Space Safety Regulations and Standards

Space Safety Regulations and Standards

Joseph N. Pelton

Ram S. Jakhu

ELSEVIER

AMSTERDAM • BOSTON • HEIDELBERG • LONDON • NEW YORK • OXFORD
PARIS • SAN DIEGO • SAN FRANCISCO • SINGAPORE • SYDNEY • TOKYO

Butterworth-Heinemann is an imprint of Elsevier

Butterworth-Heinemann is an imprint of Elsevier
The Boulevard, Langford Lane, Kidlington, Oxford, OX5 1GB, UK
30 Corporate Drive, Suite 400, Burlington, MA 01803, USA

First published 2010

British Library Cataloguing in Publication Data
Pelton, Joseph N.
Space safety regulations and standards.
1. Space flight–Safety measures–Standards. 2. Space flight–Safety regulations. 3. Astronautics–Safety measures–Standards. 4. Astronautics–Safety regulations.
I. Title II. Jakhu, Ram.
363.1'2472-dc22

Library of Congress Control Number: 2010924063

ISBN: 978-1-85617-752-8

For information on all Butterworth-Heinemann publications visit our website at elsevierdirect.com

Printed and bound in China

10 11 12 11 10 9 8 7 6 5 4 3 2 1

The IAASS Book Series on Space Safety - Dr Joseph N. Pelton. Executive Editor: G. Musgrave, A. Larsen, and T. Sgobba, *Safety Design for Space Systems* (2009) J. Pelton and R. Jakhu, *Space Safety Regulations and Standards* (2010) and pending publication in 2012 *Safety Design for Space Operations.*

Contents

PART 2 SAFETY REGULATIONS AND STANDARDS FOR UNMANNED SPACE SYSTEMS

PART 6 CONCLUSIONS AND NEXT STEPS

PART 7 APPENDICES

Acknowledgements

The Editors are especially grateful to the International Association for the Advancement of Space Safety (IAASS) for their strong support in bringing this project to fruition. Many IAASS members and fellows generously contributed to the success of this project, but we would like to particularly thank Alex Soons, the Chairman of the Board of the IAASS for his strong support to the project over the past 2 years as this book was conceived, outlined, written, and edited. We would also like to thank the people of Elsevier, namely Jonathan Simpson, Lyndsey Dixon, Hayley Salter, and Irene Hosey for their timely and always useful advice and support. Finally, we would like to thank Dr Paul Dempsey, Director of the Institute of Air and Space Law at McGill University, for writing the foreword to this book and especially to Diane Howard, Arsenault Doctoral Fellow in Space Governance at McGill, who rather miraculously did the final edits on the entire book under great pressure while traveling back and forth between Canada and her home in Florida. Finally, we express our thanks to Alejandro Restrepo, a graduate student at the McGill Institute of Air and Space Law, for his assistance in putting together and synchronizing all the authors' résumés and copyright releases. Finally we would like to express our special gratitude to the Erin J. C. Arsenault Fund at McGill University for providing financial support for assistance in editing this book.

Joseph N. Pelton and Ram S. Jakhu
Co-Editors, *Space Safety Regulations and Standards*

Foreword

Stephen Hawking observed: "It is important for the human race to spread out into space for the survival of the species. Life on Earth is at the ever-increasing risk of being wiped out by a disaster, such as sudden global warming, nuclear war, a genetically engineered virus or other dangers we have not yet thought of" [1].

Survival is a primordial instinct. We, as a species, in relatively brief time, have moved from being tree primates, to hunting primates, to farming primates, to industrial primates, and now technological primates. *Homo sapiens* gradually are coming to the collective realization that we need an inhabitable environment in which we, and our children, can survive. Space always has been an enigmatic challenge, and in Hawking's view, becomes the ultimate opportunity for survival.

There are three major catalysts for space exploration and development:

- scientific discovery, pursued predominantly by governmental institutions;
- military development, pursued by defense ministries; and
- commercial development.

Each of the three requires safety in order to enjoy the benefits of space on a sustainable basis.

In its early years, commercial development of space was dominated by satellite communications, particularly telephone and television communications. More recent commercial activities have focused on satellite imaging and global positioning. Mining of asteroids and other near-Earth celestial bodies has not yet begun. Space tourism and the transportation of passengers in space are but embryonic.

Inspired by a prize, Charles Lindbergh made the first solo piloted flight across the Atlantic in the *Spirit of St Louis* in 1927, and was a catalyst for the development of international Air Law. That event inspired the development of intercontinental aviation. Motivated by a similar prize, in 2004, *SpaceShipOne* became the first privately designed, financed, and developed spacecraft to fly humans into sub-orbital space. As did the *Spirit of St Louis*, *SpaceShipOne* inspired the development of efficient and cost-effective technology, and hopefully the development of Space Law.

For a dozen years commencing in 1967, the world community drafted five major multilateral conventions establishing the basic principles of Space Law—the "Outer Space Treaty" [2], the "Rescue Agreement" [3], the "Liability Convention" [4], the "Registration Convention" [5], and the "Moon Agreement" [6]. They were drafted in a time when governments, rather than businesses, were the principal users of space, and in the shadow of the Cold War. While they provide a skeletal framework of law, they are not enough. They were drafted in different times to deal with different issues. As commercial use of space has grown, these rules appear increasingly anachronistic, and insufficiently comprehensive. This book addresses one of the black holes that threaten space development and investment—the absence of an international governance structure to ensure safety.

The international Air Law legal regime governing air transport on issues such as safety, security, navigation and air traffic management are well developed, and set forth in the Chicago Convention [7] and the Annexes thereto promulgated by the International Civil Aviation Organization (ICAO). The ICAO promulgates standards and recommended practices (SARPs) in many of these areas for commercial aviation, facilitating international harmonization of law and enhancing the development of commercial aviation. Yet it is unclear whether aerospace vehicles fall under established principles of Air Law and, if they do, whether these laws follow them into space. Indeed, it remains unclear where the regime of Air Law ends and the regime of Space Law begins. Both the existing Air Law and Space Law regimes were developed at a time when the technology was embryonic, and commercial activity in space was nil. Thus, there is not yet a unified or integrated regime of Aerospace Law, and there appears to be much overlap and inconsistency between the regimes of Air Law and Space Law.

In instances where both regimes apply, there will be a certain amount of inevitable inconsistency. As commercial launches become more numerous, their use of airspace also inhabited by aircraft will proliferate, creating a need for defined rules of safety, security, navigation, and traffic control. The result of the absence of effective "rules of the road" will be collision, and a proliferation of space debris—the largest environmental threat to the full development of space.

In centuries when our population was smaller, our species could blithely pollute the air and water of the Earth and remain unscathed by the consequences. So, too, for the last several decades, we thought we left our space debris in orbits unlikely to collide. We are only slowly realizing the error of our ways. As terrestrial Earth became more populated, it became apparent that we were fouling our own nests. Climate change and global warming are evidence of our irresponsibility. And, as the use of space becomes more widespread, we increasingly risk the destruction of enormously expensive hardware, and thereby jeopardize the full use of space resources and their benefit for mankind. The collision of the *Iridium-33* and *Kosmos-2251* satellites in early 2009 is a canary in a coal mine. We need to come to grips with this problem before Earth begins to resemble Jupiter.

Today, space is among the fastest growing industries, accounting for more than $60 billion in annual spending by governments alone, and generating more than $180 billion a year in revenue. Space investment is a major part of the infrastructure of communications—both telecommunications and broadcast, of weather and geological monitoring, and of defense.

As the global economy grows, so too will commercial activities in space. As is the case with multinational corporations in other trade sectors, the ability of any individual nation to regulate them is circumscribed. Moreover, a fragmented and unharmonious patchwork of national laws governing such issues as safety, security, environmental harm, and liability will impede the ability of space commerce to reach its full potential.

The day may come when the world community realizes that an international body is needed to address common issues in the commercial development of space. Global harmonization of Space Law could come through creation of a new international organization, or by amending the Chicago Convention of 1944 to include space within the ICAO's jurisdiction [8].

This book fills an important void in the literature, and in the law. Our species desperately needs solutions to the problems identified herein, and policymakers should give serious consideration to the thoughtful remedies proposed in these pages.

Finally, I commend my colleagues, Professors Joseph Pelton and Ram Jakhu, the editors of this collection, who have provided invaluable leadership on the subject of safety in space, and without whom this project would never have reached your eyes.

Dr Paul Stephen Dempsey
Tomlinson Professor of Global Governance in Air and Space Law
Director, Institute of Air and Space Law
McGill University

References

[1] *Hawking: Humans Must Go Into Space*, Associated Press, 14 June 2006.

[2] Treaty on Principles Governing the Activities of States in the Exploration and Use of Outer Space, Including the Moon and Other Celestial Bodies, opened for signature 27 January 1967, 19 UST 2410, TIAS 6347, 610 UNTS 205, 6 ILM 386, GA Res. 2222 (XXI), opened for signature on 27 January 1967, entered into force on 10 October 1967.

[3] Agreement on the Rescue of Astronauts, the Return of Astronauts and the Return of Objects Launched into Outer Space, opened for signature 22 April 1968, 19 UST 7570, 672 UNTS 119, GA Res. 2345 (XXII), entered into force on 3 December 1968.

[4] Convention on International Liability for Damage Caused by Space Objects, opened for signature 29 March 1972, 24 UST 2389, TIAS 7762, 961 UNTS 187, 10 ILM 965, GA Res. 2777 (XXVI), opened for signature on 29 March 1972, entered into force on 1 September 1972.

[5] Convention on Registration of Objects Launched into Outer Space, opened for signature 14 January 1975, 28 UST 695, TIAS 8480, 1023 UNTS 15, GA Res. 3235 (XXIX), entered into force on 15 September 1976.

[6] Agreement Governing the Activities of States on the Moon and Other Celestial Bodies, opened for signature 18 December 1979, 1986 ATS 14, 18 ILM 1434; GA Res. 34/68, entered into force on 11 July 1984.

[7] Convention on International Civil Aviation, 61 Stat. 1180 (1944).

[8] The ICAO's membership of 190 nations comprises virtually the entire world community. The ICAO has more than half a century of experience in addressing safety of international commercial aviation.

About the Editors and Authors

EDITORS

Dr Joseph N. Pelton, former Dean of the International Space University, is Director Emeritus of the Space and Advanced Communications Research Institute (SACRI) at George Washington University. Dr Pelton also served as Director of the Accelerated Masters Program in Telecommunications and Computers at the George Washington University from 1998 to 2004. From 1988 to 1996, Dr Pelton served as Director of the Interdisciplinary Telecommunications Program at the University of Colorado Boulder, which at that time was the world's largest graduate level telecommunications program. Dr Pelton was the founder of the Arthur C. Clarke Foundation and remains as the Vice Chairman of its Board of Directors. Dr Pelton is a Fellow of the International Association for the Advancement of Space Safety (IAASS), and a member of its Executive Board and Chairman of its Academic Committee. He is the author of some 30 books in the field of space and applied technologies and Executive Editor of the IAASS publication series. This includes *Safety Design for Space Systems*, published by Elsevier in 2009. The follow-on book *Space Safety Regulations and Standards* is to be published in 2010. He is also Vice President of the International Space Safety Foundation of the USA as well as the former President of the Global Legal Information Network. He was Director of Strategic Policy for Intelsat in the early to mid 1980s. He is the widely read author of 30 books, including *Global Talk*, which won a Pulitzer Prize nomination. He received his degrees from the University of Tulsa, New York University, and Georgetown University.

Professor Dr Ram S. Jakhu is Associate Professor at the Institute of Air and Space Law, Faculty of Law, McGill University, Montreal, Canada, where he teaches and conducts research in international space law, law of space applications, law of space commercialization, government regulation of space activities, law of telecommunications and Canadian communications law, and public international law. He is the Managing Editor of the Space Regulations Library Series, a member of the Editorial Boards of the *Annals of Air and Space Law* and the *German Journal of Air and Space Law*, a member of the Board of Directors of the International Institute of Space Law, Research Director for Space Security Index Project, and the Chairman of the Legal and Regulatory Committee of International Association for the Advancement of Space Safety. He has been an Associate Professor at McGill University since 1998. He served as Director, Centre for the Study of Regulated Industries, McGill University, 1999–2004. He served as the First Director of the Masters Program of the International Space University, Strasbourg, France, 1995–1998. His degrees are as follows: BA Panjab University, LLB Panjab University, LLM Panjab University, International Law, LLM McGill University, Air and Space Law, Doctor of Civil Law, McGill University, Law of Outer Space and Telecommunications.

AUTHORS

Samuel Black is a Research Associate with the Stimson Center's Space Security and South Asia programs. Prior to joining the Center, he worked as a Research Assistant at the Center for Defense Information, where he focused on space security and missile defense issues. He holds a BA in Government and Politics and a Master's degree in Public Policy, both from the University of Maryland. He was a Presidential Management Fellowship Finalist for 2008 and was awarded the Capt. William P. Cole III Peace Fellowship by the University of Maryland's School of Public Policy.

Meghan Buchanan is an Aerospace Engineer with Lockheed Martin in Denver, CO, USA. She is an expert on space safety and spacecraft survivability, and has contributed papers related to space safety to past IAASS Conferences. She currently is working on the NASA Orion CEV contract, where she is supporting the spacecraft survivability analysis. She has completed her Masters in engineering.

Joseph Chan is senior manager of the Flight Dynamics Department at Intelsat Corporation. In this position he oversees the flight dynamics operations for a fleet of 64 satellites. He has been involved in the conjunction monitoring program at Intelsat since 1998. Back in 2006, he and his colleagues at Intelsat proposed the concept and worked on a prototype of data sharing among commercial operators to improve space situation awareness and safety of flight. Prior to joining Intelsat, he worked for NASA on the Mars Observer and the Topex/Poseidon program at Goddard Space Flight Center.

Sandra Coleman joined ATK in 2006 and is the Director, NASA Exploration Programs, for Alliant Techsystems' (ATK) Washington, DC operations. ATK is a $4.1 billion advanced weapon and space systems company with approximately 16,500 employees. Sandy leads the business development activities related to NASA with specific emphasis on the Exploration Program. She is a 40-year veteran of NASA, where she was Manager of the Space Shuttle External Tank (ET) Project after the *Columbia* accident responsible for overall management of the redesign, development, production, test, integration, and flight of the ET. This involved responsibility for over 100 civil servants, 2000 contractor employees, and oversight of the 43-acre government-owned, contractor-operated Michoud Assembly

Facility. Other positions held include: Transition Project Manager for the Marshall Space Flight Center (MSFC) with overall responsibility for a smooth transition of resources from the Space Shuttle Propulsion (SSP) Projects to the Crew Launch Vehicle and Cargo Launch Vehicle; Interim Manager of the SSP Office with the overall responsibility for the manufacture, assembly, and operation of the ET, Space Shuttle Main Engines (SSME), Solid Rocket Boosters (SRB), and Reusable Solid Rocket Motors (RSRM); Implementation Manager and Chief Operating Officer of the National Space Science and Technology Center at MSFC that consists of government, academia, and industry collaboration in an environment that enables cutting-edge research; Deputy Chief Financial Officer at MSFC. Her career began at NASA in the Saturn V Program Office. After the 1969 Moon landing, she became a charter member of the Space Shuttle Task Team and worked in three of the four main project offices—ET, SRB, and RSRM—and was given increasingly challenging positions in the SSP Office, including Deputy RSRM Project Manager. Sandy has received numerous awards, including the Astronaut Silver Snoopy, the NASA Outstanding Leadership Medal, and the NASA Exceptional Achievement Medal. Sandy holds a Master of Science degree in Industrial Engineering from the University of Alabama in Tuscaloosa and a Bachelor of Science degree in accounting from the University of Alabama in Huntsville.

Richard DalBello is vice president of government relations for the Intelsat General Corporation (IGC). In this position, Richard is responsible for liaison with US decision-makers at the federal, state, and local levels with regard to Intelsat General Corporation services. One of his special responsibilities is to explain how the safety and reliability of IGS satellite services are effectively ensured. Richard has previously served as president of the Satellite Broadcasting and Communications Association, and for more than 3 years he was the president of the Satellite Industry Association, the voice of the US commercial satellite industry on policy, regulatory, and legislative matters. Richard, who has over 25 years of service in the satellite industry, has served as general counsel for Spotcast Communications Inc., and as vice president of government affairs, North America, for ICO Global Communication. He has served as assistant director for aeronautics and space in the US White House Office of Science and Technology Policy (OSTP). He started his career in Washington in the Office of Technology Assessment for the US Congress. Later, he was director of commercial communications at NASA, where he was responsible for private sector experiments on the Advanced Communications Technology Satellite (ACTS). Richard has degrees from the University of Illinois, the University of San Francisco, and McGill University.

Michael E. Davis is a founding partner of Adelta Legal, a specialist commercial law firm based in Adelaide, Australia. He has practiced law in Australia for 35 years. He holds the degrees of Bachelor of Laws from the University of Adelaide, Australia and Master of Space Studies from the International Space University in Strasbourg, France. Michael has a strong interest in technology commercialization and, as a lawyer, has unique knowledge and experience of the international space and ICT industries. He has been actively involved in a number of complex projects in the areas of software commercialization, telecommunications, international telemedicine, and the space industry. Michael currently serves as Secretary of the Australian Space Industry Chamber of Commerce and in that capacity is involved in a range of space-related policy and legal issues. For 14 years, Michael served on the Advisory Board of the Institute of Telecommunications Research of the University of South Australia. He has also served as Senior Regional Representative in the Asia-Pacific Region of the International Space University. In that role, he has traveled extensively and has worked closely with space agencies, universities, telecommunications carriers and satellite operators in Asia, the USA, and Europe. He was also Chair of the Host Organizing Committee of the International Space University Space Studies Program held in Adelaide in 2004. In 2007, Michael was appointed a part-time Faculty member of the International Space University. In 2009, he was Co-Chair of the Policy and Law Department of the ISU Summer Session Program in San Francisco. He has researched and authored many papers on policy and regulatory topics and has regularly presented papers at international telecommunications and space conferences.

Professor Dr Paul Stephen Dempsey is Tomlinson Professor of Global Governance in Air and Space Law and Director of the Institute of Air and Space Law at McGill University, in Montreal, Canada. From 1979 to 2002, he held the chair as Professor of Transportation Law, and was Director of the Transportation Law Program at the University of Denver. He was also Director of the National Center for Intermodal Transportation. Earlier, he served as an attorney with the Civil Aeronautics Board and the Interstate Commerce Commission in Washington, DC, and was Legal Advisor to the Chairman of the ICC. Dr Dempsey holds the following degrees: Bachelor of Arts (1972), Juris Doctor (1975), University of Georgia; Master of Laws (1978), George Washington University; Doctor of Civil Laws (1986), McGill University. He is admitted to practice law in Colorado, Georgia, and the District of Columbia.

Dr Mohamed S. El-Genk is Regents' Professor of Chemical, Nuclear and Mechanical Engineering, and founding Director of the Institute for Space and Nuclear Power Studies at the University of New Mexico (UNM). Prior to joining UNM in 1981, he worked in industry for 13 years. He is a world authority in space nuclear power and propulsion, nuclear reactor design, safety and thermal hydraulics, thermal management of space systems, boiling heat transfer, heat pipes, advanced cooling of electronics, energy conversion, nuclear fuel and materials, nuclear fuel cycle, and radiation effects. He has published more than 275 refereed articles, 235 full articles in conference proceedings, four book chapters, and has edited a book and more than 50 volumes of proceedings that comprised more than 5000 technical papers. He has four patents to his credit and two pending. He is a Fellow of the American Nuclear Society, the American Society of Mechanical Engineers, the American Institute of Chemical Engineers and the International Association for the Advancement of Space Safety (IAASS), and an Associate Fellow of the American Institute of Aeronautics and Astronautics (AIAA). He received the US DOE Certificate of Appreciation for his Outstanding

Contributions to the Field of Space Nuclear Power and Propulsion, and served on the NASA Space Exploration of the Solar System Technology Assessment Group in 2001–2002 and the NASA Advanced Radioisotope System (ARPS) Team—2001 Technology Assessment and Recommended Roadmap for Potential NASA Code S Missions Beyond 2011. In 2001, he was named the 46th Annual Research Lecturer, the highest honor bestowed upon a member of the faculty at the University of New Mexico, and received the School of Engineering's Research and Teaching Excellence Awards.

Dr Zsuzsanna Erdélyi graduated from the University of Szeged, Hungary in 1999. After being awarded a Degree in Law and Public Administration, as well as a Certificate in European Studies, she worked as the coordinating officer of the Law Harmonization Program for Transport in the Ministry of Justice in Budapest. As part of her duties, she acted as legal adviser specializing in transport matters during the negotiations for the accession of Hungary to the European Union. In 2003, after being transferred to the Ministry of Transport to follow-up the preparatory phase of Hungary's accession to the EU, she was posted by the Ministry of Foreign Affairs as Transport Attaché to the Permanent Representation of Hungary to the EU in Brussels, Belgium, where she was a member of the European Council's Aviation Working Group until 2008. Ms Erdélyi joined the Rulemaking Directorate of the European Aviation Safety Agency (EASA) in March 2008, where she actively contributed to the regulatory work on the extension of the EASA's competences to the safety regulation of Air Traffic Management and aerodromes. She is currently working in the EASA's ATM and Airport Safety Department as Rulemaking Officer. She co-authored the EASA article on "Accommodating Sub-orbital Flights into the EASA regulatory system".

Dr David Finkleman has been Senior Scientist in the Center for Space Standards and Innovation of Analytical Graphics, Inc., since 2004. He is a retired AF Colonel and a retired member of the Federal Senior Executive Service. Dr Finkleman is a Fellow of the American Institute of Aeronautics and Astronautics (member of the AIAA Standards Executive Council), a Fellow of the American Astronautical Society, a Fellow of the American Association for the Advancement of Science, and an Emeritus in the Scientific Research Society. He is a five-time Paul Harris Fellow of Rotary International. He chairs the International Organization for Standardization (ISO) space operations working group and serves on the Consultative Committee for Space Data Standards (CCSDS). Dr Finkleman earned his PhD from the Department of Aeronautics and Astronautics at MIT. His career includes: having served as an Associate Professor of Aeronautics at USAFA; Deputy Program Manager, Navy Directed Energy Weapons; Director of Technology and Systems, Army Ballistic Missile Defense; first Director of Kinetic Energy Weapons, SDI; and Director of Analysis and Chief Technical Officer, NORAD, United States Space Command, and United States Northern Command. He served on the Battle Staff in Cheyenne Mountain during 9/11.

Dr Michael Gerhard is legal adviser to the European Aviation Safety Agency (EASA). He joined the Agency in 2008 and is working in the field of international aviation law, EU law, European administrative law and enforcement. From 1999 to 2008, Dr Gerhard worked at the German Aerospace Center (DLR). As a legal adviser he was mainly occupied with matters of public international law and administrative law. The main emphasis of his work was to advise the DLR and the federal ministries in matters of national space legislation, insurance and registration of space objects, liability issues, intellectual property issues, security aspects regarding the distribution of remote sensing data, and budget law. Furthermore, Dr Gerhard teaches Space Law to students studying Aerospace Technologies at the Aachen University of Applied Sciences. He has also given lectures at the Technical University of Berlin, as well as Summer Courses for the European Center for Space Law (ECSL). Dr Gerhard has published a book on the topic of national space legislation and more than 30 articles on space law and policy.

Dr Gérardine Meishan Goh was Legal Advisor and Project Manager at the Business and Legal Support Department of the German Aerospace Center (DLR) until August 2009, and concurrently was Lecturer at the Institute of Air and Space Law, University of Cologne, Germany, where she taught the courses International Dispute Settlement and the Law of the United Nations. She was formerly Legal and Business Development Specialist at a satellite-based geo-information services company. Gérardine completed her education in law at University College London, UK and the National University of Singapore. She received her doctorate in law from the International Institute of Air and Space Law, Leiden University, the Netherlands. In 2001, she formed half of the team that won the World Championships of the Manfred Lachs Space Law Moot Court Competition in Toulouse, France. In 2003, she won the Diederiks-Verschoor Award for the best space law paper written by a young author. Gérardine has authored more than 40 publications on international and national dispute settlement, public international law, human rights law, international space law and outer space studies, as well as psychology, sociology, genetics, and zoological physiology. She recently wrote the book *International Dispute Settlement: A Multi-Door Courthouse for Outer Space* (2007, Martinus Nijhoff/Leiden, Boston). The Assistant Secretary to the Board of Directors of the International Institute of Space Law (IISL), Gérardine is also a Fellow of the International Association for the Advancement of Space Safety (IAASS) and of the Committee on Space Research (COSPAR). She is a member of the International Law Association (Headquarters) and the European Centre for Space Law (ECSL). She is a contributing member to the Study Group on Space Debris Remediation of the International Academy of Astronautics (IAA).

Jerold Haber is Operations Manager and Program Manager for Range Safety at the Pacific Ranges at ACTA. He is a key member of the Range Commanders Risk Committee, instrumental in the definition of standard practices for quantifying risk from launch and re-entry operations and tolerable risk levels. He has served as an expert witness for US Senate subcommittee hearings on risks from natural disasters and as part of the US Non-Lethal Weapons Effect expert panel on risk analysis. Prior to joining ACTA, he served as Manager of the Risk and Systems Analysis Department at National Technical Systems Engineering. Mr Haber is an Associate Fellow of the IAASS, a Senior Member of the AIAA, and a member of the US National Research Honorary, Sigma Xi. He is listed in *Who's Who in Risk Management.* He performed his undergraduate and graduate studies at the University of California, graduating with honors.

Theresa Hitchens was appointed Director of the United Institute for Disarmament Research (UNIDIR) on 14 November 2008 and took up her duties in Geneva on 26 January 2009. She came to UNIDIR from 8 years with the Center for Defense Information in Washington, DC, where she most recently served as Director. Besides her duties as CDI director, Theresa leads the CDI's Space Security Project. Editor of *Defense News* from 1998 to 2000, she has had a long career in journalism, with a focus on military, defense industry, and NATO affairs. Her time at *Defense News* included 5 years as the newspaper's first Brussels bureau chief, from 1989 to 1993. From 1983 to 1988, she worked at Inside Washington Publishers on the group's environmental and defense-related newsletters, covering issues from nuclear waste to electronic warfare to military space. Theresa has had a long interest in security policy and politics, having served internships with Senator John Glenn, D-Ohio, and with the NATO Parliamentary Assembly in Brussels. Most recently, she was director of research at the British American Security Information Council, a thinktank based in Washington and London. The author of *Future Security In Space: Charting a Cooperative Course*, she also continues to write on space and nuclear arms control issues for a number of outside publications. Theresa serves on the editorial board of the *Bulletin of the Atomic Scientists*, and is a member of Women in International Security and the International Institute for Strategic Studies.

Dr Jeffrey A. Hoffman is Professor of Aerospace Engineering in the Department of Aeronautics and Astronautics at MIT. Dr Hoffman received a BA (summa cum laude) from Amherst College in 1966 and a PhD in astrophysics from Harvard University in 1971. He subsequently received an MSc in Materials Science from Rice University in 1988. He spent a year as a post-doctoral fellow at the Smithsonian Astrophysical Observatory, after which he worked on the research staff of the Physics Department at Leicester University in the UK (1972–1975) and MIT's Center for Space Research (1975–1978). He was a NASA astronaut from 1978 to 1997, making five space flights, and became the first astronaut to log 1000 hours of flight time aboard the space shuttle. Dr Hoffman was Payload Commander of *STS 46*, the first flight of the US–Italian Tethered Satellite System. He played a key role in coordinating the scientific and operational teams working on this project. Dr Hoffman has performed four spacewalks, including the first unplanned, contingency spacewalk in NASA's history (*STS 51D*, April 1985) and the initial repair/rescue mission for the Hubble Space Telescope (*STS 61*, December 1993). Following his astronaut career, Dr Hoffman spent 4 years as NASA's European Representative, based at the US Embassy in Paris, where his principal duties were to keep NASA and NASA's European partners informed about each other's activities, try to resolve problems in US–European space projects, search for new areas of US–European space cooperation, and represent NASA in European media. In August 2001, Dr Hoffman joined the MIT faculty, where he teaches courses on space operations, design, and space policy. Dr Hoffman is director of the Massachusetts Space Grant Alliance, responsible for statewide space-related educational activities designed to increase public understanding of space and to attract students into aerospace careers. His principal areas of research are advanced EVA systems, space radiation protection, management of space science projects, and space systems architecture. He was a member of MIT's team for NASA's Concept Evaluation and Refinement Study, and is a member of MIT's Space Architecture working group, which has been doing studies over the past several years for ESMD on lunar exploration.

Dean Hope graduated from Sydney University, BSc, BE (Elec) in 1980, and began work at British Aerospace (BAe) in Stevenage, UK in the RF department. In 1983, he joined Aussat in Los Angeles to train as an Orbital Dynamicist with the Hughes Aircraft Corp. before transferring to Aussat's ground station in Sydney to support Flight Dynamics operations. In 1989, he initially joined Inmarsat in London as a Satellite Engineer, later moving to their Flight Dynamics group in 1995. In 1999, he became Flight Dynamics manager and continues to lead a dedicated team currently responsible for station-keeping 11 geostationary satellites.

Diane Howard is an Arsenault Doctoral Fellow in Space Governance at McGill University's Institute of Air and Space Law. She has a BSc, and earned her JD (cum laude) at Nova Southeastern Shepard Broad Law Center in Florida. Ms Howard achieved Dean's Honor List for her LLM in International Air and Space Law from McGill University. She is a researcher for the Space Security Index and assists in various research projects at the Institute. To date, her research and publications have focused on aspects of private and commercial space law. She was a part of the citizens' group that lobbied Capitol Hill for the Commercial Space Law Amendments Act of 2004. Her doctoral work concerns spaceports and issues of interoperability and harmonization. Ms Howard is a licensed member of the Florida Bar.

Professor Dr Li Juquian is Associate Professor of China University of Political Science and Law (CUPL). He is also Director of the Public International Law Research Institute at the International Law School at CUPL. He currently serves as a Council Member of the China Institute of Space Law and as a Standing Council Member of Beijing International Law Society. His academic interests are mainly in the areas of international law, especially space law and WTO law. He has published more than 10 books, including four books in translation from the Chinese language. He has also published 10 articles on international law and international economic law. His education includes a Doctor of International Law from the Graduate Law School of the China University of Politics and Law (CUPL), a Masters of International and Economic Law also from CUPL, and a Baccalaureate Law degree from the Chinese South West University of Politics and Law.

Michael Krepon is co-founder of the Henry L. Stimson Center and the author or editor of 13 books and over 350 articles. Prior to co-founding the Stimson Center, Michael worked at the Carnegie Endowment for International Peace, the US Arms Control and Disarmament Agency during the Carter administration, and in the US House of Representatives, assisting Congressman Norm Dicks. He received an MA from the School of Advanced International Studies, Johns Hopkins University and a BA from Franklin and Marshall College. He also studied Arabic at the American University in Cairo, Egypt. He divides his time between Stimson's South Asia and Space Security projects. The South Asia project concentrates on escalation control, nuclear risk reduction, confidence building, and peace-making between India and Pakistan. This project entails field work, publications, and Washington-based programming, including a visiting fellowship program. The Space Security project seeks to promote a Code of Conduct for responsible spacefaring nations and works toward stronger international norms for the peaceful uses of outer space. He also teaches in the Politics Department at the University of Virginia.

Bruno Lazare graduated from Ecole Nationale Supérieure d'Arts et Métiers (ENSAM). His industrial experience includes: Head of Guyana Space Center Quality Department, 1987–1993; in charge of safety studies for GSC Ariane Launch Pads 2 and 3, 1982–1987. He now works at Central Safety Office (headquarters level) in charge of space vehicles, stratospheric balloons, and implementation of the French Space Operations Act.

Jean-Bruno Marciacq has worked for the European Aviation Safety Agency (EASA) since 2007 as Project Certification Manager. In parallel to aircraft certification, he coordinates the work on Sub-orbital Aeroplanes (SoA) for the Agency. From 2001 to 2007, he worked for the European Space Agency (ESA) as ESA Crew Safety Officer within the Astronaut Division (HME-AA) of the Human Spaceflight Department (HME-A). Within these functions, he directly contributed to the safety of seven manned space flights, and supported the ATV and Columbus projects. He was responsible for the safety of Astronaut Training and Medical Experiments performed on the ground, including EVA training operations in the ESA Neutral Buoyancy Facility (NBF). He also acted as EAC Safety Manager. From 1995 to 2001, he worked respectively as Falcon 50 Deputy Programme Manager and Customer Service Manager within the Civil Aircraft Division of Dassault Aviation, ensuring the airworthiness of the worldwide fleet of Falcon Business Jets. From 1994 to 1995, he served as Naval Aviation Officer on the French Aircraft Carrier *Foch*, for which he was awarded four medals. Jean-Bruno graduated from ESTACA University for Aerospace and Automotive Engineering, Paris and holds a Certificate of Aircraft Design from the Moscow Aviation Institute (MAI). As a complement to his studies, he has worked for various aerospace companies such as British Airways, Aérospatiale, Dornier/Deutsche Aerospace, Aero/Sagem Group, and Mudry Aircraft Co. He is a glider pilot, sailor and diver, and enjoys flying both as part of his professional duties and for leisure.

Dr Martha Mejía-Kaiser was born in Mexico City in 1957 and has lived in Germany since 1989. She received a Bachelor's degree in International Relations, a Master's degree in International Law, and a Doctoral degree in Political and Social Sciences from the National Autonomous University of Mexico (UNAM). She holds a Diploma in Air and Space Law from McGill University and is an ISU graduate (summer session). She worked in the Mexican Foreign Affairs Ministry and the Geophysics Institute (UNAM). She has served as a judge in the Manfred Lachs (Space Law) and Telders (International Law) Moot Court Competitions. She is an IISL member and is a visiting lecturer at the International Institute of Air and Space Law at Leiden University, the Netherlands.

Michael C. Mineiro is a Boeing Doctoral Fellow at the Institute of Air and Space Law, McGill University. He has published papers on a variety of subjects relevant to space law and policy. His current research interests include satellite technology export controls and commercial space transportation. He earned his LLM (2008) from the Institute of Air and Space Law, JD (2005) from the University of North Carolina, and a BA (2001) from North Carolina State University. He is a licensed member of the North Carolina State Bar.

Yves Morier is Head of Product Safety Department, Rulemaking Directorate, European Aviation Safety Agency, Cologne, Germany. He was born in 1956 and graduated in 1978 from the French civil aviation academy (ENAC) as an Air Transport Engineer. After military service, he joined the French Civil Aviation Authority (DGAC-F) as deputy head of a regional office. In 1986, he moved to the rulemaking office of DGAC-F in charge of airworthiness and operation requirements. In 1991, he was appointed by the Joint Aviation Authorities (JAA) as Regulations Director. In 2004, he joined the EASA Rulemaking Directorate as head of the product safety department. He is married and has two daughters.

Mr Yaw Otu Mankata Nyampong holds a Master of Law (LLM) degree in Air and Space Law from the Institute of Air and Space Law, McGill University, Montreal, Canada (2005), a Qualifying Certificate in Professional Law from the Ghana School of Law (2000), and a Bachelor of Law (LLB) degree from the Faculty of Law, University of Ghana, Legon (1998). He is a member in good standing of the Ghana Bar Association. Following his call to the bar in Ghana in October 2000, Mr Nyampong practiced law with the Ghanaian law firm G. A. Sarpong and Co. for 3 years, first as a pupil (student-at-law) and later as an associate. As part of his schedule of responsibilities, Mr Nyampong worked as external counsel for the Ghana Civil Aviation Authority, as a result of which he became very interested in aviation law. It was in pursuit of this new-found interest in air law that Mr Nyampong sought and gained admission to the Master of Law (LLM) program at the Institute of Air and Space Law, McGill University in September 2003. He graduated from this program in 2005 and his LLM thesis on the environmental regulation of aircraft engine emissions from international civil aviation was named on the coveted Dean's Honour List that year. At the institute of Air and Space Law, Mr Nyampong is currently conducting research for his doctoral thesis on aviation war risk insurance under the supervision of Professor Paul Stephen Dempsey. During the time that he has been with the Institute of Air and Space Law, Mr Nyampong has, on several occasions, worked as a research and teaching assistant under the supervision of Professors Paul Dempsey and Ram Jakhu. In April 2007, Mr Nyampong was appointed as Editor of *Annals of Air and Space Law*, a peer-reviewed scholarly journal published annually by the Institute and Center for Research of Air and Space Law.

Joseph Pellegrino joined ATK in 2006 and is the Director of Business Development Engineering Services. ATK is a $4.1 billion advanced weapon and space systems company with approximately 16,500 employees. Joseph leads the business development activities related to engineering services at 10 NASA centers, the Naval Research Laboratory, and the Applied Physics Laboratory. Joseph recently served as the Director of the ATK Pasadena, CA office, which provides support to the NASA Jet Propulsion Laboratory. He led program teams that supported the Mars Science Laboratory (US Mars rover), ExoMars (ESA Mars rover), and the Jason Satellite (ocean wave height and direction observation). He has 20 years of exceptional space systems experience gained at three NASA centers and the Canadian Space Agency. During his 12 years working for Boeing, Joseph trained the crews of three space shuttle crews at the NASA Johnson Space Center in Houston, TX. Joseph served as the Spacecraft Manager for three satellite programs (Anik-F1, NASA TRDS I and J) at Boeing Satellite Systems located in El Segundo, CA. During this assignment, Joseph led all of the Integration and Test activities from initial component integration to on-orbit delivery. He served as the Spacecraft Leader for numerous launch campaigns at Cape Canaveral, FL, Sea Launch, French Guyana, and the Baikonur cosmodrome. While serving as the NASA TDRS I Spacecraft Manager, Joseph led the team that recovered this satellite after a significant propulsion system failure during transfer orbit. While working at the NASA Kennedy Space Center, Joseph led the payload integration on the *STS 114* Space Shuttle Return to Flight mission. He also served as Integration Lead for the *Copula* (a space station module provided by Italy). At the Canadian Space Agency, he supported the Strategic Technologies for

Automation and Robotics (STEAR) program. He also served as the Program Manager for the teleoperation testbed that was used to develop systems to manage time delays associated with long-distance remote operations. Joseph attended the International Space University in 1995, and currently serves as a faculty member. Joseph holds a Bachelor of Science degree in Mechanical Engineering from Texas A&M University located in College Station, TX.

Thomas Pfitzer began his professional career in the 1970s as a safety officer at the Kwajalein Missile Range, where he was responsible for ensuring the safety of over 200 short and mid-range launches. There he compiled the first edition of the *Kwajalein Range Safety Manual* (1976). After 19 years as a government safety engineer, he established APT Research (1990), a company specializing in test and safety support. In 1996, the first Risk and Lethality Commonality Team (RALCT) was formed and the contract to serve as the Technical Secretariat to compile the standard was led by Pfitzer and the APT staff. In 2004, he founded the Safety Engineering and Analysis Center (SEAC), an organization providing launch safety analysis services to many of the US national ranges. He has been a key contributor to multiple safety standards, including RCC 321 for Launch Safety, ANSI Standard 0010 for System Safety, and the DDESB Standard for Managing Explosives Risk. He serves as the chair of the Launch Safety Committee of the International Association for the Advancement of Space Safety (IAASS), as a senior and active member of the International System Safety Society (ISSC), and the president of APT Research, Inc.

Michael K. Saemisch is currently the Manager for Safety and Mission Assurance on the NASA Orion program at Lockheed Martin in Denver, CO. Thus, he provides critical support for the safety, reliability, and quality assurance on the program. He has over 30 years of experience in systems safety and product assurance supporting both NASA and US Air Force space missions. He played a key role with regard to payload safety for the Space Shuttle Program and served as member of the US Department of Defense Space Shuttle Payload Safety Review Team. He has contributed papers to the IAASS international conferences and was a contributor to the IAASS book, *Space Design for Space Systems*.

Professor Dr Kai-Uwe Schrogl has been Director of the European Space Policy Institute (ESPI) in Vienna, Austria since 1 September 2007. Before this, he was Head of Corporate Development and External Relations Department in the German Aerospace Center (DLR). In his previous career, he worked with the German Ministry for Post and Telecommunications and the German Space Agency (DARA). He has been delegate to numerous international forums and recently served as the chairman of various European and global committees (ESA International Relations Committee, UN COPUOS working groups). Kai-Uwe has published 11 books and more than 100 articles, reports and papers in the fields of space policy and law, as well as telecommunications policy. He is Editor of the *Yearbook on Space Policy* and the book series *Studies in Space Policy*, both published by Springer, as well as a member of the editorial boards of international journals in the field of space policy and law (*Acta*

Astronautica, Space Policy, Zeitschrift für Luft- und Weltraumrecht, Studies in Space Law/Nijhoff). Kai-Uwe is a member of the Board of Directors of the International Institute of Space Law, a member of the International Academy of Astronautics (chairing its Commission on policy, economics, and law), and the Russian Academy for Cosmonautics. He holds a doctorate degree in political science, lectures on international relations at Tübingen University, Germany (as an honorary professor) and has been a regular guest lecturer at the International Space University and the Summer Courses of the European Centre for Space Law.

Dr Ryuichi Sekita is Associate Senior Engineer of Japan Aerospace Exploration Agency, Safety and Mission Assurance Department, a senior member of the American Institute of Aeronautics and Astronautics, and also a member of Reliability Engineering Association of Japan.

Tommaso Sgobba holds an MS in Aeronautical Engineering from the Polytechnic of Turin (I), where he has also been Professor of Space System Safety (1999–2001). Tommaso has over 30 years of experience in the aerospace industry. He is currently a staff member of the European Space Agency in charge of flight safety for manned systems, spacecraft re-entry safety, space debris, use of nuclear power sources, and planetary protection. He joined the European Space Agency in 1989, after 13 years in the aeronautical industry. Initially, he supported the development of the *Ariane 5* launcher, and of Earth observation and meteorological satellites, and the early *Hermes* space plane phase. Later he became product assurance and safety manager for all European manned missions on the shuttle, MIR station, and for the European research facilities for the International Space Station. During his long and close cooperation with the NASA Shuttle/ISS Payload Safety Review Panel, Tommaso developed at ESA the safety technical and organizational capabilities that eventually led in 2002 to the establishment of the first ESA formal safety review panel and first International Partner ISS Payload Safety Review Panel. He was also instrumental in setting up the ESA ATV Re-entry Safety Panel and in organizing the first scientific observation campaign of a destructively re-entering spacecraft (*ATV Jules Verne*). He has published several articles and papers on space safety, and co-edited the textbook *Safety Design for Space Systems*, published in 2008 by Elsevier, which is the first of its kind worldwide. Tommaso received NASA recognition for outstanding contribution to the International Space Station in 2004, and the prestigious NASA Space Flight Awareness (SFA) Award in 2007. He is president and co-founder of the International Association for the Advancement of Space Safety (IAASS), which gathers the top space safety experts worldwide. He is also vice-president of the US-based International Space Safety Foundation (ISSF).

Dr Charles H. "Herb" Shivers, deputy director of the Safety and Mission Assurance Directorate at NASA's Marshall Space Flight Center in Huntsville, AL, is responsible for safety, reliability, and quality assurance of the full range of Marshall Center programs, projects, and institutional services in support of NASA mission goals, including space shuttle, space station, space exploration, and Marshall facility safety and quality activities. He also provides technical and managerial guidance associated with related engineering, scientific and program management activities. Dr Shivers was appointed by the NASA Chief Engineer as the NASA System Safety Engineering Technical Warrant Holder from March 2005 to July 2006, ensuring certain safety requirements on the space shuttle were met for Shuttle Return to Flight. In October 2006, Dr Shivers was appointed to the Senior Executive Service, the personnel system covering top managerial positions in approximately 75 federal agencies. Dr Shivers earned a Bachelor's Degree in Industrial Engineering from Auburn University, a Master's Degree in Industrial and Safety Engineering from Texas A&M University, and a Doctorate in Industrial and Systems Engineering and Engineering Management from the University of Alabama in Huntsville. Dr Shivers also graduated from the US Army's Graduate Safety Engineering Intern Program.

Filippo Tomasello was flight test engineer in the Italian Air Force until 1984. Subsequently in ENAV (Italian provider of Air Navigation Services) he was responsible for R&D and for a number of projects, including consolidation of the Upper Area Control Centres from four to three, achieved in 2000. He then joined EUROCONTROL as manager for Northern Europe, coordinating the related medium-term plans and some projects. In 2005 he joined the European Commission (DG-TREN) on aviation safety, dealing with accident investigation, safety data collection, and extension of the competencies of the European Aviation Safety Agency (EASA). On 16 February 2007 he entered the EASA to progress the development of common safety rules for aerodromes, ATM and ANS, starting from the extension of the mandate of that Agency to these new domains. In Italy he has been, since 1991, visiting Professor at State University "Parthenope" in Naples and a member of the Italian Institute of Navigation. The EASA, established in 2002, is a Commission Agency. Per regulation 216/2008 it has responsibility on safety regulation of airworthiness, air operations, flight crews, and third country aircraft. In addition to the 27 EU Member States, the common rules also apply to Iceland, Liechtenstein, Norway, and Switzerland.

Jean-Pierre Trinchero graduated as a pyrotechnics scientific engineer from the French DoD. After industrial experience in pyrotechnics and propulsion activities for the French DoD from 1978 to 1990, he held different jobs in R&D, RAMS, and program management. From 1991 to 1996 he worked for CNES in space activities, in the European spaceport in Kourou, in charge of chemicals propellants and payload preparation (including operational safety). From 1997 to 2000, he served as a payload safety officer, and from 2001 to 2006 he was Head of Range Safety in the European spaceport (ground and flight safety). From late 2006 to the present, he has worked at the Central Safety Office (headquarters level) in charge of launch vehicles and implementation of the French Space Operations Act.

Dr Paul Wilde is an IAASS Fellow with 20 years' experience in safety standards development, launch and re-entry safety evaluations, explosive safety analysis, and operations. As a senior scientist at ACTA Inc. and the Manager of Texas Operations, he currently provides management and technical leadership for domestic and international clients. He performed leading roles for multi-organization projects in high-profile situations, such as principal investigator of public safety issues and the foam impact tests for the Columbia Accident Investigation Board, technical leadership roles in flight safety evaluations of maiden flights of the ATV, *Atlas V*, *Delta IV*, *Falcon 9-Dragon*, *SpaceShipOne*, and *Titan IVB*, and the development of US standards on launch and re-entry risk management. He received the NASA Exceptional Achievement Medal in 2004 and other awards. He has published some 100 technical reports and papers. His degrees in Mechanical Engineering are from the University of California.

Kenneth Wong is the Manager of the Licensing and Safety Division of the US Federal Aviation Administration's office of Commercial Space Transportation (AST). Prior to his present position at the FAA, he worked for private industry supporting NASA as a contractor in the areas of safety, reliability, maintainability, and quality assurance for both manned and unmanned spacecraft (space shuttle, launch vehicles, and payloads/satellites). In that position, he conducted independent technical assessments and analyses. Mr Wong has been with the FAA/AST since 1996, and has worked on several unique projects involving the safety evaluation of launch activities associated with expendable launch vehicles (ELVs) and reusable launch vehicles (RLVs). This included leading project teams to evaluate license applications, which led to the issuance of a launch-specific license for *Sea Launch*'s first inaugural mission in 1999 and an RLV mission-specific license to Scaled Composites for its *SpaceShipOne* missions in 2004. Mr Wong was an FAA/AST safety inspector for the two *SpaceShipOne* missions conducted by Scaled Composites to win the X Prize. Since 2001, Mr Wong has been leading a Human Flight Safety Team within AST to identify and assess commercial human space-flight-related issues. Mr Wong led this team in the development and drafting of guidelines for flight crew and space-flight participants on commercial sub-orbital RLV missions. He has both BS and MS degrees in Mechanical Engineering from the University of Maryland.

Olga Zhdanovich is an Increment and Mission Integration Consultant at RheaTech for the European Space Agency. In 1990 she graduated with honours from the Moscow Institute of Engineers in Geodesy, Aerial Surveying and Cartography as Engineer of Cartography. Seven years later she received an MSc in Environmental Science and Policy from Central European University, Hungary/University of Manchester, UK. She also holds a degree in Financial Management from the Moscow Aviation Institute. Although her principal area is remote sensing, for 20 years of her professional career she worked as a consultant on commercial application of space technologies in Earth observation, satellite navigation and telecommunication, and educational projects. Olga is an active member of national and international committees involved in aerospace. She was the co-coordinator of the

Forum on Space Activities in the 21st Century at UNISPACE III organized by the International Astronautical Federation and the International Space University. She has been an Associate of Athena Global Since 2001, and she was a member of the Russian Federation Delegation to UN COPOUS. She is a member of the Faculty of the International Space University and Co-Chair of the IAF Subcommittee on Global Workforce Development. Olga has received a number of awards and scholarships that includes a scholarship from the Royal Dutch National Academy of Sciences in 1995 for the International Institute of Applied Systems Analysis Young Scientist Program and holds an award from the European Space Agency for her research paper from 1998. She has authored a number of publications as chapters in books and conference papers on various applications of space technology, as well as the Russian Space Program. Olga was one of the three founding editors of the *Novosti Kosmonavtiki* magazine in Russia in 1991.

Introduction to space safety regulations and standards

Joseph N. Pelton* and Ram S. Jakhu†

* *Former Dean, International Space University,*

† *Associate Professor, Institute of Air and Space Law, McGill University*

INTRODUCTION

The term "space safety", as applied throughout this book, is quite broad. It includes the protection of human life and/or spacecraft during all phases of a space mission, regardless of whether this is a "manned" or "unmanned" activity. The concept of space safety covers:

- all aspects from pre-launch, launch, orbital or sub-orbital operations, through re-entry and landing;
- the protection of ground and flight facilities and surrounding population and buildings in proximity to launch sites; and
- the protection of space-based services, infrastructure and unmanned satellites such as communications satellite networks, global navigation systems, and remote sensing and surveillance systems, as well as scientific satellites.

In view of this broad concept of space safety, the space safety regulations and standards extend to the environmental effects of space operations, e.g. the impact on the safety of the public at large when it comes to such aspects as the longer-term negative impact of solid-fueled rockets on the protective ozone layer and the effects of perchlorate contaminants from rocket fuels that can end up in drinking water supplies.

A prerequisite for a safe and sustainable use of space therefore implies, at least, the following:

- The control of space debris and improved space situational awareness.
- International rules for space traffic.
- Internationally agreed standards and regulations to achieve compatibility among various space systems and facilities, as well as improved safety and reliability.
- Systematic elimination of weapons and other threats from space.
- Protection of people and facilities on the ground from both direct hazards and indirect environmental effects.
- And, in time, international management and controls to ensure safety for spacecraft and all types of aircraft.

There is a growing awareness that space has become much like international sea and airspace, the other realms where it is in the interest of the global community to operate in accordance with increasingly common and harmonized regulations and standards. Ultimately, space safety is a shared interest requiring international cooperative efforts. Developing increasingly effective "rules of the road" and "common standards" for space safety—at the national, regional, and international levels—need to, and will, continue and improve. These safety regulations and standards will thus evolve as the need for such measures is increasingly recognized, and an increased international participation in space makes them necessary.

THE GREAT CHALLENGE OF ACHIEVING SPACE SAFETY

Space activities remain a high-risk enterprise. As of 2010, at least 200 people have been killed by rocket explosions during processing, testing, launch preparations, and launch operations. Thirty-five fatalities have already occurred in this century. In the last 10 years at least six launches were terminated by a launch range safety officer to prevent risk to the public. Several more cases involved launches where the vehicle did not make it to orbit and came crashing back to Earth. A total of 22 astronauts and cosmonauts have lost their lives since the beginning of human spaceflight; four of these occurred on the ground during training—one Russian, plus the *Apollo 1* crew. Thus, about 4% of those who have flown into space have died. The shuttle *Columbia* accident in 2003 also posed a serious risk to civil aviation traffic when a "curtain" of debris "covered the southwestern portion of the United States". After this catastrophe, it was estimated that the risk of a collision between the debris from *Columbia* and commercial aircraft was of the order of 1 in 1000 and for general aviation it was as high as 1 in 100. As of today, there have been 10 cases of space system failures that have resulted in the dispersal of radioactive material. The orbital debris population, a major threat to both manned and unmanned vehicles, will continue to grow if not controlled. This spread of space debris could threaten the safe use of valuable regions of outer space by future generations.

Non-functional satellites, spent launch vehicle upper stages and other hardware, especially in low Earth orbit, do not remain in orbit indefinitely but gradually return to Earth. Between 60% and 90% of so-called returning "space junk" will burn up because of friction with the atmosphere at high velocity. Some components and parts can and do survive the re-entry heating and return to Earth. Tons of such space debris rain down every year. On average, one non-functional satellite, launch vehicle orbital stage, or other piece of cataloged debris has re-entered every day for more than 40 years. In February 2008, the USA shot down a malfunctioning satellite (*USA-193*) because of alleged public safety concern. The increase in worldwide space activities will only serve to increase the public safety risk, but no common acceptable risk level has yet been defined.

THE INCREASING CHALLENGE

The future of space safety presents an ever-greater challenge. We see a pattern of widening space activities, new stakeholders, increasing commercial initiatives, more space applications and networks, growing levels of space debris, and a lack of agreed international space safety standards or common regulatory approach to space systems.

The space age started in the middle of the twentieth century. For more than two decades, space capability would be an almost complete monopoly by the competing Cold-War powers of the USA and the USSR. Europe initially lagged behind, but established a substantial presence later in the century, both in commercial launch services and space sciences.

Today, space access capabilities have widened substantially worldwide. Some 50 countries have satellites on-orbit and over 100 countries utilize satellite communications and other commercial space services. Over 10 countries have unmanned orbital launch capability. A further 20 countries have sub-orbital flight capabilities—from Argentina to Syria. In 2003, China became the third country capable of sending humans into space and in 2008 even staged a successful spacewalk. India has announced plans for a human space-flight program leading to a planned first mission in 2014 and even a landing on the Moon, projected for 2020.

THE START OF AN INTERNATIONAL SPACE SAFETY STANDARDS PROCESS

Today, the creation of international space safety standards is at an embryonic stage. The International Standards Organization (ISO) has adopted a handful of standards that create a foundation for further efforts. These ISO standards include the basic space policy standards found in ISO 14300 and detailed as follows:

- ISO 14620 that covers Space Systems Safety Requirements Parts 1–3, including System Safety, Launch Site Operations, and Flight Safety Systems.
- ISO 17666 that addresses Space Systems Risk Management.
- ISO 14624 Parts 1–7 that cover Space Systems in terms of Safety and Compatibility of Materials.

Also, the United Nations' Orbital Debris Co-Ordination Working Group (ISO TC 20/SC 14) has developed and adopted the Space Debris Mitigation Principles and Management procedures (N318 (WD24113)) and the Re-entry Safety Control for Unmanned Spacecraft and Launch Vehicles Upper Stages Safety (WD 27875). While the ISO standards are both desirable and useful in establishing international coordination, the significant problem is that ISO standards are voluntary. National laws and regulations may supersede and override ISO standards. In order for international space regulations and standards to have enforcement power, they must be supported by governments as well as voluntary standards bodies.

National space programs and their oversight are currently controlled by government agencies. These controlling agencies include the Federal Aviation Administration (FAA) Office of Space Commercialization, NASA and the US Air Force in the United States, the European Space Agency (ESA) and the European Aviation Safety Agency (EASA) in Europe, and Roscosmos in Russia. The effective implementation of international space safety standards thus requires support by the relevant national or regional space agencies, otherwise they could be overridden by national regulatory bodies. In short, the ultimate objective must be standards and regulations backed by international treaties that are fully agreed by all the nations involved in space activities and implemented through national regulatory mechanisms.

To this end, the International Association for the Advancement of Space Safety (IAASS) has developed a series of proposals to create an international space safety institute that can develop globally accepted space standards, to independently assess and evaluate various space systems designs and subsystem performance. The IAASS has also assessed such issues as whether an organization equivalent to the International Civil Aviation Organization (ICAO) that regulates international aviation commerce and its safety should be established for the regulation of space safety. In addition, the IAASS has developed a draft Memorandum of Understanding that might be agreed upon by States to establish a framework for setting enforceable space safety standards and agreeing on space safety regulations.

Further, the IAASS has proposed four principles under which future goals for international space safety standards might be effectively pursued. These are to:

- ensure that citizens of all nations are equally protected from the risk of overflying rockets, space vehicles, and returning spacecraft;
- ensure that any spacecraft (manned or unmanned) is developed, built, and operated according to uniform minimum safety standards that reasonably reflect the status of knowledge and the accumulated experience;

- prevent the risk of collision or interference during transit in the airspace and on-orbit operations;
- allow mutual assistance and cooperation in case of emergency.

It is further believed that the pursuit of these principles should be carried out in a systematic manner with regard to the following five space safety standards groupings:

- The Public Safety Risk of Space Missions
- Ground Processing Safety
- On-Orbit Space Traffic Control
- Safety and Rescue
- Space Debris.

It is recognized that four of these five areas are well understood and agreed upon as significant areas to address in terms of internationally coordinated practices. The issue of on-orbit space traffic control remains controversial at this time. As in all matters involving international coordination and agreement, such matters take time and careful give and take before a global consensus is reached.

THE SCOPE AND AIM OF SPACE SAFETY REGULATIONS AND STANDARDS

The purpose of this book is to provide as much information and insight as possible as to where the current status of standards and regulations related to space safety stand today and to understand how different nations and international bodies—particularly the ISO and the United Nations—are approaching these matters at this time. The "Draft Memorandum of Understanding (MOU)" as provided in Appendix 1 of this book provides a broad-brush framework under which nations might agree to work together on a systematic basis in the field of space safety.

Throughout this book, there has been an attempt to capture and present the latest information with regard to the development of national, regional, and international space safety regulations and standards. This book has been divided into various parts to assist readers and professional users. These parts are: Introduction; (1) Developing and Improving Space Safety Regulations and Standards for Manned and Unmanned Space Systems; (2) Safety Regulations and Standards for Unmanned Space Systems; (3) Regulating Commercial Space Flight; (4) International Regulatory and Treaty Issues; (5) Creating Technical and Regulatory Standards for the Future; (6) Conclusions and Next Steps; (7) Various detailed appendices providing useful information with regard to creating space safety certification standards, draft "rules of the roads" for space activities, and more.

This book also outlines the current status of international treaties and agreements in force with regard to the exploration and use of outer space, and addresses how various spacefaring nations are addressing their own approaches to space and space safety.

In short, this book seeks to be as comprehensive as possible in addressing international and national space safety regulation and standards—both as these agreements exist today and as they may evolve tomorrow.

Over two dozen authors from around the world have contributed to this cause. These include governmental officials, academic scholars, and representatives from the aerospace industry, as well as those seeking to develop commercial space launch capabilities.

PART

1

Developing and Improving Space Safety Regulations and Standards for Manned and Unmanned Space Systems

CHAPTER 1

NASA space safety standards and procedures for human-rating requirements[1]

Charles Herbert Shivers

Deputy Director, Safety and Mission Assurance Directorate, Marshall Space Flight Center, NASA

CHAPTER OUTLINE

INTRODUCTION

This chapter identifies the standards and requirements the National Aeronautics and Space Administration (NASA) uses for major programs, including some detailed explanation for the Space Shuttle Program. Certainly, the International Space Station and Constellation are the two other large programs of NASA, but including the same level of detail herein for those major programs, as well as for major scientific endeavors being conducted via other organizations within the Agency, would likely be minimally instructive. Instead, one can obtain relevant information on these programs via simple internet searches, and safety regulations and standards in all cases follow parallel processes and decision-making flows. In short, the exhaustive listing of all the safety regulations and standards documents would be tedious and not very helpful to understanding the underlying process. Instead, it is far more important to provide a description of the flow of requirements, to identify the major policy and requirements documents, and to explain these documents' tiered relationship to lower-level programs, along with general pointers to detailed specific standards and requirements that constitute

[1]Information presented herein is taken from NASA publications and is available publicly and no information herein is protected by copyright or security regulations. While NASA documents are the source of information presented herein, any and all views expressed herein and any misrepresentations of NASA data that may occur herein are those of the author and should not be considered NASA official positions or statements, nor should NASA endorsement of anything presented in this work be assumed.

Space Safety Regulations and Standards. DOI: 10.1016/B978-1-85617-752-8.10001-7

the overall safety program within NASA. In addition, a top-level document tree for the Constellation Program is included for illustrative purposes for the planned future work of the Agency in Space Exploration. This document tree provides a useful and simple illustration of the "flow down" of requirements for a non-crewed science mission launched on an expendable launch vehicle. Some discussion of the NASA Human Ratings process is also provided.

1.1 NASA TOP-LEVEL DOCUMENTATION

Where to start is as challenging as is the enumeration of the multitude of documents that constitutes the overall NASA safety standards and regulation. NASA's governance policy is as likely a place to start as any other. It is from this document that much of the roles and responsibilities are authorized and enumerated. NASA governance is described in NASA Policy Directive—NPD 1000.0, "Strategic Management and Governance Handbook", which is a responsibility of the NASA Administrator's Office.

This NPD sets forth NASA's governance framework—principles and structures through which the Agency manages mission, roles, and responsibilities. This policy directive describes NASA's strategic management system. It sets forth the specific processes by which the Agency manages strategy and its implementation through planning, performance, and assessment of results. In addition, as a United States Federal Agency, NASA has a public responsibility prescribed in law. "NASA must meet the intent of the National Aeronautics and Space Act of 1958, which established the Agency for the purpose of expanding human knowledge in aeronautical and space activities for the benefit of all humankind. NPD 1000.0A conveys NASA's strategic approach to achieving the Agency's Mission" [1].

In NPD 1000.0, one finds reference to other top-level Agency documents pertaining to strategic management and to organization. One also finds a description of the NASA core values, NASA's governance principles, and the strategic management system that defines how the Agency establishes and conducts its missions. NASA's governance provides a check and balance system providing separate and specific authority to programs and institutional entities as they pursue the common goal of mission safety and success. Technical Authority (TA) is a particular item of interest and is delineated separately from programmatic authority such as institutional authority feeding into the Administrator. The programs hold certain authorities of risk acceptance and decision-making, while Institutional TA is provided via Center Directors, Mission Support Authority, Engineering TA, Safety and Mission Assurance TA, and Health and Medical TA. Each of these TA entities has a specific realm of authority that provides the checks and balances of Agency decision-making in executing missions. Authorities are exercised in review processes, requirements tailoring, dissenting opinions, etc. Proper implementation of the tenets in NPD 1000.0 demands a proper balance of authority, responsibility, and accountability, all of which are described in the NPD.

Perhaps the most useful of NASA's technical guidance documents to provide understanding of program and project execution is NPR 7120.5D, "NASA Program and Project Management Processes and Requirements", which establishes the requirements by which NASA will formulate and implement space-flight programs and projects, consistent with the governance model contained in the NASA Strategic Management and Governance Handbook (NPD 1000.0). Table 1.1 shows the hierarchy of NASA programmatic requirements. Figure 1.1 shows the document hierarchy and flowdown methodology.

Table 1.1 NASA Programmatic Requirements Hierarchy, taken from NPR 7120.5D, "NASA Program and Project Management Processes and Requirements"

Direction	Content	Governing Document	Approver	Originator
Needs, Goals, Objectives	Agency strategic direction based on higher-level direction	Strategic Plan and Strategic Planning Guidance	Administrator	Support Organizations
Agency Requirements	Structure, relationships, principles governing design and evolution of cross-Agency Mission Directorate systems linked to accomplishing Agency needs, goals, and objectives	Architectural Control Document (ACD)	Administrator	Host MDAA with inputs from other affected MDAAs
Mission Directorate Requirements	High-level requirements levied on a Program to carry out strategic and architectural direction, including programmatic direction for initiating specific projects	Program Commitment Agreement (PCA)	AA	MDAA
Program Requirements	Detailed requirements levied on a Program to implement the PCA and high-level programmatic requirements allocated from the Program to its projects			
Project Requirements	Detailed requirements levied on a Project to implement the Program Plan and flowdown programmatic requirements allocated from the Program to the Project	Project Plan	Program Manager	Project Manager
System Requirements	Detailed requirements allocated from the Project to the next lower level of the Project	Systems Requirements Documentation	Project Manager	Responsible System Lead

AA, NASA Associate Administrator; MDAA, Mission Directorate Associate Administrator

Direction	Content	Governing Document	Approver	Originator
Needs, Goals, Objectives	Agency strategic direction based on higher-level direction	Strategic Plan and Strategic Planning Guidance	Administrator	Support Organizations
Agency Requirements	Structure, relationships, principles governing design and evolution of cross-Agency systems linked in accomplishing Agency needs, goals, and objectives	Architectural Control Document (ACD)	Administrator	Host MDAA with Inputs from Other Affected MDAAs
Mission Directorate Requirements	High-level requirements levied on a Program to carry out strategic and architectural direction including programmatic direction for initiating specific projects	Program Commitment Agreement (PCA)	AA	MDAA
Program Requirements	Detailed requirements levied on a Program to implement the PCA and high-level programmatic requirements allocated from the Program to its projects	Program Plan	MDAA	Program Manager
Project Requirements	Detailed requirements levied on a Project to implement the Program Plan and flow-down programmatic requirements allocated from the Program to the Project	Project Plan	Program Manager	Project Manager
System Requirements	Detailed requirements allocated from the Project to the next lower level of the Project	System Requirements Documentation	Project Manager	Responsible System Lead

MDAA = Mission Directorate Associate Administrator
AA = NASA Associate Administrator

FIGURE 1.1

Program/Project Management Document Hierarchy, taken from NPR 7120.5D, "NASA Program and Project Management Processes and Requirements".

Also useful to understanding the details of programmatic execution is NPR 7123.1, "Systems Engineering Procedural Requirements", which clearly articulates and establishes the requirements on the implementing organization for performing, supporting, and evaluating systems engineering. Systems engineering is a logical systems approach performed by multidisciplinary teams to engineer and integrate NASA's systems to ensure NASA products meet customers' needs (Figure 1.2). Implementation of this systems approach enhances NASA's core engineering, management, and scientific capabilities and processes to ensure safety and mission success, increase performance, and reduce cost. This systems approach is applied to all elements of a system and all hierarchical levels of a system over the complete project life cycle [2].

Another useful illustration from NPR 7123.1, "Systems Engineering Procedural Requirements", is the logical decomposition process shown in Figure 1.3. The process is used to improve understanding of the defined technical requirements and the relationships among the requirements (e.g. functional, behavioral, and temporal) and to transform the defined set of technical requirements into a set of logical decomposition models and their associated set of derived technical requirements for input to the design solution definition process.

1.2 NASA TOP-LEVEL SAFETY STANDARDS AND REQUIREMENTS

The NASA Office of Safety and Mission Assurance is responsible for, among others, NPD 8700.1E, "NASA Policy for Safety and Mission Success", which provides top-level policy and responsibilities,

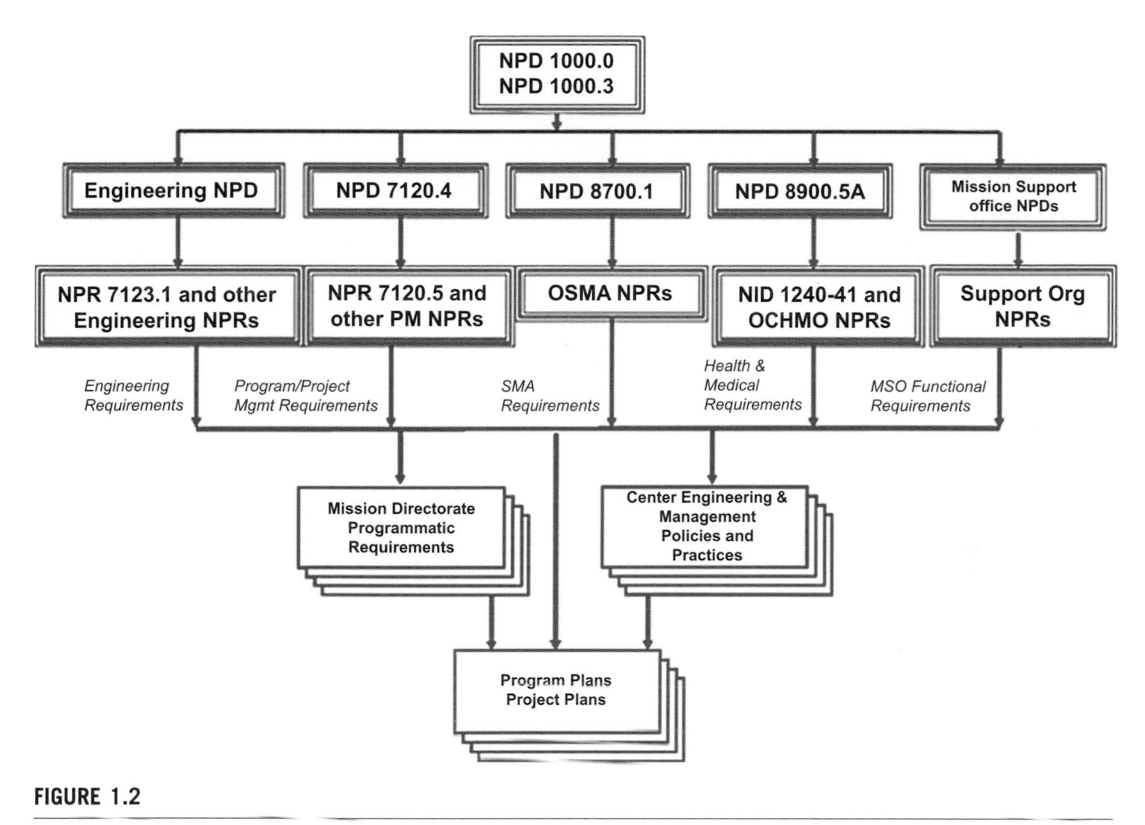

FIGURE 1.2

The Systems Engineering Engine, taken from NPR 7123.1, "Systems Engineering Procedural Requirements".

and NPR 8715.3 "General Safety Program Requirements", which provides the top-level safety requirements for all Agency activities. "This NASA Procedural Requirements (NPR) provides the basis for the NASA Safety Program and serves as a general framework to structure more specific and detailed requirements for NASA Headquarters, Programs, and Centers" [3]. The document is not a standalone document, but is used in conjunction with the references contained within the document. In a broad sense, that reference section is an exhaustive bibliography of NASA's safety documentation. Thirty-one documents are referred to in the "Authority" section, and 99 are included in the "Applicable Documents" section.

Within those references are the Federal Laws and Directives with which NASA must comply, including those specific to activities in space exploration. Some items of particular interest to this subject are (the original numbering from the parent document is intentional):

j. NPD 8700.1, "NASA Policy for Safety and Mission Success."
k. NPD 8700.3, "Safety and Mission Assurance (SMA) Policy for Spacecraft, Instruments, and Launch Services."
m. NPD 8710.3, "NASA Policy for Limiting Orbital Debris Generation."

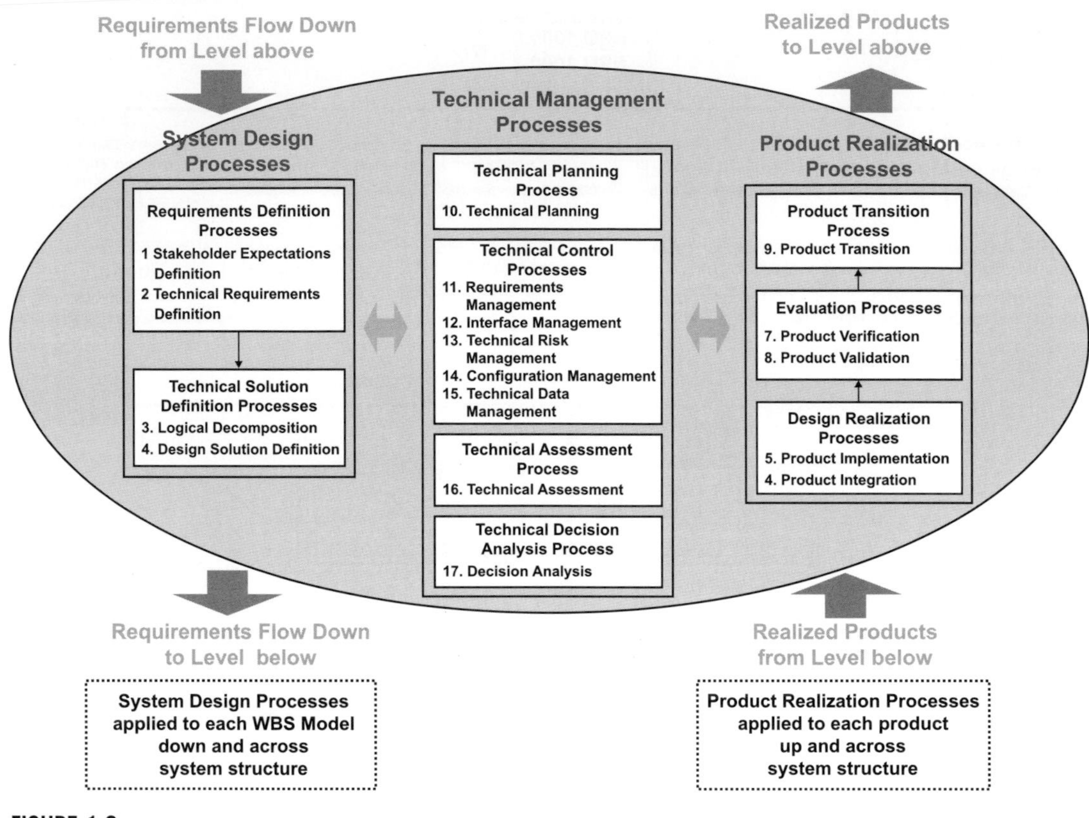

FIGURE 1.3

Logical decomposition process.

- **n.** NPD 8710.5, "NASA Safety Policy for Pressure Vessels and Pressurized Systems."
- **o.** NPR 8715.7, "Expendable Launch Vehicle Payload Safety Program."
- **p.** NPD 8720.1, "NASA Reliability and Maintainability (R&M) Program Policy."
- **q.** NPD 8730.5, "NASA Quality Assurance Program Policy."
- **aa.** NPR 7120.5, "NASA Program and Project Management Processes and Requirements."
- **ab.** NPR 7120.6, "Lessons Learned Process."
- **ac.** NPR 7123.1, "Systems Engineering Procedural Requirements."
- **ad.** NPR 7150.2, "NASA Software Engineering Requirements."
- **af.** NPR 8000.4, "Risk Management Procedural Requirements."
- **ai.** NPR 8705.2, "Human-Rating Requirements for Space Systems."
- **aj.** NPR 8705.4, "Risk Classification for NASA Payloads."
- **ak.** NPR 8705.5, "Probabilistic Risk Assessment (PRA) Procedures for NASA Programs and Projects."
- **al.** NPR 8705.6, "Safety and Mission Assurance Audits, Reviews, and Assessments."
- **ao.** NPR 8715.5, "Range Safety Program."

ap. NPR 8715.6, "NASA Procedural Requirements for Limiting Orbital Debris."
ar. NASA-STD-8709.2, "NASA Safety and Mission Assurance Roles and Responsibilities for Expendable Launch Vehicle Services."
at. NASA-STD-8719.8, "Expendable Launch Vehicle Payload Safety Review Process Standard."
aw. NASA-STD-8719.13, "Software Safety Standard."
ax. NASA-STD-8739.8, "Software Assurance Standard."
ba. NSS 1740.14, "Guidelines and Assessment Procedures for Limiting Orbital Debris."
bb. MIL-STD-882, "Standard Practice for Safety Systems."
bd. SSP 50021, "Safety Requirements Document."
bi. "Wallops Flight Facility Range Safety Manual."
bj. AFSPCMAN 91710, "Licensing and Safety Requirements for Launch."
cp. "Eastern and Western Range (EWR) 127-1, "Range Safety Requirements."
cq. NASA SP 8013, "NASA Micrometeoroid Environment Model (Near Earth to Lunar Surface)."
cr. NASA SP 8038, "Micrometeoroid Environment Model (Interplanetary and Planetary)."
cs. SSP 30425, "Space Station Program Natural Environment Definition for Design."
ct. NASA TM 4527, "Natural Orbital Environment Guidelines for Use in Aerospace Vehicle Development."
cu. McNamara, H., Suggs, R., Kauffman, B., Jones, J., Cooke, W. and Smith, S. (2004). Meteoroid Engineering Model (MEM): A Meteoroid Model for the Inner Solar System. *Earth Moon and Planets*, 95:123−139.

NASA uses these and other requirements to meet its stated goal: "NASA's goal is to maintain a world-class safety program based on management and employee commitment and involvement; system and worksite safety and risk assessment; hazard and risk prevention, mitigation, and control; and safety and health training" [4].

NASA's Safety and Mission Assurance Requirements Tree is shown in Figure 1.4. In this figure, one sees the relationship of the top-level documents and the flow to S&MA disciplines and programs and projects. In addition, reference is made to the NASA Technical Standards Program and to the NASA Directives System where the documents listed above and others may be found. Again, a simple internet search will yield links to these systems or documents, but being designed primarily for the use of NASA employees, there are specific requirements for system access, left to the reader to explore.

1.3 NASA HUMAN RATINGS PROCESS

Of particular interest are the Human Rating requirements imposed by NASA on select systems. Many NASA systems require "Human Rating". Systems requiring Human Rating must implement additional processes, procedures, and requirements necessary to produce human-rated space systems that protect the safety of crew members and passengers on NASA space missions.

Human-rated systems accommodate human needs, effectively utilize human capabilities, control hazards, and manage safety risk associated with human space flight, and provide, to the maximum extent practical, the capability to safely recover the crew from hazardous situations. Human rating is an integral part of all program activities throughout the life cycle of the system, including: design and development; test and verification; program management and control; flight readiness certification; mission operations; sustaining engineering; maintenance, upgrades, and disposal.

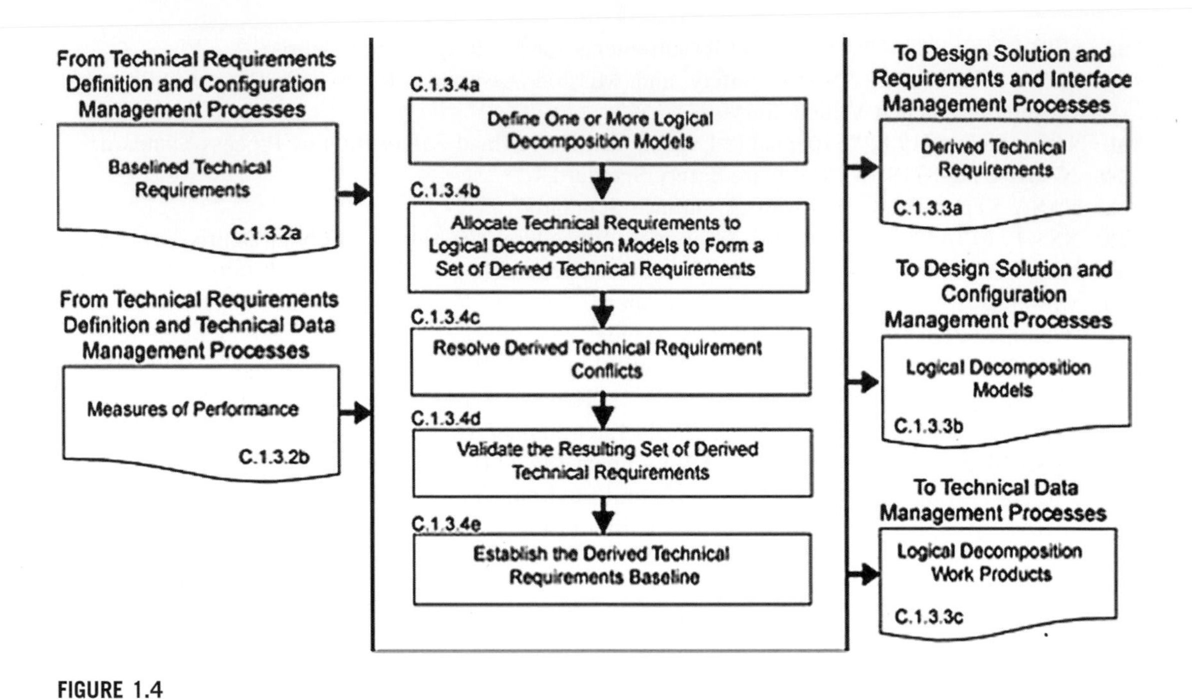

FIGURE 1.4

NASA Safety and Mission Assurance Requirements Tree.

The Human-Rating Certification is granted to the crewed space system but the certification process and requirements affect functions and elements of other mission systems, such as control centers, launch pads, and communication systems. The types of crewed space systems that require a Human-Rating Certification include, but are not limited to, spacecraft and their launch vehicles, planetary bases and other planetary surface mobility systems that provide life support functions, and Extravehicular Activity (EVA) suits. A crewed space system consists of all the system elements that are occupied by the crew during the mission and provide life support functions for the crew. The crewed space system also includes all system elements that are physically attached to the crew-occupied element during the mission, while the crew is in the vehicle/system.

Verification of program compliance with the Human Ratings requirements is performed in conjunction with selected milestone reviews (System Requirements Review (SRR), System Definition Review (SDR), Preliminary Design Review (PDR), Critical Design Review (CDR), System Integration Review (SIR) and the Operational Readiness Review (ORR)) conducted in accordance with the requirements of NPR 7120.5, "NASA Space Flight Program and Project Management Requirements", and NPR 7123.1, "NASA Systems Engineering Processes and Requirements". NPR 8705.2, "Human-Rating Requirements for Space Systems", specifies development of products that are reviewed at each of the selected milestone reviews. The adequacy of those products and the acceptability of progress toward Human-Rating Certification are used to verify compliance. In addition, the Human-Rating requirements and processes are subject to audit and assessment in accordance with the requirements contained within NPR 8705.6, "Safety and Mission Assurance Audits, Reviews, and Assessments".

NPR 8705.2, "Human-Rating Requirements for Space Systems", also defines and delineates specific responsibilities for Human Rating including overall authority assigned to the NASA Associate Administrator and assurance of implementation assigned to the Chief, Safety and Mission Assurance, and the NASA Chief Engineer as Technical Authorities within their realms of responsibility. Additional responsibilities and authorities are described as appropriate.

The Human-Rating Certification Process is linked to five major program milestones: System Requirements Review, System Definition Review, Preliminary Design Review, Critical Design Review, and Operational Readiness Review. The program's compliance with the Human-Rating requirements and the contents of the Human-Rating Certification Package are endorsed and approved by all three Technical Authorities (Safety, Engineering, and Health and Medical) at each of the five milestones. Since it is not the intention of this article to restate the documented requirements and processes required for human rating, a summary of major certification elements from NPR 8705.2 is hereby provided:

- **a.** The definition of reference missions for certification.
- **b.** The incorporation of system capabilities to implement crew survival strategies for each phase of the reference missions.
- **c.** The implementation of capabilities from the applicable technical requirements.
- **d.** The utilization of safety analyses to influence system development and design.
- **e.** The integration of the human into the system and human error management.
- **f.** The verification, validation, and testing of critical system performance.
- **g.** The flight test program and test objectives.
- **h.** The system configuration management and related maintenance of the Human-Rating Certification.

> *A human-rated system accommodates human needs, effectively utilizes human capabilities, controls hazards with sufficient certainty to be considered safe for human operations, and provides, to the maximum extent practical, the capability to safely recover the crew from hazardous situations. Human-rating consists of three fundamental tenets: (i) Human-rating is the process of designing, evaluating, and assuring that the total system can safely conduct the required human missions; (ii) Human-rating includes the incorporation of design features and capabilities that accommodate human interaction with the system to enhance overall safety and mission success; and (iii) Human-rating includes the incorporation of design features and capabilities to enable safe recovery of the crew from hazardous situations. Human-rating is an integral part of all program activities throughout the life cycle of the system, including design and development; test and verification; program management and control; flight readiness certification; mission operations; sustaining engineering; maintenance, upgrades, and disposal [5].*

Specific requirements, including specific technical requirements, are described in detail in Chapter 3 of NPR 8705.2 and are publicly available for review.

1.4 FLOWDOWN METHODOLOGY

NASA requirements are designated by level and flow down to lower work levels gaining more specificity as the levels change. All programs and projects are required to specifically identify those requirements that are applicable and track and verify status of completion in verification plans before mission execution. In conjunction with the Technical Authorities, agreement is reached for the complement of requirements for specific projects. Generally the levels can be described as follows:

- Level 0—Top-Level Agency Requirements controlled by the Administrator and Associate Administrators.
- Level 1 Requirements—Mission Drivers controlled by a NASA Mission Directorate and serving as the basis for mission assessment during development. These are NASA requirements and standards (latest versions apply).
- Level 2 Requirements—System/Segment (Mission Requirements Document—to be baselined at System Requirements Review).
- Level 3 Requirements—Element (Instruments, Spacecraft, etc.).
- Level 4 Requirements—Subsystem (Instrument Subsystem, Spacecraft Subsystem, etc.).
- Level 5 Requirements—Component (Instrument Component, Spacecraft Component, etc.).

1.5 SPACE SHUTTLE PROGRAM

The Space Shuttle Program (SSP) Manager documents and controls program requirements in Volumes I–XVIII of NSTS 07700, "Program Definition and Requirements Document" [6]. This document is supported by more than 225 subordinate NSTS documents, more than 400 applicable documents, and thousands of project- and element-level documents. Specific content topics of NSTS 07700 are structured as shown in Table 1.2. In addition, there is a Shuttle Master Verification Plan supported by subordinate verification plans, Shuttle System Interface Control Documents, Payload Interface Control Documents, and Operations and Maintenance Documents. NSTS 08171, "Operations and Maintenance Requirements and Specifications Document (OMRSD)", is the location for on-line operations and maintenance tasks to support Space Shuttle turnaround. NSTS 08151, "Intermediate and Depot Maintenance Requirements Document (IDMRD)", is the location for off-line operations and maintenance tasks in support of Space Shuttle turnaround. The Certification of Flight Readiness (CoFR) process, documented in NSTS 08117, "Space Shuttle Requirements and Procedures for Certification of Flight Readiness", constitutes the main part of the SSP risk management review process. NSTS 16007, "Shuttle Launch Commit Criteria and Background Document", is the Space Shuttle Program baseline document that provides the launch support system and the launch team with the Shuttle launch commit criteria and background information.

NSTS 5300.4(1D-2), "Space Shuttle Safety, Reliability, Maintainability and Quality Provisions for the Space Shuttle Program", establishes common safety, reliability, maintainability, and quality provisions for the Space Shuttle Program. NSTS 22206, "Requirements for Preparation and Approval of Failure Modes and Effects Analysis (FMEA) and Critical Items List (CIL)", provides detailed instructions for the preparation of Failure Modes and Effects Analyses (FMEAs) and Critical Items Lists (CILs). NSTS 22254, "Methodology for Conduct of Space Shuttle Program Hazard Analyses", provides the methodology required for the preparation of SSP hazard analyses, hazard reports, safety analysis reports, and Management Safety Assessments. NSTS 08209, "Shuttle Systems Design Criteria", is a seven-volume set of design criteria and performance requirements. The Shuttle Operational Data Book, "NSTS 08934 (JSC 08934)", also a seven-volume set, is the single authoritative source of properly validated data, which most accurately and completely describe the Shuttle operational performance capabilities and limitations. Other critical Shuttle Program documents include:

Table 1.2 NSTS 07700 Contents

NSTS 07700 Documents (Current Issue)	Document Title
Volume I	Program Description and Requirements Baseline
Volume II, Book 2	Program Structure and Responsibilities—Book 2, Space Shuttle Program Directives
Volume II, Book 3	Program Structure and Responsibilities—Book 3, Space Shuttle Program Interface Agreements
Volume III	Flight Definition and Requirements Directive
Volume IV, Book 1	Configuration Management Requirements—Book 1, Requirements
Volume IV, Book 2	Configuration Management Requirements—Book 2, Configuration Deviations/Waivers
Volume V	Information Management Requirements
Volume VI	Flight Support Equipment (FSE) Management
Volume VIII	Operations
Volume IX	Ground Systems Integration and Operations
Volume X, Book 1	Flight and Ground System Specification—Book 1, Requirements (Sections 1.0–6.0)
Volume X, Book 2	Flight and Ground System Specification—Book 2, Environment Design, Weight and Performance, and Avionics Events (Apx 10.3–10.16)
Volume X, Book 3	Flight and Ground System Specification—Book 3, Requirements for Runways and Navigation Aids (Apx 10.17)
Volume X, Book 4	Flight and Ground System Specification—Book 4, Active Deviations/Waivers
Volume X, Book 6	Flight and Ground System Specification—Book 6, Retired Deviations/Waivers (Apx 10.1)
Volume XI	System Integrity Assurance Program Plan
Volume XII	Program Logistics and Supportability Requirements
Volume XIV	Space Shuttle System Payload Accommodations
Volume XV	Resource Management Policy and Requirements
Volume XVIII, Book 1	Computer Systems and Software Requirements—Book 1, Allocation of Computational Functions
Volume XVIII, Book 2	Computer Systems and Software Requirements—Book 2, Allocation of Simulation Functions
Volume XVIII, Book 3	Computer Systems and Software Requirements—Book 3, Software Management and Control
NSTS 07700-10-MVP-01	Shuttle Master Verification Plan—Volume I, General Approach and Guidelines
NSTS 07700-10-MVP-02	Shuttle Master Verification Plan—Volume II, Combined Element Verification Plan
NSTS 07700-10-MVP-09, Part 1	Shuttle Master Verification Plan—Volume IX, Computer Systems and Software Verification Plan—Part 1, Guidelines and Standards
NSTS 07700-10-MVP-09, Part 2	Shuttle Master Verification Plan—Volume IX, Computer Systems and Software Verification Plan—Part 2, Verification Requirements

- NSTS 1700.7B, "Safety Policy and Requirements for Payloads Using the Space Transportation System".
- NSTS 08080-1, "Manned Spacecraft Criteria and Standards".
- NSTS 12820, "STS Operational Flight Rules".
- NSTS 08126, "Problem Reporting and Corrective Action (PRACA) System Requirements".
- NSTS 17462, "Flight Requirements Document (FRD) and Other Payload/Flight Related Documents".
- JSC 17481A, "JSC Safety Requirements Document for Space Shuttle Flight Equipment".

1.6 SPACE EXPLORATION (CONSTELLATION) PROGRAM

Constellation is the program NASA has developed to meet the United States' Vision for Space Exploration. The Program includes developing new launch and crew vehicles for journeys to the Earth's Moon, to Mars, and beyond. The Constellation Program Document Tree is shown in Figure 1.5.

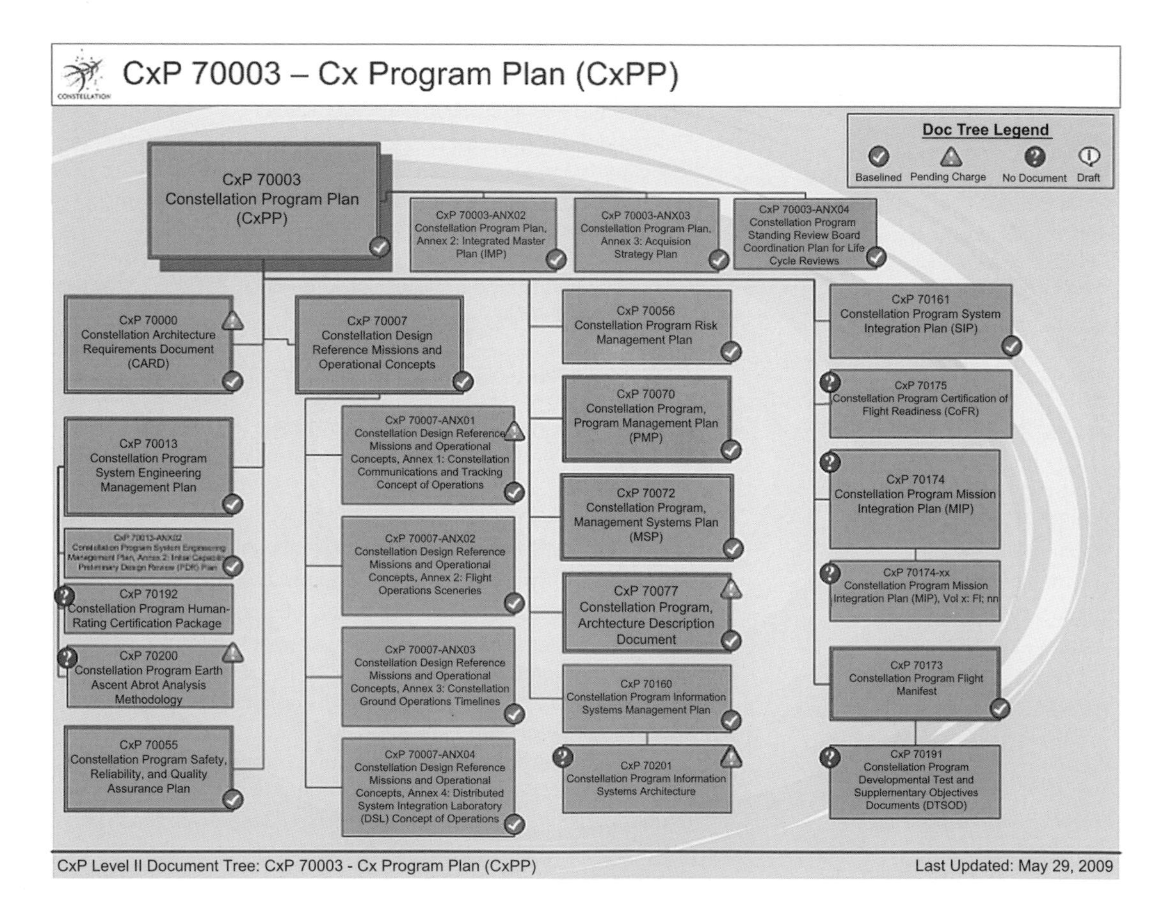

FIGURE 1.5

The Constellation Program Document Tree.

1.7 A TYPICAL NON-CREWED MISSION LAUNCHED ON AN EXPENDABLE LAUNCH VEHICLE

Program Level Requirements for the typical science program project launched on an ELV:

- Level 1 Requirements—Mission Drivers.
  - Program Level 1 Requirements—baselined after Key Decision Point, KDP-A:
    - Highest unique requirements for the project
    - Controlled by Science Mission Directorate, documented in the Program Plan Program Level Requirements Appendix
    - Imposed for the development and operation of the project
    - Serve as the basis for mission assessment during development
    - Provide the baseline for determination of science mission success
    - NASA requirements and standards—latest versions apply.
- Level 2 Requirements—System/Segment (project Mission Requirements Document).
  - Baseline at Systems Requirements Review.
- Level 3 Requirements—Element (Instruments, Spacecraft, etc.).
- Level 4 Requirements—Subsystem (Instrument Subsystem, Spacecraft Subsystem, etc.).
- Level 5 Requirements—Component (Instrument Component, Spacecraft Component).

CONCLUSIONS

Protecting the health and safety of humans involved in or exposed to space activities, specifically the public, crew, passengers, and ground personnel, is NASA's primary policy. The policy is implemented through the application of NASA directives and standards through a rigorous process of identification, allocation and verification. As the NASA activities are carried out, general and specific requirements are identified and allocated as appropriate to the programs and projects and verified to have been met before mission execution. The process is exhaustive and deliberate to ensure that safety and mission success are achieved. This article cannot cover the entirety of NASA's safety regulations and standards that are intended to execute its safety goals because of their sheer size and great complexity. It is hoped that this article and the various graphics provided herein can provide useful insight into the intricate inner workings of that painstaking process.

References

[1] NPD 1000.0, Strategic Management and Governance Handbook.
[2] NPR 7123.1, Systems Engineering Procedural Requirements.
[3] NPR 8715.3, General Safety Program Requirements, p. 17.
[4] NPR 8715.3, General Safety Program Requirements, p. 17.
[5] NPR 8705.2, Human-Rating Requirements for Space Systems.
[6] NSTS 07700, Program Definition and Requirements Document.

CHAPTER

US Department of Defense launch safety standards

2

Thomas Pfitzer
APT Research

CHAPTER OUTLINE

INTRODUCTION

The US Department of Defense (DoD) launch safety standards had their origins in the 1950s. These standards have evolved and improved as risks have become better understood, lessons have been learned, and launch systems have become more complex. Risk management practices have become the norm. From the outset, protection of the public has been a high priority. As the space launch industry becomes global and commercial organizations become responsible for ensuring safety, the experience gained in developing and improving standards within the US DoD becomes a valuable resource, serving as a foundation for managing future risks. This chapter provides a short history of the development of US DoD standards, followed by discussions of the current policies and standards now in use.

This article discusses how US DoD launch ranges address the cultural safety questions of "How safe is this launch?" and "How safe is safe enough?"

Space Safety Regulations and Standards. DOI: 10.1016/B978-1-85617-752-8.10002-9

2.1 LAUNCH SAFETY HISTORY

Within the US DoD, launches are performed by a system of national ranges established by public law. Each of these ranges operates as a semi-autonomous organization with a high degree of independence and freedom to establish its own policies regarding safety.

Following the Second World War, guided missile technology in the USA became a very popular topic with the US Congress which, in 1949, established requirements for joint long-range proving grounds. In developing these requirements, Congress established the first launch safety policy. The policy established that: "From a safety standpoint, they [launched vehicles] will be no more dangerous than conventional airplanes flying overhead." This policy established a baseline risk level for safety standards [1]. Until the late 1990s, national ranges had full latitude to interpret these requirements for their own use.

While there were many similarities in the interpretation, there were also differences. Because of these differences, a Risk and Lethality Commonality Team (RALCT) was formed in 1996 by the Range Commander's Council (RCC). The RALCT was formed to reach a consensus on reasonable common standards for falling debris protection criteria and analytical methods. The initial version, RCC 321-97, was a ground-breaking document in that, for the first time, a common basis and rationale was established for use by all ranges. The RALCT also developed approaches that were later adopted by other agencies and organizations, including the Federal Aviation Administration (FAA), the DoD Explosives Safety Board (DDESB), and the North Atlantic Treaty Organization (NATO). Following that initial effort, there have been numerous upgrades and additions. Currently, the RCC 321 consists of a standard with approximately 20 pages of text and a supplement containing more than 200 pages of analytical approaches and rationale supporting the criteria. The focus of this chapter is the RCC 321 Standard. The supplement contains background information that, although arcane in its detail, is nevertheless a valuable resource in tracing the history of US launch safety standards.

2.2 ASSIGNED RESPONSIBILITIES

The US DoD Directive 3200.11, Major Range and Test Facility Base (MRTFB) [2], assigns responsibility to each "Range Commander" for ensuring that all launch missions are conducted safely, consistent with operational requirements. Therefore, an important aspect of the range safety responsibility is to ensure that the risk is properly managed within prescribed limits. To accomplish this, each Range Commander (or designee) must:

1. Establish risk management procedures (including hazard containment) to implement the risk management process described in RCC 321.
2. Establish acceptable risk criteria appropriate to each type of mission flown in consideration of the guidance provided in RCC 321.
3. Accept any risks, including those that exceed the established risk criteria when warranted for a mission in consideration of the operational requirements and national need.
   a. Make such decisions based on a thorough understanding of any additional risk that exceeds the risk criteria and the benefits to be derived from taking the additional risk.
   b. Ensure such decisions are documented in a formal waiver process (or equivalent), preferably in advance of the mission.

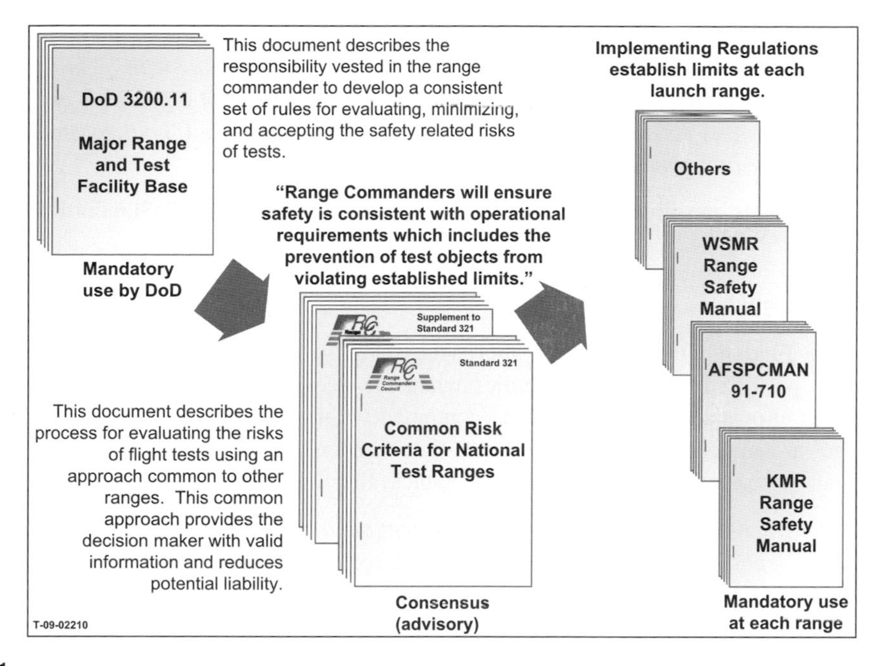

FIGURE 2.1

Overview of US Department of Defense Regulatory Process for Range Safety.

4. Maintain related range policy and requirements documents.
5. Maintain records of risk assessments and waivers to established risk criteria.

Implementation of the assigned responsibility is documented in a hierarchy of documents beginning with DoD 3200.11, detailed in RCC 321 Standard, and implemented at each launch agency with local regulations as shown in Figure 2.1.

2.3 GUIDING PRINCIPLES

The guiding principles which are the foundation of the RCC 321 Standard are (1) minimize risk, (2) contain all risks where possible, and (3) manage risks where containment is not possible. The first two of these are longstanding policies that can be traced to the earliest policy documents from the 1960s. The third, "risk management", has emerged to the forefront during the last decade. The implementation of each is discussed below.

2.3.1 Minimizing risks

The concept of minimizing risk is applied as a guiding philosophy. The standard says:

> *In planning any operation, risk must be reduced to the extent that is practical in keeping with operational objectives. Safety should be balanced with operational objectives by cooperative*

interaction between the range and the range user. To maximize achievement of mission objectives within safety constraints, the range user should consider overall risk along with other factors that affect mission acceptability. These factors include criticality of mission objectives, protection of life and property, the potential for high consequence mishaps, local political factors, and governing range or programmatic environmental requirements.

(All indented italics are direct quotations from Chapters 2 and 3 of RCC Standard 321 [3].)

2.3.2 Containment

There exist two fundamentally different approaches to ensuring launch safety. The containment approach is the preferred method and applies when the geography of the launch facility and its surrounding area allows all of the risks to be contained within boundaries having restricted access. Most US ranges are adjacent to oceans or deserts. Surveillance of these unpopulated areas can ensure a large area that is void of people at the time of launch, thus implementing containment.

All ranges must strive to achieve complete containment of hazards resulting from both normal and malfunctioning flights. If a planned mission cannot be accomplished using a containment approach, a risk management approach may be authorized by the Range Commander or the designated representative. The risk management approach should conform to the guidelines presented in this document or otherwise demonstrate compliance with the objectives presented.

For most space launches, the containment approach cannot be fully implemented. This leads to the need to manage risks.

2.3.3 Managing risks

While the first principle of minimizing risk defines an operational philosophy, the third, managing risk, implies an *active* role to ensure that reasonable steps are taken to reduce risks. These actions are considered necessary to demonstrate an effective program of managing risks.

Compliance with RCC 321 leads to defensible launch support and launch commit decisions. Employing a sound basis for accuracy and repeatability in risk assessments leads to consistent risk acceptance decisions, thereby fostering public confidence that the ranges are operated with appropriate regard for safety. Thus, individuals living or working at or near a range may go about their daily lives without concern for their proximity to range activities. Moreover, compliance with these guidelines provides assurance that flights near or over communities by space boosters or weapon systems does not significantly increase the risk to these communities.

In defining objectives for risk assessment and risk management, the RCC goals are to:

a. Create a uniform process among the ranges that will achieve the stated risk management goals.
b. Promote accurate, repeatable risk assessments by minimizing errors in estimating and ensuring their scientific validity.
c. Create a process that fosters innovation to support challenging missions.
d. Nurture openness and trustworthiness among the ranges, range users and the public.

To implement the risk management program, the standard defines objectives. The first and most important of these is to protect the general public.

> *General Public. The general public includes all people located on and off base that are not essential to a specific launch. This definition applies to all people, regardless of whether they are in some mode of transportation (such as airplanes, ships, and buses), are within a structure, or are unsheltered. The general public should not be exposed, individually or collectively, to a risk level greater than the background risk in comparable involuntary activities, and the risk of a catastrophic mishap should be mitigated.*

In the above context, the RCC considers "comparable involuntary activities" as those where the risk arises from man-made activities that:

a. Are subject to government regulations or are otherwise controlled by a government agency, and
b. Are of vital interest to the USA, and
c. Impose involuntary risk of serious injury or worse on the public.

Other objectives in managing risk include protection of mission essential personnel, operational personnel, ships, aircraft, and spacecraft.

2.4 THE RISK MANAGEMENT PROCESS

The dominant thrust of the DoD Launch Safety Standard is to manage risks. This is viewed as a management process that includes critical elements of: (1) risk identification, (2) assessment and understanding of risk, and (3) making informed decisions to accept or reject the risks. The process used to manage risk is well defined and is highly similar to risk management processes used by the DoD in Systems Safety and Explosives Safety.

Risk management is a systematic and logical process to identify hazards and control the risk they pose. This process should include the following elements (phases), which are identified as:

Phase I Mission Definition and Hazard Identification
Phase II Risk Assessment
Phase III Criteria Comparison and Risk Reduction
Phase IV Risk Acceptance.
(See Figure 2.1).

The initial goal of the risk management approach is to contain the hazards and isolate them from populated areas wherever practical. An alternative to hazard isolation is to define hazard containment areas so as to minimize the population exposed or be able to evacuate persons not associated with the hazard-generating event. This is in accordance with the primary policy that no hazardous condition is acceptable if mission objectives can be attained from a safer approach, methodology, or position, i.e. minimizing the hazards and conducting the mission as safely as reasonably possible. When hazards cannot be contained or minimized to an insignificant level, then assessments that are more detailed are performed to determine if the remaining risk is acceptable. An additional benefit of hazard containment is that this process is typically less costly than risk assessments and can be evaluated relatively quickly with straightforward assumptions and with fewer required data.

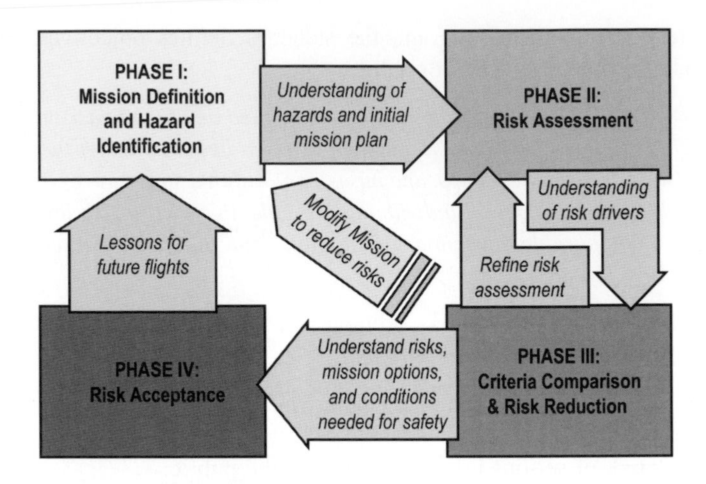

FIGURE 2.2

The Four Phases of US Department of Defense Risk Management.

The RCC 321 Standard offers significant guidance concerning each of the four phases of the risk management process. This helps ensure that a consistent and repeatable process is applied at all US ranges.

Phase I: Mission Definition and Hazard Identification. Phase I is the "problem definition" step of the process. Information is assembled to identify mission characteristics, objectives, and constraints. Potential hazard sources must be identified by evaluating the system to be flown and the range safety constraints. Information sources typically include:

a. Range safety data packages.
b. System description documents.
c. Mission essential and critical operations personnel locations.
d. Surrounding population data to include public and commercial facilities and public and commercial transportation assets (including aircraft corridors and shipping lanes).
e. Seasonal meteorological data.
f. The range safety system used.
g. Lessons learned on similar missions.

Further details of information sources are in Chapters 2 and 7 of the RCC 321 supplement. The output of this step provides a basis for hazard analysis and risk assessment, and for use in evaluating options for mitigating risks in ways that will minimize adverse mission impact.

Phase II: Risk Assessment. This step provides information needed to determine whether further risk reduction measures are necessary. Risk levels for identified hazards are expressed using qualitative and quantitative methods. This step produces basic measures of the risks posed by hazards. These hazards include inert, explosive, and flammable debris dispersions, explosive overpressure fields, exposure to toxic substances, and exposure to ionizing and non-ionizing radiation. In some cases, this step will provide sufficient information to support the decision-making without further analysis.

A valid risk assessment must account for all potential hazards posed by the range activity to personnel, facilities, and other assets. The assessment must be based on accurate data, scientific principles, and an application of appropriate mathematics. The assessment must be consistent with the range safety control that is planned for the mission. Valid calculations to assess risk can be made using the methods presented in the supplement. These typically produce conservative estimates; i.e. they produce a scientifically plausible result that characteristically overestimates risk given existing uncertainties. In all cases, the analyst is responsible for ensuring that the application of the methods in the supplement produces reasonable results. This assessment leads to mitigation measures needed to protect individuals and groups of people; this topic is discussed more fully in Chapter 3.

In general, risk is expressed as the product of the probability of occurrence of an event and the consequences of that event. Total risk is the combination of the products, over all possible events, of the probability of each event and its associated consequence. The probability of an event is always between zero and one; however, the consequences of that event can be any value. Risk can be relatively high if the probability is high, or the consequence is great, or a combination of the two.

Simple risk models are often employed to make an initial determination of risk. They are also used when the identified hazards are known to result in low risks and the analyst is assured that the estimated risk is conservative. For example, simple models can be used when only inert debris occurs and the debris is fairly limited in size and weight, with relatively low values of kinetic energy or ballistic coefficients, and shelters would provide protection from debris. These models are generally less costly, minimize schedule impacts, and have the following characteristics:

a. Simplified application of input parameters and assumptions.
b. Simplified measures of population estimation utilized.
c. A basic injury model and associated casualty areas.
d. Conservative assumptions of debris fragmentation and survivability.

If the resulting risk estimate is conservative and well within acceptable limits, then models that are more costly and time consuming, more complex, or of higher fidelity, will not be necessary.

When the identified hazards are significant or the initial risk estimate shows that acceptance criteria are, or may be, exceeded, then more complex risk models are typically used. Use of these models may be more costly, be time consuming to execute, and require a higher fidelity and more sophisticated application of input data and assumptions. The assessment may require detailed population and sheltering models, more complex human vulnerability models, and more realistic debris fragmentation and survivability models. This may require input parameters and assumptions to be supported by empirical evidence or expert elicitation. The complex risk assessment models are typically used when significant size debris or explosive debris impacts are present that could compromise shelters and the associated population.

The product of Phase II is a set of numerical estimates of the risk that address the question "How safe is this mission?" The actual metrics used are addressed later in this chapter.

Phase III: Criteria Comparison and Risk Reduction. Risk measures are compared with criteria to determine the need or desirability for risk reduction. If the risk is initially unacceptable, measures should be considered to eliminate or mitigate it. Elimination is achieved by design or system changes that remove the hazard source, such as replacing a hazardous material with a non-hazardous one or moving a trajectory to achieve containment. Mitigation is achieved by reducing the consequences of an event or the probability of an event happening. For example, increasing system reliability of a launch vehicle or test article will increase the probability of success, thereby lowering risk. Alternatively, designing a mission to avoid flight over densely populated areas will decrease consequences of casualties and thereby reduce the risk. Mitigation measures may include elements in the operation plan that reduce risk and are consistent with operational objectives, flight termination systems, containment policies, evacuation, sheltering, and other measures to protect assets from the hazards. Flight termination criteria should be optimized by balancing the risk given a failure and termination against the risk given a failure and no flight termination. To evaluate the effectiveness of mitigation measures, risk must be reassessed assuming they have been implemented. These risk reduction procedures should be followed until risk levels are as low as reasonably practical.

In general, no one wants to accept known risks. Therefore, the risk acceptance phase is of key legal importance. It integrates unique elements of US law (Federal Tort Claims Act) to provide for an individual, named by their position (Range Commander), to accept risks. The essential elements of this law are that if *a properly designated individual* makes an *informed decision* in support of a *documented decision process*, that individual is not liable, even if the information (i.e. risks) upon which the decision was based later proves to have been faulty.

Phase IV: Risk Acceptance. Presentations to the decision authority must be sufficient to support an informed decision. The presentations should include all range-mandated risk control measures, residual risks, measures of catastrophic loss potential (such as maximum collective risk given a flight termination action, maximum collective risk given failure of a flight termination system, and risk profiles), key analysis assumptions, and the protective measures that have been considered and implemented. The decision authority must approve proposed mission rules and should compare the operational risk to the criteria defined in this document and to other applicable mission requirements. When local agreements are in place and the range has adequately communicated the content and rationale of RCC Document 321 to the representatives of local government, local agreements should govern. This shall not be interpreted as overriding any Federal or state laws or regulations. The three-tiered hierarchy of requirements is:

a. Federal and state laws and regulations.
b. Local agreements.
c. RCC Document 321.

In general, higher-risk operations require a higher level of approval. The Range Commander may tolerate risk levels within criteria given herein to secure certain benefits from a range activity with the confidence that the risk is properly managed and consistent with "best practices". The outcome of these presentations to the decision authority is the acceptance of operational risks by a properly informed decision authority. This acceptance includes a determination that the residual operational risk is within tolerable limits. By doing so, it avers/justifies that the proposed conditions for allowing the operation to be initiated and the rules to allow the mission to continue to completion comply with "best practices" for ensuring that risk falls within accepted levels.

The terms of this acceptance and required implementation conditions must be documented. The responsible safety office should document a risk assessment to demonstrate compliance with the risk management policy applied.

2.5 RISK UNCERTAINTY

A necessary part of risk management is gaining an understanding of the uncertainty in the assessment computations. Some calculations of risk have little uncertainty. For example, playing Russian roulette with a six-shooter has a probability of one "win" in six trials, with very little uncertainty. Range safety calculations, however, are much more complex. Nevertheless, providing information about the inherent elements of the uncertainty to the decision-maker is considered an essential element of the informed decision principle.

There is significant uncertainty in the computed risks of launches. Ninety percent confidence bounds describing the uncertainty in the computed risk can have a range of several orders of magnitude. For this reason, uncertainty cannot be ignored. On the other hand, most of the current risk computation tools are not at a level where they compute uncertainty. This is a significant area to be addressed, but the schedule of the current RCC Document 321 standard did not allow for the considerable time required for the launch risk community to produce adequate modeling approaches that can respond to uncertainty requirements in risk acceptability. For this reason, this standard does not include uncertainty in the risk acceptability requirements, but it is understood that uncertainty will be addressed in future versions. In the meantime, the Risk Committee encourages the community to develop uncertainty models that can eventually be used with risk acceptability standards that require the use of uncertainty.

2.6 CRITERIA FOR ACCEPTABLE RISK

In meeting the obligations to manage risks and protect the public, launch agencies must (1) *execute* the risk management process and, in doing so, (2) *apply* protection standards. RCC Standard 321 achieves this in the following manner.

There are two major components of the risk acceptability criteria: a set of performance standards for establishing and implementing appropriate risk criteria at a range, followed by a set of quantitative standards. The quantitative risk criteria contained in this chapter prescribe limits on a per mission and an annual basis.

2.6.1 Proper execution

To ensure the risk management process is properly executed, the standard requires that each range must:

a. Assess the risk to all people from launch and re-entry activities in terms of hazard severity and mishap probability. Note: Hazardous operations that can be contained within a controlled area may not require a risk assessment.

b. Estimate the expected casualties associated with each activity that falls within the scope of this document. Additional risk measurements may be useful for range operations that are dominated by fatality to ensure fatality risks do not exceed acceptable limits.

c. Document its measure(s) of risk and risk acceptability policy in local requirements and policy documentation.

d. Maintain documentation to demonstrate that its risk measures provide a complete and accurate assessment of the risks, to include documentation needed to demonstrate that its risk measures:

  (1) Clearly convey the risk for decision makers.

  (2) Are consistent with the measures used by other scientific or regulatory communities involved in "comparable involuntary activities" (as described in section 2.1).

e. Estimate the risk on a per mission basis, except under special conditions where risk management on an annual basis is justified as described below.
f. Periodically conduct a formal review to ensure that its activities and its mission risk acceptability policy are consistent with the annual risk acceptability criteria.

2.6.2 Adequate protection

Protection of personnel (or assets) to adequate levels involves establishing criteria for "how safe is safe enough?" To answer this question, the RCC 321 Standard defines 17 criteria associated with personnel and assets to be applied in different situations. They are contained in Chapter 3 of the standard. While many of the categories are relatively new in their application and are not universally used, two standards are widely used and have strong historical foundations. They are:

- **Maximum acceptable risk to a single person (public) for a single mission is 1×10^{-6}.** This "one in a million" standard has widespread use for numerous safety applications and many precedents for protection of the public.
- **The maximum total expected risk from a single mission is 3×10^{-5}.** "Thirty in a million" has become the most common basis for risk acceptance decisions for launches from all US ranges. In this case, the total risk is the summation of the risk to all persons (public) exposed to risks from the launch.

In the process of defining and refining these acceptable standards over the last decade [4], substantial research has been conducted providing supporting rationale. Details of this research may be obtained from the authors or other participants involved in developing RCC Standard 321.

2.6.3 Summary of Criteria from RCC 321

The US DoD Document RCC 321 (Table 3.2) summarizes the 17 criteria defined by the US DoD for use at launch ranges.

2.7 EFFECTIVE USE OF STANDARDS

Any application of quantitative risk criteria to protect people or assets will come under scrutiny and critical review. Perfection and total agreement are not possible because different interpretations of risk measures are inevitable. Risk calculations are particularly vulnerable to critique, as noted by the mathematician, Blaise Pascal, centuries ago [5]. The proof of an effective risk standard, then, cannot be clearly defined because differing expert opinions will always exist. Instead, the proof comes from its successful and effective application. By this measure, the set of standards used by the US DoD to protect population on the ground is highly successful. The composite safety record achieved by the US national ranges in applying the processes and criteria herein compare very favorably to the safety record of most other activities with obvious hazards. The US DoD ranges have achieved a 60-year record that has experienced numerous mishaps, but, by applying the standards of protection for the public, this industry sets a very high standard of success. Statistics clearly show that the original goal of the US Congress of "being no more dangerous than airplanes flying overhead" has been met and substantially exceeded.

ACKNOWLEDGEMENTS

This chapter is based on work done by many individuals, who deserve credit for the ideas contained herein. The primary foundational work was a collaborative effort of all members of the RALCT. Of particular note is Alice Correa, who has consistently served to keep the development team on focus through numerous changes, updates, and rewrites. As the primary "book boss" for the RCC 321 manuscript, she has excelled in merging disparate perspectives from the national ranges into a useful standard. The compilation of this chapter was greatly assisted by J. P. Rogers and Max Tomlin, each with substantial background and expertise in launch safety. Finally, thanks to Pat Clemens and Bob Baker for editing the chapter, and to Heather French for typing, graphics, and final preparation for publication.

Notes and references

[1] EWR 127-1, *Range Users Handbook*, 21 August 1995, p. 1-1.

[2] US Department of Defense Directive 3200.11, Major Range and Test Facility Base (MRTFB), 27 December 2007.

[3] RCC Standard 321-07, Common Risk Criteria Standards for National Test Ranges, June 2007, Chapters 2 and 3.

[4] One of the topics of greatest interest and research was the selection of the specific metric to use when making risk decisions. Historically, the metric of "expected casualty" has been used. Within the USA, the term "casualty" is most often used as a euphemism for the more concise term, "fatality". It was adopted in the mid-1900s to minimize public concern. However, as the science of risk management has improved, the ability to evaluate various levels of injury has also improved, and today the analytical definition of casualty is "serious injury or fatality". Currently the US DoD has separate standards for both casualty and fatality, with casualty generally being approximately three times more likely than fatality.

[5] In 1662, Blaise Pascal first defined a set of rules for use in quantitative risk management, many of which are adopted in RCC 321.

CHAPTER

Space safety standards in Europe[1]

3

Gérardine Meishan Goh

CHAPTER OUTLINE

[1] The opinions expressed in this chapter are entirely those of the author and do not in any way represent the organizations with which they are affiliated.

Space Safety Regulations and Standards. DOI: 10.1016/B978-1-85617-752-8.10003-0

INTRODUCTION

Political furbelows and the "better, cheaper, faster" economic paradigm do not disguise the fact that, today, ground and mission safety is the cynosure of the European space programs. This is in direct contrast to the military origins of European rocketry, as well as the mission- and budget-driven national programs of Europe's initial decades in space exploration. Two paradigm changes were to have significant impact on space safety standards in Europe: regional cooperation and commercialization.

Despite national security concerns and the initial inclination to "go it alone", regional cooperation for civil space exploration in Europe developed in parallel with purely national space programs. The scientific community in Europe admitted early on that, following the intellectual exodus that took place during and after the Second World War, space programs centered on solely national initiatives would not prove competitive against the major superpowers of the time. This led to the foundation of the European Space Research Organization (ESRO) and the European Launcher Development Organization (ELDO) in the 1960s, which would merge in 1975 to form the present European Space Agency (ESA). With the advent of a European-wide involvement in launch technology, human space flight, and civil and commercial space exploration, it has been acknowledged that the development of a standardized space safety culture is necessary, if not crucial, to Europe's continued involvement in the exploration of outer space.

On a separate stratum, the commercialization of activities related to space design, manufacturing, launch, and mission control meant increasing numbers and categories of actors involved in space exploration in Europe. Varied national safety standards applicable in different countries meant that work sourced from different companies would not be interoperable or compatible. Of particular concern to safety was that the final product would essentially only be certifiable to the *lowest* of the motley safety standards.

The harmonization of a Europe-wide framework for space safety standards addresses both regional cooperation and safety in commercialization. Work on a common European safety regime, compliant with recommended international standards and applicable both to public entities and to industry, not only makes European space activities safer, but also more economically competitive. The constant review of published safety standards by technical experts means that, although the European safety standards continue to be a work in progress, they also remain up to date and feasibly stringent [1].

This chapter provides an overview of the European space safety standards for manned and unmanned space systems [2]. It surveys the context and historical background of space safety standards in Europe and the current institutionalized push for harmonization. Focus is placed on the European Cooperation for Space Standardization (ECSS), which aims to provide standards relating to space project management, engineering, and product assurance.

3.1 BACKGROUND

Space safety standards in Europe are closely linked with the concepts of product assurance, technical and quality management, and dependability [3]. In recent years, safety has become an important issue in the provision of satisfactory space products and services. Safety encompasses the protection of human life and well-being, and avoidance of damage to public and private property, as well as the protection of the environment.

3.1.1 National safety standards

Safety standards in Europe were administered at the national level before 1993, with the national space agencies and national standards bodies of spacefaring States drafting and implementing their own safety standards. This meant that safety standards were applicable on a national basis, in some cases on a project-by-project basis, which created incompatibility across borders in Europe. In terms of safety, this practice also led to the undesirable outcome that the lowest denominating standard of safety would ultimately be effective in cross-border European space projects. Further, this also led to inconsistent safety and product assurance standards depending on the nationality of the product and service providers and of the contracting agency or organization.

3.1.2 Changing paradigms: regionalization and commercialization

The dual advent of regionalization and commercialization meant that a sea change was necessary with regard to safety standards in Europe.

Through regionalization, the phenomenon of a common European Space Agency, working in parallel and cooperation with its counterpart national space agencies, meant that issues of compatibility and minimum international standards were brought to the forefront. Together with developments with the European Union for a common market, harmonization of safety and other standards became increasingly necessary to ensure interoperability and improving safety standards.

Commercialization provided further impetus for the harmonization of safety and other standards in the European space industry. The geographic return policy of the European Space Agency allowed technology transfer among Member States of the Agency while kick-starting the space sector in various countries in Europe. The plethora of commercial space manufacturing, operations, and launch companies that has arisen as a result of commercialization has also led to international and, indeed, intercontinental cooperation. In order to remain economically competitive, players in the European space industry needed an up-to-date, objective set of standards on which they could base their space products and services. Increasing the economic viability of the European space industry in the face of global competition was to become one of the objectives to be achieved through the standardization of space safety frameworks across Europe.

3.2 HARMONIZATION OF SAFETY STANDARDS AT THE EUROPEAN LEVEL

The phenomenon of regionalization in space activities is nowhere better demonstrated than through the European participation in space activities. An unprecedented amount of regional cooperation, through various contractual and institutionalized means, has meant a sweeping need for harmonization and compatibility at the European level. This section looks at the institutions and mechanisms that are involved in the harmonization of safety standards at the European level. Further to the safety standards developed within the ESA, other European standardization bodies are also surveyed.

3.2.1 Safety standards and culture within the European Space Agency (ESA)

3.2.1.1 History

The interaction of ESA with the various national space agencies and with industry meant that where national industry players responded to Invitations to Tender by the ESA, ESA safety and standards were to be complied with. The importance of safety standards within the ESA can be seen from its early implementation in space projects, and from the Agency's initial attempt at standardization through the ESA Procedures, Specifications and Standards (PSS) documents. The PSS architecture was as follows [4]:

- PSS-01-0 Product Assurance and Safety
- PSS-02-0 Electrical Power and EMC
- PSS-03-0 Mechanical Engineering and Human Factors
- PSS-04-0 Space Data Communications
- PSS-05-0 Software Engineering
- PSS-06-0 Management and Project Control
- PSS-07-0 Operations and EGSE
- PSS-08-0 Natural and Induced Environment
- PSS-09-0 Control Systems
- PSS-10-0 Ground Communications and Computer Networking.

Many of the PSS standards are presently obsolete or have been replaced by more up-to-date standards from the European Cooperation for Space Standardization framework [5].

3.2.1.2 ESA List of Approved Standards

The ESA periodically issues the List of "ESA Approved Standards" [6], which lists the titles of documents that provide the standards in management, product assurance, and engineering to be used in implementing all ESA project space activities. This list is divided into two sections: "Applicable Standards" and "Reference Documents". Documents listed under "Applicable Standards" are compulsory requirements for each project, and requirements listed under those standards must be applied as appropriate. This means that safety standards listed in the List of ESA Approved Standards apply also, via contractual clauses, to products and services provided by industry partners and national space agencies and companies that contract work from the ESA. The application of standards in documents listed under the section "Reference Documents" is discretionary.

Updates of the List of ESA Approved Standards generally illustrate the current status of standards endorsed by ESA in the context of its activities within the European Cooperation for Space Standardization and the Consultative Committee for Space Data Systems. The development of standards to facilitate European space activities within the common market has been recognized as an agreed common objective of the ESA, the various national space agencies, and players in the European space industry. In particular, all three of these sectors concur that the development of a complete set of harmonized standards for direct application, including safety standards, is crucial to the European space sector.

Standards are usually drafted by a working group of experts, in an iterative and consultative process. Upon endorsement by the ESA Standardization Boards and the ESA Standardization Steering Board, the standards are released to the user community for feedback. Continual dialog between the various stakeholders ensures that any new list of approved standards will be effective. This is of special importance with regard to the application of safety standards.

3.2.1.3 Product assurance and safety at the ESA

Safety at the ESA is inextricably linked to product assurance [7]. The Product Assurance and Safety (PA&S) group comprises more than 80 experts tasked with the identification and mitigation of all aspects that could have adverse consequences on the ESA's space missions. These include issues relating to human safety, including astronauts, ground personnel and the public, property, and the environment. The group also works to consider the impact of such safety considerations on the budget, operations, and schedule of a space project. The scope of the PA&S group includes purview over safety standards for hardware and software quality, product reliability, safety, electrical and electronic engineering, materials, industrial fabrication processes, and configuration control. Of particular interest in the context of this chapter is the fact that the PA&S group controls the safety standards of the components and parts supplied by industry partners to the ESA, and ensures that the contractually agreed standards are met. Through a series of reviews, audits, and inspections, the PA&S group enforces safety standards throughout ESA projects, controlling and reducing risks and initiating any necessary corrective actions. The PA&S group is further subdivided into five divisions, focusing on:

1. Quality, Dependability, and Safety
2. Components
3. Materials and Processes
4. Requirements and Standards
5. Projects Product Assurance.

In the context of this chapter, the work of the Quality, Dependability, and Safety division and that of the Requirements and Standards division are of special interest.

Within the ESA, quality, dependability, and safety are the three criteria identified as crucial to mission success. The ESA has a designated Quality, Dependability, and Safety division that is in overall charge of the quality of ESA products and services [8]. It establishes quality management systems within the ESA to enable control over system behavior in space missions. In particular, for safety-critical systems, the division aims to provide a sound understanding of the risks posed to human life, public property, and the environment by space missions undertaken by the ESA. The Quality, Dependability, and Safety division comprises three sections that work in tandem to develop and implement techniques and processes to achieve safe and reliable systems design, manufacturing, operations, and disposal. These three sections are: Quality Assurance and Management, Dependability and Safety, and Software Product Assurance.

Quality assurance, failure investigations, and the management of product assurance are the three cornerstones of the Quality and Assurance section. Providing expert support in quality assurance to all ESA space missions, the section also collates, analyzes, and distributes information relating to past and present failures of existing products and processes both to the ESA and to its contractors and partners, so as to disseminate lessons learnt that may be useful for ongoing and future missions. The section provides direct support for project product assurance, and participates in project reviews and failure investigations. It also provides certification of European-coordinated test centers, internal quality management of the laboratory equipment pool, and internal audit information services. It had a significant role in the three ISO-9001 certification projects completed by the ESA in relation to ground segment development and satellite operations, Earth observation satellite data processing, and computer support services.

The Dependability and Safety section is the seat of responsibility for Reliability, Availability, Maintainability, and Safety (RAMS), the engineering discipline applied to hardware, software, and the human components of space missions [9]. Aside from being crucial to mission success, safety considerations and standards ensure environmental protection and the protection of humans from dangers associated with space flight. Recognizing the interdisciplinarity of the RAMS concept, the ESA Dependability and Safety section integrates spacecraft as well as ground segments and operations, taking an end-to-end approach throughout all the life-cycle phases of the space mission in question. The mandate of the section focuses on the development, implementation, and maintenance of expertise and standards relating to RAMS. This includes:

1. Evaluating proposed concepts, designs, and operations for their compliance with defined dependability and safety requirements
2. Assessing system robustness and weaknesses
3. Evaluating risk assessments
4. Defining and controlling the implementation of applicable dependability and safety-related acceptance requirements
5. Reviewing projects to assess compliance with safety criteria and formulating accept/reject recommendations
6. Participating in flight safety reviews and reviewing hazard reports for the relevant authorities and safety review panels.

The holistic approach taken by the section sees its responsibilities also extend to:

1. The definition of programmatic and technical dependability and safety requirements as per accepted standards (e.g. those of the European Cooperation for Space Standardization [10])
2. The identification of project-specific safety requirements
3. The evaluation of contractors' plans, technical specifications, and other issues related to the execution of a space mission
4. The monitoring of contractors' dependability and safety programs to ensure compliance with baseline European safety standards
5. The direction and controlling of risk-reduction processes and undertaking of necessary preventive measures.

Further, the section also conducts research into risk management, safety processes and standards, and methodologies to support hazard analysis, failure modes, effects, criticality analysis, and risk assessment and reduction.

The third section within the Quality, Dependability, and Safety division is the Software Product Assurance section [11]. Aside from research into methodologies and tools necessary for software product assurance, the section also develops training modules to implement an effective software product assurance culture within the ESA. The section successfully developed the following tools, which are presently offered as services both internally and externally to the ESA:

1. The ISO 15504-compliant "Spice for Space" (S4S) process assessment
2. Software Product Evaluation and Certification (SPEC), which evaluates software products on an agreed, objective quality model
3. A software dependability evaluation that applies the RAMS techniques to software systems and products.

Apart from the work of the Quality, Dependability, and Safety division, another relevant division is the Requirements and Standards division [12]. As a regional space agency comprised of Member States in cooperation with their respective national space agencies, the ESA also works closely with many industry players from various European countries. Naturally, the ESA has for many years been working in association with several international organizations to create and promote common safety standards for space projects. In this vein, the Requirements and Standards division works at close quarters with the European Cooperation for Space Standardization (ECSS) to construct a single European-wide set of standards for space projects. The Requirements and Standards division is the mechanism through which the ESA provides technical support for the drafting of safety and other standards and administrative support for the flow of crucial information, both of which are essential for the work of the ECSS. The Requirements and Standards division also maintains the list of standards approved for application by ESA space projects, and supports the work of the ESA Standardization Steering Board (ESSB).

3.3 HUMAN SPACE-FLIGHT SAFETY STANDARDS

The ESA Product Assurance and Safety Office (PASO) is also responsible for safety standards for human space flight. The Office administers the Requirements and Certification processes for payloads and cargo items that are to be transported on the ESA's Automated Transfer Vehicle (ATV), or that are to be used on board the International Space Station (ISS). While ISS safety standards utilized by the ESA are generally those in use by the American and Russian partners of the ISS, the ESA PASO is in overview charge of two categories of cargo and human safety standards: ATV Requirements, and Kourou Ground Safety Requirements.

ATV Requirements comprise the specific set of safety standards applicable to payloads and cargos on board the ATV [13]. It should be noted that ISS payloads and cargo must further also be compliant with the applicable on-orbit requirements of NSTS/ISS 1700.7B Addendum and SSP 50021 [14]. When jointly used, the documents aim to provide a safe design for the ATV transportation phase and the ISS on-orbit phase. The ATV Requirements are intended to "protect the general public, flight and ground personnel, the ATV, other payloads, public—private property, and the environment from payload-related hazards" [15]. ATV Requirements include the following standards and forms:

1. Safety Policy and Requirements for Payloads/Cargo Items to be Transported on ATV, Issue 1, November 2004, ESA Document ESA-ATV-1700.7b [16]
2. ATV Pressurized Payload/Cargo Safety Certification Process, Issue 1, July 2004, ESA Document ESA-ATV-PR-13830 [17]

3. ATV Payload/Cargo Item Ground Safety Checklist, ESA Document ATV Form 476 [18]
4. ATV Payload/Cargo Hazard Report, ESA Document ATV Form 542 [19]
5. ATV Payload/Cargo Item Flight Safety Certificate, ESA Document ATV Form 879 [20].

Additional safety standards unique to ground operations and ground support equipment are in the Centre Spatial Guyanais (Kourou) Safety Regulations Document.

The Kourou Ground Safety Requirements are a unique set of documents issued by the French CNES, which is the operating launch authority of the Centre Spatial Guyanais in Kourou, French Guyana. This set of Safety Requirements is particularly distinctive in that it is issued by a national space agency (in this case, CNES of France) and adopted by the ESA without an attempt at revision thus far. The reason for this could be that the Kourou launch site is on sovereign French territory, making France a State liable in international law for any damage arising from launch activities undertaken from Kourou [21]. The Kourou Ground Safety Requirements include the following safety regulations and standards:

1. CSG Safety Regulations: General Rules Volume 1, Centre Spatial Guyanais Safety Regulations Document CSG-RS-10A-CN [22]
2. CSG Safety Regulations: Specific Rules Volume 2—Part 1 (Ground Installations), Centre Spatial Guyanais Safety Regulations Document CSG-RS-21A-CN [23]
3. CSG Safety Regulations: Specific Rules Volume 2—Part 2 (Spacecraft), Centre Spatial Guyanais Safety Regulations Document CSG-RS-22A-CN [24].

3.3.1 Current institutions for the harmonization of safety standards in Europe

Currently, there are several institutions working to harmonize standards in Europe. All but one of these organizations have mandates that extend beyond space safety standards. However, it is clear that there is a significant amount of dialog and exchange that takes place between these organizations. This is useful to prevent duplication of standards and, in particular, to provide for an iterative process. This section will briefly look at the main institutions relevant to the harmonization of safety standards in Europe.

3.3.1.1 The European Commission (EC)

The European Commission, under its Enterprise and Industry Directorate-General, has a specific section that deals with Space Policy and Coordination under its Aerospace, Security, Defense, and Equipment wing. On 25 June 2007, the EC addressed a mandate to the Comité Européen de Normalisation (CEN, the European Committee for Standardization), the Comité Européen de Normalisation Electrotechnique (CENELEC, the European Committee for Electrotechnical Standardization), and the European Telecommunications Standards Institute (ETSI) to establish space industry standards [25]. The scope of the mandate was to establish a program for space-related standards "to

- ensure an adequate safety level for space hardware and services;
- foster European Union projects such as the *Galileo* satellite navigation system, the Global Monitoring for Environment and Security (GMES) and projects in the satellite telecommunications field;
- stimulate the emergence of European end-user terminals;

- mitigate space-related threats such as debris; and
- support the international competitiveness of the European space industry" [26].

This mandate followed on the European Council Resolution of 7 May 1985 [27] on taking a new approach in harmonizing technical standards in line with the Directive 98/34/EC of the European Parliament and of the Council of 22 June 1998, as twice amended [28]. The "new approach" provides a clear delineation between the responsibilities of the EC and those of the European standards bodies. EC Directives would define the essential requirements that goods must meet before being placed on the market, while European standards bodies would draft the corresponding technical specifications. The mandate was also an essential component of the European Space Programme, which intends to integrate a variety of space systems from the European Union, ESA, and individual Member States into a European space infrastructure for the implementation of EU policies.

The 2003 White Paper on Space Policy [29] delineated a multi-year European Space Programme that will determine the priorities, objectives, roles, responsibilities, and budgets of the European space effort. A demand-driven, economically inclined policy, it aims to ensure "faster economic growth, job creation and industrial competitiveness, enlargement and cohesion, sustainable development, security and defence of the European citizen" [30] through exploiting the benefits of space technology and exploration. Initiatives on regulation and standardization, including initiatives specific to space safety, would be developed through cooperation between the ESA and the EC, in line with their Framework Agreement on collaboration [31]. Amongst the issues slated for standardization were criteria for performance, accuracy, interoperability and compatibility, and safety and user-friendliness. The standardization program was also intended to deal with issues of "design and manufacture of equipment, environmental aspects, services, quality, safety and interoperability" and the "design and manufacture of equipment, environmental aspects, services, quality, safety and interoperability" [32]. For the purposes of this chapter, the CEN, CENELEC and ETSI were specifically instructed to execute the mandate in cooperation with the ECSS and ESA in particular, international and European standards bodies, the European space industry through Eurospace, national space agencies and regulatory authorities of Member States.

3.3.1.2 Comité Européen de Normalisation (CEN)

The European Committee for Standardization (Comité Européen de Normalisation, CEN) is a federation of the national standards bodies of the 27 European Union countries and the three European Free Trade Area countries. It generates standards through a consensus-based process. Formal European standards (ENs) adopted by the CEN also become national standards in every one of the CEN's 30 member countries.

The CEN works at close quarters with the ECSS for the development of ENs in the space sector. Documents prepared by the ECSS are submitted to the CEN's procedures for adoption as ENs in fields such as "project management requirements, product assurance and safety requirements, technical requirements for equipments, mechanisms, assemblies, systems, the application of materials and components in space missions, as well as interface requirements for information relating to space systems and activities and transmitted between organizations" [33].

To date, 45 standards relating to space systems have been published by the CEN, of which the following relate directly to product assurance and safety:

1. Space product assurance—General requirements, Part 1: policy and principles, Document EN 13291-1:1999
2. Space systems—Risk management (ISO 17666:2003), Document EN ISO 17666:2003
3. Space product assurance—Quality assurance for test centres, Document EN 14736:2004
4. Space product assurance—Non-conformance control system, Document EN 14097:2001
5. Space systems—Safety requirements, Part 1: System safety (ISO 14620-1:2002), Document EN ISO 14620-1:2002
6. Space product assurance—General requirements, Part 3: Materials, mechanical parts and processes, Document EN 13291-3:2003
7. Space product assurance—Thermal vacuum outgassing test for the screening of space materials, Document EN 14091:2002
8. Space product assurance—Thermal cycling test for the screening of space materials and processes, Document EN 14098:2001
9. Space product assurance—Measurement of the peel and pull-off strength of coatings and finishes using pressure-sensitive tapes, Document EN 14099:2001
10. Space product assurance—Flammability testing for the screening of space materials, Document EN 14090:2002
11. Space product assurance—The control of limited shelf-life materials, Document EN 14089:2002
12. Space product assurance—The determination of offgassing products from materials and assembled articles to be used in a manned space vehicle crew compartment, Document EN 14100:2001
13. Space product assurance—Material selection for controlling stress-corrosion cracking, Document EN 14101:2001.

3.3.1.3 European Aviation Safety Agency (EASA)

The EASA is the linchpin of the European Union's aviation safety policy. It aims to promote the highest European-wide standards of safety and environmental protection in the European aviation industry. Recently, the EASA has taken note of the two-stage-to-orbit space vehicle slated for commercial space-flight purposes. The EASA regulates the safety criteria and standards of a very specific type of aerospace vehicle for specific uses, as detailed in the chapter on "Regulation and Licensing of European Commercial Aerospace Vehicles" in Part 4.

3.3.1.4 European Cooperation for Space Standardization (ECSS)

The most significant effort for the production of space safety standards in Europe is the initiative known as the European Cooperation for Space Standardization (ECSS). Due to the importance of including safety standards in all stages of a space project, although the ECSS standards cover diverse branches ranging from project management to engineering and to product assurance, the harmonization of these standards permits the inclusion of important minimum safety standards throughout the European space industry. Thus, further to the dedicated discipline on safety in the ECSS framework, the ECSS standards should be studied in their entirety in the context of this chapter. The next section focuses on the ECSS standards, with a particular focus on the safety and safety-related standards therein.

3.4 EUROPEAN COOPERATION FOR SPACE STANDARDIZATION (ECSS)

3.4.1 History

The initiative for the ECSS began in Fall 1993 with the adoption of its Terms of Reference (TOR). All partner signatories to the ECSS TOR jointly undertook the development of a single set of coherent, user-friendly standards for all European space activities. Interestingly, the ECSS initiative followed on requests by actors in the European space industry through the Eurospace trade association for the harmonization of the ESA and CNES product assurance standards. The harmonization was subsequently extended to include those related to project management, product assurance, and engineering activities for the entire European space community. Participating national space agencies include: the Italian Agenzia Speziale Italiana (ASI); the Belgian Office for Scientific, Technical and Cultural Affairs (OSTC); the British National Space Centre (BNSC); the French Centre National d'Etudes Spatiales (CNES); the German Deutsches Zentrum für Luft- und Raumfahrt e.V. (DLR); the European Space Agency (ESA); the Dutch Nederlands Instituut voor Vliegtuigontwikkeling en Ruimtevaart (NIVR); and the Norwegian Space Centre. Further, all 18 ESA Member States support the ECSS initiative, as evident in ESA Council Resolution 3 [34].

With the objective of developing standards to improve industrial efficiency and market competitiveness in the European space industry, various standards that had been unique to particular contractors, companies, and space agencies were reviewed and revised into a single coherent set. With the active participation of the ESA, the major European national space agencies and the majority of the contractors and companies in the European space sector, the first major step was undertaken with the draft of the Standardization Policy [35], which addressed the scope, structure, implementation, authority, organization, and documentation of the ECSS standards to be proposed.

3.4.2 Impact and significance of the ECSS standards to safety standards in Europe

In the scope of this chapter, the ECSS standards are particularly crucial for the following reasons:

1. Safety as an element throughout the design, manufacturing, launch, and mission. Although the scope of the ECSS standards extends beyond safety standards, safety is a basic element in the project management, engineering, and product assurance of a space mission or product. Further to having a specific section on Safety Standards categorized under the Product Assurance Branch, there are various standards peppered throughout the framework that are crucial to space safety.
2. Specific focus on space safety. The ECSS standards feature a dedicated discipline in safety under its Product Assurance Branch. Entitled "Q-40: Safety", the discipline addresses all aspects related to safety risks associated with design, development, production, and operations of space products, including the implementation of a safety assurance program and a risk assessment based on qualitative and quantitative analyses.
3. Compliance with international safety standards. One of the policies of the ECSS is not to duplicate standards. There is a conscious effort to harmonize all ECSS standards with international standards and generally accepted working practices in the European space industry. Moreover, the first standard on safety, the ECSS Q-40A Standard, formed the basis for the International Standards

Organization (ISO) in their draft ISO DIS 14620-1 Standard. Work was then completed on the ECSS level to meet and exceed the ISO standard, while addressing unique European needs. The ECSS remains a liaison A partner of the ISO TC20/SC14 entities.

4. Iterative development of stringent yet feasible safety standards. The ECSS standards are in continual review. At the time of writing, the third revised batch of ECSS standards had been released [36]. The sustained, iterative review of standards by the ECSS working groups, which are comprised of experts and industry players in the field, ensures the progressive evolution of rigorous, up-to-date, and yet workable safety standards for the industry.
5. General adoption and use in the European space sector. The ECSS initiative was prompted by calls for harmonization by the European space industry through its trade association. The support of the ECSS standards by the ESA, the various national agencies, and the main stakeholders in the European space industry ensures the general adoption and use of these standards in the European space sector. The upshot is that the ECSS standards relating to space safety would become the most widely adopted and used standards in the European space sector. This is especially so given the partnership between the ECSS and the CEN, which is the main standardization organ of the European Community.
6. Addressing novel issues related to safety. The ECSS initiative also engages in the further development of other standards related to novel and pressing issues related to safety. Examples include risk management, debris mitigation, and situational awareness.

3.4.3 Architecture of the ECSS standards on safety

The ECSS is structured on a three-pronged approach: Space Project Management, Space Product Assurance, and Space Engineering. The overview ECSS-S-ST-00 document provides a system description of the ECSS standards, in conjunction with a glossary of terms as listed in the document ECSS-S-ST-00-01. The ECSS standards are intended to be applied as a consistent set rather than individually. As such, all safety standards within the ECSS framework should be considered in the matrix of the other ECSS standards. The list of ESA-approved standards indicates which of the ECSS standards are presently in force [37].

Documentation within the ECSS system makes a distinction between standards, handbooks, and technical memoranda. It was deemed that the ECSS standards had reached critical mass to be relevant as a consistent set of standards in space projects in 2006. Since that time, revision and re-release of the standards has occurred three times to ensure their continued applicability in space projects and in space business agreements. The schematic in Figure 3.1 provides an overview of the ECSS architecture [38].

At first glance it is clear that an entire discipline under the space product assurance branch is dedicated to safety standards. Designated discipline Q-40, the "Safety" discipline should be read in the context of the other Q disciplines, in particular those related to product assurance (Q-10), quality assurance (Q-20), and dependability (Q-30). Moreover, of direct importance to the Q-40 safety discipline are the M-10 ("Project planning and implementation") and M-80 ("Risk management") disciplines of the space project management branch.

Under the Q-40 "Safety" discipline, three further categories of standards have been identified: Safety (Q-ST-40), Hazard Analysis (Q-ST-40-02), and Fault-tree Analysis (Q-ST-40-12), as seen

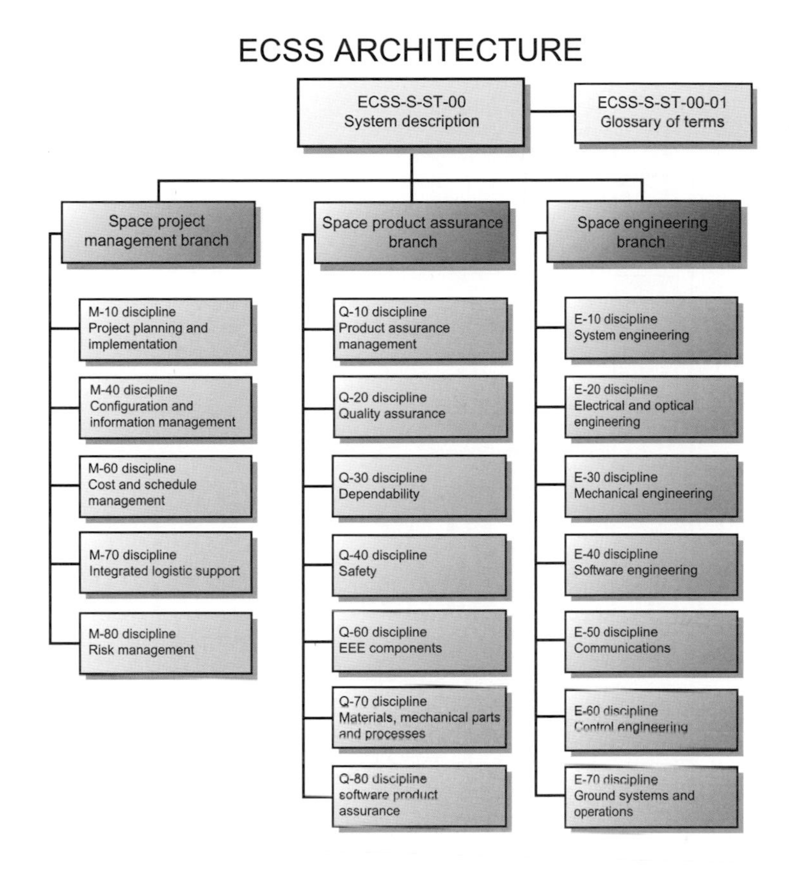

FIGURE 3.1

ECSS architecture.

below. Of these, the ESA has approved of the standards under safety assurance in its third release for general use in ESA missions and projects:

1. ECSS-Q-ST-40C "Safety" on 6 March 2009
2. ECSS-Q-ST-40-02C "Hazard analysis" on 15 November 2008
3. ECSS-Q-ST-40-12C "Fault tree analysis—Adoption notice ECSS/IEC 61025" on 31 July 2008 [39].

3.4.3.1 Content

Each standard from the ECSS framework, including those related to safety standards, includes:

1. A change log identifying relevant changes between the new standards, and the previous, now superseded, standards.
2. A description of the scope, coverage, and applicability of the standard.

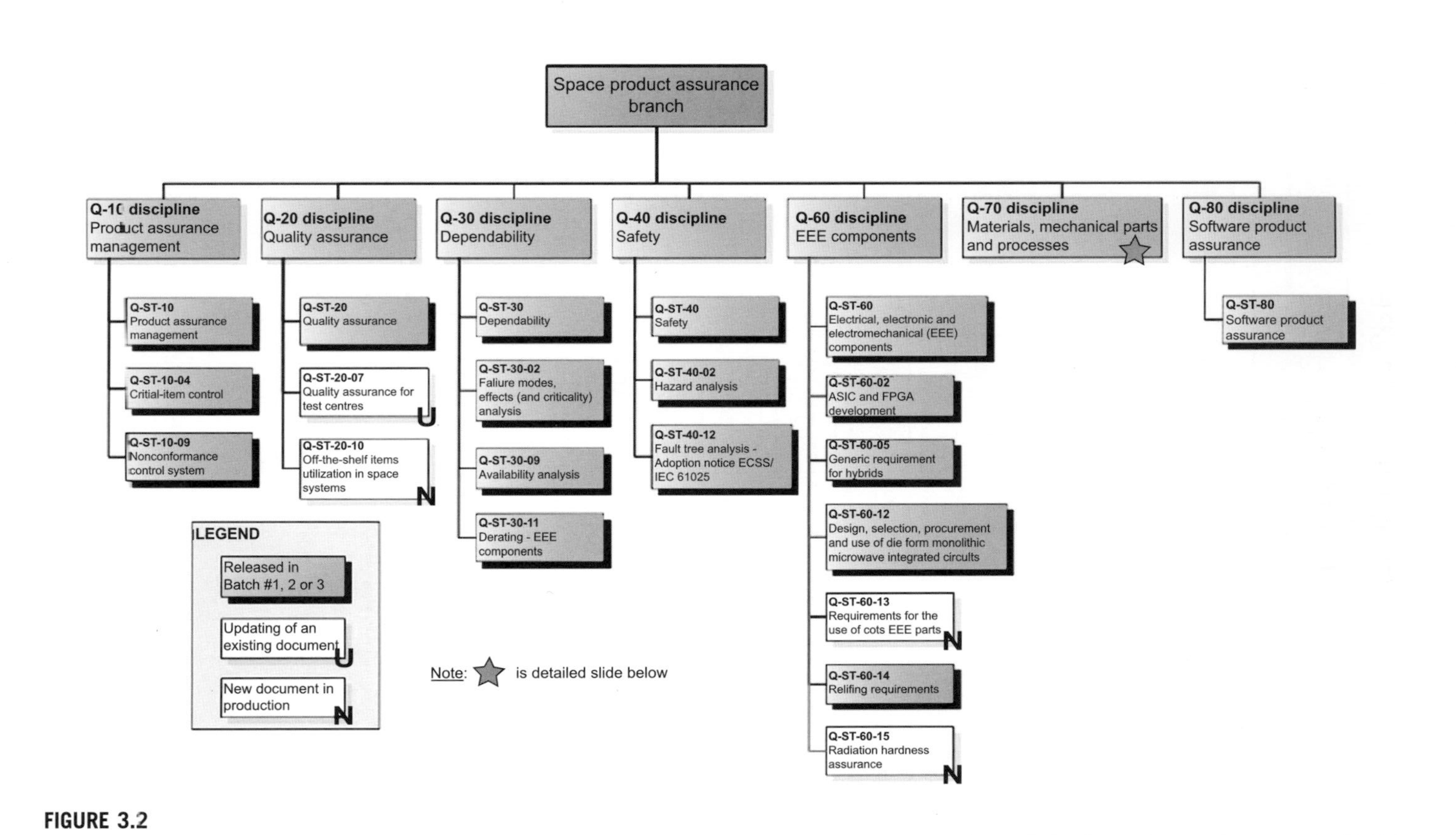

FIGURE 3.2

ESA Space Product Assurance Branch.

3. The list of documents referred to in the standard, including technical memoranda and handbooks.
4. The terminology used, including relevant definitions.
5. The requirements of the standard and a clearly separated, descriptive elaboration.
6. The required content of deliverable documents.

3.4.3.2 Published standards relating to safety

The ECSS safety policy aims to ensure the safety of flight and ground personnel, the launch vehicle, associated payloads, ground support equipment, the general public, public and private property, and the environment from hazards associated with European space systems [40].

Document ECSS-Q-ST-40C, released on 6 March 2009, is the newest incarnation of the safety standards under the ECSS framework. It builds upon the first and second issues respectively, ECSS-Q-40A of 19 April 1996 and ECSS-Q-40B of 17 May 2002, which formed the basis for work done by the CEN and the ISO on space safety standards. The newest document exceeds the international standards recommended by both the CEN and ISO. The following is the structure of the 75-page document:

1. Foreword
2. Log of Changes
3. Scope
4. Normative references
5. Terms, definitions and abbreviated terms, including terms from other standards, those specific to the present standard, and abbreviations
6. Safety principles. Objectives, policies and implementation of overall safety principles
7. Safety program. A system safety program incorporating a risk assessment is applied for all European space systems under the auspices of European public institutions such as the ESA and the EC. The risk assessment involves a five-fold checklist, including:
   a. the identification and evaluation of hazardous characteristics and functions through the iterative performance of systematic safety analyses;
   b. the minimization and elimination of potential hazardous consequences associated with such characteristics and functions through a hazard reduction sequence;
   c. the progressive assessment of remaining risks;
   d. the formal verification of the adequacy of the hazard and the risk control measures applied so as to support safety validation and risk acceptance; and
   e. the obtainment of approval from the relevant authorities.
8. Safety engineering. The safety requirements identification and traceability are inserted into the safety design objectives, which include considerations of safety policy and principles, design selection, hazard reduction precedence, environmental compatibility, external services, hazard detection—signaling and safing, space debris mitigation, atmospheric re-entry, safety of Earth return missions, safety of human space-flight missions and access. A safety risk reduction and control mechanism is also detailed, with a standard on the severity of possible hazardous events, failure tolerance requirements, design for minimum risk, and probabilistic safety targets. Safety-critical functions are to be identified and controlled via mechanisms for identification, status information, safe shutdown and failure tolerance requirements. Basic requirements for operational safety in flight operations, mission control and ground operations are also elaborated upon.

9. Safety analysis requirements and techniques. Safety requirements for functions and subsystems of a spacecraft are defined and specified. A description of the standards to be complied with throughout the safety analyses in a project's life cycle, including hazard analyses and safety risk assessments, are explained.
10. Safety verification. A five-part safety verification process is envisaged. Starting with hazard report and review, the document also details the methods to be used for safety verification. Further, the procedure and standards for the verification of safety-critical functions, such as validation, qualification, failure tests, verification of design or operational characteristics, and safety verification testing, are described. Standards relating to hazard close-outs and safety assurance verification are detailed. The final part comprises standards for the declaration of the conformity of ground equipment with the safety standards.

The Document also contains seven Annexes, three of which are normative:

Annex A (informative) Analyses applicability matrix
Annex B (normative) Safety programme plan—DRD
Annex C (normative) Safety verification tracking log (SVTL) DRD
Annex D (normative) Safety analysis report including hazard reports—DRD
Annex E (informative) Criteria for probabilistic safety targets
Annex F (informative) Applicability guidelines
Annex G (informative) European legislation and "CE" marking.

3.4.3.3 Implementation and enforcement of safety standards

Europe's pioneering pursuit of harmonizing safety standards has established the first and only international space safety standards to date [41]. It is important to note that the ECSS standards themselves are not legally binding, and that it has always been the policy of the ECSS that the standards would be made applicable to contracts and projects by way of inclusion in contractual clauses. The contractual party imposing the use of the ECSS standard is also the responsible party for its correct application and enforcement [42].

CONCLUSIONS

European space safety standards constitute the first truly international set of space safety standards. Developed with the consensus and active participation of the various stakeholders, the standards continue to be reviewed and reissued so as to remain relevant and competitive in a rapidly evolving technical and economic matrix. With the full support of the European Commission, European Space Agency, national space agencies and industry players, the published standards relating to space safety not only benefit from the expertise of practitioners and policymakers in the field, but also from the overt support thrown behind their effective implementation.

An attitudinal change towards safety standards can also be observed in Europe. Together with the increase in complex unmanned space programs, safety standards in Europe are no longer targeted solely for the benefit of the space system or the astronaut. Rather, a broad and robust approach has been undertaken to consider the safety of ground personnel and private individuals, avoidance of property damage, and environmental protection. The European Cooperation for Space Standardization and the space safety standards that result from this initiative are cornerstones in the common strategy undertaken by agencies and industry stakeholders to establish a sound, user-friendly, and effective framework for European space safety standards.

Bibliography

Official European Documents

Directive 98/34/EC of the European Parliament and of the Council of 22 June 1998 laying down a procedure for the provision of information in the field of technical standards and regulations, Official Journal L 204 (21 July 1998), pp. 37–48, as twice amended by Directive 98/48/EC of the European Parliament and of the Council of 20 July 1998 amending Directive 98/34/EC laying down a procedure for the provision of information in the field of technical standards and regulations, Official Journal L 217 (5 August 1998), pp. 18–26 and Council Directive 2006/96/EC of 20 November 2006 adapting certain Directives in the field of free movement of goods, by reason of the accession of Bulgaria and Romania, Official Journal L 236 (23 September 2003), pp. 81–106.

European Commission, Enterprise and Industry Directorate-General, Aerospace, security, defence and equipment, Space policy and coordination, "Programming Mandate Addressed to CEN, CENELEC and ETSI to Establish Space Industry Standards" (Brussels, 25 June 2007), Document M /415 EN.

European Council Resolution of 7 May 1985 on a new approach to technical harmonization and standards, Official Journal C 136 (4 June 1985), pp. 1–9.

European Council, Space: a new European frontier for an expanding Union—an action plan for implementing the European Space Policy (2003), COM (2003) 673.

Technical Safety Standards

ATV Pressurised Payload/Cargo Safety Certification Process, Issue 1, July 2004, ESA Document ESA-ATV-PR-13830.

CSG Safety Regulations: General Rules Volume 1, Centre Spatial Guyanais Safety Regulations Document CSG-RS-10A-CN.

CSG Safety Regulations: Specific Rules Volume 2—Part 1 (Ground Installations), Centre Spatial Guyanais Safety Regulations Document CSG-RS-21A-CN.

CSG Safety Regulations: Specific Rules Volume 2—Part 2 (Spacecraft), Centre Spatial Guyanais Safety Regulations Document CSG-RS-22A-CN.

ESA, List of "ESA approved standards", Version 3.1 (8 June 2009), ESSB-AS V3.1, as applicable through ESA Document ESA/ADMIN/IPOL(2007)11 (20 July 2007).

European Cooperation for Space Standards (ECSS), in three Batches (Batch 1 published on 4 August 2008, Batch 2 published on 15 November 2008 and Batch 3 published on 18 March 2009), all available from the website of the ECSS, online at http://www.ecss.nl/ (last accessed 20 August 2009).

Safety Policy and Requirements for Payloads/Cargo Items to be Transported on ATV, Issue 1, November 2004, ESA Document ESA-ATV-1700.7b.

Notes and references

[1] European Cooperation for Space Standardization (ECSS), online at its website http://www.ecss.nl/ (last accessed 20 August 2009).

[2] The scope of this chapter is limited to the 18 Member States of the European Space Agency (Austria, Belgium, Czech Republic, Denmark, Finland, France, Germany, Greece, Ireland, Italy, Luxembourg, the Netherlands, Norway, Portugal, Spain, Sweden, Switzerland, and the UK), and the States participating in the ESA Plan for European Cooperating States (PECS) (Hungary, Poland, and Romania), as well as Estonia and Latvia, which have both signed Cooperation Agreements with the ESA without joining the ESA PECS initiative. It specifically excludes Canada, which operates under a Cooperation Agreement with the ESA.

[3] ESA, "Feature: About Product Assurance and Safety", online at http://www.esa.int/esaTQM/1069167508574_productassurance_0.html (last accessed 20 August 2009).

[4] ESA, "PSS Documents", online at http://www.esa.int/TEC/Microelectronics/SEMMYUU681F_0.html (26 June 2007) (last accessed 20 August 2009).

[5] The ESA PSS Standards catalog can be obtained from the ESA Publications website and office; see information from their website at http://www.esa.int/SPECIALS/ESA_Publications/SEMMI5LTYRF_0.html (last accessed 20 August 2009).

[6] ESA, List of "ESA approved standards", Version 3.1 (8 June 2009), ESSB-AS V3.1. For the application of this list, refer to ESA Document ESA/ADMIN/IPOL(2007)11 (20 July 2007). The list is regularly updated and can be downloaded from http://ice.sso.esa.int/intranet/public/standards (last accessed 20 August 2009).

[7] ESA, "Feature: About Product Assurance and Safety", online at http://www.esa.int/esaTQM/1069167508574_productassurance_0.html (last accessed 20 August 2009).

[8] ESA, "Feature: ESA Quality, Dependability and Safety Division", online at http://www.esa.int/esaTQM/1090331584093_productassurance_0.html (last accessed 20 August 2009).

[9] ESA, "Feature: Dependability and Safety section", online at http://www.esa.int/esaTQM/1069167510085_productassurance_0.html (last accessed 20 August 2009).

[10] See section 3.4 below.

[11] ESA, "Feature: Software Product Assurance Section", online at http://www.esa.int/esaTQM/1069167510077_productassurance_0.html (last accessed 20 August 2009).

[12] ESA, "Feature: About Requirements and Standards", online at http://www.esa.int/esaTQM/1090331582852_productassurance_0.html (last accessed 20 August 2009).

[13] ESA, "Safety requirements for payloads/cargos on board the ATV" (November 2004) ESA Document ESA-ATV-1700.7b, Issue 1 (Revision 0), available online at http://paso.esa.int/8_Additional/ESA_ATV_1700_7B_Issue1.pdf (last accessed 20 August 2009).

[14] A list of the payload safety requirements for the space shuttle and the ISS is available online at http://paso.esa.int/Safety_Requirements.htm (last accessed 20 August 2009), with further links to specific Payload Safety Requirements Documents.

[15] *Ibid.*, p. 7.

[16] Document available online at http://paso.esa.int/8_Additional/ESA_ATV_1700_7B_Issue1.pdf (last accessed 20 August 2009).

[17] Document available online at http://paso.esa.int/8_Additional/REQ-ESA-ATV-PR-13830-Signed.pdf (last accessed 20 August 2009).

[18] Document available online at http://paso.esa.int/8_Additional/ATV%20Form%20476.pdf (last accessed 20 August 2009).

[19] Document available online at http://paso.esa.int/8_Additional/ATV%20Form%20542.pdf (last accessed 20 August 2009).

[20] Document available online at http://paso.esa.int/8_Additional/ATV%20Form%20879.pdf (last accessed 20 August 2009).

[21] Article VII, 1967 Treaty on Principles Governing the Activities of States in the Exploration and Use of Outer Space, including the Moon and Other Celestial Bodies, adopted by the United Nations General Assembly in its resolution 2222 (XXI) and opened for signature on 27 January 1967. It entered into force on 10 October 1967, and as of 1 July 2009, has received 100 ratifications and 27 signatures. The same responsibility is also placed on France pursuant to Articles I–III of the 1972 Convention on International Liability for Damage Caused by Space Objects. The Convention opened for signature on 29 March 1972, after it was adopted by the UN General Assembly in its resolution 2777 (XXVI). After its opening for signature on 28 March 1972, it has to date received 84 ratifications, 24 signatures, and three acceptances of rights and obligations. The Liability Convention entered into force on 1 September 1972.

[22] Document available online at http://paso.esa.int/8_Additional/REQ-CSG-RS-10A-CN.pdf (last accessed 20 August 2009).

[23] Document available online at http://paso.esa.int/8_Additional/REQ-CSG-RS-21A-CN.pdf (last accessed 20 August 2009).

[24] Document available online at http://paso.esa.int/8_Additional/REQ-CSG-RS-22A-CN.pdf (last accessed 20 August 2009).

[25] European Commission, Enterprise and Industry Directorate-General, Aerospace, security, defence and equipment, Space policy and coordination, "Programming Mandate Addressed to CEN, CENELEC and ETSI to Establish Space Industry Standards" (Brussels, 25 June 2007), Document M /415 EN.

[26] *Ibid.*, p. 1.

[27] European Council Resolution of 7 May 1985 on a new approach to technical harmonization and standards, Official Journal C 136 (4 June 1985), pp. 1–9.

[28] Directive 98/34/EC of the European Parliament and of the Council of 22 June 1998 laying down a procedure for the provision of information in the field of technical standards and regulations, Official Journal L 204 (21 July 1998), pp. 37–48. This Directive was amended twice: Directive 98/48/EC of the European Parliament and of the Council of 20 July 1998 amending Directive 98/34/EC laying down a procedure for the provision of information in the field of technical standards and regulations, Official Journal L 217 (5 August 1998), pp. 18–26 and Council Directive 2006/96/EC of 20 November 2006 adapting certain Directives in the field of free movement of goods, by reason of the accession of Bulgaria and Romania, Official Journal L 236 (23 September 2003), pp. 81–106.

[29] European Council, Space: a new European frontier for an expanding Union An action plan for implementing the European Space Policy (2003) COM(2003), 673.

[30] See note 23, pp. 1–2.

[31] European Council, Official Journal L 261 (6 August 2004), pp. 64–68.

[32] See note 23, pp. 5–6.

[33] Information on the CEN, its work, and the available standards can be found online at its website, http://www.cen.eu (last accessed 20 August 2009).

[34] ESA Council Resolution 3.

[35] ECSS-P-00, see note 1.

[36] Batch 1 of ECSS Standards was released on 4 August 2008, with Batch 2 published on 15 November 2008 and Batch 3 on 18 March 2009. All Batches of ECSS standards can be obtained online at its website, http://www.ecss.nl (last accessed 20 August 2009).

[37] ESA, List of "ESA approved standards", Version 3.1 (8 June 2009), ESSB-AS V3.1. For the application of this list, refer to ESA Document ESA/ADMIN/IPOL(2007)11 (20 July 2007).

[38] ECSS, "Release of Batch #3: Hot News from ECSS" (6 March 2009), available from its website, see note 1.

[39] See note 35, p. 6.
[40] ECSS Overview Document, see note 1.
[41] Bohle, D. H. K. et al. The First International Space Safety Standard, Proceedings of Joint ESA-NASA Space Flight Safety Conference, ESTEC, Noordwijk, Netherlands, 11–14 June 2002, ESA SP-486 (August 2002), pp. 393–396.
[42] Kriedte, W. "ECSS—A Single Set of European Space Standards", ECSS, online at http://www.ecss.nl/ (last accessed 20 August 2009).

CHAPTER

4 Space safety standards in Japan

Ryuichi Sekita
Safety and Mission Assurance Department, JAXA

Japan's space safety standard has its roots in one engineering document, "CF-86001 System Safety Standard for National Space Development Agency of Japan", which was established in 1986. Now, more than 20 years after the establishment of the original standard, the Japanese Aerospace Exploration Agency (JAXA) has two kinds of standards and six documents, as shown below:

- JAXA Management Requirement (JMR)
  - JMR-001, System Safety Standard
  - JMR-002, Launch Vehicle Payload Safety Standard
  - JMR-003, Space Debris Mitigation Standard
- JAXA Engineering Requirement Guideline (JERG)
  - JERG-0-001, Technical Standard for High Pressure Gas Equipment for Space Use
  - JERG-1-006, Launch Vehicle Development Safety Technical Standard
  - JERG-1-007, Launch Site Operation Safety Requirements.

JAXA's safety standards are to specify integrated requirements for safety management and safety design that the unmanned payload organization should conduct to protect human life, properties, and environments from any mishap or accident occurring associated with payload and its Ground Support Equipment. Also, note that JAXA's manned space projects are compliant with NASA's safety standards.

JAXA's safety standards set forth the following basic requirements:

1. Establish management organization. Effectively implements the system safety program.
2. Establish safety design requirements for hazard control.
3. Identify and control system hazards throughout the life cycle.
4. Verify safety design by tests or other means.
5. Ensure operation procedures and other relevant documents.
6. Record the results of all activities.
7. Maintain and manage safety data.
8. Prepare an effective system safety program plan.
   a. Milestone. Schedule for performing hazard analysis and safety review.

Figure 4.1 shows the contents of JMR-002 Launch Vehicle Payload Safety Standard as an example of Japan's safety standards outline.

Space Safety Regulations and Standards. DOI: 10.1016/B978-1-85617-752-8.10004-2

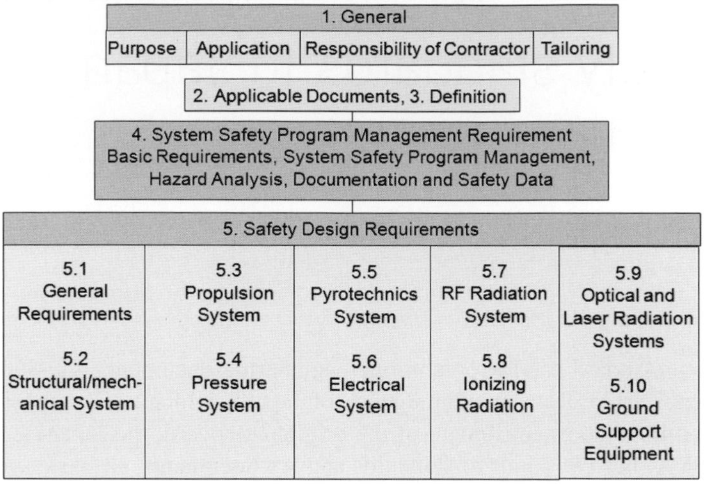

1. General

Purpose | Application | Responsibility of Contractor | Tailoring

2. Applicable Documents, 3. Definition

4. System Safety Program Management Requirement
Basic Requirements, System Safety Program Management,
Hazard Analysis, Documentation and Safety Data

5. Safety Design Requirements

5.1 General Requirements 5.2 Structural/mechanical System	5.3 Propulsion System 5.4 Pressure System	5.5 Pyrotechnics System 5.6 Electrical System	5.7 RF Radiation System 5.8 Ionizing Radiation	5.9 Optical and Laser Radiation Systems 5.10 Ground Support Equipment

FIGURE 4.1

Contents of JMR-002.

CHAPTER

Russian national space safety standards and related laws

Olga Zhdanovich

CHAPTER OUTLINE

INTRODUCTION

The Russian Federation inherited the national standardization system from the Union of Soviet Socialist Republics (USSR). In July 2003, Federal Law No. 184-FZ, titled "On technical regulation", was enacted by the Russian parliament and made the Standardization Law of 1993 invalid. The law has changed the established structure of standards and technical regulations in Russia, and the system of their development, approval, and control. The primary goals of this law were to eliminate barriers to trade and to encourage free movement of goods, supporting Russia's entrance into the World Trade Organization. The law allowed for a 7-year transitional period during which earlier standards and regulations relating to standardization, certification, metrology, and compliance with conformity would remain in full force. Implementation of the Technical Regulation Law causes a number of discrepancies in the field of a national standardization process.

Space Safety Regulations and Standards. DOI: 10.1016/B978-1-85617-752-8.10005-4

The modern Russian Space Program is influenced by a hybrid set of standards. This includes some of the standards and regulations that were developed in the Soviet Union, plus standards that were developed following the collapse of the USSR. Today, new space standards are introduced based on compliance with the international principles and Russian legislation on technical regulation. Safety is an important topic for the Russian Space Program. There are two main clusters of safety issues: (a) safety of space technology and human space flight, and (b) safety of the population of the Russian Federation living under launch trajectories of Russian boosters.

Although the adopted Technical Regulation Law makes conformity to technical regulation mandatory and compliance to standards voluntary, no Space Technology Technical Regulation has been approved to date. In addition, the Russian Space Program has retained a mandatory requirement for compliance to standards.

This chapter considers recent legal development of a national technical regulation in the Russian Federation. It describes the organization of the Russian Space Program, enacted space safety-related legal documents, and the process of mandatory certification for Russian space technology. The chapter also considers the system of development of Russian space standards. It examines the space safety-related legislation under preparation and public discussion, and discusses standards that relate to safety of space technology, as well as safety requirements for human space flight.

5.1 NATIONAL STANDARDIZATION SYSTEM OF THE RUSSIAN FEDERATION: HISTORICAL OVERVIEW

The Russian National State Standards are named GOST R (*ГОСТ Р* in Russian), which in English means "State Standard of Russia" (*Gosudarstvennyi Standard Rossii*).

The development of a comprehensive national standards system in Russia has occurred over a period of more than 80 years. The Soviet governmental standardization agency was founded in 1925 and was responsible for writing, publishing, and dissemination of national standards [1]. After the Second World War the USSR national standardization program was transformed to provide the necessary methodological and technological support for the ambitious industrial and technological development programs undertaken by the Soviet government. The Cold War and Space Race were the main driving forces for that technological development.

In 1968, the first standard GOST 1 State Standardization System was published—GOST (in Russian *Gosudarstvennyi Standard*) stands for state standard [2]. Two years later, the Committee of Standards, Measures and Measuring Instruments, working under the Council of Ministers, was renamed as the State Committee of Standards under the Council of Ministers of the USSR [1]. Since that time, the Committee has undergone many transformations, and for a long time in Soviet Union and Russian Federation history the Committee was known as the State Standard Organization or *GosStandard*. *GosStandard* was the main body responsible for the development of the national standardization system.

In the Soviet Union, standards were mandatory and obligatory for all activities. The USSR State Standardization Ministry together with other ministries and departments introduced national standards. For example, the following categories of standards were developed and applied: a USSR state national standard—GOST (*Gosudarstvennyi Standard*) was a standard applicable to the national economy on the territory of the Soviet Union; OST (*Otraslevoi Standard* in Russian) was the standard of a specific branch of industry; a standard of the specific enterprise/plant—STP (*Standard Predpriyatiya*)—was a standard of a specific company that was mandatory only for that specific entity. In addition, documents that provided technical specifications—TUs (*Technicheskie Ukazaniya*), Regulations (*Polozheniya*)—were introduced.

Standards and norms related to public safety were introduced by specific agencies that dealt with public health, such as the Ministry of Health, Sanitary and Epidemiology Service, and the Agency for Fire Safety, etc. Norms and standards relating to human health were abbreviated as *SanPiNs* in Russian language (*Sanitarnye Pravila i Normy*), which stands for sanitary rules and norms. Also, hygienic norms (*Gigienicheskie Normativy* in Russian) were important to human health. Occupational health problems were defined by the System of Standards for Labour Safety, abbreviated as SSBT (in Russian, *Sistema Standartov Bezopasnosti Truda*). For the environment, a level of pollution was controlled by PDK norms (in Russian, *Predelno-Dopuctimaya Koncentracia*; PDK indicates maximum exposure to a physical or chemical agent allowed in an 8-hour working day to prevent death or injury), which in English means "maximum admissible concentration of toxic substances to the environment". The USSR Ministry of Environmental Protection developed these norms.

During the Cold War, the ministry in charge of the national Space Program was the Ministry of General Machine-building. This entity, together with the Ministry of Defense and *GosStandard*, was responsible for the development, introduction, and control of implementation of space technology-related standards among other activities. However, from the very beginning, safety standards relating to the safety of cosmonauts on board spacecraft were developed by institutions within the USSR Ministry of Health and Soviet Academy of Sciences (Institute of Biomedical Problems). Space industrial enterprises developed and implemented their own STPs. Historically, within the Ministry of General Machine-building, the Central Institute of Machine-building (TsNIIMASH) was responsible for the overall scientific coordination of the development of standards, norms, and regulations for USSR space technology.

In 1991, following the collapse of the Soviet Union, the Russian State Committee for Standardization, Metrology, and Certification became the main body responsible for standards on the territory of independent Russia. It also inherited the state standard system from the Soviet Union. As a result, the Russian Federation acquired approximately 500,000 regulations officially approved by the government agencies [3].

On 24 February 1992, the Russian Space Agency (Russian abbreviation RKA, *Rossiiskoe Kosmicheskoe Agentstvo*) was founded by the Act of the Russian Government No. 85, "On structure of management of space activity in the Russian Federation". The Agency was specifically responsible for the development of a national civil space program. Seven years later, the Agency incorporated the field of aviation and became the Russian Aviation and Space Agency (RAKA). In 2004 the Agency was separated into the Russian Space Agency (*Roscosmos*) and the aviation branch had its own agency.

Since 2004, the *GosStandard* of Russia has had the new name *Rostehregulirovanie* [1]. Russian national standards are now developed under the auspices of the Russian Federal Agency for Technical Regulation and Metrology (*Rostehregulirovanie* in Russian).

The modern Russian national standards are developed through the system of the Standardization Technical Committees that covers all aspects of the functioning of the national economy, including military and national defense. *Rostehregulirovanie* consists of around 400 Technical Committees. These Technical Committees develop new national standards, which are approved and then disseminated by the Agency.

Today, technical regulations in the Russian Federation are enacted by means of federal law, and by the orders of the President of Russia or the Cabinet. Under law, technical regulations are mandatory whilst conformity by industry to standards is voluntary. The change of mandatory compliance to standards to mandatory conformity with technical regulation creates a contradictory situation in the standardization process in Russia because only 11 technical regulations have been approved so far and they are unable to replace thousands of Russian standards. The 11 approved technical regulations are available on the official website of *Rostehregulirovanie* [4].

5.2 TECHNICAL REGULATION LEGISLATION

There are two main laws that are important for the systems of the Russian National Standards and for the Russian Space Program: Federal Law "On technical regulation" 184-FZ (FZ means "Federal Law" in Russian, *Federalnyi Zakon*) of 27 December 2002 and Federal Law "Providing unity of measurements", No. 102-FZ of 26 June 2008.

Federal Law No. 5154-1 "Standardization Law" from 10 June 1993 has become invalid after the enactment of the Technical Regulation Law. This has created a difficulty because the majority of national economic activities are not covered by the technical regulations that have been approved during the last 7 years. These areas still operate on a principle of mandatory compliance to standards. Since March 2009, a new draft of "Standardization Law" is under public discussion and is available at the official website of *Rostehregulirovanie* [5].

The main goal of the Law "On technical regulation" is to bring the Russian approach in the field of technical regulation closer to international practice and regulations, as well as to facilitate the entrance of the Russian Federation into the World Trade Organization.

Federal Law No. 184-FZ "On technical regulation" has been amended four times through Laws 45-FZ (May 2005), 65-FZ (December 2007), 309-FZ (July 2008), and 160-FZ (July 2009). The enactment of the Law has led to the following changes in the national standardization process [6]:

- Replacement of mandatory compliance with standards by mandatory compliance with technical regulations, intended to ensure protection of life and health of people, and protection of actions misleading consumers.
- Voluntary applications of standards. The concept "code of practice" has been introduced as a standardization document, which can be used as a basis of proof of conformity.
- In terms of mandatory conformity assessment procedures, more declaration of conformity (subject to official registration) and less certification depending on risk level.
- Introduction of specific certification labeling.
- Separation of the following functions: certification, accreditation, and surveillance.

A technical regulation is normally a federal law, a government resolution, or an order by the President of the Russian Federation or the government. A technical regulation is a legal document; it has the status of a federal law. A Technical Regulation document (TR), *Tehnicheskiy Reglament* in Russian, places a mandatory requirement on operation, production, service, and processes (as objects of technical regulations). The technical regulation outlines the requirements for public safety and health, personal and public property safety, environmental safety, safety and protection of animals and plants, and prevention of abusive activities that might affect customers [7].

There are two types of TR: general and special. The general technical regulation requirements are mandatory for any kind of commercial products, production technologies, operation, and use. A special technical regulation has requirements for specific manufacture technologies and products, their utilization, storage, transportation, dumping, and for environmental hazards not regulated by the general technical regulations.

Pursuant to this Law, technical regulations, which are compulsory, are developed and approved through the following four processes before they are enacted:

- A draft of a technical regulation is developed.
- The draft is subject to public discussion.
- A public open administrative review of the draft is undertaken by an expert commission involving government agencies, the scientific community, trade associations, and consumer advocacy groups.
- Following this review, a further review and legislative approval by the Lower Chamber of the Russian Parliament, called the State Duma, is conducted prior to enactment.

Technical regulations provide the basis of public safety at the national level. TR establishes minimal necessary requirements that provide for the following specific safety issues listed in the Law: (1) safety of emissions, (2) biological safety, (3) safety from explosions, (4) mechanical safety, (5) fire safety, (6) industrial safety, (7) thermal safety, (8) chemical safety, (9) electrical safety, (10) nuclear and radiation safety, (11) electromagnetic compliance for the safety of operation of instruments and equipment, and (12) unity of measurements [7].

Rostehregulirovanie publishes lists of standards and codes of practice related to test, methods, and sampling rules to be used for conformity assessment according to technical regulations.

The following categories of standards are used in the Russian Federation and in the Russian Space Program:

- Russian National standard, **GOST R**—mandatory in the territory of the Russian Federation.
- Standards for the former Soviet Union states, **GOST**—mandatory in the territory of Commonwealth of Independent State (CIS) countries. (Standards adopted during the time of the former USSR have the same abbreviation, GOST. The last two numbers in a GOST number indicate the year of standard approval. It is possible to avoid confusion and discern if a GOST is a USSR standard or a CIS standard by reading the year of approval; before 1991 denotes a USSR standard and after 1991 a CIS standard.)
- Standards for branches of the Russian national economy, **OST**—mandatory for use by *Roscosmos*, for example, in the Russian Space Program.
- Standards of the specific plant/corporation, **STP**—mandatory to use only for the specific enterprise, for example, STP of Rocket and Space Corporation "Energiya".

Russian standards also incorporate international standards in the following ways: (a) direct acceptance of the ISO standard without any changes as a Russian state standard, for example GOST R ISO 9000-2001; and (b) acceptance of an ISO standard with changes, after which the name of a standard will look, for example, like GOST R 50231-92 (ISO 7173-89).

There is a special logic, or hierarchical structure, with GOST R, GOST, OST, and STP. *Rostehregulirovanie*, the main federal body for standardization, approves a Russian National Standard "GOST R". This is a fairly general standard, based on common knowledge, and can be applied to various activities of the national economy. "GOST" standards are approved by CIS national standardization bodies. After the collapse of the Soviet Union, a number of cooperative projects were implemented by CIS countries together, including a common marketplace organized by Russia with a few CIS countries. A specific governmental agency/ministry approves "OST" standards. In reality, OST standards are GOST R standards applied to a very specific field of the national economy, and then a company of specific industry develops its own "STP" standards to implement the GOST R, GOST, and OST standards.

Today, there are a few discrepancies in the whole standardization process because of the technical regulation law that made the previous standardization law invalid. The original idea behind technical regulation was the elimination of barriers to trade of industrial and agricultural products for Russia's membership in the World Trade Organization. The implementation of technical regulation has left behind, without any legal basis for a standardization process, a number of areas that are not involved in Russia's WTO activities, such as federal needs, the state communication system, the state medical system, the labor safety system, the defense industry, and manufacturing of dual-use systems. As a result, the standardization process occurring in Russia today is sometimes in contradiction to the technical regulation law. For example, compliance with standards for the Russian Space Program and other defense-related programs is still mandatory. Another problem arises in that, although technical regulations should be in full force by 2010, there are still only a dozen technical regulations that have been approved and implemented, and so far none has been approved for space technology. The vice-head of *Rostehregulirovanie*, S. Pugachev, when describing the need in Russia for a new standardization law, explained that the current technical regulation law is unable to replace the standardization law and related activities fully because the nature and scope of standardization are far broader than in the technical regulation [8].

The Federal Law "Providing unity of measurements", No. 102-FZ of 26 June 2008, is building a legal basis for uniform metrological support for technological development in Russia.

5.3 RUSSIAN SPACE PROGRAM: ORGANIZATION, CERTIFICATION, LEGISLATION, AND STANDARDIZATION

This section discusses the organization of the Russian Space Program through its main elements, such as the Russian Space Agency and the Federal Space Program. The section considers the certification process of space technology, describes enacted laws relating to space safety, and outlines newly enacted space safety acts that are under development and public discussion. It also discusses the process of technical regulation and standardization of Russian space technology, and outlines space safety standards and safety procedures for human space flight on board the International Space Station.

5.3.1 Organization of the Russian Space Program

The Russian Space Program has three cornerstone elements: the Federal Space Agency, the Federal Space Program, and federal space activity law. In other words, the Russian Space Program is run by the Russian Space Agency that executes the Federal Space Program on the basis of the Federal Law "On space activity", No. 5663-1.

The Federal Space Agency (*Roscosmos*), in its current setting, operates on the basis of Provisions approved by the Russian government on 26 June 2004, No. 314, which included some additions from

30 July 2007. The Russian Space Agency is the body of the federal executive power that is responsible for the following: implementation of a state policy, legal regulation, providing state services and management of state property in the field of space activity including dual-use purposes, as well as international cooperation [9]. Paragraph 5.3.7.3 states that *Roscosmos* is organizing certification of space technology developed for scientific and social economic purposes (the certification process is discussed in section 5.3.2).

The Federal Space Program is the national space program developed over 10 years and approved by the government of the Russian Federation. The current Federal Space Program covers a decade, 2006–2015, and was approved by decree of the government of the Russian Federation on 22 October 2005 (No. 635).

The Federal Space Program is a plan for a decade of space activities in Russia with given funding. Every year, the Russian Parliament confirms funding for the Russian Space Program as part of the national state budget [10].

The current Federal Space Program, for example, has a section dealing with cosmodromes and maintenance of ground test facilities. There is a special action listed toward the development of a system of environmental monitoring of territories/administrative regions affected by rocket and space technologies (those regions located along the trajectory of a flight of a booster above Russian territory, where first and second stages land, and/or remnants of launchers fall, during accidents).

Federal Law on Space Activity, No. 5663-1 of 20 August 1993, created a legal basis for performing space activity in the Russian Federation [11]. The law is described in detail in section 5.3.3.

The Russian Space Program inherited its type of organization from industry: *Roscosmos* consists of a number of state industrial enterprises while at the same time acting as a state customer to the companies that constitute the agency, as well as to external entities. This is also reflected in the Provisions of the Russian Space Agency.

Historically, the roles within the Russian Space Program were demarcated, with each having a specific title. In other words, chain types of customer–contractor relationships in the Russian Space Program are [12]:

- State customer (mainly *Roscosmos*), who develops technical specifications for a proposed space complex under development.
- Prime system contractor of a space complex, the leading development organization that is responsible for development of a space complex based on technical specifications given by a state customer.
- Prime subsystem contractor, which is an organization developing space complex subsystems, based on technical specifications given by a state customer and prime system contractor for a whole space complex.
- Subcontractor company, which develops a specific component part, prepares component parts given by technical specifications provided by the prime system contractor and the prime subsystem contractor.
- General contractor, a company that is responsible for the manufacture of a space complex.

The state customer–contractor chain of relationships is an important consideration when further describing safety space laws under development.

A study made by the European Space Agency (ESA) in 1995 revealed that the essential differences between the internal structure of the Russian space industry and the European space industry are [13]:

- Separation of design and production processes (original units).
- Contact with different ministries and consideration of administration, organization, and sanctioning of expenditure.
- Space systems as series products and not as individual parts, like European space technology.
- Decisions regarding problems of development, tests, or operations often made by government authorities or committees appointed by them.
- Differentiation into prime contractors and subcontractors for development and manufacture.

In addition to the Russian Space Agency, the following ministries have been involved in the Russian Space Program: Ministry for Liquidation of Emergency Situations and Civil Defense; Ministry of Industry; Ministry of Energy; Ministry of Transport; Ministry of Defense; Ministry of Natural Resources; Ministry of Information Technologies and Communications; Federal Agency on Geodesy and Cartography; the Federal Agency for Hydrometeorology and Environmental Monitoring; and Agency for Fishery, as well as a few others [10].

5.3.2 Federal Certification System of space technology

In the Russian Federation, certification can be done either on a mandatory or voluntary basis. For a state mandatory certification, there are 16 independent systems of certification for specific branches of industry. Mandatory certification is performed for a number of legislative acts, such as federal laws, for the certification of goods and services, and protection of the rights of consumers. There is a list of goods and services that require mandatory certification, and general requirements of the procedure for mandatory certification are defined and described by Russian national standards.

In Russia, certification of space technology is mandatory. The Russian Space Agency is the state entity responsible for mandatory certification of rocket and space technology within the federal system of certification of space technologies, which are developed for scientific and national economic purposes. The system, which does not have a special label, was registered on 22 June 1995.

The Central Institute of Machine-building (TsNIIMASH) is responsible for the overall scientific and methodological management of the certification of the Russian Space Program. The Center of the Certification of RKT of the TsNIIMASH has the role of the central body of the Federal Certification System of the RKT (the Russian abbreviation RKT stands for *Raketno-Kosmicheskaya Technika*, which in English means rocket and space technology). Also, a number of research institutions and manufacturing plants have their own certification units and test laboratories. These certification units and test laboratories are officially registered with the Federal Certification System of RKT [14].

The main functions of the RKT Certification Center of TsNIIMASH are:

- Certification of RKT:
  - Certification of quality management systems
  - Attestation of production facilities
  - Certification of space-based services
  - Scientific and technical expertise
- Control:
  - Quality of certified RKT articles/goods
  - Stability of production

- Accreditation:
  - Bodies/entities responsible for certification
  - Test centers and laboratories of the Federal Certification System of RKT
  - Development of the normative and technical documentation of the Federal Certification System of RKT.

The main difference between the Russian mandatory certification system for aviation and space programs and those elsewhere is that certification begins with project definition and design and is applied to all stages of the life cycle of a space system (e.g. ground tests, flight tests, serial manufacturing, operation, and damping) [15].

5.3.3 Russian space safety legislation

In total, there are more then 400 legal documents that define the Russian Space Program [16]. Among them, legal acts constitute only 10%. The other 90% are decisions of administrative executive bodies. The majority of legal acts within that 10% do not relate directly to a space program; they are significant but define other types of activities. For example, the following legal acts related to the Russian Space Program are important: the Federal Law on Technical Regulations, No. 184-FZ; the Federal Law on Supply for Federal State Needs, No. 60-FZ of December 1993; and the Federal Law on Industrial Safety of Dangerous Industrial Objects, No. 116-FZ of July 2007.

5.3.3.1 Enacted space-related laws

Russia has a hierarchy of norms governing space activities. The Russian Space Program is governed by the Constitution of the Russian Federation, general principles and norms of international law, and international treaties signed by the Russian Federation, as well as other legal documents mentioned above [17].

Current civilian and military space activities are regulated by the Law of the Russian Federation on Space Activity, No. 5663-1 of 20 August 1993. As mentioned earlier, the Russian Space Agency performs its activities on the basis of the "Provisions on Federal Space Agency" (*Polozhenie* in Russian) issued by the government of Russia on 26 June 2004, No. 314. Any company involved in the development or application of space technology should have a license from the Russian Space Agency on the basis of "Resolution of the government of the Russian Federation on adoption of space activity licensing" from 30 June 2006, No. 403.

It is important to mention that a number of legal documents were signed with the Republic of Kazakhstan for the rental of Baikonur cosmodrome and clean-up of the territory of Kazakhstan in case of explosion of boosters that create toxic pollution of its territory and a threat to human health and safety.

This section looks into safety, certification, and liability issues from the Space Activities Law and discusses legal issues applied to the safety of Russian regions and Republic of Kazakhstan, for protection against falling stages of boosters, or their remnants, in the case of an accident.

Safety in Space Activity Law

The Space Activity Law (No. 5663-1) is the main federal law defining space activity in Russia. This Law has been amended eight times: 147-FZ in November 1996, 15-FZ in January 2003, 8-FZ in March 2004, 122-FZ in August 2004, 19-FZ in February 2006, 231-FZ in December 2006, 309-FZ in December 2008, and 313-FZ in December 2008 [11].

The amendment in November 1996, 147-FZ, was the main amendment to the Space Activity Law, effective 1993. For example, the names of the Federal Space Agency and Ministry of Defense were taken out and were phrased as federal executive bodies for space activity and for defense. Federal Law 8-FZ introduced the invalidity of some statements in the Space Activity Law from 1993. Law 19-FZ outlined changes in a few legislative acts of the Russian Federation and/or made invalid a few acts because of the adoption on 21 July 2005 of the new Law 94-FZ, "On placement orders for supply, doing works, providing services for state and municipal needs". The 313-FZ amendment of the Space Activity Law replaced the word "certification" with the phrase "mandatory certification and declaration of compliance" due to the enactment of the Law on Technical Regulation.

The Space Activity Law (as amended) consists of seven main sections: section 1 considers general issues, section 2 provides organization of space activity in the Russian Federation, section 3 discusses economics, section 4 looks into space infrastructure, section 5 describes safety of space activity, section 6 considers international cooperation, and section 7 describes liability aspects.

In Russia, space activities are carried out in accordance with the Constitution of the Russian Federation. The President of the Russian Federation has overall responsibility for space activities. The government of the Russian Federation coordinates space-related activities. The Federal Space Agency and Ministry of Defense are federal agencies that execute the Russian Space Program. The law refers to the Russian Space Agency and the Ministry of Defense as the federal executive body for space and the federal executive body for defense. The Space Activity Law mentions principles of international law as a binding source of law governing space activities of and in the Russian Federation [17].

Detailed descriptions of Articles of the Space Activity Law important to standardization, safety, and liability of the Russian Space Program are described below.

Section 1, "General issues", has four Articles. Article 4, "Principles of space activity", contains the safety principle "*Provision of Safety of Space Activity and Environmental Protection*", which is among the main principles of space activity in the Russian Federation.

Section 2, "Organization of space activity", has seven Articles. Article 10, "Mandatory certification and declaration of compliance of space technics", has three paragraphs. Space technics is the Russian term for space technology.

1. Space technics, including space objects and objects of space infrastructure created for scientific and national economic purposes, should be checked for compliance with the requirements established by the legislation of the Russian Federation (mandatory certification or declaration of compliance; amendments No. 147-FZ, November 1996; No. 313-FZ, December 2008). Equipment that is used for the development and use of space technics might be required for mandatory certification and declaration of compliance (version No. 313-FZ, December 2008).
2. Mandatory certification or declaration of compliance of space technics is done in an order established by the legislation of the Russian Federation for the technical regulation (amendment of No. 313-FZ, December 2008).
3. The certification agencies, manufacturers of space technics, and corresponding governmental officials found guilty of the violation of rules of certification or declaration of compliance of space technics are liable by virtue of the legislation of the Russian Federation (No. 313-FZ, December 2008).

The safety of space activity has a separate section in the main legal document; section 5 consists of four articles (22–25) that relate to the safety of space activity. Article 22 describes the safety of space activity, investigation of space incidents is provided by Art. 23, search-and-rescue and clean-up of accidents is considered in Art. 24, and insurance of space activity is discussed in Art. 25.

The safety of space activity covered under Art. 22 provides that any space activity must comply with the safety requirements established by the laws and other legal regulations of the Russian Federation. There are two main bodies that are responsible for overall guidance and execution of safety of space activities—the federal executive body for space activities and the federal executive body for defense. Space activity should be done in compliance within allowed anthropogenic pressure to the environment (No. 309-FZ, December 2008).

By requirement of the organizations involved or citizens, the federal executive body for space activities and federal executive body for defense are obliged to give information about threats that can arise from space activity. In the case of any real threat to the public safety and environment, the federal executive body for space activity and federal executive body for defense are obliged to immediately inform governmental agencies and organizations, as well as citizens.

Under Art. 23 ("Investigation of space incidents"), all accidents and disasters while carrying out space activity shall be subject to investigation with a procedure that should be set up in the legislation of the Russian Federation or other legal regulations of the Russian Federation. The procedure and validity of results of investigation of space incidents, including accidents and catastrophes, may be appealed in court.

"Search-and-rescue, clean-up from incidents" (Art. 24) shall be done by appropriate state agencies if necessary with participation of regional governments, and local administrations, as well as organizations and citizens. This article covers two aspects: clean-up and search-and-rescue works.

Clean-up of accidents with space activity involved shall consist of restoration and reconstruction of the industrial and other plants suffering and necessary environmental measures, and compensation for damage for regions, organizations, and citizens (the paragraph became invalid by No. 122-FZ, July 2004).

Search-and-rescue works, as well as the clean-up of an accident on a territory of a foreign state, shall be performed by agreement with the competent authorities of this state using funds of organizations and citizens and funds from the Federal State Budget (No. 147-FZ, November 1996; No. 122-FZ, August 2004).

"Insurance of space activity" (Art. 25) states that while performing any space activity on a territory of Russian Federation, insurance of space activity is compulsory. The amount is determined by legislation of the Russian Federation. Compulsory insurance should be effected against damage to the life and health of the cosmonauts and the personnel on the ground and the objects of space infrastructure, as well as against property damage to third parties. Organizations and citizens performing space activity may voluntarily insure space technics (risks of losses, missing or damage to space technics).

Section 7, on "Liability", comprises Articles 29 and 30. Article 29, "Liability of officials, organizations and citizens", asserts that state organizations and their officials, other organizations and their officials, and citizens liable for the breaking of that law and other legislation acts that regulate space activity are liable in accordance with the Russian legislation.

Article 30, "Liability for damage while performing space activity" (No. 147-FZ from 29 November 1996), has three paragraphs:

1. Liability for damages inflicted by space objects of the Russian Federation within the territory of the Russian Federation or outside the jurisdiction of any state except outer space shall arise regardless of the fault of the inflictor thereof.

2. If in any place, except from the Earth' surface, damage has been inflicted on a space object of the Russian Federation or on property onboard such an object by another space object, the liability of the organization and citizen owners of another space object shall emerge with their being at fault and in proportion to their fault.
3. Damage inflicted to a person or property of a citizen, as well as damage inflicted on a property of an organization by a space object of the Russian Federation while performing space activity on a territory of the Russian Federation or outside its territory, shall be compensated by the organization or citizen that insured their liability for damage in a size and order foreseen by the Civil Codex of the Russian Federation.

Safety issues at Baikonur through agreements with the Republic of Kazakhstan

On 25 May 1992, the Russian Federation and Republic of Kazakhstan signed an agreement about order of use of the Baikonur cosmodrome. On 28 March 1994, both governments agreed on the main principles and conditions of use of Baikonur. By mutual agreement from 1995, Baikonur received the status of federal city of the Russian Federation under a special regime of operation. On 23 December 2004, Russia and Kazakhstan agreed on rent of Baikonur up to 2050 in the framework of a new agreement on cooperation for effective use of the complex "Baikonur" [18].

Search and rescue of crews from the International Space Station is carried out based on the special agreement between Russian and Kazakhstan for the procedure of utilization of a search-and-rescue system of the Russian Federation in the territory of the Republic of Kazakhstan (agreement signed on 25 December 1993).

On 18 November 1999, the governments of Kazakhstan and the Russian Federation signed a special agreement on the order of cooperation in the case of an accident during launch activity from "Baikonur" cosmodrome. Article 5 states that, in case of another space object, the liability shall derive from their being at fault and in proportion to their fault. The Russian party informs the Kazakhstan party about the reasons that caused the accident, provides a list of measurers for safety of the next launches of that specific booster type, and, by coordination with the Kazakhstan party, restarts launches. In Kazakhstan, around 4 million hectares are affected by launch activities from Baikonur cosmodrome.

Financial compensations to the Republic of Kazakhstan by the Russian Federation are foreseen by the agreement of 10 December 1994 on the rent of Baikonur complex, where it is stated that in the case of damage the Russian Federation is liable due to the international convention on international liability for damage done by space activity of 29 March 1972. For example, on 27 July 2006 there was an accident of a booster "Dnepr" that primarily carried university microsatellites. That launcher blew up 86 seconds after take-off. Toxic propellants affected Kazakhstan's environment. The damage was estimated as US $1.1 million [18].

Safety regulations for Russian regions along launch trajectories

A number of administrative regions of the Russian Federation are located along trajectories of a space orbit of Russian boosters launched from Baikonur and Plesetsk cosmodromes. For more than 50 years of space activity, these regions and population have been impacted directly by first the Soviet and then the Russian Space Program.

The Russian government issued the "Order and conditions of random utilization of regions where separated parts of rockets fall" on 31 May 1995. Three years later, a special technique for calculation of payment to the administrative regions where rocket parts have fallen was adopted. The Russian regions (subjects of the Russian Federation) started to receive additional state funding for damage from space activities [19].

On 12 November 1998, the Upper Chamber of the Russian Parliament adopted a decree "On development of a temperate commission of Soviet Federation for protection of interests of Subjects of Russian Federation, organizations and citizens from the dangerous consequences of rocket and a space activity". In December 1998, the government of the Russian Federation issued a provision for that commission and approved a plan of activities.

In 1998, the development of a unified system of danger warning and liquidation of emergency situation caused by space technology was started. This was initiated because the Russian government adopted the "Rules on informing bodies of executive power on launch of a spacecraft with a nuclear power source as well as informing local municipalities and providing protection to population in a case of emergency landing of a spacecraft with nuclear power source" on 15 August 1998 [19].

The Ministry of Ecology developed a special ecological passport for administrative regions with fall of rocket parts with annual environmental monitoring of air, water, soil, and human health. Today, serious attention is given to the environmental impact of space technology. Standards on ecological safety for space technology that became effective from 2008 have been introduced recently. Technical specifications, given in tenders by *Roscosmos* for all newly designed space systems, contain mandatory requirements for protection of the environment and humans from harmful impacts of space technology. Safety issues in the regions where parts of rockets may fall require annually: (1) keeping special workforce units; (2) renting of aviation; (3) special medical and biological research of space technology-related pollution on people and the environment. In 2005, these three items were estimated as 55 million roubles to be funded from the state budget of the Russian Federation [20].

5.3.3.2 New space safety law under development

There are two space safety federal level legislation documents currently under discussion and preparation. The first is a special technical regulation law on the safety of space technology. The second is a draft law applicable to the administrative regions of the Russian Federation where launcher stages or remnants of launchers together with spacecraft in the case of an accident may fall along the trajectory paths over the territory of the Russian Federation.

Proposed Special Technical Regulation on Space Technology Safety

The draft of the Federal Law "On special technical regulation on safety of space complexes produced for scientific and social economic purposes" (hereafter Special TR on Space Technology Safety) has been under discussion since September 2006. The Special TR on Space Technology Safety has two attachments: Attachment A lists terms of space activities where special TR is applicable; Attachment B describes "Rules and forms of assessment of compliance for mandatory requirements of space complexes produced for scientific and social economic purposes" [21].

Drafts of the Special TR on Space Technology Safety with two attachments and the results of public discussion are available at websites of various entities involved in the process of public discussion of new technical regulations, such as the Non-Governmental Council for Technical Regulations and the Institute of Standardization and Certification of Innovations.

The draft of the Special TR on Space Technology Safety has six chapters and 20 articles [21]. Chapter 1 considers general issues, chapter 2 describes general requirements for the organization of safety during the development and operational phases of space complexes, chapter 3 discusses the essential requirements that define the necessary levels of the safety of space complexes, chapter 4 provides an assessment of compliance and a form of confirmation of compliance due to risk levels, chapter 5 outlines state control for compliance with safety requirements and use of results of control, and chapter 6 discusses enactment of the special technical regulation.

Chapter 1 has eight articles (1–8). More detailed description of Articles 4, 6, and 8 are given below. The main terms used for the Special TR on Space Technology Safety (Art. 4) are:

- Active space activity—manufacturing, test, launch, and operation of goods of rocket and space technics.
- Safety program—a document that contains requirements both for safety program content and safety program implementation, as well as a list of types of safety activities that should be performed during all phases of the life cycle of a space system. The aim of a safety program is to provide, to confirm, and to control a safety level given in the technical specifications document.
- Reliability program—a document that contains requirements both for reliability program content and reliability program implementation and a list of types of works that should be performed at all stages of a life cycle. The aim of a reliability program is to provide, to confirm, and to control the level of reliability given in the technical specifications document.

Classification of danger produced by a space activity and objects of technical regulation are described in Art. 6, chapter 1 of the Special TR on Space Technology Safety.

The different classes of danger produced by space activity are divided into three categories: risk objects, environment, and property. The risk object category includes cosmonauts, personnel from cosmodromes and other ground infrastructures, local population of the Russian Federation where the cosmodromes are located, employees of the rocket and space industry, the population of Russia, and populations of other countries. The environment is considered as two classes: Earth and outer space. The property category is divided into space objects property, space infrastructure property, and property of third parties.

Possible maximum consequences of danger that might occur are divided into four categories: (1) damage to life and health, (2) damage to nature, (3) significant property damage, and (4) global catastrophes.

Classes of dangers of space activity are differentiated as follows:

- Accident of a launcher followed by a fall to the ground of the booster itself, any component parts of the launcher or parts of a payload carried by this launcher beyond special zones that are allocated along the trajectory of a booster on the territory of the Russian Federation.
- Fall to the Earth of a spacecraft, or of a return capsule of a spacecraft or any component part beyond special zones.
- Fall of a booster at the launch pad.
- Accident of a launcher with fall to the ground of a booster itself, or any component parts of the launcher or fall of a payload carried by this launcher inside special zones.
- Landing of a first stage and second stage with remnants of propellants inside special zones.

- Damage to health of cosmonauts due to off-nominal situations caused by unforeseen external harmful impacts.
- Death of cosmonauts because of malfunctions of a technology, software of a spacecraft, or mistakes of crew and ground controllers.
- Collision of automatic spacecraft with space objects or space infrastructure object beyond Earth's surface.
- Explosion or fire at a launch pad.
- Accident or off-nominal flow of technology activities by exceeding allowed norms of impacts on the personnel and the environment.

Technical regulations cover:

- Space systems (process of its development and operations)
- Boosters and acceleration units
- Human spacecraft and orbital stations
- Automatic spacecraft
- Launch pads and ground control infrastructure
- Rocket engines
- Spacecraft with nuclear power sources
- Processes of utilization
- Transport and damping of propellants
- Means and processes of transportation of rocket and space technics.

Article 8 of chapter 1 considers implementation of safety requirements as follows:

- The basis of safety of a space activity is created by high reliability of space technics. High reliability is planned during the design phase, tested during the test phase, and supported and analyzed during the operation phase. A check of the possibility of implementation of required reliability characteristics is done on the basis of the experience accumulated in the Russian space industry and on the basis of analysis of corresponding data on design, manufacture, and operation from similar space complexes.
- *A reliability program* is a thorough analysis of possible effectiveness from increasing reliability measures at the expense of simplification of construction, improvement of technology, application of functional reservation, optimization of structural reservation, maintenance and repair, and use of controlled (that can be monitored) processes.
- An analysis of necessity and sufficiency of safety procedures is done in a *safety program*. This is done through analysis of types, consequences and criticality of failures, real-time control of parameters of the state of a space complex, diagnostics and warning of off-nominal situations, utilization of an emergency protection system, and a danger-warning system, implementation of alarm rescue procedures and/or crew evacuation, and search-and-rescue procedures, as well as evacuation procedures of personnel from a ground infrastructure.

Chapter 2 has four articles (9–12). Article 9 considers the requirements for planning of a space activity by taking into consideration the space technology safety issues. All types of launchers and spacecraft, as well as objects of space infrastructure, have safety passports. Safety passports record in chronological order all accidents and dangerous events that happened to the holder of the passport, consequences that were caused by accidents, findings about reasons that cause accidents, measures taken against accidents, and assessment of implementation of safety requirements. In Russia, for all cosmodromes including Baikonur, there is a requirement for all space objects to be registered and risk logs to be filled in for all types of launchers, for all chosen trajectories of launch, for all defined regions of fall of parts of boosters, and for all types of payloads.

Chapter 3 provides essential requirements that define different levels of space safety of space complexes. Chapter 3 has two articles (13 and 14). Article 14 outlines the design and technological solutions used during the implementation of requirements for a limitation of risk from the main classes of rocket and space technics as follows:

1. Identification of critical elements, functions, and processes for every class of danger with definition of sufficient depth of functional and structural reservation, maintenance and repair. Real-time control and operation with the goal of achieving best results of every flight with guaranteed risk limits.
2. Definition of necessity and sufficiency of safety, emergency, search-and-rescue, protection, and evacuation systems for every class of space technology product. This includes:
   - For launchers: (a) control in a regime of retention at a launch pad; (b) regime of withdrawal from launch pad; (c) liquidation before exit beyond borders of special zones; (d) withdrawal of last stages from orbit; (e) minimization of remnants of fuels in first and second stages.
   - For a manned spacecraft: (a) shelter from unforeseen impacts; (b) participation of crew in the operation of a spacecraft; (c) means of rescue at all stages of a flight; (d) means for disaster evacuation; (d) control from Earth whenever possible.
   - For an automatic spacecraft: (a) protection from any impact and influence of a process of preparation to launch and a process of launch; (b) prevention from spontaneous fall to Earth (withdrawal from orbit after termination of work).
   - For ground infrastructure: (a) system of disaster protection; (b) safety against a fire and an explosion.

Chapter 6 on enactment of the technical regulation consists of two articles (20 and 21). Article 20 states that technical regulations are used for mandatory implementation only for the protection of health and life of citizens, and for protection of property of physical and legal entities, state and municipal property, as well as for environmental protection, life and health of animals and plants. Article 20 identifies that the technical regulation will be effective 2 years from the day of its official publication [21].

Rules and forms of assessment of compliance of the Special TR on Space Technology Safety

Attachment B of the draft of the Special TR on Space Technology Safety describes rules and forms of assessment of compliance for mandatory requirements of space complexes produced for scientific and social economic purposes [22].

The document has six sections. Section 1 discusses terms and definitions, section 2 outlines general issues, section 3 considers forms of assessment of compliance, section 4 identifies rules of compliance assessment, section 5 defines forms and schemes of confirmation of compliance, while section 6 describes an order of confirmation of compliance of space complexes.

Chapter 3 identifies forms of assessment of compliance that include: (a) testing and research, (b) examination, (c) mandatory confirmation of compliance, and (d) state control.

Chapter 4 discusses rules of assessment of compliance. The assessment of compliance of space complexes is done within the following 11 aspects of the following life-cycle phases:

1. **Conceptual design phase.** A research institute of a state customer performs scientific and technical examination of a project prepared by a prime system contractor. The aim of examination is to reach conclusions on compliance or non-compliance of a conceptual design phase of a project with technical specifications given originally by a state customer; an important element of this stage is coordination of reliability and safety programs.
2. **Development of documentation for tests of a space complex prototype and mock-ups.** Assessment of compliance is done by scientific and technical examination of completeness and sufficiency of the experimental test program for a space complex and its components parts. A research establishment of a state customer or independent expert organization performs an examination of a test program followed by transfer of results of examination to a prime system contractor.
3. **Manufacturing of prototypes and mock-ups for autonomous tests.** Correction of documentation involves an assessment of compliance by autonomous tests of a space complex and its component parts, followed by the writing of autonomous tests reports.
4. **Manufacturing of prototypes for integration and inter-agency tests.** Corrections of documentation are made. Assessment of compliance is done by:
   - Integration and inter-agency ground tests followed by a report. Contractor companies prepare this report. The aim of the report is to conclude the readiness of a space complex under development for flight tests.
   - An examination by a research establishment of a state customer of results of ground tests, and preparation of a report. The report concludes on the readiness of a space complex under development for a flight test program.
   - A confirmation of compliance with technical specifications on the basis of results of ground tests. Compliance is confirmed by: (a) flight test readiness declaration of a prime system contractor and subcontractors; (b) certificate for a flight test technical readiness that can be issued by an independent accredited body of mandatory Federal Certification System of Space Technics; (c) examination and agreement on a flight test program by a research establishment of a state customer. An examination decides on completeness and sufficiency of the flight experimental test program and its conformity to documentation; (d) control by a state customer of the readiness of a space complex for a flight test program. Control is performed by creation of a Flight Test State Commission. Later, the Russian government approves both the Flight Test State Commission and Flight Test Program.
5. **Flight test program.** Evaluation of compliance is done by:
   - State control through a decision of a State Commission on a startup of the flight test program. This is done by consideration of results of a flight test readiness declaration prepared in written form by a prime system contractor, subsystem contractors, and organizers of flight test programs.
   - Analysis of each launch by preparation of a report of each launch according to the flight test program.

- Submission by a State Commission of data on the main results of flight tests to the government of the Russian Federation and the Russian Space Agency.
- Decision of a State Commission on termination of a flight test program with analysis of results of flight tests. An analysis of flight test results includes evaluation of implementation of a flight test program, compliance of characteristics of a space complex with given technical specifications, and preparation of a report with a decision to accept a space complex into operation and to start full-scale production if necessary.
- A certificate of compliance aimed to confirm compliance of the space complex through design, ground tests, and flight tests to technical specifications. A certificate of compliance is given by an organization accredited in the mandatory Federal Certification System on Space Technology.

6. **Preparation of documentation for full-scale series production.** Evaluation of compliance of documentation for full-scale series production is done in the form of state control by creation of an inter-agency commission by a state customer. The purpose of the inter-agency commission is to consider coordination and acceptance of designer documentation for full-scale series production.
7. **Preparation and mastering of full-scale series production, manufacturing, and tests of full-scale products.** Corrections of documentation for full-scale series production of a space complex are made. Assessment of compliance is done as follows:
   - Declaration of an implementation plan for a preparation for full-scale series production by manufacturers of a spacecraft and its component parts.
   - Confirmation of compliance of quality management systems together with confirmation of compliance of a space complex and its component parts by certificates of compliance given by certified bodies accredited with *Roscosmos*.
   - Tests of samples with preparation of acts on a readiness of production facility for full-scale series production of a given space complex.
   - Conclusions of a prime system contractor and subsystem contractors on suitability of designer documentation for full-scale series production.
   - Examination of implementation of requirements of designer documentation during manufacture and correction of documentation.
   - Examination of acceptance test program of full-scale samples according to designer documentation.
   - Incoming inspection of component parts.
   - Examination of quality of produced samples and stability of a production process with preparation of reports by manufacturers.
   - Examination of factual level of reliability of a space complex in the process of full-scale series manufacture with preparation of a report by prime contractors.
   - Examination of annual programs of quality and reliability by research establishments of a state customer with preparation of reports on coordination of quality and reliability programs.
8. **Acceptance of a space complex into operation.** Assessment of compliance is done in the form of: (a) A draft of a decree of the government of the Russian Federation on acceptance of a space complex into operation and start of full-scale series production. The draft of a decree is prepared on the basis of a report prepared by a State Commission together with a state customer and a prime system contractor. (b) A special State Acceptance Commission for acceptance of the subsystems

of a space complex into operation created by the government of Russia. The State Acceptance Commission issues an act in a form of a legal document that confirms the start of an operation of a space complex. (c) A certificate of compliance issued by a certification entity of *Roscosmos*.

9. **Acceptance of a spacecraft or orbital system into a flight operation.** Assessment of compliance is done by: (a) An examination of the technical state of a spacecraft by a technical commission formed from various entities responsible for spacecraft operation, for space data and services acquisition, development, and manufacture of a spacecraft. Results of the examination are fixed in the technical act approved by entities that assigned the technical commission. (b) A confirmation of compliance with certificate of compliance.
10. **Operation of a space complex.** Assessment of compliance is done as follows: (a) Ground and flight tests. (b) Examination of data analysis received from preparation to launch, launch, and flight operation. An organization responsible for operation prepares a report with results of each launch. (c) Examination of results of operation and evaluation of the technical state of a spacecraft by developers and manufacturers. (d) Confirmation of compliance by annual inspections of a certification body with the aim to confirm the validity of a certificate of compliance.
11. **Termination of full-scale series production and dumping.** Assessment of compliance is done: (a) by state control with a decree of the Russian Federation government on termination of industrial series production and start of dumping; (b) by examination of the technical state of articles designated for dumping; (c) by confirmation of compliance of articles designated for dumping and conversion according to a space technology dumping and conversion program in the form of certificates of compliance.

Chapter 5 outlines forms and procedures of confirmation of compliance. It has two paragraphs:

- Confirmation of compliance of space technology on the territory of the Russian Federation is mandatory. Mandatory confirmation of compliance of space complexes is done in a form that includes declaration of compliance as well as mandatory certification. A declaration of compliance takes the form of a decision of organizations (developers and manufacturers) on compliance of their space technology products with given technical specifications. A declaration is made on the basis of their own actual proof, confirmed by expert examination of the main research institute of a state customer.
- Mandatory certification of space complexes is done according to the scheme envisaged: (a) conduction of tests of space complexes, their component parts according to a valid procedures, and examination of test results by a third party (certification body) with the aim of confirmation of compliance; (b) certification of quality management systems of organizations (developers and manufacturers) of a space complex regarding compliance with valid scientific and technical documentation; (c) inspection control of certified space technics and quality systems; (d) confirmation with a certificate of compliance of space complexes and its component parts with established technical specifications, and confirmation of compliance by a certificate of compliance of quality management systems. Certificates are given by an entity accredited in the mandatory Federal Certification System of Space Technology.

The Special TR on Space Technology Safety has been under public discussion for nearly 3 years [21,22].

Draft of a federal law on safety of territories along launch trajectories

As discussed earlier, a number of administrative regions of the Russian Federation are located along the trajectory paths of Russian launchers to space. Parts of boosters with remnants of toxic propellants affect the environment and population of these regions.

Space technology pollution of Kazakhstan's territory from Baikonur cosmodrome, which is rented by Russia for a period of 50 years, is subject to the intergovernmental agreement between the Russian Federation and Republic of Kazakhstan (see above).

For Russian domestic territory, there are a number of activities performed for environmental protection of territories. These activities involve the mitigation of risks to human health, flora, and fauna.

In September 2009, all activities resulted in the draft of a federal law "On regions of fall of space objects and implementation of changes in a some legal acts" presented on the official website of *Roscosmos* [23]. The draft of this law has four chapters. Chapter 1 considers the main terms (Articles 1–3), chapter 2 outlines the basics of state regulation in the field of use of regions with fall of rockets (Articles 4 and 5), chapter 3 identifies rules of creation, use, and liquidation of areas with fall of rockets (Articles 6–8), and chapter 4 concludes and also considers the exact parts of specific legal acts where changes should be made due to enactment of the law (Articles 9–13).

Article 8 of chapter 3 has two sub-articles and outlines safety in regions with falling space objects:

- Safety measures in the regions with falling space objects include:
  - Establishment of a special procedure of land use described in Article 7, chapter 2.
  - Control of flight of a space object with the aim to locate the position of fall.
  - Search for and rescue of fallen space objects.
  - Liquidation of the impact of a fall of a space object includes: detoxification of places of fall of a space object in the case of leakages of toxic propellants; land re-cultivation of areas with fallen space objects; control of ecological situations; other safety measures necessary for elimination of impact caused by a fallen space object.
- An order of safety activities coordinating federal bodies of executive power is performed by the government of the Russian Federation, in the form of a decree of the Cabinet. Financing of the safety measures in these regions is the financial obligation of the government of the Russian Federation.

Article 8.2 outlines liability for damage caused by the fall of space objects. Damage to a landed property, land use, and land rent shall be compensated in accordance with the land use legislation of the Russian Federation.

There are 17 administrative regions of the Russian Federation that are affected by the fall of rockets. The proposed law should register regions that have falling rocket parts in a Russian state land cadastre as special zones with regulated economic activity and special conditions for all types of transportation. Regional administrations will receive extra state funding as compensation for the direct impact of space technology. Compensation will be differentiated due to natural geographic conditions and demographic situation in each administrative region of the Russian Federation [23].

5.3.4 Technical regulation and standardization of the Russian Space Program

This section discusses technical regulation of the standardization process in Russia, types of standards used in the Russian Space Program, and the procedure for adoption and update of space system standards, and also considers safety standards for human space flight, as well as safety documentation and procedures for payloads onboard the Russian segment of the International Space Station.

5.3.4.1 Technical regulation of space technology

Rostehregulirovanie has Technical Committee No. 321: Rocket and Rocket-and-Space Technics to oversee technical regulation of the Russian Space Program. Rocket-and-Space Technics is the translation of the Russian abbreviation RKT, which stands for *Raketno-Kosmicheskaya Technika*; in English it means rocket and space technology. The Committee is responsible for the development of the Russian National Standards for the Russian Space Program. It consists of 22 subcommittees as shown at the website of *Rostehregulirovanie* covering the categories listed in Table 5.1.

These 22 subcommittees form thematic groups for space technology standards development. Traditionally, safety in the Russian Space Program comes under the Reliability category. Human safety on board a spacecraft is not included in Technical Committee No. 321 because human safety standards are developed by Ministry of Health-related institutions.

The Central Institute of Machine Building (TsNIIMASH) of the Russian Space Agency (*Roscosmos*) is responsible for general scientific coordination of the development of standards for today's Russian Space Program. In TsNIIMASH, there is a special department that has historically

Table 5.1 Subcommittees of Technical Committee No. 321, Rocket-and-Space Technics

Subcommittee Number	Category
1	Space launcher means
2	Orbital means
6	Terminology in rocket and rocket-and-space technics (RKT)
7	Aero-gas dynamics and thermal regime of RKT
8	Assurance RKT
9	Reliability of RKT
10	Liquid-fuel rocket engines
11	Solid-fuel rocket engines
12	Control systems
13	Radio technical systems
14	Telemetry
15	Hydroscopic instruments
16	Ground technological equipment
17	Production technology of RKT
18	Technology of instrument making
19	Materials for RKT
20	Design regulations, requirements, and manufacture of general use
21	Metrology and metrological provision of production of RKT
22	Management of production quality and certification of the RKT

Subcommittees 3–5 are not shown on the *Rostehregulirovanie* website, http://www.gost.ru/wps/portal/pages.TechCom (in Russian).

been responsible for reliability, quality, safety standardization, and certification of the Russian Space Program. In 1975, the system of development and serial manufacturing of rocket and space technology was created and it contains around 300 normative technical documents [24].

Rostehregulirovanie has it own state research entities related to machine-building, known as VNIINMASH (Russian Research Institute of Standardization and Certification in Machine-building) and VNIIStandard (Russian Research Institute for Standardization). In total, *Rostehregulirovanie* comprises around 15 state entities responsible for development of national standards.

In Russia, when new space standards are developed they can be developed by any organization involved in the Russian Space Program (involvement of any organization or company requires a license to perform space activity given and approved by *Roscosmos*). The national standards under development will usually be verified by one of the above institutions from the national standardization agency, but not always. As part of their function, VNIINMASH performs checks on the compliance of Russian safety standards with international standards in the area of machine-building [25].

Technical Committee No. 321 will suggest new standards for approval by the Russian national standardization agency (*Rostehregulirovanie*). Following approval, the Agency will publish and disseminate a newly adopted standard. The Russian Space Agency then has the main responsibility for control of compliance of space activities with existing and new standards.

5.3.4.2 Types of space technology standard

The Russian Space Program uses the following system of standards: (a) GOST R are state standards that are approved by *Rostehregulirovanie*; (b) OST are standards developed for the Russian space industry, approved and controlled by *Roscosmos*; (c) STP are standards of industrial companies developed for their own needs, such as STP of the Rocket and Space Corporation (RSC) "Energiya". In addition, such documents as technical specifications (TU or TY—*Tehnicheskie Ukazaniya* in Russian) and regulations (P—*Polozheniya* in Russian) are important.

Branch industry standards are state standards modified according to the needs and requirements of a specific branch of industry in Russia. In the Russian space program, OST standards are developed for unified products, manufacturing processes, services, and metrological support used in the national space industry. OST also outlines specific rules of standardization inside the space industry. Before adoption of the OST standard, the standardization document is checked for compliance with standardization rules and procedures of the Russian Federation. The OST standard never contradicts the GOST standard.

Standards of a given company, known as STP, are developed for: (a) compliance with international, national, and branch industry standards; (b) products, processes, and services used at that company; (c) technological processes and safety issues; (d) management of production processes. Standards of entities should not break mandatory requirements of state standards.

For example, GOST, OST, STP standards, and P regulation used in the Russian Space Program are: GOST 18353-73, "Non-destructive control: methods of classification"; OST 92-4272-86, "Non-destructive dye penetration control: methods of control"; STP 351-54-86, "Organization and implementation of non-destructive control at the enterprise" (the standard of RSC Energiya); P 17375-082, "Baseline data for the selection of non-metallic materials used in habitable pressurized module compartments of products based on their hazard factors" [26].

There are GOST standards in Russia that cover different categories of technological development; these are known in Russia as common systems of technical standards. Common systems of technical standards are created as standards applicable to similar processes and stages as engineering design and production in various branches of industry. Common systems of technical standards are also known in Russia as inter-branch industry standards.

In the Russian Space Program, the following systems of state standards (GOST) are used: (1) Common System of Design Documentation; (2) Common System of Technological Documentation; (3) System of Standards for Development and Production Preparation; (4) Common System of General Technical Requirements on Special Equipment; (5) Common System of Special Equipment Quality Assurance; (6) State System of Insurance of Uniform Metrology; (7) Common System of Materials and Article Protection against Corrosion and Aging; (8) Common System of Technological Preparation for Production; (9) Common System of Software Documentation; (10) State Standards System on Space Factors and Protection Against Their Influence; (11) System of Standards of Labor Safety; and (12) System of Standards for Ergonomics [13].

Systems of standards for Russian space industry (OST) are created as extensions of state common technical standards and regulate the following issues: (a) development of space systems; (b) space system design documentation requirements; (c) space system engineering documentation requirements; (d) space system metrological quality assurance requirements; (e) space system quality requirements; (f) space system safety and reliability requirements; and (g) space system test specifications.

In addition, a system of standards for product assurance and safety of space systems includes tasks on standardization of all space system life-cycle phases starting from conceptual design, project documentation development, tests, and full-scale production and operation, and ends with worn product utilization [13].

5.3.4.3 Space standards adoption and update procedure

In Russia, state standards (GOST), space industry standards (OST), and company standards (STP) concerning specifications on product assurance and safety, design, testing, manufacture, and maintenance for products are still compulsory for the Russian space industry.

The procedure of standard adoption at a given space industrial entity requires the fulfillment of a number of organizational and engineering actions to put a standard into practice and adhere to it at an entity. The standards are considered as adopted if a procedure is established in an entity, organization and engineering actions are planned and fully completed, and a report on standard adoption is drawn up [13].

Every standard (GOST and OST) is reviewed regularly every 5 years; for human space flight, standards can be updated immediately if necessary. About 20% of all space standards are checked every year. Obsolete standards, which are not applicable at review time, are canceled. When a standard is under review, it is assessed against modern requirements of engineering and scientific level with the following conclusions: further application with/without corrections, restriction/non-restriction of the standard's scope and validity period, revision, and cancelation if necessary. About 30% of changes to standards are aimed at improving the scientific and engineering level of standards (adoption of new technologies, materials, procedures, and processes, extension of a standard application scope, etc.) [13].

5.3.4.4 Space safety standards

The nature and scope of safety standards and procedures in Russia are different from those used at, for example, the ESA.

As mentioned above, differences in nature and scope of Russian safety standards are caused by the difference in organization of the Russian space industry compared to the European space industry (see section 5.3.1), as well as differences in approaches of development of safety standards related to occupational health issues. The organization of the Russian space industry is different mainly because of the separation of the design and production processes into different entities and manufacture of products of space technology series rather than as individual parts as in European space technology [13]. Occupational health problems were traditionally covered by different systems of standards, mainly through the System of Standards of Labor Safety, System of Ergonomics standards.

The major difference regarding safety in the Russian Space Program is that safety is not a stand-alone separate program or activity of the Russian Space Program. Safety measures appear as part of various programs mandatory for space technology development and operation, such as reliability programs, quality control, ecological safety issues, complex programs of tests of space technology items, programs of flight tests, complex plans of space technology operation, utilization of space infrastructure. Safety is considered at every phase of the life cycle of space technology development, starting with design, through to tests, manufacture, and operation. This is clearly shown above, in the discussion on rules and forms of assessment of compliance of space technology due to phases of the life cycle as part of the special technical regulation on space technology safety. For example, during research and development of new space technology, reliability and safety programs make up around 55–65% of total costs of the project at that stage. In total, safety activities make up about 70% of the total budget of a full life-cycle implementation of a space project from design to operation [20].

An ESA study of 1995 showed that although both ESA and Russian standards on reliability assurance require reliability engineering and assurance activities with the aim of meeting the specified reliability objectives of space systems, the approaches for performance of a reliability program are different (Table 5.2) [13].

Human space-flight systems are developed under a unified set of space standards. Some of the important GOST R and OST standards deal directly with reliability and safety of space systems, and contain special requirements concerning human space flight. General requirements for GOST R and OST are complemented by special standards developed by the prime system contractor for human space flight—the Rocket and Space Corporation "Energiya". At the same time, RSC Energiya is able to develop GOST R and OST standards through the established procedures of *Rostehregulirovanie* and *Roscosmos*. Of course, there are a number of other documents that make up technical specifications and regulations (TU and P documents).

Table 5.2 Comparison of ESA and Russian Reliability Programs

ESA Reliability Program	Russian Reliability Program
Analytical design work	Experimental design work
Qualitative analysis	Quantitative analysis
Emphasis on development phase	Distributed over all phases, including development, production, operation, and maintenance
Failure prevention by design	Curing and evaluation of test failures

Russian space standards for human space flight, including safety issues, are given in the document SSP 50094, revision A6/20/97 TIM # 20. These standards are applicable to the Russian segment of the International Space Station. Actually, this document contains the main groups of space standards—GOST, OST, STP, TU, P—and others. The major disadvantage of this document is that all standards are listed as a continuous flow of information [26].

Safety has its own chapter (6.5) and deals with the following issues: electrical hazards design requirements, acoustic condition design requirements, and crew radiation safety that has been listed but not considered. Also, safety issues are distributed among other chapters inside the document. A good example is the issue of safety in Structures and Mechanisms, Chapter 7.1. Safety factors are given in a table on page 390 titled "Russian minimum factors of safety for metallic and composite space station structure", where safety factors are listed against external loads for the following activities: ground operations, pre-launch preparation, launch, injection to orbit, landing, orbital flight, docking in orbit [26].

At the same time there are industrial standards devoted to safety directly, for example [21,22]:

- OST 134-1021-99, "Space systems and complexes: general safety requirements".
- OST-134-1021-99, "Manned spacecraft: general requirements for safety".

Traditionally, the Product Assurance and Safety Department of RSC "Energiya" is responsible for safety issues [13]. Below are a few standards listed that are used for a human space flight [27]:

- GOST 25645.103-84, "Physical conditions of space: terms and definitions".
- GOST R 25645.167-2005, "Space environment (natural and artificial): model and time distribution".
- GOST R 52017-2003, "Spacecraft: preparation and realization of space experiments".
- GOST R 52925-2008, "Space technology items: general requirements for limitation of technological pollution of near Earth space".
- GOST R 52985-2008, "Ecological safety of rocket and space technics".

5.3.4.5 Human-related safety standards

Beside space technology technical standards, safety standards in the Russian Space Program related to humans (this might be the crew of a spacecraft and/or personnel of ground infrastructure and/or the population of Russia or other countries) are subject to two main safety groupings: (a) safety and protection of humans; (b) safety and protection of environment.

Safety standards related to the safety of humans can be grouped as follows (not an exhaustive list) [28]:

- System of Standards for Labor Safety
  - Standards for drinking water
  - Standards for protection from noise
  - Standards for protection from emissions
  - Standards from protection from vibrations
  - Standards for fire and explosions
  - Standards for quality of air in the working areas
- System of Standards for Ergonomics

- System of Sanitary Rules and Norms
- System of Hygienic Norms
- System of Standards for Safety during Emergency Situations and Accidents.

Any GOST state standard text in Russia in paragraph 2 gives reference to related important standards that create the basis of the standard. The basis of two important safety standards for human space flight and environmental safety are considered below as an example: safety standards on habitable environments on board manned spacecraft that were developed in 1995 and the ecological safety standard on space technology that was adopted in 2008. This was done with the purpose of showing the basis of development of human-related safety standards that stand alone from space technical standards.

State Standard GOST R 50804-95, “Cosmonaut habitable environments onboard a manned spacecraft: General medical and technical requirements”, was developed by the Institute of Biomedical Problems, together with the Russian Space Corporation Energiya and the Joint Stock Scientific Research and Production Company “Zvezda” (the main producer of Russian spacesuits). This standard is based upon 49 standards that are listed in section 2 of the standard and 10 important methodological literature references at the end [29]. A few of these are:

- GOST 12.1.012-90, “Vibration safety: general requirements”.
- GOST 12.1.014-84, “Air of the working zone: method of measurement of concentrations of dangerous substances by indicator tubes”.
- GOST 2874-82, “Drinking water: hygienic norms and quality control”.
- GOST 21829-76, “System ‘human machine’: coding of visual information and general ergonomic rules”.
- GOST 25645.203-83, “Radiation safety of the crew of a spacecraft during space flight: requirements for individual and onboard dosimeter control”.
- GOST 12.4.026-76, “SSBT: signal colors and safety signs”.

State Standard GOST R 52885-2008 “Ecological safety of rocket-space technique: general technical requirements” is based on seven standards given in section 2:

- GOST 12.1.004-91, “System of standards for labor safety: fire safety and general requirements”.
- GOST 12.1.007-76, “System of standards for labor safety: toxic substances, classification and general safety requirements”.
- GOST 12.1.018-93, “System of standards for labor safety: fire and explosion of static electricity, general requirements”.
- GOST 12.1.008-78, “System of standards for labor safety: biological safety, general requirements”.
- GOST 12.1.019-79, “System of standards for labor safety: electric safety, general requirements and nomenclature of protection”.
- GOST 28496-90, “System quality assessment and certification of mutually released production: label of compliance, forms, sizes, and order of application”.
- GOST 30333-2007, “Passport of safety of chemical production: general requirements”.

The ecological safety space technology standard gives 22 references on SanPiNs and hygienic norms and other important methodological literature. For example: (a) SanPiN 4015-85, “Maximum content of toxic substances in industrial waste in a storage place located outside territory of an enterprise”; (b) hygienic norms 2.1.6.696-98, "Approximate safety levels of impact of pollutants in the atmospheric air of populated locations: the Letter of Commission on State Sanitary Epidemic Regulation No. FK-13-4-155 from 30 March 1998"; and (c) PDK 3210-85, "Maximum allowed concentration in chemical substances in soil" [30].

Below are listed important safety standards for a crew during space flight [27]:

- GOST 25645.202-83, "Radiation safety of a crew of a spacecraft during space flight: requirements for the individual and onboard dosimeter control".
- GOST 25645.203-83, "Radiation safety of the crew of a spacecraft during space flight: model of a human body for calculation of a tissue dose".
- GOST 25645.204-83, "Radiation safety of the crew of a spacecraft during space flight: techniques of calculation of points of shielding inside a phantom".
- GOST 25645.211-83, "Radiation safety of the crew of a spacecraft during space flight: characteristics of nuclear interactions of protons".
- GOST 25645.215-83, "Radiation safety of the crew of a spacecraft during space flight: safety norms for a space flight with duration up to 3 years".
- GOST 28040-89, "Complexity of life-support system of a cosmonaut in a manned spacecraft: terms and definitions".
- GOST R 50804-95, "Cosmonauts' habitable environment on board a manned spacecraft: general medical and technical requirements".
- GOST R 25645-99, "Radiation safety during space flight: test of shield efficiency of crew, general requirements".

5.3.4.6 Safety procedures for payload on board the International Space Station

The preparation of documentation for payloads operating on the Russian segment of the International Space Station (ISS) is different from the part operated by the USA.

The process begins with the Flight and Ground Experimental Program Definition Kick-off meeting, where the definition of the individual activities is presented to the Russian party (this is an established process for ESA payloads and experiments in the Russian segment). The project preparation phase is supported by the following five scheduled reviews with the goal of assessing the technical and programmatic aspects of proposed experiments: (a) Mission Feasibility Review; (b) Mission Integration Review; (c) Mission Verifications Review (done two times); (d) Final Project Review [31].

For confirmation of conformity to the agreed technical requirements, the proposed equipment is subjected to a series of pre-flight tests. The test program includes qualification tests and acceptance tests, check of interfaces with other payloads (if necessary), incoming inspection after transportation, and in special cases integrated tests as part of the spacecraft, as well as tests at the engineering complex of Baikonur before installation into the launch vehicle. The flight model, training models, and baseline data collection models are subject to a safety inspection prior to any issue by the cosmonauts.

Two acceptance tests will be carried out during the integration process of equipment for a specific mission. Flight and training equipment is only admitted to the acceptance tests if it has successfully passed the qualification tests. Acceptance Test 1 (AT-1) verifies whether or not the equipment conforms to the requirements that are specified in the technical specification for equipment (TS-EQ) document. Upon completion of AT-1 the consent to ship the equipment to the Russian party will be given. A separate AT-1 test will take place 2–3 months before the Progress and Soyuz launch. Acceptance Test 2 applies to the equipment with active interfaces to the spacecraft and to the ISS. It allows the equipment to be tested with the vehicle and station simulators in order to provide the final verification of those interfaces. The equipment, without active interfaces, undergoes a simplified AT-2, which includes post-delivery inspection. The successful conclusion of a review allows for the acceptance of a flight [31].

After transportation of the equipment to Moscow and then to Baikonur, the equipment is inspected in accordance with the requirements, including a visual inspection of the equipment, external inspection and completeness check, and functional check. Flight readiness check and safety inspection of the equipment are conducted prior to or immediately after its installation into the spacecraft; completion and signing of the documentation reflects the incoming inspection results.

The required documents for an experiment at the Russian segment are: (1) technical specification for the experiment (TS-EX_100); (2) technical specification for the equipment (TS-EQ_200); (3) technical description (TD_201); (4) operations and maintenance manual (OM_202); (5) acceptance test program (ATP_203), equipment incoming inspection manual (204); (6) qualification test program (205); (7) qualification test report (206); (8) safety assessments and certificates (207); (9) crew training documentation and inputs to crew procedures (208); (10) ground test equipment/checkout equipment (209); (11) passport (logbook) for the equipment (200PS); (12) requirements of the contents of electrical circuit diagrams (210—Annexes 1 and 2); and (13) requirements of the contents of outline installation drawings (211) [31].

The most important documents for safety are: (a) safety assessment reports and certificates (207); and (b) qualification test reports (206).

The safety assessment reports and certificates (207) document does not have a strictly regulated form. However, it should contain all necessary data that will allow for conclusions about the safety of the equipment relative to the ISS and to the crew members in all phases of equipment operation to be reached. If the equipment being used has data on toxicological safety, the appropriate test certificates or reports should be submitted with this document. The following sections should be considered while preparing the 207 document:

- Description of the equipment unit by unit
- Materials safety
- Electrical safety
- Structural safety
- Electromagnetic and other types of radiation
- Gas bottle handling systems if necessary
- Noise level
- Temperature
- Appendices, including safety check results, equipment toxicological test data
- List of non-metallic materials
- Materials certificates.

The draft of the safety assessment reports and certificates (207) document is submitted to the mission integration review, initial version at the mission verification review and baseline version at Acceptance Test 1 [31].

The qualification test reports (206) document contains reports on the results of qualification tests of the equipment conducted by the hardware developers. The documents generally consist of the following sections:

- Introduction
- General provisions
- Conditions and order of tests
- Scope of qualification tests
- Test procedures and results
- Appendices including:
  - Testing record sheets
  - Reports on equipment characteristics non-conformity to TS-EQ requirements
  - Defect/failure analysis/checklist
  - Detailed test procedures
  - Warranties.

The draft version of the qualification test reports (206) document is submitted to Acceptance Test 1 and baseline version at Acceptance Test 2 [31].

CONCLUDING REMARKS

Russian safety standards are different by nature and scope from European and US safety standards. This has happened because of the different organization of the Russian space industry and the historically different approaches for space safety standards development and implementation that date back to the time of the former Soviet Union.

Today, Russian space safety standards form a part of the reliability and quality control programs distributed along all phases of the life cycle of space technology, starting with the development phase, followed by testing and continuing through the operation phase. There are not that many special standards for crew health and safety on board spacecraft. Various systems of state national standards such as the System of Standards of Labour Safety, Systems of Hygienic Rules, System of Environmental Protection Standards regulate human health and safety-related issues for the crew of a spacecraft and personnel on the ground.

The positive outcome of the proposed Special Technical Regulation on Space Technology Safety is that for the first time in Russian history a special space safety document has been prepared and publicly discussed. Hopefully, the forthcoming new standardization law will make clear the legal distinction between technical regulation and standardization, especially in the area where and when compliance to standards is mandatory and where it is not.

ACKNOWLEDGEMENTS

The author wishes to acknowledge the support of Mr Mikhail Azeev from RSC Energiya of Russia for his valuable contribution to this paper and Dr Noel Simon, a space policy expert from Australia, for useful comments and help in editing the paper. All errors, as always, are the responsibility of the author.

References

[1] Official website of *Rostehregulirovanie* Agency, http://www.gost.ru/wps/portal/, Agency's history page (in Russian).
[2] GOST page at Wikipedia encyclopedia, http://en.wikipedia.org/wiki/GOST (in English).
[3] The collection of technical standards in Russia and Kazakhstan, http://www.snip.com.ru/ (in English).
[4] Official website of *Rostehregulirovanie* Agency, http://www.gost.ru/wps/portal/ Technical Regulations page (in Russian).
[5] Official website of *Rostehregulirovanie* Agency, http://www.gost.ru/wps/portal/, Page with a draft of a new "Standardization Law", and public discussion of the proposed law (in Russian).
[6] Information on Technical Regulation Law by the Association of European Business to Russian Federation, http://www.aebrus.ru/files/File/CommitteeFiles/TRSTF2007.07.19_1IHeller.pdf (in English).
[7] Technical Regulation Law available at the official website of *Rostehregulirovanie*, http://www.gost.ru/wps/portal/, Federal Law page (in Russian).
[8] Pugachev, S. (2008). Russia Needs Standardisation Law. *Automobile Engineers Magazine*, No. 2. Article available at the official website of *Rostehregulirovanie*, http://www.gost.ru/wps/portal/, Discussion of a new Standardization Law page (in Russian).
[9] Provisions of the Federal Space Agency at the official website of the Russian Space Agency, http://www.federalspace.ru/, Legal documents page (in Russian).
[10] Federal Space Program at the official website of the Russian Space Agency, http://www.federalspace.ru, Federal Space Program page (in Russian).
[11] Space Activities Law at the official website of the Russian Space Agency, http://www.federalspace.ru/, Legal documents page (in Russian).
[12] GOST R 51143-98, "Launch and technical complexes of rocket and space complexes: general technical requirements for tests and acceptance". Parts of the GOST R 51143-98 standard are available on the web page of the *Novosti Kosmonavtiki* magazine, http://www.novosti-kosmonavtiki.ru/, Terms and Definitions Forum page (in Russian).
[13] "ESA-CIS Standards Identification and Comparison", Study Final Report by Daimler-Benz Aerospace and RST Rostock, September 1995 (in English).
[14] Alexandrovskaya, L. et al. (2001). *Certification of Complex Technical Systems*. Logos (in Russian).
[15] Standardization and Certification Basics prepared by Department of Automation of Electronic Systems and Control Systems of Tomsk State University for Control Systems and Radio Electronics. Lecture course available at the website, http://iit.tusur.ru/docs/standart.doc (in Russian).
[16] Moiseev, I. (1998). Normative and Legal Regulation of Space Activity in Russia. *Novosti Kosmonavtiki*, No. 3. Article available on the website of the *Novosti Kosmonavtiki* magazine, http://www.novosti-kosmonavtiki.ru/content/numbers/170/15.shtml (in Russian).
[17] Malkov, S. and Doldirina, C. (2010). Regulation of Space Activities in the Russian Federation. In *National Regulation of Space Activities* (Jakhu, R., ed.).
[18] Uryngaliev, A. (2008). Kazakhstan–Russia: Perspectives of Cooperation in Space. *Kazakh–Russian Relations at Modern Stage: Summary and Perspectives of Cooperation* conference, 6 March. Paper available at http://www.zakon.kz/61489-kazakhstan-rossija-perspektivy.html (in Russian).
[19] Ministry of Ecology (1998). State Report on State of Environment. Report available at the website of the Center for Geology and Objects of Nuclear Power Cycle at http://www.wdcb.rssi.ru/mining/Welcome.htm (in Russian).
[20] "Financial and Economic Provisions of Safety of Space Activity". Discussion by Lower Chamber of the Russian Parliament, 13 September 2005. Discussion available at http://asozd.duma.gov.ru/arhiv/a_dz.nsf/ (in Russian).

[21] Technical regulations that are under public discussion, see web page at Institute of Standardization and Certification of Innovations, Section Technical Regulations at website http://www.isci-gost.ru/index.files/48.htm (in Russian).
[22] Non-governmental Council for Technical Regulation, http://www.texsovet.ru/, Web page with technical regulations under discussion (in Russian).
[23] Draft of a Federal Law, "On regions of fall of space objects and implementation of changes in some legal acts" at the official website of the Russian Space Agency, http://www.federalspace.ru, Legal documents page (in Russian).
[24] Official website of the Central Institute for Machine-building, http://www.tsniimash.ru/struct/CAR/br_RKT/, RKT Reliability Department page (in Russian).
[25] Official website of VNIINMASH, http://www.vniinmash.ru/ (in Russian).
[26] SSP 50094—revision A 6/20/97 TIM#20 (in English).
[27] Library of Russian standards website, "GOSThelp.ru", http://www.gosthelp.ru/ (in Russian).
[28] GOST documents for download, at the website of the Institute of Standardization and Certification of Innovations, http://www.isci-gost.ru/index.files/48.htm (in Russian).
[29] State Standard GOST R 50804-95, "Cosmonaut habitable environments on board of a manned spacecraft: general medical and technical requirements", at the Library of Russian standards, "Gosthelp.ru", http://www.gosthelp.ru/ (in Russian).
[30] State Standard GOST R 52885-2008, "Ecological safety of rocket-space technique: general technical requirements", at the Library of Russian standards, "Gosthelp.ru", http://www.gosthelp.ru/ (in Russian).
[31] ESA (2005). *European Users Guide to Low Gravity Platforms*, Erasmus User Centre and Communications Office, ESA (in English).

CHAPTER

6

Space safety regulations and standards in China

Li Juquian

CHAPTER OUTLINE

INTRODUCTION

Space activities in China were developed in the 1950s after the first satellite launched in Russia. For nearly 50 years, particularly during the 1990s, China has achieved great things in many important aspects of space activities, including launching vehicles, satellite systems, and manned spacecraft flight. Though these space activities have developed very quickly, in comparison space law has developed relatively slowly in this period. China still lacks special space law, and many of the activities are regulated by ministerial regulations and regulatory documents that need to be systematized and codified further.

During this period, space safety has not only been a concern of practical organizations in space activities, but has also been keenly discussed in academic circles, most especially those including space law scholars. The satellite collision between a US satellite and a non-functional Russian satellite raised many concerns regarding space safety regulations and/or standards in the international arena.

This chapter is composed of four parts. The first part surveys the background of space safety regulations in China via a panoramic view. The second part researches current safety controls and restriction regulations relating to pre-launch stage activities. The third part discusses issues

Space Safety Regulations and Standards. DOI: 10.1016/B978-1-85617-752-8.10006-6

relating to on-orbit operation. The last part includes the author's proposals for further development of China's regulations and standards in the field.

6.1 CURRENT STATUS OF SPACE LAW IN CHINA

China began its space activity in 1956, just one year before the first man-made satellite launched into outer space, and achieved great progress during more than 50 years of effort. China successfully launched its first satellite *Dongfanghong-I* in 1970, its first spacecraft *Shenzhou-I* in 1999, its first manned spacecraft *Shenzhou-V* in 2003, and the first spacewalk by a Chinese astronaut in 2008. All of these achievements were focused upon by and witnessed by the world.

However, the law in this important field has lagged far behind China's accomplishments. The most noticeable phenomenon is that there is no special space law at national level. The law in this field can be described mainly as ministerial regulations or ministerial orders, which are at the very lowest level in the hierarchy of law in China. This also means that these ministerial regulations and orders can be easily changed, modified, or canceled.

The hierarchy of law in China is mainly regulated by the 2000 Legislation Law, which is based on the constitution and years of legislation practice. According to this law, law is defined broadly in China, and could be:

- National law, which is passed by the National People's Congress and its Standing Committee [1].
- Administrative regulation, which is passed by the central government, known as the State Council [2].
- Local decree, which is passed by the provincial-level local congress and its standing committee, or by the major city-level local congress and its standing committee; or autonomous decree, which is passed by the local autonomous area congress [3].
- Ministerial regulation, which is passed by any of the ministries, committees, the central bank, the central auditing organ, or the administration directly under the State Council; or local regulation, which is passed by the provincial-level local government, or major city government [4].

In the hierarchy of Chinese law, the constitution always comes first if a law is in conflict with it. National law comes after the constitution and can override the other types of law if conflict exists between it and the others. An administrative regulation can override any law except the constitution and national law. If a local decree conflicts with the ministerial regulation, then the matter should be reported to the central government to decide. When the central government decides the ministerial regulation should override the local decree, the matter should be submitted to the Standing Committee of the National People's Congress to decide [5].

A simple survey of the current space law in China reveals the following characteristics:

1. Currently, there is no exclusive space law at national level, nor is it anticipated in the near future. Space activity is an important field, but less so than basic civil and criminal institutions. The draft of national space legislation has been discussed several times in academic circles, and was planned to be put on the legislative agenda of the National People's Congress. It seems more research has to be done to convince the relevant organs of the Standing Committee that the time to legislate a national space law is ripe. According to the legislation agenda of the 11th Session

National People's Congress announced by its Standing Committee last year, space law was not on the list of the 64 laws to be legislated in the next 5 years [6]. Though one may hope it could be put on the list as a special consideration, this is not likely to be the case considering the factual situation.

Some national laws relating to some field of space activity exist, but their primary aims are not space activities.

2. There is no administrative regulation particular to space activities.
Not only has the National People's Congress not introduced space legislation, but the State Council has also not promulgated any special administrative regulation of space activities. Though the administrative regulation, compared to national law, is a kind of lower-level law, it can be applied all over China and has binding force. It is easier to pass an administrative regulation by the State Council without following a complicated procedure. It seems practical to pass an administrative regulation before promulgating a special space law at the national law level.
3. Various ministerial regulations exist, mainly relating to space activities.
All the ministries under the central government have the power to make a regulation and put it into effect within the scope of the ministry's function, provided it does not conflict with any upper-level regulation or law. It is easier to pass a ministerial regulation than to pass an administrative regulation. Each ministry can pass a ministerial regulation relating to space activities if it is within the function of the said ministry. Though there are only two specific ministerial regulations that focus on the specific matter of space activities, there are many different ministerial regulations relating to space activities. This situation creates the likely possibility of conflicts between or among different ministerial regulations.

In addition, each ministry can announce an order, notification, or regulatory document. All of these can be put into effect, though with a lower level of binding force.

This factual situation makes it difficult to research space law in China, and also makes it difficult to enact a brand new space law at national level. But there are some advantages for space activities, such as efficiency, flexibility, and creativity, especially when the government needs to carry out a new policy to adapt to a changing situation. These advantages explain the progress of space activities through the years, despite the lack of national space law in China.

6.2 PRE-LAUNCH SAFETY ADMINISTRATION

Four out of five of the United Nations space law treaties were approved by the Standing Committee of the National People's Congress, namely the Outer Space Treaty, the Rescue Treaty, the Liability Convention, and the Registration Treaty [7]. All these treaties established the basic principles and rules pertaining to space activities internationally, and can be applied as law in China. Of course, the provisions of the treaties relating to space safety are inevitably part of Chinese law.

Though China has no special space law, it does have some regulations relating to and emphasizing the issue of space safety. These regulations can be mainly divided into two different categories: (1) the approval requirement before launching a space object and (2) the insurance requirement before launching a space object. In addition, there are some regulations relating to commercial use of space activities, but the main aims are not necessarily encouragement of commercial use.

6.2.1 Approval requirement before launching a space object

In order to fulfill China's obligation under international treaties, regulate civilian launching projects, promote the healthy development of the civilian space industry, and protect the national safety and the public interest, the Commission of Science, Technology and Industry for National Defense (COSTIND), established in 1982 after the merger of several organs in relating fields, in 2002 promulgated a ministerial regulation named "Provisional Methods for Administering Permit of Civilian Space Launching Project" (hereafter referred to as "Provisional Methods"). COSTIND itself was canceled in March 2008 during the governmental reform, and some of its functions were allocated to the State Administration of Science, Technology and Industry for National Defense, which is a national bureau administered by the Ministry of Industry and Information Technology. Though some terms used in the Provisional Methods need to be modified because of the cancellation of COSTIND, its provisions still have effect.

Looking at the Provisional Methods, it can be observed that some systems relating to space safety have been established.

6.2.1.1 Permit requirement

The permit requirement means the person who wants to launch a space object must get governmental approval in advance, and those who obtain a permit must carry out the project within the scope of permission.

Article 3 of the Provisional Methods dictates that a civilian space launch project is subject to the permit administering system. This means that anyone who wants to launch a space object needs to submit an application to COSTIND to get the permit before the actual launch of the space object, regardless of whether the applicant is a naturalized person, legal person, or other type of organization.

It is illegal for anyone to attempt a space launch project without a permit from the government and under the Provisional Methods there is legal responsibility for such action. COSTIND has the power to order the termination of the project and impose administrative sanction; in the case of a crime committed, criminal punishment shall be imposed according to the law [8].

It is also illegal for an entity to have a permit but carry out the project outside the scope of permission. COSTIND has the power to sanction, or to order the permit holder to adjust; in the case of a serious situation, COSTIND may cancel the permit [9]. If the permit has been canceled, then the applicant cannot apply for the same project within 2 years [10]. This is a severe sanction for civilian projects because they will lose the opportunity to make profits within 2 years.

6.2.1.2 Qualification requirements

The permit system itself involves the examination of qualifications. Only those who meet the qualification requirements can be approved, which is not all applicants. The applicant should submit as many documents as necessary to convince COSTIND that they are qualified to launch the space project. Among the necessary documents, the following are very important as regards space safety:

1. A report of safety design and materials to show how the project will protect public safety, the reliability of key safety systems, the impact on property and personal safety near the launch site and within the range of the launch track in the case of normal and abnormal launch processes of the launch vehicle, how to avoid pollution and space debris, and other supplementary material [11].

These requirements mean that only those civilian space projects meeting the conditions listed here may be approved by COSTIND. All the requirements emphasize safety in space and on the Earth.

2. For a project executed at the launch site within China's territory, the applicant needs to submit: the planned launch time; the technical requirements of the satellite, launch vehicle, and launch, track, measure, and communication system; detailed orbital parameters of the launch vehicle; detailed satellite orbital parameters; and use of spectrum resources [12].
   All these requirements relate to space safety in technical aspects. The concern about space debris and its inclusion in the regulations is an important signal that China deals seriously with space safety issues and hopes to control them. Needless to say, if the applicant does not provide the required information, it is not possible to obtain a permit from COSTIND.
3. The material relating to the project complies with national environmental protection laws and regulations. This requirement does not directly relate to space safety, but mainly concerns environmental protection on the Earth. In a broad sense, space is a field of environmental protection, but it is not the focus of these regulations [13].

6.2.1.3 Insurance requirement before launching a space object

Strictly speaking, the insurance requirement does not directly relate to space safety. However, if we consider it in conjunction with the permit system, this requirement does relate to safety in space.

According to the Provisional Methods, the permit holder must purchase third-party insurance and other relevant insurance in compliance with the relevant regulations [14]. This is a very strict requirement, and the regulations use the word "must". Before the applicant progresses to the launch site procedure, it shall submit to COSTIND the effective third party insurance policy; if the launch site is located in a foreign country, the applicant shall submit the effective third-party policy 60 days before the planned launch date [15]. Then, on approval from COSTIND, the launch project can proceed to the launch site procedure.

Space activity insurance is promulgated in other ministerial regulations, including the Ministry of Finance Notification of Publishing the Administering Methods of Satellite Insurance Fund 1997 (hereafter referred to as "Fund Methods"), the Ministry of Finance and State Administration of Taxation's Letter Relating to the Satellite Launching Insurance 1997 (hereafter referred to as "Taxation Letter"), and the Ministry of Finance and State Administration of Taxation's Notification on the Taxation Policy on the Satellite Launching Insurance 1997 (hereafter referred to as "Taxation Policy"). In addition, the China Insurance Regulatory Commission promulgated in 2006 The Index of Category Methods of Risk Unit Relating to Property Insurance (hereafter referred to as "Risk Unit Index"). According to these documents, satellite launch insurance is required and encouraged; the government takes all necessary measures to promote this insurance.

In the 1990s, international insurance corporations were unwilling to underwrite insurance policies for China's foreign launch projects. In order to promote the ability of the launch service, a robust market of insurance must be established. The Ministry of Finance promulgated the Fund Methods based on the documents and opinions of the State Council and the Central Military Commission. According to the Fund Methods, the financial resources of the satellite launch insurance fund are from the insurance premiums of what is called the Space Insurance Combination [16]. The Space Insurance Combination comprises 10 insurance corporations, which are all state-owned enterprises (SOEs) [17]. The satellite launch insurance premium is entirely collected by the Satellite Insurance Combination and deposited separately in the RMB general and foreign currency account, before it is transferred to the Space Insurance Combination members' accounts [18]. The Taxation Letter clarifies that the part of insurance income allocated to the satellite launch fund is exempt from income tax, but not business tax. This is a proactive measure from the State Administration of Taxation.

Space activity is so unique that many special measures are applied by the organs of central government. According to the Risk Unit Index, space risk is an independent category. The Risk Unit Index states that space activity is a kind of "high-value, high-technology, and high-risk" activity, and cannot be further divided into sub-risk units.

6.2.1.4 Commercialization and space safety

Commercialization of space activity is a new phenomenon, and may be a trend in future space activity. Commercialization of space activity does not necessarily mean space tourism. It also refers to satellite remote sensing, telecommunications, satellite broadcasting, and the entire launch vehicle process. China encourages the satellite application industry. For example, the State Development Plan Commission promulgated an announcement in 2002, "The Announcement Concerning the Organizing Special Program of Industrialization of Satellite Navigation Positioning Application", with the goal of establishing a special market of more than 10 billion RMB, a satellite navigation positioning infrastructure system, backbone enterprises, and dominant products [19]. The Ministry of Transport announced a ministerial regulatory document in 2006 to encourage the application and regulation of a satellite navigation positioning system [20]. With more and more persons and industries involved in space activity, it is very important to assure the safety of such activity.

China promulgated many ministerial regulations and regulatory documents concerning the issue of space safety.

- Certain profitable carrying is prohibited.
  In 2006, COSTIND announced a notification pointing out that the seed-breeding satellite *Shijian-VIII* performed a national basic scientific experiment. The seed material was collected by the Ministry of Agriculture without remuneration; no material from any other unit can be carried on the basis of commercial service [21].
- The satellite utilization industry is encouraged and space safety is emphasized.
  In 2007, two organs of the central government, the State Development and Reform Commission and COSTIND, announced a notification recounting the encouragement of the development of satellite application industry [22]. In the notification, one of the medium- and long-term policy goals is to further promote the performance of nationally manufactured telecommunication satellites in various aspects, including safety, reliability, capacity, lifespan, precision, etc. The safety performance of the satellite surely means, *inter alia*, space communication safety.
- Production and business is subject to general safety regulations.
  General safety law and regulation can be applied in the space launch process if it is a commercial service. The Standing Committee of the NPC promulgated the Safe Production Law in 2002. The law requires that all the units performing production and business activity obey the obligation of safe production. Those who breach this legal obligation are subject to administrative sanction or criminal punishment accordingly [23]. It is apparent that there are thousands of regulatory documents relating to the general issue of safe production. In 2008, the State Council

announced a notification on safe production, and emphasized safe production in the military, industry, mining, fireworks, etc.

An important aspect that should be mentioned here is that of military personnel used in space activity safety. There is a military regulation promulgated by the Central Military Commission relating to this topic, "Safety Regulation of the People's Liberation Army" (hereafter referred to as the PLA Safety Regulation) [24]. Safety training is a major issue in this regulation. Article 30 of the regulation states that military personnel shall perform adaptive training prior to performance of an important mission. A space launch is one of the important missions clearly listed in the regulation.

Although all of the above-mentioned regulations, with the exception of the PLA Safety Regulation, primarily relate to civilian space projects, the PLA Safety Regulation itself does not distinguish between civilian space activity and military space activity.

China's emphasis on the safety of space activity can also be observed in the announced governmental policy. Both the 2000 white paper and 2006 white paper on "China's Space Activities" reiterate the policy of peaceful use of space. Further, the 2005 white paper on "China's Efforts on Arms Control, Disarmament and Non-proliferation" asserts prevention of a space arms race, and calls on negotiations to conclude an international legal document to prohibit deployment of weapons in space, the use of force, or the threatened use of force against a space object.

6.3 ON-ORBIT OPERATION SAFETY ADMINISTRATION

After a space object is launched safely, it will soon operate on its orbit. On-orbit safety of a space object is a topic of great concern. The collision of a US satellite and a Russian satellite in 2009 was a warning, showing the importance of the telemetry, tracking and command (TT&C) of on-orbit satellites. China's regulations in this field mainly relate to the satellite [25].

The regulations or standards relating to on-orbit safety can be broken down into three categories: (1) TT&C, (2) telecommunications, and (3) satellite application and safety.

6.3.1 TT&C

No ministerial regulations exist on this issue. The reason may be that, to a great extent, it is a technical issue and can be resolved via technology. In fact, TT&C is a vital issue each spacefaring country must face seriously. The current technology cannot trace all existing space debris, allowing for the possibility of a collision between space objects, however unlikely according to expert calculation [26].

China first established a TT&C team in May 1956 [27]. Until 2006, the TT&C system in China was composed of a launch vehicle TT&C system, a C-band satellite TT&C network, and an S-band satellite TT&C network, providing TT&C support for launch vehicles and various orbits' space vehicles [28]. Some companies developed advanced TT&C technology that was easily applied to space activities. For example, Beijing Aerospace Measurement & Control Technology Company, established in 1982, developed "Automatic Computer Measurement and Control System and its Standards Appling Generally to Missile and Satellite" in 1986 [29]. Xi'an TT&C center successfully used this TT&C technology to repair several malfunctioning satellites and to trace spacecraft including *Shenzhou-V* [30].

6.3.2 Telecommunications

Within the field of space safety, the role of radio waves cannot be ignored. Many regulations and regulatory documents were promulgated to administer the use of the radio wave. As early as 1963, China promulgated "Administering Provisions of Use of Radio Waves". The latest regulation of radio waves is "The Administering Regulation of Radio Waves", promulgated jointly by the State Council and the Central Military Commission in September 1993 (hereafter referred to as Radio Regulations). According to that regulation, the Headquarters of the General Staff of the People's Liberation Army promulgated "The People's Liberation Army Regulation on Administering Radio Waves" in December 1994.

The aim and purpose of the Radio Regulations is to enhance the administration of the use of radio waves, maintain the order of radio waves in the air, use radio wave resources efficiently, and guarantee the normal operation of radio business [31]. All units or individuals who want to establish and use radio stations must submit a written application to obtain a permit for a radio station [32]. Any establishment and use of a radio station, as well as the research, production, and importation of radio emission equipment, must obey the regulations.

All radio wave use is under the control of the state. These measures can assure that the use of radio waves will not unduly affect the normal operation of radio waves in TT&C for space activities, and will minimize the risk of malpractice, which could probably cause danger to space activities.

Besides the general rules relating to the use of radio waves, the government also promulgates some special orders for specific matters or specific situations. For example, the Ministry of Industry and Information Technology announced a notification in 2008, emphasizing the safe use of radio for the Olympic Games [33]. This ministry promulgated in March 2009 "Administrative Provisions on Radio Station License" to enhance the administration of radio stations [34]. The CNCA made an announcement in 2009 relating to the GB9254-2008 standard, "The Radio Disturbance Limit Value of Information Technology Equipment and its Measurement" [35].

6.3.3 Satellite application and safety

Satellite applications cover a very wide range, and include *inter alia* remote sensing, navigation positioning, and television broadcasting. The greater the number of applications of a satellite, the more complex the function and mission required, and hence the more risks to which the satellite is exposed. Thus, satellite applications also necessitate satellite safety.

The regulations on this issue vary because of the multiple ministries involved, as well as the various businesses covered.

- Data service and satellite safety.
  COSTIND announced a notification of "Domestic Data Administering Rules of the Chinese-Brazil Earth Resources *Satellites* 01/01/03 (Trial)" in 2007 [36]. This notification deals with the relationship between data service and satellite safety, and puts forward a principle on the priority of satellite safety. "The resources satellite center shall to the utmost extent satisfy the demand of data by the customer provided the safety of the satellites operation is guaranteed. If it is necessary to adjust the normal receiving and dealing of the satellites data, a notification to the customer shall be made in time" [37].

- Personnel stability and equipment safety.
  The State Oceanic Administration announced a notification in 1991, "Administering Rules of Satellite Telecommunication Network of Oceanic Live Data Transfer" [38]. In the notification, the State Oceanic Administration requires that personnel ought to be absolutely stable in order to assure the normal operation of the satellite network, and the equipment room shall be safe and neat, and the equipment in good condition [39].
- Satellite television broadcasting and safety.
  Satellite television broadcasting has become widely used in recent years. It, too, concerns the safety of satellites. The State Administration of Radio and Television announced a notification in 30 May 2007 on the safety of broadcasting as well as the integrity of the satellite system [40]. The administration requires, in order to assure the safety of radio and television signal transfer, that the satellite transponder is rented or used systemically, and the radio station and television station shall not rent or use the satellite transponder themselves. The channel and the frequency approved by the administration must be complied with, and the equipment and safety guarantee measure must satisfy the administration's relevant requirements.
- Prohibition of a signal sent to a telecommunications satellite without permission.
  In order to maintain order in satellite telecommunications, the Ministry of Post and Telecommunication promulgated "The Provisional Administering Rules of Domestic Civilian Telecommunication Satellite Transponder" in 1991. In the following year, the ministry announced a notification "Prohibition Sending Signal without Permission to Telecommunication Satellite". Both of these emphasize that, without permission, sending signals to the telecommunication satellite is strictly prohibited. The reason for this provision is that some units sent signals without authorization from the government and caused disorder, seriously interfering with satellite telecommunication, including telephone, data, fax, television, and radio program broadcasting.

CONCLUSIONS

Survey and analysis of the regulations relating to space safety clearly shows the status quo of China's law and its relationship with space activities. Some aspects of Chinese law are the same as at the international level:

1. Regulations are complicated; there is a lack of systematized, national law.
2. Different ministries deal with different issues in the field. Some issues are covered and regulated concurrently, which may cause conflicts; some issues are omitted, which may leave room for increasing risk.
3. Many regulations are provisional and unstable and may cause long-term problems.
4. No safety regulation exists to deal with space stations, which surely will exist in the future when China develops these.
5. Commercialization of space activity is currently an important phenomenon, if not a trend. There are more and more regulations relating to commercialization, but not enough.
6. International cooperation is only limited in less important fields, and no international standard for space communication safety exists although it is a very important field.

It is clear that space safety regulations and standards are not sufficient at either national or international level. However, with the development of space activities, space safety is becoming more and more important, challenging the rules and standards currently existing. The following proposals are probably feasible ways of dealing with the problem:

1. China needs a systematized mechanism of national space law, for the reasons listed above.
2. International standards are needed for international society. International standards can give all participants in space activity a useful tool to deal with space safety. It is the duty of every country to produce an international standard as soon as possible.
3. Registry of information should be in greater detail, making tracking easier. Though the Registration Convention has been in existence for more than 30 years, the registration information is not always enough for tracking. So, it is necessary to either amend the Registration Convention making it possible to track efficiently, or to make a new international document promoting better tracking.
4. A tracking cooperation mechanism and necessary information disclosures system should both be established. This measure can further promote tracking, and will make space safer. The existing space treaties all recognize international cooperation. The proposals mentioned above are reasonable results.
5. Precautionary principles should be followed. This principle is a widely accepted concept in the environmental law field. It can be incorporated and applied to the issue of space safety, imposing a legal obligation to the country participating in space activities.
6. An early-warning system at both national and international levels should be established. The system is mainly a technical one, and every country may create an international cooperation mechanism for the early-warning system.

Notes and references

[1] See the Legislation Law, Article 7.
[2] See the Legislation Law, Article 56.
[3] See the Legislation Law, Articles 63–66.
[4] See the Legislation Law, Articles 71–73.
[5] See the Legislation Law, Articles 78–82, 86.
[6] See NPC website, http://www.npc.gov.cn/npc/xinwen/dbgz/yajy/2008-10/29/content_1455984.htm (last accessed 22 August 2009).
[7] See the 2000 White Paper, "China's Space Activities".
[8] See the Provisional Methods, Article 25.
[9] See the Provisional Methods, Article 16.
[10] See the Provisional Methods, Article 17.
[11] See the Provisional Methods, Article 6.
[12] See the Provisional Methods, Article 6.
[13] See Article 2 of the Environmental Protection Law of the People's Republic of China, 1989.
[14] See the Provisional Methods, Article 19.
[15] See the Provisional Methods, Article 21.
[16] There is a complicated calculation to find the exact number of the fund. According to Article 2 of the Fund methods, it is the insurance premium minus the business tax, management fee and reinsurance premium, plus the net income and interest of the reinsurance recoverable.

[17] According to the Insurance methods, the 10 insurance corporations are China Reinsurance Company, PICC Property and Casualty Company limited, China Pacific Insurance Company, Ping'an Insurance Company of China, Huatai Property Insurance Company, Tian'an Insurance Company, Dazhong Insurance Company, Sinosafe Insurance Company, Yong'an Insurance Company, and Xinjiang Corps Insurance Company.

[18] See the Fund methods, Article 4. All the accounts can only be opened in four state-owned enterprise banks: Commercial and Industrial Bank of China, Agriculture Bank of China, Bank of China, and China Construction Bank.

[19] The commission has now been reformed into the State Development and Reform Commission, absorbing the former's functions.

[20] MOT (2006). No. 617.

[21] COSTIND 1st Division (2006). No. 555.

[22] SDRC (2007). No. 3057.

[23] See 2002 Safe Production Law, Article 2, Articles 77–95.

[24] This military regulation came into effect on 1 August 2008, replacing the military regulation of the same name in 2003.

[25] China will establish a space station in the future according to its long-term space activity plan. So it can be concluded that the regulations relating to space stations will be promulgated, but not in the near future.

[26] Scientifically Treat the Satellite Collision. *Guangming Daily* (in Chinese), 14 February 2009.

[27] Tracking Satellite, Devoting Spaceflight. *Scientific Chinese* (in Chinese), October 2007, p. 94.

[28] Yu Zhijina (2006). *Development Tendency of TT&C System in China*. Science Association Annual Conference.

[29] See the website of the company, http://www.casic-amc.com/company/index.jsp?subCid1=credit (last accessed 22 August 2009).

[30] See Xinhua News Agency, http://news.xinhuanet.com/newscenter/2003-06/11/content_913300.htm (last accessed 22 August 2009).

[31] See the Radio Regulations, Article 1.

[32] See the Radio Regulations, Article 11.

[33] SRC Letter (2008). No. 12.

[34] See MIIT (2009). Order No. 6.

[35] See CNCA Announcement (2009). No. 31.

[36] See COSTIND 1st Division, No. 1417.

[37] See the notification, §10.

[38] See SOA (1991) No. 140.

[39] Articles 30–33.

[40] See "Notification of Enhancing the Administering the Satellite Broadcasting Technology", 30 May 2007.

CHAPTER

Space launch safety in Australia

7

Michael E. Davis

CHAPTER OUTLINE

7.1 BACKGROUND

In 1996, Kistler Aerospace Corporation entered discussions with the Australian government to use the Woomera range for the testing of its new *K-1* reusable launch vehicle. Kistler was developing a two-stage reusable launch vehicle for the low earth orbit satellite launch market. Both stages were designed to be returned to the launch site in order to be reused within a short time frame and the Woomera range in South Australia was an ideal return site. A test program was scheduled to take place from Woomera commencing in 2000. In April 1998 an agreement was signed between the Australian government and Kistler providing for the establishment of the Kistler test program and detailing the government's regulatory requirements.

Other companies also saw commercial opportunities to provide launch services in Australia to meet the surging demand for new telecommunications satellites. Asia-Pacific Space Centre Pty Ltd proposed to establish a launch facility on Christmas Island, an Australian territory to the north-west of Western Australia. The company planned to use *Soyuz* or the new Russian *Angara* rockets to launch satellites into low earth and geostationary orbits. United Launch Systems International Pty Ltd was an Australian company with a major investment by the Thai Satellite Telecommunications Company. The company proposed to establish a low earth orbit space launch facility using the newly developed, Russian-built *Unity* launch vehicle. The proposed launch site was at Hummock Hill Island, south of Gladstone on the eastern coast of Australia. Test launches were anticipated in 2002 with commercial

Space Safety Regulations and Standards. DOI: 10.1016/B978-1-85617-752-8.10007-8

operations commencing in 2003. Spacelift Australia Limited, an Australian company, announced in September 1999 that it had entered an agreement with a Russian company, STC Complex – MIHT, to establish a low earth orbit launch service from Woomera commencing in early 2001. The agreement involved the use of the Russian *Start* launch vehicle.

7.2 THE SPACE ACTIVITIES ACT

Prompted by these developments, legislation, to be known as the Space Activities Act, was introduced into the Australian Parliament on 12 November 1998. The government decided that a clear legislative and regulatory framework was essential if the commercial space launch projects proposed for Australia were to be successful. It would allow potential investors in such projects to know what government requirements must be met. The Space Activities Act became law on 21 December 1998.

Australia is a signatory to the five space treaties [1] and is therefore required to monitor and regulate space activities within its territory or under its control, and to register with the United Nations any space objects for which Australia is a Launching State. Because Australia has responsibility under the treaties for any loss or damage caused outside of Australia by objects launched from Australian territory, the Australian government has an unlimited liability under international law. Subject to certain limitations, the legislation also imposed an obligation, under domestic law, on the part of launch operators, to indemnify the government against its international liability for loss and damage.

The objectives of the legislation were to establish a regulatory regime for commercial space activities carried out in Australia and by Australian nationals, to establish a compensation regime for third party damage caused by space launch activities, and to make provision for the government to implement its obligations under the UN Space Treaties [2]. The scheme of the legislation was to prohibit space activities such as launches from and returns to Australia, and launches by Australian nationals, unless certain approvals are obtained [3].

7.2.1 Licenses

Space licenses cover a particular launch facility in Australia and a particular kind of launch vehicle. To obtain a space license, the Minister responsible for the legislation must be satisfied that:

- the proposed licensee must be competent to operate the launch facility and launch vehicle;
- all necessary environmental approvals under Australian law have been obtained and an adequate environmental plan has been made for the construction and operation of the launch facility;
- the proposed licensee has sufficient funding to construct and operate the launch facility;
- the probability of the construction and operation of the launch facility causing substantial harm to public health or public safety or causing substantial damage to property is as low as is reasonably practicable;
- there are no reasons relevant to Australia's national security, foreign policy, or international obligations why the space license should not be granted; and
- any other requirements prescribed in regulations are satisfied [4].

7.2.2 Launch permits

Launches in, and connected returns to, Australia require a launch permit or exemption certificate. To obtain a launch permit:

- the proposed licensee must hold a space license for the launch facility and the kind of launch vehicle concerned;
- the proposed licensee must be competent to carry out the launch and any connected return;
- the proposed licensee must satisfy the insurance/financial requirements of the Act;
- the probability of the launch or launches, or any connected return, causing substantial harm to public health or public safety, or causing substantial damage to property, should be as low as is reasonably practicable;
- the space object will not contain a nuclear weapon or a weapon of mass destruction; and
- there are no reasons relevant to Australia's national security, foreign policy, or international obligations why the launch permit should not be granted [5].

7.2.3 Overseas launch permits

Where an Australian national is a responsible party for an overseas launch, an overseas launch certificate is required. The requirements for an overseas launch certificate are that:

- the proposed certificate holder must have the ability to satisfy the insurance/financial requirements for each launch (there is power to waive this requirement having regard to the nature and purpose of the space objects);
- the probability of the launch causing substantial harm to public health or public safety or substantial harm to property is sufficiently low; and
- there are no reasons relevant to Australia's national security, foreign policy, or international obligations why the launch permit should not be granted [6].

The return of an overseas-launched space object must be authorized. This may be satisfied by permission or agreement where:

- the person who is to carry out the return is competent to do so;
- the insurance/financial requirements are satisfied;
- the probability of the return causing substantial harm to public health or public safety or substantial harm to property is as low as is reasonably practicable;
- the space object does not contain a nuclear weapon or a weapon of mass destruction; and
- there are no reasons relevant to Australia's national security, foreign policy, or international obligations why the launch permit should not be granted [7].

7.2.4 Register of Space Objects

The Space Activities Act also establishes a Register of Space Objects, contains provisions relating to the powers of the Launch Safety Officer (LSO), and requires and authorizes the conduct of investigations in the event of launch accidents [8].

7.3 SPACE LICENSING AND SAFETY OFFICE

In June 2001, the Australian government established the Space Licensing and Safety Office (SLASO). Its role was to assess applications for approval under the Act, ensuring that space activities do not jeopardize public safety, property, the environment, Australia's national security, foreign policy, or international obligations [9]. Although SLASO oversees compliance with a wide range of regulatory obligations such as occupational health and safety legislation, explosives and dangerous goods legislation, transport of dangerous goods legislation, and environmental protection, its prime responsibility is the protection of public safety and property from risks arising from the flight of the launch vehicle.

In addition to the space license and launch permit required under the Space Activities Act, a Launch Safety Officer must monitor operations at each launch facility. The Launch Safety Officer may give any directions necessary to avoid danger to public health, or to persons or property, including directions to stop the launch or destroy the space object.

An accident involving the space object must be investigated to establish the circumstances surrounding the accident and prevent other accidents occurring. Immediately after an accident occurs, the launch permit under which the launch was conducted is automatically suspended. An incident involving the space object may also be investigated for these purposes [10].

7.3.1 "As low as reasonably practicable"

Originally, the Act and the Regulations required the government to be satisfied that the risk to public health and safety was "sufficiently low" in connection with a particular launch. In 2002, the Act and the Regulations were amended to raise the risk standard to being "as low as reasonably practicable", requiring launch operators to show that the lowest practicable risk is achieved within reasonable cost. This was seen as a significant increase on the regulatory burden for launch operators, though there may have been an indirect benefit in reducing the cost of insurance because of the resulting risk reduction.

SLASO uses a systems approach to assess launch safety involving the interconnection of the following elements of launch capability: launch facility, space object comprising the launch vehicle and payload spacecraft, organizational environment and launch operations, and flight of the space object [11]. According to SLASO, "as low as reasonably practicable" means "a level of risk that is not intolerable, and cannot be reduced further without the expenditure of costs that are grossly disproportionate to the benefit gained" [12].

7.3.2 Risk Hazard Analysis Methodology

In addition to applying qualitative "as low as reasonably practicable", operators must use a Risk Hazard Analysis Methodology to quantify the risks posed to such persons and property and to ensure that exposure levels are very low. These are set out in the Flight Safety Code and a launch permit will not be issued unless the Launch Safety Standards set out in the code are met. The standards comprise Third Party Casualty Safety Standards, Asset Safety Standards, and other mandatory standards in respect of drop zones, landing sites, and flight safety systems.

The Third Party Casualty Safety Standards are as follows:

- The maximum third party collective risk (the sum of casualty risks to all individuals in the general public) on a per-launch basis is one in 10,000 per launch.
- The maximum third party individual casualty risk on a per-launch basis is one in 10 million per launch.
- The maximum third party individual casualty risk on a per-year basis is one in 1 million per year [13].

The Asset Safety Standards are unique to the Australian space safety regime. They apply to particular facilities, termed Designated Assets and Protected Assets, which have been determined to warrant special protection from the risk of impact because of the potential for such impacts to trigger a catastrophic chain of events. These facilities include remotely located energy production facilities, such as offshore oil or natural gas platforms, and onshore natural gas facilities that are in the vicinity of a proposed space vehicle flight path.

The probability of trigger debris impacting a Designated Asset must not be higher than one in 10 million per launch and must not be higher than one in 1 million per year. There are also safety standards set in relation to the probability of non-trigger debris impacts on a Designated Asset. A Protected Asset has a further 10-kilometer buffer zone over and above the protection from trigger debris it would enjoy as a Designated Asset. Trigger debris is space debris of a particular shape, weight, velocity, or explosive potential that is capable of triggering a catastrophic chain of events on a Designated Asset or Protected Asset.

CONCLUSIONS

In conclusion, the combination of the qualitative and quantitative standards described in this article amount to a safety regime that is very much geared to protecting public safety and property from risks arising from the flight of a launch vehicle. In fact, according to SLASO "the standards are significantly lower than the upper tolerance limits used in quantitative risk assessment in other industries, including the offshore petroleum industry" [14]. None of the commercial launch projects mentioned at the beginning of this article that gave rise to the Space Activities Act has been realized and there is no longer a strong economic or political case for the establishment of a commercial launch industry in Australia. However, the safety regime is well established and ensures that the highest standards of public safety are achieved in any Australian space launch activity.

Notes and references

[1] The Treaty on Principles Governing the Activities of States in the Exploration and Use of Outer Space, including the Moon and other Celestial Bodies, 1967 (commonly known as the Outer Space Treaty); the Agreement on the Rescue of Astronauts, the Return of Astronauts and the Return of Objects Launched into Outer Space, 1968 (commonly known as the Rescue Agreement); the Convention on International Liability for Damage Caused by Space Objects, 1972 (commonly known as the Liability Convention); the Convention on Registration of Objects Launched into Outer Space, 1975 (commonly known as the

Registration Convention); and the Agreement Governing the Activities of States on the Moon and other Celestial Bodies, 1979 (commonly known as the Moon Agreement).

[2] See Second Reading Speech, Australian Senate Hansard, 12 November 1998, p 148.

[3] See http://www.spacelaw.com.au/

[4] Space Activities Act, s18.

[5] Space Activities Act, s26 (3).

[6] Space Activities Act, s35 (2).

[7] Space Activities Act, s43.

[8] Space Activities Act, s50 and Parts 5 and 7.

[9] Since the establishment of the SLASO in June 2001, six space activities have been approved—two HyShot launches at Woomera, the FedSat micro-satellite, the Optus C1 and two Optus D-series satellites (see http://www.innovation.gov.au/Industry/Space/Pages/SpaceLicensingandSafetyOfficeFactSheet.aspx).

[10] The only investigation to date in Australia under the Space Activities Act was in relation to an incident on 30 October 2001 when an anomaly during a HyShot experimental launch to test a scramjet engine from the Woomera range caused debris to land near a main highway. The report of the investigation is available at http://www.atsb.gov.au/publications/2002/sir200206_001.aspx

[11] Introduction to the Australian Space Safety Regime, Version 2.2, 18 May 2009. Available at http://www.innovation.gov.au/General/MEC-SLASO/Documents/Australian_Space_Safety_Regime_Overview.pdf

[12] Introduction to the Australian Space Safety Regime, *ibid.*

[13] The Australian standards for individual casualty risk are identical to those adopted in the USA. The Australian collective risk standard is less conservative than the equivalent US standard of 3 in 100,000 per launch; see Introduction to the Australian Space Safety Regime, *ibid.*

[14] Introduction to the Australian Space Safety Regime, *ibid.*

CHAPTER

Certification of new experimental commercial human space-flight vehicles

8

Tommaso Sgobba*, Joseph N. Pelton†

* *President, International Association for the Advancement of Space Safety,* † *Former Dean, International Space University*

CHAPTER OUTLINE

8.1 COMMERCIAL HUMAN SPACE-FLIGHT INITIATIVES

We live in a new commercial space era. It started with the XPrize competition, followed by the dramatic winning of the Ansari Prize by Burt Rutan and Paul Allen. The successful flights of *SpaceShipOne* in the summer of 2004 heralded the beginning of a new commercial space age. This innovative era generated novel opportunities for commercial system innovation, but also stimulated new space safety concerns and the need for an additional type of regulatory regime as well. Today, there is almost what might be called a tug of war between the "new space entrepreneurs" who are all for innovation and change, and the well-established "space establishment" of governmental space agencies. The official space agencies have been "man rating" human space vehicles for many decades and feel they have a great deal of hard-won knowledge about what works and what does not in the harsh and unforgiving world of outer space.

The new space entrepreneurs are eager to bring large numbers of different people into space in safe yet innovative ways that simplify some of the real, as well as the perceived, complexities that surround large and established official space programs. Their thesis is that totally original design concepts, as opposed to following time-tested space safety concepts, will be the key to breakthroughs in space safety. Something will have to give between these dramatically different views of the way forward. However, it is possible that the new space entrepreneurs can learn from the space establishment, and the space establishment can find fresh inspiration from the latest generation of space innovators. No matter what, the next few decades of space travel should be both exciting and eventful as the "rules of game" and the "space rules of the road" unfold. There is much merit in the viewpoints of the new and old space ventures.

Care needs to be taken to ensure that the best ideas from past experience are preserved while we are also searching for innovation that can lower the cost of space travel while also making it safer. The

Space Safety Regulations and Standards. DOI: 10.1016/B978-1-85617-752-8.10008-X

ideas presented in this article and the detailed document presented in Appendix 1 could perhaps provide a basis for "initial certification" of commercial human space-flight vehicles (CHSV). This process would allow entrepreneurs to be innovative in their design, but such a "preliminary certification" would allow governmental officials, passengers, and those who are concerned about their safety the knowledge that some form of "preliminary certification" had been independently executed. We believe that a process of "certification" by an independent space safety board that follows a systematic methodology to review safety measures—in design, fabrication, and operation—is a prudent and viable way forward.

8.2 REGULATORY OVERSIGHT OF COMMERCIAL SPACE TRAVEL

On 23 December 2004, President George W. Bush signed into law the Commercial Space Launch Amendments Act of 2004 (CSLAA). The CSLAA promotes the development of the emerging commercial space-flight industry and makes the Department of Transportation and the Federal Aviation Administration (DOT/FAA) responsible for regulating commercial human space flight under 49 USC Subtitle IX, Chapter 701 (hereafter Chapter 701). The CSLAA requires that before receiving compensation from a space-flight participant or making an agreement to fly, a Reusable Launch Vehicle (RLV) operator must inform the space-flight participant in writing that the US government has not certified the launch vehicle as safe for carrying crew or space-flight participants (49 USC §70105 (b)(5)(B)). In addition, the RLV operator should have a placard displayed in the launch vehicle in full view of all space-flight participants to warn that the launch vehicle does not meet certification standards required of a commercial aircraft. Thus, the operation of this type of craft is under an experimental license, at least through 2012.

There are further indications, based on the independent report provided to the FAA and the US Congress under the provisions of the CSLAA, that the period under which this new type of "experimental commercial human space-flight" license could be granted might be extended additional years into the future. This report, as prepared by the Aerospace Corporation, George Washington University, and the Massachusetts Institute of Technology (MIT), indicated that the progress that had been anticipated in 2004 had not been achieved in the 2004–2008 period and, thus, experimental licenses might reasonably be continued after 2012.

The CSLAA acts to protect the people and property on the ground but places the risk of those who might wish to fly on "experimental craft" on to the passengers. The problem that many perceive is that the average "space tourist" will not likely have the necessary background or the technical experience to truly grasp the risk that "experimental" space flight might entail.

8.3 REGULATORY CONTROL OF EXPERIMENTAL COMMERCIAL HUMAN SPACE FLIGHTS AND THE ISSUE OF "INFORMED CONSENT" ON THE PART OF PASSENGERS

The future issue or dilemma that may need to be faced in coming years is precisely what the "written consent" of passengers on experimental commercial spacecraft really means. Although this written consent is required by law, it may very well be interpreted by the US government (DOT/FAA) in one way, the commercial organization providing the experimental flights in another, and passengers and crew yet another. Indeed, the greatest challenge to interpretation is most likely to come after a fatal accident of an "experimental craft" has occurred. The estates of those concerned might claim damages based not only on the written consent—but on whether or not that consent was "informed" or "uninformed". The issue of what is "informed consent" may therefore not be resolved until there are legal claims made and the issue is resolved in a court of law.

Meanwhile the European Aviation Safety Agency (EASA) is currently seeking to develop rules and regulations with regard to the flight of "winged space planes" with crew and passengers aboard that may indeed seek to establish some form of certification for such vehicles. If it should evolve that the US government would authorize "experimental flights" without certification while the European regulatory agency was requiring some form of certification, this could lead to problems not only of business operation and regulatory oversight but ultimately to legal complexities and complications with regard to future compensatory claims as well.

The bottom line is that the lack of an independent safety certification for commercial human space flight appears likely to cause difficulty going forward. This is not only because there might not be a seamless global regulatory approach, but because of other practical issues as well. Certainly, the lack of an independent safety certification may deter the participation of at least some potential customers. Furthermore, the current CSLAA regime does not relieve the RLV operator of any responsibility for gross negligence. In this regard, obtaining an independent safety certification would be very much in the interest of the RLV operator in case of future litigation. The issue of certification is, of itself, a complex and problematic issue. The question is whether one would start with the certification of key subsystems to certain standards; whether certification only applies to the entire vehicle; whether, in this case, certification applies to certain types of technical, performance, or quality standards for the vehicle and its subsystems; or whether it would apply only to testing of "in-flight performance". This, of course, is quite different for aircraft—that are designed for very long duration flights and can be tested for long endurance reliability—versus the case of "space planes" where we are referring to vehicles designed to operate more like a rocket with short explosive performance rather than long-duration operations.

The FAA guidelines for "Commercial Sub-orbital Reusable Launch Vehicle Operations with Space Flight Participants", issued in February 2005, provide details about the overall process of written informed consent. They require that an RLV operator should describe to each space-flight participant the safety record of all launch or re-entry vehicles that have carried one or more persons on board, including both US government and private sector vehicles. The safety record should not be limited to the vehicles of a particular RLV operator. An RLV operator should also describe the safety record of its vehicle to each space-flight participant. The RLV operator's safety record should include vehicle ground-test and flight-test information. This information should describe all safety-related anomalies or failures that occurred and corrective actions taken to resolve the anomalies or failures. The FAA guidelines consider that the development of commercial launch vehicles to carry space-flight participants is in the embryonic or early stages.

Consequently, newly developed launch vehicles will not have the extensive flight-test history or operational experience that exists for commercial airplanes. Because of the lack of flight-test and operational experience, the risks with the RLV operator's particular launch vehicle and with vehicles like it, including both government and private sector vehicles, should be disclosed. The US House

Committee on Science report, H. Rep. 108-429, clarifies that Congress intended that all government and private sector vehicles be included in this description. Because most human space flight to date has taken place under government auspices, the government safety record provides the most data. The RLV operator should provide a record of all vehicles that have carried a person because they are the most relevant to what the RLV operators propose. Likewise, because they were intended for a human on board, greater care was likely to have been taken in their design and construction. The same should be expected for commercial human space flight. Furthermore the RLV operator is required to provide space-flight participants an opportunity to ask questions orally to acquire a better understanding of the hazards and risks of the mission.

8.4 "PRELIMINARY CERTIFICATION" BY AN INTERNATIONAL SPACE SAFETY BOARD

The requirements set forth in Appendix 1 of this book have been established on the basis of the safety experience accumulated in manned space flight to date. These requirements, and the proposal, attempt to use an independent International Space Safety Board to assist in establishing internationally accepted technical guidelines that might be used as a preliminary step toward "certification" of vehicles for Commercial Human Space Flight and would be a key step forward. These principles and technical recommendations are set forth in detail at the end of the book. These requirements are clearly not definitive and certain elements will need to be added after further international technical, policy, and regulatory review. Yet, the materials presented in Appendix 1 do, nevertheless, represent a good deal of careful thought and reflect the best practices from around the world over the past four decades. The verification processes, safety specifications, materials, and other information presented in Appendix 1 specifically represent the technical and operational experience of safety experts from the International Association for the Advancement of Space Safety (IAASS). This is the international association formed in 2003 to champion the cause of both governmental and commercial space safety.

By demonstrating design compliance with these requirements, the commercial manned space-flight operators would be able to show that they have taken into due consideration past experiences and best practices from all relevant data for the sake of making their spacecraft design and operations safe.

The International Association for the Advancement of Space Safety (IAASS) has developed a quite detailed description of what an Independent Space Safety Board (ISSB) would do to provide flight safety certification services to the emerging commercial manned space industry in terms of its competence, operations, verification testing, and document requirements.

The ISSB would be composed of a multidisciplinary team of space safety experts and would be available to all RSV providers or would-be providers on a non-discriminatory and a not-for-profit fee basis. The capabilities of the ISSB would, of course, evolve over time as experience and knowledge in the field expands. Because of the great diversity in the design of both sub-orbital and orbital flight systems that exist today, the approach to certification may tend to start with the key subsystems in terms of their design quality, durability, and redundancy. Over time the ability to achieve certification on a more systematic basis is expected to improve, much as was the case with aircraft design and certification.

CONCLUSIONS

The detailed information presented in Appendix 1 is seen as the first completely synoptic overview of a set of safety specifications and safety testing procedures for the new commercial space travel industry. Although it is comprehensive, it is not entirely complete and several items of these specifications are yet to be finalized. Further, these specifications are to be completely available according to both metric and US units of measurement. Also, with time and experience, these might be amended for reasons that include the need for greater specificity, to address new technology or flight concepts, or even reasons of obsolescence. Even if the ISSB does not function as envisioned in the document set forth in Appendix 1, the existence of these various specifications and safety verification tests, and this information, can have considerable value to the developers of commercial human space-flight vehicles (CHSV) and the providers of commercial human space-flight (CHSS) services.

PART

Safety Regulations and Standards for Unmanned Space Systems

2

CHAPTER

9

Regulatory procedures and standards for launch range safety for manned and unmanned launches

Jeff Hoffman*, **Joseph N. Pelton**[†]

* *Massachusetts Institute of Technology,* [†] *Former Dean, International Space University*

CHAPTER OUTLINE

INTRODUCTION

Ground operations are an essential component of any space launch system. In the particular case of commercial human space flight, some of the important questions are: Should ground systems be owned and/or operated by the government, the private sector, or a combination of both? Who should maintain regulatory supervision and who should set the safety standards for these ground systems? If there were to be a combination of public/private ownership and regulatory control, then what specific functions should be exercised by whom and do these vary if a manned versus an unmanned space flight is concerned and does it matter if the launch operation is an orbital or sub-orbital flight?

What is clear today with regard to commercial space flight involving humans is that there is, and will continue to be for some time, a diversity of launch vehicle design concepts. At least the next 5–10 years will continue to be the embryonic stage of the commercial human space-flight industry. This characteristic of the new commercial human launch vehicle industry proceeding with a great diversity of designs not only in their launch systems, but also their safety systems and ground operations, clearly inhibits any systematic attempt to define standards for these systems or even a single set of standards for mission control, range operations, and other launch and re-entry support functions. In contrast, both

Space Safety Regulations and Standards. DOI: 10.1016/B978-1-85617-752-8.10009-1

commercial and governmental unmanned launch operations have decades of experience. Launch range operations and launch range safety standards and practices are well established and fail-safe protections and elements such as flight termination systems (FTSs) have served to minimize ground operation accidents and harm to adjacent properties.

Even after the industry matures, there is a difference in regulatory perspective, with the USA tending toward the encouragement of commercial space operations and even "industrial safety enforcement processes" to a greater extent than is the case with European, and possibly other, governments. Respondents to a questionnaire administered to the US commercial space industry, that was conducted during 2008 by the Aerospace Corporation in response to the mandate of the Commercial Space Launch Amendment Act (CSLAA) of 2004, concluded that there was a nearly consensus viewpoint among the US space industry that ground operations will need to be very system-specific. This led to the widespread opinion that associated ground operations should largely, if not exclusively, be the responsibility of the vehicle operator. In contrast, the European Aviation Safety Agency (EASA) is proceeding toward the licensing of manned winged space planes and their ground control facilities.

The Federal Aviation Administration Office of Commercial Space Transportation (FAA/AST) has developed a detailed concept of operations for integrating commercial space transportation (CST) into the aviation-based National Airspace System (NAS) [1]. This has specifically led to the concept of "mission planning" with respect to commercial space transportation operation. This is considered by the FAA/AST to be a more apt label than "mission control" with regard to CST operations in that it seeks to achieve the desired seamless integration of all aspects to be considered from pre-launch through flight, re-entry, launch and post-flight checkout.

The FAA/AST is authorized to license and regulate US commercial space launch and re-entry activities and the operation of non-federal launch and re-entry sites. Various relevant sections of the statute are included below:

Title 49 USC, Section 70101(a); Congress finds that:

> *"... private industry has begun to develop commercial launch vehicles capable of carrying human beings into space, and greater private investment in these efforts will stimulate the Nation's commercial space transportation industry as a whole ...*
>
> *"(8) space transportation, including the establishment and operation of launch sites, re-entry sites, and complementary facilities, the providing of launch services and re-entry services, the establishment of support facilities, and the providing of support services, is an important element of the transportation system of the United States, and in connection with the commerce of the United States there is a need to develop a strong space transportation infrastructure with significant private sector involvement ..."*
>
> **"(10) the goal of safely opening space to the American people and their private commercial, scientific, and cultural enterprises should guide Federal space investments, policies, and regulations ..."** [2].

The FAA has perceived, within this guidance from the US Congress, that commercial space transport could evolve as an important new part of the overall US transportation system. This, of course, may take many years to become reality. The long-term objective of the FAA is therefore to develop a Space and Air Traffic Management System (SATMS). Within this SATMS, space and aviation operations would become seamless and fully integrated in a modernized, efficient National Airspace System (NAS). The even grander concept beyond SATMS is a World Airspace System (WAS) where space and air management is also integrated. This concept clearly is decades in the future and, because of national security and political constraints, the USA is focusing on the NAS and not the WAS in its current effort. Planning both in the USA and around the world will be needed to evolve into an integrated world-based system.

Since 2003, the FAA/AST has been supporting the efforts of the Next Generation Air Transportation System (known as "NextGen"). A NextGen Joint Planning and Development Office was officially established on 12 December 2003 in response to a US Congressional FAA reauthorization (under Vision 100, Public Law 108-176). The objective of this office is to develop an integrated national plan for air and space traffic control [3].

9.1 THE CURRENT DEVELOPMENT OF REGULATION OF GROUND OPERATION FOR COMMERCIAL SPACE TRANSPORTATION

The possible commercial space transportation vehicles of the near and longer-term future as seen from the US perspective, including ground operations, are described in a recent report of the FAA [4].

The FAA's approach continues to be to envision a full scope of potential commercial launch vehicles. These options range from conventional expendable to partially reusable to fully reusable. Thus, the options include vertical take-off and landing through horizontal take-off and landing, and every combination thereof, plus options such as balloon ascent and a variety of carrier or towing vehicles. In short, there is an attempt to envision every combination of aircraft, balloon, rocket, space plane, or other ascent vehicle with a view to seeking a seamless integration of CST and aviation demands on future airspace.

The future of space transportation, as seen from the USA, anticipates a significant degree of commercial involvement, just as was the case with US aviation, railroads, and maritime systems. These commercial precedents indicate that this evolution of CST will depend on development of a market for commercial services. This future market may well lie in both developing and satisfying the commercial business interests for cargo movement and public transport in new forms of point-to-point operations. Critical factors in new market development may well involve not only safety and reliability, but also cost, convenience, reduction in environmental gas and sound pollution, and even global regulatory homogeneity, such as now provided by the International Civil Aviation Organization (ICAO) [5].

9.2 THE NATIONAL AND REGIONAL AIRSPACE SYSTEMS

The US National Airspace System (NAS) comprises the entire network of interconnected systems and infrastructures, the people who operate those systems, the vehicle operators who rely on those systems to be operable, and the myriad of operational procedures and processes and certifications. In the USA, the NAS includes some 19,000 airfields and airports, extensive air traffic control facilities, and a plethora of equipment that operates and interacts continuously to keep the NAS operating efficiently and safely. Further, the NAS represents a continually evolving system as dictated by technology enhancement of equipment (such as GPS, automatic landing systems, advanced radar and imaging systems, and satellite-based telemetry systems). It also includes improvements in operating procedures and processes, growing airspace demands, and the addition of new flight requirements, such as represented by commercial space transportation. The implementation process will ensure that each aspect of the NAS infrastructure is addressed and readied for satellite navigation. Although the US NAS is perhaps the most complex, the European Airspace now managed by the EASA is comparable in terms of size, complexity, and difficulty to manage, with the further challenge of integrating what was, until recently, a collection of national air traffic control systems [6].

Neither the US NAS nor other national or regional airspace systems currently have set an upper limit. The practical limit is established by the application of the "controlled airspace", which encompasses that volume of air from mean sea level to 60,000 feet (or 18,250 meters) in elevation. That includes the airspace overlying the waters within 12 nautical miles of the coasts of the 48 contiguous states and Alaska. The legal and regulatory competence for air or space operations outside of this region is less clear. Past incidents such as the shooting down of a US U-2 spy plane in the 1950s strongly suggest that national airspace and national airspace systems are, to a degree, defined by the ability to defend these regions.

But beyond the issue of defense of national airspace, there is the practical question of coordinated regulation of airspace safety. Clearly one of the open questions today is what national, regional, or international agency will in the future regulate flight services in flights that reach above the 60,000-foot (or 18,250-meter) level as commercial space flight—or possibly hypersonic flights into "near space" or "protospace"— grows in the future. This also generates additional questions of "ground control or ground operations" with regard to international overflight, return from orbit over another country, abort to another country, and which agency or agencies will control and monitor the various vehicles, which may be internationally owned. The FAA, EASA, and other aviation safety agencies will need to examine their policies in these respects as flight envelopes expand and commercial space transportation diversifies [7].

9.3 DEVELOPMENT OF TRAFFIC MANAGEMENT TOOLS FOR INTEGRATED OPERATIONS

New methods and procedures allow rapid assessment and reconfiguration of airspace structures and traffic. Thus, sector configurations are unconstrained by current boundaries. Dynamic reconfiguration of airspace within and between facilities increases operational flexibility. Two key tools can be employed to support the integration task with regard to air and space traffic management.

9.3.1 Space transition corridors

Space transition corridors (STCs) provide dynamically reserved and released airspace in the NAS for space vehicles launched from spaceports to fly over populated areas and through commercial airways to reach orbit or achieve sub-orbital trajectories. STCs are selected and determined based on performance characteristics of the vehicle and overall safety considerations. STCs may be tailored as mission needs or Air Traffic Control (ATC) needs dictate. These STCs provide more flexibility than today's special-use airspace (SUA).

9.3.2 Flexible spaceways

Flexible spaceways are similar to today's airways and jet routes, but they serve the purpose of routing traffic transitioning to and from space. These are dynamically designated to meet specific mission objectives, such as transitioning to airborne launch points, and aerial refueling. Depending on the mission and vehicle profile, spaceways may be used in conjunction with an STC, to segregate different types of missions, to concurrently accommodate different mission phases (e.g. launches vs. re-entries), and to ensure safety in the case of contingencies [8].

9.4 THE CURRENT REGULATORY FRAMEWORK FOR CST IN THE USA IN CONTRAST TO EUROPE

A review of current air transport regulations on the support and operation of human/cargo space-flight systems suggests that these current regulations and associated guidelines are directly applicable to, and adequate to support, the CST operations. To that end, Congress directed the FAA to adopt an active role in CST and authorized the FAA:

- to directly support the CST: "... to promote public–private partnerships involving the United States government, State governments, and the private sector to build, expand, modernize, or operate a space launch and re-entry infrastructure" (Title 49 USC Section 70103(b));
- to regulate the CST: "(13) a critical area of responsibility for the Department of Transportation is to regulate the operations and safety of the emerging commercial human space-flight industry" (Title 49 USC, Section 70101); but
- to regulate with due caution and consideration: "(15) the regulatory standards governing human space flight must evolve as the industry matures so that regulations neither stifle technology development nor expose crew or space-flight participants to avoidable risks as the public comes to expect greater safety for crew and space-flight participants from the industry" (Title 49 USC, Section 70101).

The need for further regulation will emerge as the Concept of Operations is being implemented principally to provide the mechanism to fully integrate aviation and commercial space into the same airspace. For the most part, near-term requirements for regulatory oversight in the USA may be satisfied by modifying appropriate sections of 14 CFR to explicitly incorporate the operating procedures and processes of the Concept of Operations. Subsequent regulatory needs will evolve as operating experience, technology enhancement, and industry growth necessitate. On the other hand, the EASA is moving ahead with an integrated policy with regard to the take-off and landing of commercial aircraft and commercial "space planes" (i.e. winged aircraft that "fly" to and from outer space). This will entail the "licensing" of such craft and an integrated approach to their operations.

9.5 LAUNCH RANGE SAFETY REGULATION

Launch range safety experience for various types of vehicles, including protection of public safety, is now more than 50 years old. A great deal of knowledge has been acquired about how to prevent accidents that could harm launch support crews and the uninvolved public. In short, much more is known about safety on the ground than is known about human space flight, especially its commercial variety [9].

NASA and the US Air Force in the USA, the European Space Agency (ESA) in Europe, and Rascom in Russia, in particular, have well-developed procedures for range safety, which form the basis of the criteria for commercial human space flight. Commercial satellite launches from the Eastern and Western Test Ranges (ETR, WTR) have to abide by the same range safety criteria as older, government-owned and operated vehicles. Launches of new vehicles also are subject to the same safety criteria. The procedures enforced by the ESA at their launch site in Kourou, French Guinea, as well as sounding rocket launches in Europe, are likewise well established and quite parallel to those employed in the USA. Russian launch range safety procedures were once quite different, but today, especially due to coordination arising from the International Space Station (ISS) construction, and commercial coordination that arose from joint ventures such as the International Launch Services Ltd, have served to establish a great commonality to on-the-ground safety procedures.

One of the most important safety responsibilities of the range commanders (i.e. the commanders of the United States Air Force Space Command's 30th Space Wing at Vandenberg Air Force Base, California; the 45th Space Wing at Patrick Air Force Base; and Air Force Command at Cape Canaveral, Florida) is to ensure public safety during launch and flight. Range safety personnel: (i) evaluate vehicle design, manufacture, and installation prior to launch; (ii) monitor vehicle and environmental conditions during countdown; (iii) monitor the track of vehicles during flight; and (iv) if necessary, terminate the flight of malfunctioning vehicles. The method used for flight termination depends on the vehicle, the stage of flight, and other circumstances of the failure. In all cases, propulsion is terminated. In addition, the vehicle may be destroyed to disperse propellants before surface impact, or it may be kept intact to minimize the dispersion of solid debris. Flight termination can also be initiated automatically by a break-wire or lanyard pull on the vehicle if there is a premature stage separation.

Current flight termination practices have an excellent safety record for the USA. From 1988 through November 1999, there were 427 launches at the ETR, during which 11 destruct commands were issued (two Atlas 2, one Delta 3, one Titan 4, four Trident submarine-launched ballistic missiles, and three other missiles). Over the same time period, there were 177 launches at the WTR, during which 11 destruct commands were sent (one Athena, two Pegasus, one Titan IV, and seven intercontinental ballistic missiles). Total failure of a flight termination system (FTS) is extremely rare at either range, and destruct commands are often superfluous because vehicles explode or break up because of dynamic forces before the mission flight control officer can react [10].

From a range safety standpoint, reusable vehicles (i.e. the space shuttle) are presently subject to the same constraints as expendable vehicles. The fact that a launcher is carrying humans does not lessen range safety constraints. This is a widely accepted understanding for astronauts, cosmonauts, and others that fly into space on governmentally controlled programs. The fact that there has not been a case where an FTS had to be activated prior to an actual launch malfunction on a "crewed" rocket system has not drawn active attention to this capability. The advent of commercial space transportation systems with human passengers and crew aboard and the possibility that many more people may fly on such systems in future years will undoubtedly flag this as a key regulatory policy issue going forward.

The US space shuttle has a launch FTS on the solid rocket boosters (SRBs). This was activated after the breakup of the *Challenger* stack in January 1986. Each time the shuttle is launched at Cape Canaveral, an Air Force range safety officer monitors events during the first 2 minutes. If the spaceship should veer off course and endanger a populated area, this officer would have the responsibility of flipping a pair of switches on a flight termination control panel (Figure 9.1).

The first switch arms explosives on the shuttle's two SRBs. Flipping the second switch would detonate them, destroying the shuttle and crew. The shuttle, however, has no FTS capability for re-entry on the premise that the crew could prevent the shuttle from landing in a manner that could create major loss of life.

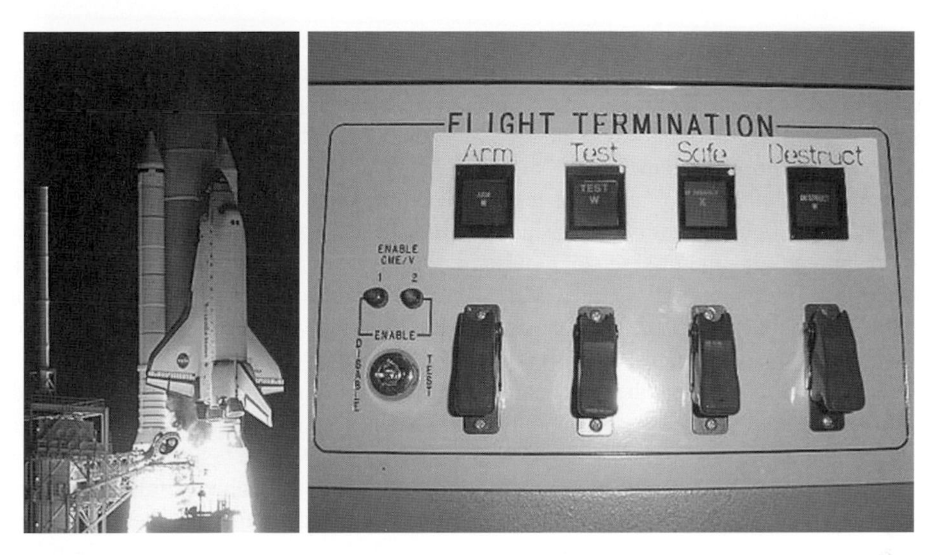

FIGURE 9.1

Graphic showing the space shuttle *Endeavor* and its flight termination system control panel.

Reprinted courtesy of operations, NASA

Critical aspects of traditional range safety philosophy, which relate to commercial human spaceflight plans involving privately developed and operated spaceports, are as follows:

- Present conventional rockets carry a large amount of explosive propellants, with the potential to cause extensive property damage and loss of life if they land in populated areas. Certain developmental Reusable Launch Vehicles (RLVs) use hybrid propellants, such as nitrous oxide (N_2O)/hydroxyl-terminated polybutadiene, which can decrease, but not eliminate, explosive potential upon impact.
- Rocket trajectories must be monitored during launch to ensure that they do not impose an undue hazard to populated areas. The FAA's review of license applications for spaceport facilities requires launch trajectories that result in acceptable "expected casualty" (Ec) analyses for flights over or near populated areas [11].
- Present standard practice for launch vehicles having sufficient propellant to cause extensive ground damage is to equip them with FTSs, operable by ground personnel, to allow timely independent response in the event of loss of guidance or control that could cause the vehicle trajectory to deviate in such a way as to threaten populated areas.
- Range safety and launch FTSs are the responsibility of the range operator (USAF for the Eastern and Western Test Ranges and NASA for the Wallops Flight Facility). Range safety is an independent operation, and range safety officers are not responsible to the launcher or the payload organizations [12].

The requirement for an independent range safety office and an FTS does not apply to airplanes. Despite the fact that airplanes carry large amounts of fuel and can cause extensive damage if they crash into populated areas, the governing principle is that with a human pilot on board (two pilots for large airplanes), loss of guidance or controllability is unlikely and precludes the need for an FTS. Also, airplanes move much slower than rockets, and air traffic controllers can detect potentially threatening situations and alert military resources in time to deal with a developing situation (at least in principle). The presence of a pilot or pilots on a number of planned developmental RLVs brings their flight operations closer to those for conventional aircraft.

Alternate flight safety systems are being tested for commercial RLVs under development. Examples include thrust termination systems (TTSs), which end the propelled stage of flight. These systems can be controlled by humans in line-of-sight or monitoring a computer screen, to prevent the rocket from leaving the designated fly zone. They can also be triggered by valves within the vehicle that terminate the thrust should the propellant pressure increase or decrease to a value outside specified safety parameters. Other safety systems include auto-landing systems and the ability to switch from horizontal flight to angling downward for controlling the flight path. Some small sounding rockets are not required to have FTSs or TTSs. Sounding rockets are generally spin-stabilized, meaning that active guidance is not required; hence the probability of loss of control taking a rocket out of the test range is extremely small.

Where in this risk spectrum will future commercial human space vehicles fall? This is still to be determined and varies greatly based on the wide range of designs currently being pursued. Some systems such as XCOR's *Lynx* rocket plane will operate much like a jet aircraft and will be piloted much like an aircraft during take-off and landing. Other systems such as Blue Origin's *New Shepard* and Armadillo Aerospace's *Black Armadillo* (as currently designed) will take off vertically and land with vertical stabilizing thrusters like a spacecraft. Many other designs are also being pursued, as previously discussed. In short, no systematic and unified set of controls for launch range safety can be applied at this time because of the great diversity of designs for space systems.

Each potential launch system seeking a license undergoes an analysis to determine the maximum probable loss in the case of loss of control, as has been done for all existing launch vehicles. The risk analysis needs to include the size of the protected range and potential inhabited regions near the perimeter. This is why there is a preference for launch facilities to be located in isolated areas or on seacoasts.

Different systems and different launch sites will have different damage potentials. Also, a hazard analysis must be conducted to determine scenarios that could lead to loss of control. Systems using an aircraft or balloon as a first stage would presumably have very different risks early in flight than those using more traditional vertical take-off rockets. During the re-entry portion of the flight, the vehicle will presumably not be carrying large quantities of propellant, so the potential hazards diminish but do not disappear. The FAA/AST takes these issues into account when issuing permits and licenses. It should be noted, however, that at this stage of CST development all commercial vehicles intended to carry human crew or passengers have been "authorized" by the FAA only as test flights under experimental permits. In short, no licensing has even been considered and, under the terms of the CSLAA legislation in the USA, the FAA cannot "license" such flights at least through 2012. Further, it currently seems unlikely that Congress will move quickly to grant the FAA such regulatory authority to "license" such flights in the post-2012 period.

There may be commercial human space-flight vehicle configurations with sufficient risk that an FTS or TTS should be required. This is not something that can be determined at this time, but designers of future commercial human launch systems should be aware of the possibility. The presence of an FTS or TTS might seriously impair the commercial viability of a company's launcher, so it is important to establish range safety criteria well enough in advance to be of use to designers and operators. New techniques for range safety, including autonomous FTSs based on GPS, have been under study for several years [13].

The USAF Space Command is considering switching to GPS satellite-based tracking and use of more automated destruct capabilities for vehicles that veer off course. There are plans to add these capabilities to the Evolved Expendable Launch Vehicle (EELV) Atlas 5 and Delta 4 rockets. This is an area where standardized equipment would be of great benefit to the emerging commercial human space-flight industry.

Assignment of responsibility for range safety operations is a crucial question. As mentioned above, for government launches, range safety is the responsibility of the range operator. The Wallops Flight Facility (WFF) has formed a partnership with the Virginia Commercial Space Flight Authority to deal with these issues at WFF [14].

The FAA licenses commercial launches, and in the future might also "license" future human launches, but for the time being will continue to issue only "experimental permits". In this respect the FAA, at a future time as the licensing agency, would be responsible for human safety. However, range safety for any launches from existing government launch facilities will almost certainly remain the responsibility of the USAF or NASA.

Part 417 of 14 CFR addresses range safety requirements for launches from both federal and non-federal sites. The required activities fall into three general categories:

- Determination of range safety requirements for different launch vehicle configurations
- Determination of the type of range safety FTSs or TTSs that should be used, if required
- Operational management of range safety once flight activities start [15].

Part 431.43 of 14 CFR requires that an applicant for RLV mission safety approval submit procedures for:

1. Monitoring and verifying the status of RLV safety-critical systems sufficiently before enabling both launch and re-entry flight to ensure public safety, and during mission flight (unless technically infeasible).
2. Human activation or initiation of a flight safety system that safely aborts the launch of an RLV if the vehicle is not operating within approved mission parameters and the vehicle poses risk to public health and safety and the safety of property in excess of acceptable flight risk [16].

Part 431.43 also specifies that any RLV that enters Earth orbit may only be operated such that the vehicle operator is able to:

1. Monitor and verify the status of safety-critical systems before enabling re-entry flight to assure the vehicle can re-enter safely to Earth.
2. Issue a command enabling re-entry flight of the vehicle. Re-entry flight cannot be initiated autonomously under nominal circumstances without prior enablement [17].

Part 437.67 of 14 CFR requires that applicants for experimental permits track reusable sub-orbital rockets. Specifically, a permittee must:

1. During permitted flight, measure in real time the position and velocity of its reusable sub-orbital rocket.
2. Provide position and velocity data to the FAA for post-flight use [18].

As mentioned above, determination of range safety requirements should be a near-term, high-priority activity. This has a potentially significant impact on launch system design; these requirements should be available to start-up launch companies early in the design process. At this time, much national expertise in range safety resides in the current range operators, although the FAA has developed its own safety expertise. The FAA should continue to draw upon this expertise, at least for initial range safety considerations. Eventually, the FAA will need to determine whether it should develop its own internal range safety expertise or continue to rely on other organizations. This decision will in part depend on the willingness of the USAF and/or NASA to become involved with range safety at private spaceports over the long run.

The licensees are ultimately responsible for public safety in cases where private spaceport operators provide range safety operations. This poses a potential issue for the independence of the public safety function. As noted above, the USAF or NASA has the responsibility for range safety for the ranges that each organization operates. The range safety officer has no programmatic connection with or responsibility for the organization launching a payload. Also, the range operators are not trying to turn a profit from their launch operations. In contrast, the commercial success of a private spaceport will depend on the number of customers that can be attracted to use the facility. Minimizing the cost of meeting range safety constraints will increase the commercial attractiveness of a spaceport. This of course remains true only up until the time when an accident occurs and then the reverse conclusion quickly applies.

A guiding principle for safety within large organizations is that the safety division must operate independently of operational programs. (The incorporation of safety functions into line program organizations at NASA was cited as an important factor in the *Columbia* accident report in weakening NASA's safety culture.) Safety personnel must answer to independent management, not to the programs whose safety they are responsible for. Ensuring adequate separation of interests will be a significant problem if range safety is the responsibility of the same organization that is attempting to make a profit from spaceport operations [19].

The US Congress has demonstrated that it recognizes a need for government involvement in the CST system and has provided direction and authority to the Department of Transportation under Title 49 USC. Congress constrained the regulatory process with this language: "... the regulatory standards governing human space flight must evolve as the industry matures ... [so as not] to stifle technology." The FAA has met that constraint as presented in Concept of Operations reports by absorbing the CST into the NAS rather than developing an entirely separate system. What remains to be done is to incorporate the Concept of Operations into the US Code (Title 49 USC) and the US Code of Federal Regulations (14 CFR) to accommodate the current state of commercial space transportation (CST). This would establish a "regulatory base from which evolutionary regulatory changes and strictures could be adopted as needed in the future".

In Europe, the development of "commercial space transport systems" and so-called space planes is, in general, occurring with a much greater degree of governmental involvement. This also means that the ESA and the EASA are working in close tandem with various development efforts and have sought to simplify their "regulation" of launch site safety by concentrating on winged vehicles that take off and land from, essentially, somewhat specialized airport facilities. Exactly how "private or commercial space launches" might be controlled or regulated in Europe remains an open question.

Regardless of the location, space launches using either expendable or reusable vehicles are complex operations that require careful technological planning and oversight at all levels of preparation, launch operations, and landing. Furthermore there are many types of launch operations and each might require some varying degree of launch range safety oversight and regulation.

In fact, an examination of four categories of current and proposed vehicles and operating modes reveals a lack of common characteristics that would suggest some difficulty with a completely common regulatory approach. The vehicle categories that were considered are discussed below.

- **Fully expendable (no components returned).** This is the classic launch vehicle, designed to function only once to deliver a payload (science experiment, satellite, etc.) into orbit, for a landing on another planet, for long-range space exploration, etc. The launch vehicle ignites and lifts off, shedding stages, fairings, and assorted other parts, generally into the ocean but sometimes on land, as it travels to its ultimate delivery point. It spends a minimum amount of time in the National Airspace System and generally is autonomous. The vehicles fly in constrained, well-defined vertical columns in the NAS.
- **Partially expendable (some components returned).** This is essentially the same as above with the exception that the payload and various other parts may be recoverable and reusable. Historical examples are the Corona class of satellites used for reconnaissance operations (film returned to Earth) and the Mercury, Gemini, and Apollo manned space programs (people returned to Earth). These vehicles were generally the same as fully expendable vehicles except for spending considerably more time in the NAS, both on launch and return to Earth and, in the case of the manned programs, having a crew on board that could fully or partially control flight (John Glenn on *Freedom 7*, Neil Armstrong on *Gemini 8*, responding to a stuck altitude control system valve, and Neil Armstrong again during the *Apollo 11* landing).
- **Partially reusable.** The prime example here is the Space Transportation System, which is reusable to the extent of returning the solid boosters and the fly-back Orbiter Vehicle. The Pegasus flight system also belongs in this class to the extent that the carrier vehicle is reused. These concepts spend considerably more time in the NAS but fly in a mix of column and horizontal box flight patterns.
- **Fully reusable.** This group represents "future space", though two examples have flown, and one (Virgin Galactic/Scaled Composites *SpaceShipTwo*) is anticipated to be introduced into the marketplace in the relatively near future. In this instance, the vehicle is intended to be reused in its entirety and, of the various concepts, has the longest loiter time in the NAS. Generally, these vehicles fly in a box/column flight pattern.

The above variations in vehicle designs and flight characteristics are substantial and do not realistically allow regulation simply on the basis of "expendable" or "reusable" categorization. The primary purpose of regulatory activity in the USA under the CSLAA is to promote and protect the safety of (1) the public and public infrastructure, (2) the other users of the NAS, and (3) the participants and crew of the commercial space transporter. Fundamentally the regulatory process should address the need to protect the above and should be based on enhancing the safety of operations in those areas that affect this need.

In light of these considerable differences, the launch of expendable vehicles, when used as a first stage to lift reusable rockets carrying crew and space-flight participants (passengers), as well as launch and re-entry of reusable launch vehicles with crew and space-flight participants aboard, should be regulated differently than the launch of expendable vehicles without humans aboard. Nevertheless range safety controls, regulation, and licensing may have substantial areas of commonality regardless of whether the vehicle is expendable or reusable, or has humans aboard. Current range safety processes can continue to apply, with pertinent upgrades as needed. Until more experience is gained with commercial private space-flight vehicles, however, the regulation of expendable and reusable vehicles for launching humans should remain on a case-by-case basis. This conclusion would seem logical whether the operations were in the USA or some other international launch site. This is essentially due to the great diversity of vehicle design, system components, and flight characteristics that exist today and presumably will be the case for some time to come. The development of a metadata system to monitor, in a systematic manner around the world, the development and actual performance of commercial launch systems, and to better identify different launch risk factors and criteria, would assist greatly in the regulatory process.

There are a large number of safety factors related to the launch of expendable vehicles, but even more stringent requirements are needed for reusable and re-entry launch systems with humans aboard, as well as for expendable vehicles that are employed to launch a reusable craft with space-flight participants [20]. These standards and their verification may be the responsibility of separate commercial enterprises, newly constituted entities from the launch insurance industry, or some other independent agency. These standards and their verification could also remain with the US government.

Although there could be a case where there are reusable vehicles that are flown with robotic controls without humans aboard, this would seem to require the same degree of safety regulation and stringent safety standards to protect the public during re-entry and landing operations as discussed previously. Regulatory planners can expect that there will be a number of launch configurations in which an expendable vehicle is used to insert a reusable vehicle into orbit that would later re-enter and land. In this case, the expendable vehicle would also be subject to more stringent safety standards. In the case of a private operation, safety inspections and safety standards enforcement would be conducted either by independent experts, government inspectors, or both. The responsibility for the standards, inspections, licensing, and oversight should be clearly established before these events occur.

Table 9.1 Comparison of Reusable Orbital Launch Systems and Sub-orbital Space Plane Systems for Human Space Flight [21]

Characteristics	Reusable Orbital Launch System	Sub-orbital Space Planes
Maximum velocities	Up to Mach 25	Mach 4–6
G forces	High *g* forces	3–5 *g* (during descent)
Thermal gradients on re-entry	Thousands of degrees C	Hundreds of degrees C
Environmental protection systems and structural strength of vehicle	Very demanding in terms of design and materials	Much lower demands in terms of structural strength, atmospheric systems, life support, etc.
Exposure to radiation	Can be high levels	Minimal exposure due to short flight duration and lower altitudes
Exposure to potential orbital debris collisions	Exposure increases as length of mission increases	Exposure risk is very low due to short duration and lower altitudes
Escape systems	Parts of the flight during high thermal gradients make escape systems extremely difficult and expensive to design	Escape systems are much easier to design due to lower thermal gradients, lower altitude, etc.
Type of flight suits required	Expensive and complex flight suits required	Simple and lower cost flight suits are required due to lower altitudes, lower thermal gradients, much shorter exposure to low oxygen atmosphere
Launch risk factors (overall)	Very high	Considerably lower and different

NASA's Commercial Orbital Transportation Services (COTS) program, whereby contractor companies would provide access to the International Space Station (ISS) or private orbital facilities, could provide the first test case of such a situation.

In the case of reusable vehicles involving the launch and re-entry of humans, there should be clearly established oversight and licensing processes that consider performance standards, performance margins, and standards verification, as well as processes that pertain to the granting of waivers to those standards. For reasons explained below, somewhat different standards, different licensing and inspection procedures, and different processes and/or authority for granting of waivers would appear appropriate for reusable vehicles making sub-orbital flights. This is in contrast to reusable vehicles actually going into orbit and de-orbiting, since such vehicles involve much more demanding technical performance requirements. The very different performance characteristics of sub-orbital and orbital launchers illustrate why this difference in regulatory approach is appropriate. In simplistic terms, reusable sub-orbital craft that operate much like experimental high-altitude jet aircraft should be regulated differently than reusable launch systems that go into orbit and have much more demanding performance characteristics. These considerable differences are set forth in Table 9.1.

CONCLUSIONS

The key public safety question that will ultimately need to be addressed in the USA, Europe, and other parts of the world is whether certain mandatory requirements will be applied. Prime among these issues is whether commercial reusable space-flight vehicles should require both mandatory escape capabilities and a destruct capability that could operate from the range safety facility. These questions require more study by the various aviation and space safety agencies around the world as the relevant technologies and safety systems mature.

Factors common to both expendable and reusable vehicles are general risk management techniques, independent validation and verification, the perceived need for oversight of launch safety as exercised by US government regulatory procedures, and any state or local oversight requirements that may apply, especially as exercised at the launch facility and environs. These also may be common for the safety inspection and oversight activities exercised by insurers of the launch, and the safety precautions and due diligence exercised by the operators of the commercial launch. The inspection by the US government, and due diligence with regard to meeting international requirements to minimize orbital debris, would also be the same. Another area requiring further study is the degree to which there should be oversight of launch training facilities against clearly established standards.

Despite these similarities and common risk management techniques, it is believed that separate procedures need to apply to reusable launch vehicles, especially those with humans aboard. Nevertheless, the presence of humans is only one factor. Other factors and considerations include: orbital flight versus sub-orbital flight, lower *g* forces (i.e. 3–6 *g*s for short duration) versus much higher *g* forces of longer duration, different environmental conditions with regard to radiation exposure, orbital debris exposure, much different temperature differentials during re-entry and landing operations, different abort and escape options, etc. The lack of commonality in these various safety factors at this stage suggests that personal commercial space flight is something that needs to continue to be regulated on a case-by-case basis.

A case-by-case approach does not imply a lack of stringent safety regulation. Nevertheless, during the experimental period, elements of risk must be considered part of the process. Historically, human space flight has resulted in about a 4% fatality rate for those that have flown (factoring in the multiple missions of some astronauts) and approximately a 1% probability of fatality per flight [22]. Clearly the commercial space transportation industry needs to aim for much higher safety standards and performance. "Space tourists" who pay large sums of money for a "space adventure" will ultimately want a much higher likelihood of their safe return than just 1%.

References

[1] Murray, D. P. (FAA/AST) and VanSuetendael, R. (FAA/ATO-P) (2006). A Tool for Integrating Commercial Space Operations Into The National Airspace System. AIAA Paper 2006-6378, August.

[2] Title 49 USC, Section 70101(a).

[3] FAA Joint Planning and Development Office (JPDO), "Concept of Operations for the Next Generation Air Transportation System (NextGen)."

[4] FAA/AST. "2008 US Commercial Space Transportation Developments and Concepts: Vehicles, Technologies, and Spaceports."

[5] FAA/AST (2005). "Suborbital Reusable Launch Vehicles and Emerging Markets," February.

[6] FAA/AST, Space and Traffic Management System, "Addendum 1: Operational Description to the Concept of Operations for Commercial Space Transportation in the National Air Space System, Narrative, Version 2.0."

[7] *Ibid.*

[8] *Ibid.*

[9] National Research Council, Aeronautics and Space Engineering Board, Committee on Space Launch Range Safety (2000). *Streamlining Space Launch Range Safety.* National Academies Press.

[10] Future Interagency Range and Spaceport Technologies, Interagency Working Group of the FAA, DoD, and NASA, "Space Vehicle Operators Concept of Operations: A Vision To Transform Ground and Launch Operations."

[11] Murray, D. P. (FAA/AST) and Ellis, R. E. (FAA Air Traffic Organization, Fort Worth Air Route Traffic Control Center) (2007). Air Traffic Considerations for Future Spaceports. *2nd International Association for the Advancement of Space Safety (IAASS) Conference*, Chicago, 14–16 May.

[12] Leung, J. S., Fay, G. L. II, Patrick, T. A., Osburn, S. L. and Seibold, R. W. (2006). Space-Based Navigation for RLVs and ELVs. Final Report, US Department of Transportation, Contract DTRS57-99-D-00062, Task 16, Aerospace Corporation Technical Report No. ATR-2006(5200)-1, 8 February.

[13] *Ibid.*

[14] Underwood, B., Kremer, S. and Woodhams, W. (2004). NASA's Wallops Flight Facility Rapid Responsive Range Operations Initiative. *AIAA 2nd Responsive Space Conference*, April.

[15] *Ibid.*

[16] *Ibid.*

[17] *Ibid.*

[18] *Ibid.*

[19] Pelton, J., Logsdon, J., Smith, D., MacDoran, P. and Caughran, P. (2005). *Space Safety: Vulnerabilities and Risk Reduction in US Space Flight Programs.* Washington, DC.

[20] *Ibid.*

[21] *Ibid.*

[22] Newman, S. Vice President, ARES Corporation, and former NASA Employee in the Office of Safety and Mission Assurance (2007). Briefing at *Reach to Space Conference*, George Washington University, 12 November.

CHAPTER

Commercial systems due diligence in the application of standards and procedures designed to avoid the creation of orbital debris in GEO

10

Dean Hope
Inmarsat Flight Dynamics

CHAPTER OUTLINE

INTRODUCTION

Geostationary communications operate by relaying radio signals sent from the surface of the Earth to a satellite seemingly fixed above one meridian of longitude, and then re-transmitting these signals to users on the ground that are either at a fixed location or who are continually changing their locations, the so-called mobile user.

By positioning satellites at three or more widely spaced longitudes around the globe, it becomes possible to create a network of satellites that can provide communications coverage to most of the Earth's surface. At every stage in the life of a geostationary Earth orbit (GEO) satellite, there are many factors taken into account to minimize the possibility of the satellite breaking up and adding to the ever growing population of debris orbiting the Earth.

Satellite manufacturers initiate the process of protecting the space environment by implementing safe satellite designs. For example, propulsion systems are designed not to allow propellants to combine in such a way as to cause an explosion that would fragment the satellite structure into a myriad of small pieces of debris. Equally, the measurement of propellant mass remaining on board ought to be of sufficient accuracy that operators can safely determine when to raise a satellite above GEO altitude at the end of life to avoid it running out of fuel and drifting through the geostationary arc as one large piece of debris.

Launch vehicle providers and mission planners play their part by designing safe mission profiles for the launch and early orbit phases (LEOP) of their customers' satellites as they pass through many

Space Safety Regulations and Standards. DOI: 10.1016/B978-1-85617-752-8.10010-8

intermediate transfer orbits avoiding other operational satellites and debris objects in order to reach their final destinations.

Having arrived on-station, geostationary satellites, by definition, are then required to maintain position within a specified latitude–longitude box over their lifetime. This lifetime is primarily determined by the amount of propellant remaining on board to execute the maneuvers necessary to remain on-station, although certain hardware failures, such as in the payload, can sometimes shorten useful life as well.

At some point, when fuel is low, the GEO operator must apply the remaining fuel to raising the orbit to an altitude higher than geostationary, i.e. higher than 35,786 km, so as not to pose a collision risk to other geostationary satellites had the satellite remained in GEO due to unexpected fuel depletion.

The origins and development of the standards and procedures observed by Inmarsat and others over the years is instructive in itself. As more and more satellites were launched, it became clear that managing these fleets would require the creation of, or better monitoring by, national and international regulatory bodies to help manage both the diminishing resources of frequency spectrum and physical space along the geostationary arc.

10.1 ORIGINS OF GEO COORDINATION

In the early days of satellite communications, the main coordination concern was the allocation of radio frequency spectrum given that transmissions from space could transgress international boundaries. The primary regulatory body for assigning radio frequencies used in space and elsewhere was, and still is, the International Telecommunications Union (ITU). The ITU was originally formed to manage changes made to the first International Telegraph Convention signed in Paris way back in 1865.

With the formation of the United Nations in 1945, the ITU became one of its specialized agencies and in 1948 moved its headquarters to Geneva, where it has been based ever since [1]. After the launch of the world's first successful geosynchronous satellite *Syncom-2*, in 1963, the ITU deemed it appropriate to hold an Extraordinary Administrative Radio Conference to address space communications and put in place regulations that would govern the radio frequency spectrum used by all future space stations (satellites) [2].

A significant consequence of this and subsequent World Administrative Radio Conferences was the introduction of the concept of orbital slots for geostationary satellites, which effectively defined the latitude and longitude station-keeping box limits that modern satellite operators use today [3].

Intelsat was the first international telecommunications satellite consortium and launched its first satellite, *Early Bird*, in April 1965 [4]. It was some 14 years later that the world's second global international satellite network, Inmarsat, was created. As with Intelsat, its name reflected its origins, in this case supporting communications at sea, hence the International Maritime Satellite organization.

The fact that Intelsat had been launching and operating GEO satellites for such a long time before Inmarsat means that inevitably it has pioneered many of today's best practices and procedures for the operation of GEO-based satellites, many of which Inmarsat and others have adopted. Two prime examples of these are, firstly, the implementation of close proximity or co-location agreements where satellites belonging to two organizations share adjacent station-keeping longitudes that are close together and, secondly, the notification to other operators of times when a satellite is planned to relocate to a new operational longitude.

10.2 PRE-LAUNCH COORDINATION ACTIVITIES

Before any communications satellite is launched, the radio frequencies to be used for transmitting and receiving signals must be agreed and coordinated through the ITU. The nature of the traffic to be passed through the satellite must also be fully characterized and assessed so it is compatible with the frequency spectrum made available and the onboard radio frequency (RF) payload designed accordingly.

The process of RF coordination for a satellite-based communications system can take several years due to the detailed negotiations required between the many interested parties and so requires the assistance of a dedicated Frequency Coordination group within the applicant communications organization or company. Ultimately, the aim of such coordination is to ensure that the signals passing to and from the satellite do not cause damaging interference to other communications systems, which could affect the safety of those operations.

Launching a satellite into GEO can also take many years to prepare. Apart from the predominant cost factor, the size and mass of the satellite largely define the range of vehicles from which to choose for launch. A detailed mission analysis is required to determine the most fuel efficient orbit sequence to launch on specific days of the year in order to reach the desired target geostationary orbit. There are many constraints taken into consideration in the mission analysis, ranging from satellite apogee or perigee motor firing attitude limitations due to Sun and Earth sensor fields of view, signal strength limits at the vast apogee distances, visibility of the satellite in transfer orbits as seen from the global tracking network stations, and many other factors. The end result is what is known as a launch window for each possible launch day. The launch window represents that period of time, or times, during which the launched satellite is able to achieve its final target orbit with an acceptable on-station lifetime after executing a series of intermediate transfer orbits during the LEOP phase.

Having developed both a mature mission analysis and a mission events timeline, it is then time to generate an RF interference prediction for those periods when the satellite being launched will pass close to other geostationary satellites whose operating frequencies overlap the telemetry and telecommand (TTC) spectra allocated to the new satellite. The Frequency Coordination group of the new satellite's operator company contacts all potentially affected operators as a common courtesy—an unwritten convention—in the weeks approaching the launch date and warns each operator of the times between which the satellite will be within $\pm1°$ of their satellite's on-station longitude.

The amount of Earth orbiting debris objects has grown steadily since the early 1960s and so the possibility of a satellite colliding with debris at the time of its launch has also increased. Many of these objects are regularly tracked by the US Space Surveillance Network (SSN) and are cataloged. Their orbit elements can be checked against the orbit elements expected for the satellite after launch and proximity assessments are made. If the estimated object versus satellite minimum separation distance is found to be below a given warning threshold, it is then possible to adjust the launch window opening or closing times or apply an intermediate window cut-out to avoid that particular conjunction.

Recent Inmarsat policy has been to contact USSTRATCOM using their Form-1 process to request a collision avoidance (COLA) analysis for the separated satellite to be performed in the days leading up to launch. Because of the dynamic nature of the forces affecting an orbiting body, a more meaningful COLA analysis can be achieved using the best orbital position and velocity data predictions available for both bodies.

Hence, the closer in time that the observational data for the debris population are to the predicted satellite transfer orbit parameters at launch, the higher the accuracy of the conjunction assessment. Typically, the first COLA request is submitted 48 hours before launch and another is submitted some 6 hours before launch, although the timing of the latter is somewhat dependent on the number of potential collision candidates appearing in the first set of results. Launch providers generally request a COLA analysis from USSTRATCOM at the same time to assess the collision risk for the launch vehicle itself, since the final rocket stage enters a similar transfer orbit to the separated satellite payload with an apogee approaching GEO altitude. Both types of COLA analyses can result in a last minute change to the nominal lift-off time.

In the months prior to launch, it is customary to submit specific satellite and mission details to a launch-licensing authority. This is usually a national quasi-government body charged with the legal responsibility for ensuring that the company whose satellite is being launched is compliant with all the regulations and policies that the Launching State has committed itself to through international treaties and obligations. In the case of Inmarsat, for example, the launch-licensing authority in the UK is the British National Space Centre (BNSC), which was established in 1985 [5]. In the USA, this role is performed by the Federal Communications Commission (FCC).

10.3 LEOP AND ON-STATION COORDINATION

During the launch and early orbit phase of a satellite mission, the main coordination activity involves refining the pre-launch warning times issued to other GEO satellite operators as to when possible RF interference may occur. This is because the actual intermediate transfer orbits achieved after each firing of the liquid apogee engine (LAE) to raise perigee may not be exactly as predicted due to slight variations in the engine thrust performance from burn to burn, or differences in the satellite pointing attitude while firing.

In the drift orbit phase, which occurs between the time of the last apogee engine firing and reaching the in-orbit test (IOT) longitude, it is possible that when the longitude drift rate is low compared to being geostationary, some operators whose satellites lie between the final apogee burn longitude and the IOT longitude will request an exchange of orbital elements for each satellite in order to independently calculate the minimum separation distance. This is a good example of commercial operator best practice—mutual due diligence in ensuring safe passage of each other's prime assets. Similarly, if the IOT location is close to another operational GEO satellite, an exchange of orbit elements is made to maximize the separation between the two satellites until the testing phase is completed.

Once the satellite arrives at its geostationary operating longitude, an ongoing regular sequence of station-keeping maneuvers must be performed in order to maintain the satellite's position within its assigned latitude and longitude box. For fixed service satellite systems, the station-keeping box is usually $\pm 0.05°$ in both the north–south (NS) direction and east–west (EW) direction; however, in the case of Inmarsat, which provides service to mobile users, the station-keeping box is $\pm 0.1°$ in the EW direction.

At several locations along the geostationary arc, Inmarsat and Intelsat satellites share operating longitudes. The satellites are not co-located at precisely the same longitude, but are positioned as close as possible within agreed limits. These agreements take the form of legally binding documents signed by each party. The separation profile is typically to have an Intelsat satellite operating $\pm 0.05°$ in the EW direction about one sub-satellite longitude, and then a $0.05°$ longitude guard band followed by an Inmarsat satellite operating within a $\pm 0.1°$ EW box about a second sub-satellite longitude. This separation profile takes account of the fact that Inmarsat satellites operate using mission-specific inclined orbit strategies whereby the inclination can be anywhere between $0°$ and $5°$. For cases of geostationary satellites operating with inclinations higher than $5°$, the ITU station-keeping box requirements relax to ensuring that only the longitude of the orbit node crossings remain within $\pm 0.1°$ [6]. This is typical of situations where satellites have reached the end of their nominal mission design life and yet still have more than the planned amount of fuel remaining. This may happen through a combination of fortuitous launch vehicle performance, LAE engine performance, minimal LAE attitude pointing errors, or prudent Flight Dynamics (FD) management of the station-keeping fuel over life.

If predicted distortion of the downlink beam footprint caused by increased ground track motion due to higher orbit inclination is deemed acceptable to users, it becomes possible to suspend NS station-keeping and so prolong the lifetime of a GEO satellite. Allowing the inclination to continue to grow conserves onboard fuel by only having to maintain the EW orbital motion of the satellite inside the station-keeping box.

Such a situation does, however, increase the need for even closer coordination with any immediate geostationary neighbors whose satellites operate with similar high-inclination orbits in order to avoid any possibility of collision. The reason for this is that the geometric EW longitude deviation from the sub-satellite longitude grows as the square of the orbit inclination, so for inclinations above approximately $4.7°$, it is not physically possible to contain the daily EW longitude motion within $\pm 0.1°$. This notwithstanding, the daily maximum longitude excursion points only occur close to the maximum and minimum latitude regions high above and below the equatorial plane due to the well-known "figure-eight" effect [7].

During the on-station life of a GEO satellite, it is possible that an operator may elect to relocate the satellite to a different geostationary longitude. In the weeks prior to the expected move date, the Frequency Coordination group contacts all the other GEO operators between the drift start and drift stop longitudes whose satellites have similar payload communication frequencies or TTC frequencies to the satellite being moved. Contact was once typically by faxed letters; however, in recent years, this has been superseded by the sending of emails via the internet.

The nature of advance details provided to other operators is another example of commercial best practice that has emerged over many years but has not yet been standardized. Typical radio frequency interference (RFI) coordination information includes the names of responsible staff and contact phone numbers usually based at the satellite control center (SCC), the names of the primary tracking stations being used for the move and their location coordinates, telemetry beacon transmitter frequency, polarization and effective isotropically radiated power (EIRP), the modulation scheme, tele-command receive frequency and polarization, ranging frequency and the payload status (normally disabled).

Inmarsat, Intelsat, and others also provide certain orbit-related information such as entry and exit time predictions for when the relocating satellite is expected to be within $\pm 1°$ of the box center longitude of any satellites being passed and which have similar operating frequencies. The drift rate relative to geostationary is included since this gives a coarse indication of physical separation assuming a near circular drift orbit. A drift rate of 1° east/day equates to 78 km below geostationary altitude and, conversely, 1° west/day equates to 78 km above the GEO arc for a circular orbit. For relocating satellites that pass by with much lower drift rates, more detailed orbital element information can be made available for close approach monitoring.

Once the relocating satellite starts drifting, real-time RFI coordination is handled directly through the SCC of the operator moving the satellite. Intermediate operators who may be affected by the move will have their particular SCC contact the relocating satellite operator's SCC if any RF interference appears at times that were not predicted.

The monitoring of one satellite passing by another satellite represents an instance of a much larger concern in the twenty-first century, space situational awareness (SSA). While commercial GEO satellite operators have managed their space traffic on the busy GEO highway for many years, the growth in satellite population and the number of orbiting debris objects poses a commensurate increase in the risk of collision and could rapidly escalate without improved awareness and avoidance of these objects.

Inmarsat has provided industry inputs to several SSA-related bodies, both national and international. In the UK, Inmarsat regularly attends meetings of the ACE68/Panel 9 Space Debris Committee held under the auspices of the British Standards Institute (BSI). This committee reviews the UK contributions to the development of standards to be issued by the International Standards Organization (ISO), which apply to satellite manufacturers and operators and aim to minimize the creation of orbital debris.

There is a single top-level standard called the Space Systems—Space Debris Mitigation standard, which calls out lower-level standards addressing topics such as the Disposal of Satellites Operating at GEO, the Process for Orbital Information Exchange, Estimating the Mass Remaining of Usable Propellant, and the End-of-Life Passivation of Unmanned Spacecraft, although the latter is still in an early draft stage.

Of particular interest to all GEO operators is the Process for Orbital Information Exchange. Most satellite operator Flight Dynamics (FD) groups recognize that in order to accurately assess proximity between two objects in space, it is important to make the necessary calculations in a common orbital reference frame. To that end, there is a growing consensus among operators to make their orbital elements and ephemerides available in a standard format that clearly identifies the particular reference frame used by that operator and thereby allows the correct transformations into a common frame to be made.

The origins of this particular ISO standard stem from an existing mature draft recommendation from the Consultative Committee for Space Data Systems (CCSDS). It defines specific formats for the two main types of orbital information exchanged between operators—namely, the Orbit Parameter Message (OPM) and the Orbit Ephemeris Message (OEM) [8]. The ISO is expected to publish a formal version of their current draft incorporating the CCSDS recommendation by the end of 2009.

The US government has been actively looking into setting up a satellite coordination body through its Department of Defense (DoD) space surveillance support to the Commercial and Foreign Entities (CFE) program. There are two key benefits of the US government promoting satellite coordination and surveillance: first, that the population of objects being tracked and monitored for conjunctions becomes much larger since it also includes military satellites as well as commercial and civil satellites;

second, the CFE program could also include higher accuracy special perturbation (SP) two-line-element set (TLE) tracking data for the population of cataloged uncontrolled debris objects for conjunction analysis.

As of mid-2009, the CFE program had expanded its capability to be able to monitor around 800 actively controlled satellites in the GEO region of space and send any close approach warnings to the SCC of affected GEO operators from its Joint Space Operations Center (JSpOC) located at Vandenberg Air Force Base, California.

In January 2008, a meeting of FD representatives from five commercial GEO operators, Intelsat, Inmarsat, SES-Astra, Telesat, and Echostar, was hosted by Intelsat in Washington, DC. At that meeting, it was agreed that in order to voluntarily improve SSA for the commercial operators, they would make their high-accuracy orbital data available for conjunction analysis on a proof-of-concept basis to a trusted third party, the Center for Space Standards and Innovation (CSSI) [9]. The CSSI offered to provide a hardware platform and its SOCRATES-GEO service to perform the required SSA conjunction analyses and issue email warnings of possible collisions to the group of participating operators.

The prototype system has proven invaluable in improving knowledge of the presence of controlled satellites passing close to other operational GEO satellites; however, in the case of uncontrolled debris objects, conjunction assessments have been hampered by having to rely on the lower accuracy general perturbation (GP) two-line-element (TLE) sets made publically available by the US government's Space Surveillance Network (SSN).

By including optical observation data for offending objects from the International Scientific Optical Network (ISON) at times when coarse-level close approaches are detected, proximity recalculations can be made against the higher accuracy ISON orbital data to refine the predictions and so improve confidence in the warnings. At the time of writing, there are now more than 150 GEO satellites actively involved with the prototype system and the process of creating a permanent Satellite Data Center working with the commercial GEO operators is well under way.

In the European SSA theater, in 2006 Inmarsat volunteered its services as a commercial GEO operator to participate in the Expert Users Representative Group which was convened by the European Space Agency (ESA) as a first step in defining the user requirements for the proposed Proof of Concept for Enabling Technologies for Space Surveillance Project. In November 2008, the ESA Council at ministerial level approved phase 1 of the project, which is expected to take 3 years to identify all of the existing space surveillance and monitoring resources present among the 11 participating Member States and essentially create a prototype system of federated national assets that can be used to test architectures for the monitoring of orbital debris and issuing warnings of potential collisions. If approved, a second phase would follow with the aim of building and installing any missing capabilities and so eventually create a fully operational European SSA monitoring system able to meet the needs of all its users [10].

A satellite that has survived the stresses of a thunderous launch and then successfully threaded its way through a cloud of natural and man-made objects at geostationary altitude for its entire lifetime eventually arrives at the day when its operator must decide if the time is right to finally decommission the satellite. This decision has always been difficult to make when operating with a commercial imperative since a delicate balance must be struck between the desire to maintain a revenue stream and the need to ensure that the satellite can at least achieve the mandatory minimum altitude above GEO stipulated in the launch license agreement for that satellite.

10.4 END-OF-LIFE DECOMMISSIONING

The operational lifetime of a satellite is determined by the maximum number of maneuvers that it can perform in order to maintain position within its assigned station-keeping box. Maneuvers are small changes made to the satellite orbital velocity (delta-V) that are applied at precise times and in specific directions by the firing of onboard thrusters. Most commercial GEO satellites use a bi-propellant propulsion system that mixes fuel with oxidant in a thruster to efficiently generate the required delta-V.

The FD group of the satellite operations department carefully monitors the amount of propellant used by each maneuver and subtracts this mass from the known amounts loaded prior to launch. This is known as the bookkeeping or dead-reckoning method. There are other methods, but the bookkeeping method is the simplest and therefore the most commonly used. Unfortunately, it is also a method that carries a high degree of uncertainty in the knowledge of how much propellant is left in the tanks.

The largest contributor to this uncertainty concerns how much propellant is consumed when the large LAE thruster fires to raise the transfer and intermediate orbits to geostationary altitude during the LEOP phase. Each LAE engine has a performance dispersion, and therefore fuel consumption dispersion, of around 0.1%, which is never quite the same from one engine to the next. Given that typically 90% of the loaded propellant is burned within the first few days of LEOP to reach the final geostationary orbit, it is clear that the fraction of fuel consumed by dispersion becomes significant when compared to the fuel remaining at the beginning-of-life (BOL) phase.

Improving the methods for estimating remaining onboard propellant would be a major step forward in reducing the risk of premature fuel depletion of satellites. This is one reason why Inmarsat has been a keen industry participant in the development of ISO standards, especially those specific to establishing requirements for the design of satellite propellant measurement methods and onboard hardware such as the standard for Disposal of Satellites Operating at GEO and the standard for Estimating the Mass Remaining of Usable Propellant.

As part of the launch-licensing process, GEO satellite operators must make clear how they intend to comply with requirements to raise the orbit of their satellites above the geostationary altitude of 35,786 km and make the satellite safe once it has reached the end of its useful life. It has long been recognized by Intelsat, Inmarsat, and other GEO operators that having an uncontrolled satellite stranded in GEO for any reason, including fuel starvation, represents a major collision risk to other satellites as it drifts away from its assigned longitude.

Early generations of Inmarsat satellites had a 7 m/s delta-V allocation in their propellant budgets to target an end-of-life (EOL) orbit raising altitude of 200 km above GEO; however, subsequent recommendations defined by the Inter-Agency Space Debris Coordination Committee (IADC) typically increased the preferred altitude to greater than 300 km for GEO satellites, the specific altitude being dependent on individual satellite geometry [11]. It should be recognized that the IADC, a body representing 11 of the world's flagship space agencies founded in 1993, has been a leading light in defining ways to minimize the growth of space debris. In February 1999, the ESA issued satellite disposal guidelines as part of their ESA Space Debris Mitigation Handbook and then, in June 2004, the ESA was again instrumental in the preparation of the European Code of Conduct for Space Debris Mitigation. Both of these documents are consistent with, and build upon, the IADC recommendations.

For satellites operating over the USA, expanded FCC regulations were released in June 2004 [12], and further expanded to cover disclosure of orbital debris mitigation plans by license applicants in a Public Notice issued in October 2005 [13]. The scope of these new regulations includes the disposal of GEO satellites and, again, in general, they endorse the recommendations of the IADC.

Having successfully raised the perigee of the final orbit to the required altitude above GEO, the satellite operator has one final task to perform and that is to make the satellite safe. The overall aim is to remove, or at least reduce where possible, all remaining energy sources present on the satellite. This entails activities such as spinning down any momentum wheels, discharging batteries, and venting any pressure vessels that may still be holding propellants or pressurant gases. These, and related actions, are outlined in the pending ISO standard, Disposal of Satellites Operating at GEO.

CONCLUSIONS

The volume of space bounded by the radius of the Earth's atmosphere all the way up to geostationary altitude has become an increasingly valuable natural resource. Already there are large numbers of discarded rocket bodies, satellites that have failed to reach their intended orbits, and other pieces of man-made debris, as well as a large population of natural objects such as meteoroids, orbiting and passing through this volume.

As we continue to launch more and more satellites, this population will inevitably increase and so will the risk of collisions, which can only breed more debris. Inmarsat and other GEO operators recognize that they have an obligation to help preserve the viability of the space environment as a medium for vital activities such as global communications and Earth resource monitoring.

In the years since the launch of *Early Bird* in 1965, the world's leading GEO satellite communications providers have actively coordinated operations of their satellite fleets, beginning with a combination of best practices and mutual coordination agreements and later becoming subject to a more formally regulated environment with the introduction of international standards, guidelines, and legal frameworks such as licensing by the Launching State.

With the latest initiatives of the US government's CFE program, the commercial GEO operators' Space Data Association and its Satellite Data Center, and, in the longer term, the European SSA monitoring system, the risk of debris creation from collisions with controlled satellites should be greatly minimized. However, this being said, for these systems to produce more meaningful and reliable conjunction assessments in the future, there is still a need for work to be done on improving the overall accuracy of orbital data currently available for all the different classes of tracked objects.

References

[1] ITU main website (history), http://www.itu.int/net/about/history.aspx
[2] Martin, J. *Communications Satellite Systems*, Chapter 6, p. 132.
[3] International Telecommunications Union (ITU) Radio Regulations, Article 22, Sect. II and III.
[4] Intelsat main website (history), http://www.intelsat.com/about-us/history/intelsat-1960s.asp
[5] BNSC main website (history), http://www.bnsc.gov.uk/About%20BNSC/How%20we%20work/8003.aspx
[6] International Telecommunications Union (ITU) Radio Regulations, Article 22, Sect. III (27).

[7] Soop, E. M. (1994). *Handbook of Geostationary Orbits*, p. 37. ESA.
[8] CCSDS recommendations, http://public.ccsds.org/review/default.aspx
[9] CSSI main website, http://www.centerforspace.com/
[10] ESA SSA website, http://www.esa.int/SPECIALS/Operations/SEMFSG6EJLF_0_iv.html
[11] IADC Space Debris Mitigation Guidelines IADC-02-01, Section 5.3.1, p. 5.
[12] FCC Second Report and Order (2004). FCC 04-130, IB Docket No. 02-54, June.
[13] FCC (2005). Public Notice, DA 05-2698, Report No. SPB-112, October.

CHAPTER

11 Data sharing to improve close approach monitoring and safety of flight

Joseph Chan*, Richard DalBello†

** Intelsat Global Services, † Intelsat General*

CHAPTER OUTLINE

11.1 COLLISION MONITORING TODAY

In 1999, Intelsat contracted with the Aerospace Corporation via the Space Operation Support Office (SOPSO) to conduct close approach monitoring. The Aerospace Corporation developed a fully automated two-tier program that determined satellite close approaches based on miss-distances and conjunction probabilities. The initial detection was based on the publicly available "two-line-element" sets (TLE). Once a potential conjunction was identified, Aerospace would request the more accurate "special perturbation" (SP) ephemeris data from the Air Force to confirm the conjunction. The Aerospace Corporation shut down the SOPSO office abruptly in November 2002. In March 2003, Intelsat contracted MIT Lincoln Lab to perform close approach analysis. It was a semi-automated system and the conjunction detection was based on miss-distances only. Because MIT had a contractual relationship with the Air Force, and therefore had direct access to the observations from the deep space surveillance network, the conjunction monitoring was based on a single-tier process. However, the monitoring was restricted to non-active or "passive" space objects. This restriction was due to the difficulties in detecting past maneuvers and predicting future maneuvers of active satellites and thus invalidated longer-term close approach predictions.

Since January 2007, Intelsat has relied on an in-house close approaches monitoring system. This system follows the two-tier model and relies on the US Joint Space Operations Center (JSpOC) to validate the potential conjunctions detected via the TLE. We routinely screen our satellites using the TLE data and, during special activities such as satellite relocations and transfer orbit missions, we also exchange data with other satellite operators whose satellites are operating in an adjacent box next to

Space Safety Regulations and Standards. DOI: 10.1016/B978-1-85617-752-8.10011-X

our satellites. Participants in the Commercial and Foreign Entities cooperative program can enroll in this program but the duration is limited to 1 year and then participants must re-enroll. The exchanged data usually consist of the latest orbital information, near-term maneuver plans, frequency information, and contact information for further discussion [1].

There are drawbacks in the current close approach monitoring process. In addition to lack of standards of TLE propagators, TLE data do not have the required accuracy for credible collision detection. An operator that is forced to rely on TLE data must increase the calculated collision margin to avoid potential close approaches. In most cases, threats identified using the basic TLE data are downgraded after coordination with other operators or further evaluation with more precise orbital data. In addition to the inaccuracies of the TLE data, this data also lacks reliable maneuver information. This limits the usefulness of the TLE for longer-term predictions, since maneuver information is necessary to properly predict the ephemeris for active satellites. The lack of this data becomes increasingly problematic as more satellites employ ionic propulsion systems and are, essentially, constantly maneuvering.

Because of the relatively imprecise nature of the TLE data, the US Air Force established the "Interim Commercial and Foreign Entities (CFE) Data/Analysis Redistribution Approval Process" (commonly referred to as the Form 1 Process) for granting operators access to information that goes beyond the basic TLEs. Through the Form 1 Process, operators can request additional information (the special perturbation, or SP, data) on specific "close approach" situations. Although helpful, it is cumbersome to rely on the Form 1 Process as an operational tool because it requires advance notice, which is often impossible in emergency situations. In addition, conjunction events often require close cooperation and interactive communication. Today, the Form 1 Process relies primarily on email as a method of communication and the US government does not guarantee the rapid turnaround necessary in most cases.

The US government is currently reviewing its policies on the distribution of TLE data. One proposal would require the negotiation of an individual "tailored agreement" between each satellite operator requesting information and the US government. Other proposals have suggested that the US government might be willing to provide additional conjunction assessment services on a reimbursable basis. At this writing, it is unclear how or whether the Commercial and Foreign Entities (CFE) cooperative program, which is scheduled to terminate in 2009, will continue [2].

There is no single approved way to represent the position of an object in space. As a result, operators generally use different presentation of their orbital position depending on the software they use for flight operations. In addition, there is no one agreed protocol for sharing information and coordinating operators must be prepared to accommodate the practices of other operators. To do this, operators must maintain redundant file transfer protocols and tools to convert and reformat information so that it can be input to the owners/operators software systems for computing close approaches.

Separate tools are necessary to exchange data with each operator. Some operators write their own software tools for monitoring and predicting the close approach of other spacecraft, while others contract with third parties for this service. The magnitude of the effort to maintain "space situational awareness" grows quickly as the number of coordinating operators increases. Unfortunately many operators are not able or willing to participate in close approach monitoring due to lack of resources or capabilities.

11.2 PROPOSAL TO CREATE A "DATA CENTER" FOR SPACE SITUATIONAL AWARENESS

In response to the shortcomings of the current TLE-based CFE program and the recognition that better interoperator communication is desirable in and of itself, satellite operators have recently begun a broad dialog on how to best ensure information sharing within the satellite communication industry. One proposal currently being discussed in the international operators' community is the so-called Data Center, which would enhance space situational awareness for commercial operators. As conceptualized, the Data Center would be an interactive repository for commercial satellite orbital, maneuver, and frequency information. Satellite operators would routinely deposit their fleet information into the Data Center and retrieve information from other member operators when necessary. The Data Center would allow operators to augment existing TLE data with precision orbit data and maneuver plans from the operator's fleets. The Data Center would also:

- perform data conversion and reformatting tasks allowing operators to share orbital element and/or ephemeris data in different formats;
- adopt common usage and definition of terminologies;
- develop common operational protocols for handling routine and emergency situations;
- exchange operator personnel contact information and protocol in advance of need.

If the Data Center were to gain acceptance, it could perform additional functions, such as the close approach monitoring tasks currently being conducted by the operators. In this phase, US government-provided TLE data could be augmented by the more precise data available from the operators. This would improve the accuracy of the Center's conjunction monitoring and could provide a standardized way for operators to share information with the US government and other governments. In the early stages, information on non-operational space objects would still be supplemented by TLE data from the Air Force Commercial and Foreign Entities (CFE) program and/or other government programs. US government, or other government, support would still be required when precise information is needed to resolve close approaches and avoidance maneuver planning.

11.3 DATA CENTER PROTOTYPE

A prototype of an active Data Center was established to study the feasibility of such an approach following a workshop of commercial owners/operators held in February 2008 in Washington, DC. Owners and satellite fleet operators, including Intelsat, Inmarsat, Telesat, Echostar, SES, and Eutelsat, participated in this workshop [3]. A majority of the operators present agreed on the need to simplify the data exchange process to minimize risk for safety of flight and on the importance of creating a common Data Center. The operators agreed to work on a prototype Data Center as a proof of concept to improve coordination for conjunction monitoring.

The Center for Space Standards and Innovation (CSSI) agreed to work with the operators to develop such a prototype. Once initiated, the prototype Data Center expanded quickly and today seven operators are participating and regularly contributing data from over 116 satellites. The participating operators receive daily close approach alerts when the miss-distances and conjunction probabilities fall below certain thresholds and a daily neighborhood watch report showing the projected separations of satellites that are flying in the adjacent control box. The participating operators provide their ephemeris data in the reference frames and time systems generated in their flight software, and the Data Center performs the transformation and reformatting to a common frame for close approach analysis.

This approach greatly simplifies the efforts and reduces the burden on individual operators to participate to the Data Center and thus encourage participation. A strict data policy has been put in place to ensure privacy of the data. The Data Center is not allowed to redistribute the data received from the owners/operators without approval from the owners of the data. While there is still significant work left to refine the processing, the initial results from the early development of the "Data Center" are very promising. Future owners/operators workshops will continue to discuss the technical issues that can improve the efficiency of the Data Center's operation as well as how to create a better organizational structure.

CONCLUSIONS

The principal goal of the "Data Center" is to promote safety in space operations by encouraging coordination and communication among commercial operators. The Data Center could also serve as a means to facilitate communication between operators and governments. Details on the implementation of the Data Center, services to be provided, usage policies, structure of the organization, and by-laws have yet to be determined and would require agreements with the member operators at the creation of the organization. The development of a Data Center could provide new visibility and awareness of the geostationary orbit, allow all satellites to be flown in a safer manner, and reduce the likelihood of an accidental international incident in space.

Within the next decade, many more countries will gain the ability to exploit space for commercial, scientific, and governmental purposes. It is essential that the world's governments provide leadership on space management issues today in order to protect the space activities of tomorrow. Bad decisions and short-term thinking will create problems that will last for generations. Wise decisions and the careful nurturing of our precious space resources will ensure that the tremendous benefits from the peaceful use and exploration of outer space are enjoyed by those who follow in our footsteps in the decades to come.

Notes and references

[1] Fact Sheets: US Joint Space Operations Center, www.vandenberg.af.mil/library/factsheets/factsheet.asp?id

[2] US Air Force Space Command—Change in Commercial and Foreign Entity Data Exchange Sparks Review, May 2009, www.stratcom.mil/news/article/62/Change_sparks_review

[3] In the second half of 2009, Intelsat, Inmarsat, and Eutelsat formally created the Space Data Association to share information on a routine basis about commercial satellite orbits in order to minimize the possibility of orbital collisions. The organization is open to all satellite operators.

CHAPTER

12

Space situational awareness and space traffic management standardization

David Finkleman

Center for Space Standards and Innovation, Analytical Graphics, Inc.

CHAPTER OUTLINE

INTRODUCTION

The continuing increase in orbital debris represents a threat to space safety for manned space missions and unmanned commercial and scientific satellites. Despite United Nations voluntary guidelines to diminish space debris, the threat remains. Space situational awareness and space traffic management standardization are key methods to address the hazards of on-orbit collisions.

There can be no collaboration or interoperability unless the operational environment exists for exchanging information. Cooperative maneuver and orbit analysis require that each satellite operator has confidence in the assessments of trusted third parties and fellow operators. Consensus standards can establish the cooperative environment and promote technical understanding and confidence.

12.1 DIVERSITY IN SPACE OPERATIONS

Newtonian two-body or Keplerian orbits that students of science and technology know well are the most simplistic representations of trajectories. They suitably apply only for massive bodies whose separation is large with respect to their sizes. They overlook gravitation of other massive bodies and

Space Safety Regulations and Standards. DOI: 10.1016/B978-1-85617-752-8.10012-1

neglect all non-conservative forces, such as momentum transfer from photons (light pressure), which are present even in sparse interplanetary space.

Considering non-conservative forces, or the fact that massive bodies are not concentrated points, leads to great diversity in trajectory estimates. None of these forces can be described precisely. They are "abstracted" at different levels of detail. Different abstractions of these forces serve different needs, for example the drag abstraction

$$D = 5\rho \bullet V^2 \bullet C_{\mathrm{d}}$$

where D is the non-conservative drag force, ρ is atmospheric density, V is the velocity tangent to the trajectory, and C_{d} is the drag coefficient, valid only where the atmosphere is dense. Much more complex abstractions that consider the properties of the satellite's surface, the interactions of individual gas particles sparsely distributed, and the orientation of the satellite are required at high altitude.

Unless an operator states what drag coefficient and what density were used to estimate an orbit, others using the same observations will arrive at a different estimate. If the operator who provided a trajectory estimate employed a more sophisticated drag abstraction than the recipient, the potential for misunderstanding is greatly amplified.

Abstractions and parameters greatly affect what one thinks the orbit of a satellite is. The most commonly used abstraction of distributed gravitation employs two parameters that signify the bi-angular mass distribution, called "tesserals" and "zonals" or "order" and "degree". Complex gravitation that predicts equatorial gravity wells that cause satellites in geostationary orbit to vibrate requires several high orders and degrees. Unless the operator who provides orbit estimates to others states what order and degree were employed, the recipient may use those estimates incorrectly. Vallado has demonstrated the differences among orbit estimates for the same satellites using different orders and degrees, as well as different abstractions of atmospheric density, sea and land tides, and other forces over a spectrum of prediction intervals [1]. The differences can be hundreds of kilometers over even relatively short propagation intervals. Even using different Earth radius or fundamental gravitational parameter leads to orbit estimate diversity that can be the difference between safe passage and a collision.

The numerous parameters and narrative statements are metadata, essential information about the data one is providing. Lack of metadata has compromised many missions. Launch providers may not have included forces that those accepting satellites on orbit have included. Even the applicable system of units has been confused for lack of metadata. It is not sufficient for operators to exchange only numbers. No matter how precise the data, the numbers cannot be used properly without accompanying metadata.

Single operators who deal only with their own satellite need orbit data consistent with their ability to control their satellite. They can use any set of abstractions, parameters, and numerical techniques that meet this need. All will be consistent and appropriate. When two organizations interact, they negotiate Interface Control Documents (ICDs) that include all parameters, abstractions, and interface descriptions required for the collaboration. These are extremely detailed documents developed with extreme diligence in due course.

Space operations now engage many different operators that have not developed ICDs for collaboration and who employ diverse, often unstated, parameters and abstractions. There must be consensus on the elements of data and accompanying metadata if we are to collaborate to avoid collisions. Such consensus is also guidance for developing ICDs. Standards are the mechanism for achieving consensus.

12.2 CONSENSUS IN SPACE SITUATIONAL AWARENESS

Space situational awareness is an abstraction. Many organizations have conceived definitions to suit their own needs, but there is no consensus on what it is. Awareness is not omniscience. We need not and cannot be aware of everything of interest, everywhere, all of the time. Space situational awareness requires only being aware of significant events in time to act, if necessary.

12.2.1 Natural phenomena

Natural phenomena are the common ground of significant events. Solar phenomenology is most important. Particulate and electromagnetic radiation during solar events lead to forces that affect interplanetary trajectories and interact with sensitive electronics and navigation systems. Interactions with the Earth's gaseous and electromagnetic environments lead to dissipation of orbital energy, changes in satellite attitude, and degraded communications. This common interest is sometimes called "space weather", although that terminology may mislead the less informed to draw a poor analogy with hurricanes, droughts, and thunderstorms.

Guidelines and recommendations of the research coordination elements of the United Nations Committee on the Peaceful Uses of Outer Space (UNCOPUOUS) have been vetted as International Standards [2]. The COPUOUS Committee for Space and Atmospheric Research (COSPAR) issues recommended solar indices for atmospheric and thermospheric density. These are reflected in a series of international standards in process. The most recent publication deals with the Earth's magnetic fields [3], an uncertain discipline with extremes of detailed theory and brute-force empiricism. Correlations of the characteristics of solar radiation in specific bands are widely used as "proxies" for variations in atmospheric density.

It is extremely important for space situational awareness that one clearly reveals which proxies and which correlations one employs to estimate satellite trajectories. This is a conundrum for estimating a satellite's lifetime. There is not enough historical data to predict well the extreme variations in solar activity through solar cycles. In order to enforce low Earth orbit lifetime guidelines, there must be consensus approaches to representing the near-Earth atmosphere over decades.

This has been accomplished by the ISO 27852 Orbit Lifetime Standard [4]. ISO member bodies worldwide have agreed on approaches to abstracting satellite attitude, orbital parameters, and atmospheric characteristics over decades. This leads to a uniformly accepted graphic that clearly delineates the regimes of satellite and atmospheric parameters within which orbit lifetime restrictions are satisfied. Standardized analysis techniques are delineated outside of these clearly satisfactory regions. Figure 12.1 shows that standard chart.

12.2.2 Astrodynamic data

Knowing the state of motion of objects in orbit, and estimating where those objects will be in the future, are central elements of space situational awareness. Traffic management demands that

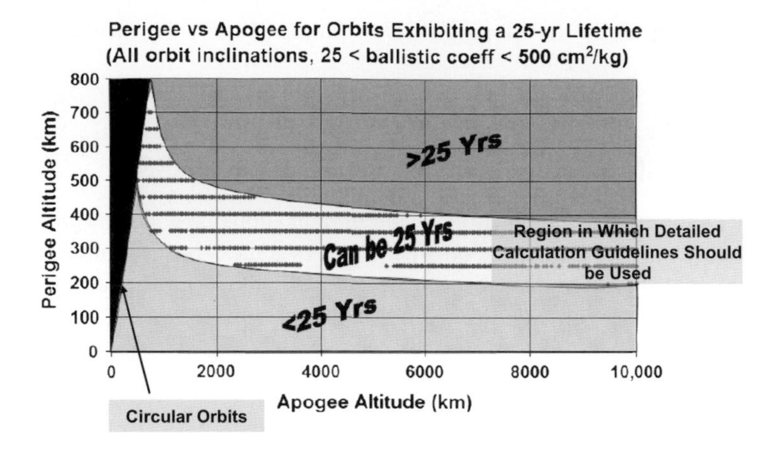

FIGURE 12.1

Standard orbit lifetime estimation guidance.

information about satellite trajectories and characteristics be communicated with sufficiently precise data and completely descriptive metadata. This has become possible only recently, and the capability is not institutionalized worldwide.

There are two distinct data types that must be exchanged among satellite operators: observational data and orbit data. Metric observations of objects in orbit are the foundation for estimating the current and future states of motion. We employ the word "estimate" consistently because there are uncertainties in every measurement (measurement noise) and physical hypotheses are always incomplete (process noise). These convolve to make the satellite states and trajectories quantifiably uncertain. We can never know precisely where satellites are or where they will be.

12.2.2.1 Observational data

There are irreducible uncertainties in the measurement of each of the observables. The uncertainties are described by probability distributions that reflect the variation in the quantity measured over the volume of measurements that are made (probability density). There is an expected value averaged over the entire probability density, and there are moments of the probability density, the lowest order, the mean value, and the variance between the two.

Each measurement affects all other measurements. For example, when we move a telescope in azimuth there are inevitable changes in elevation. The variations of the value of one observable with respect to changes in another are called covariances. All of these uncertainties form a covariance matrix whose diagonal contains all variances and whose off-diagonal terms are called covariances. For the mathematically inclined, this matrix must always be positive, semi-definite.

Observation variables depend on the nature of the observation. Telescopes measure angles relative to a fixed reference direction at a given location. Radars measure range and angles relative to corresponding references. Pulsed radars measure range rate and angles. A number of observations are required to infer the trajectory of a satellite. It is very important that the observations be statistically independent. If the observations are correlated, the uncertainty inferred for the orbit or satellite state will be too small and unrealistic. Covariance realism is assessed by comparing the uncertainty bounds of the trajectory to the actual or analytically determined trajectory. If the trajectory propagates outside the covariance bounds, the covariances are unrealistic.

There are no standards for exchanging observations. Some organizations, such as the Russian International Scientific Observation Network (ISON) and the Center for Space Standards and Innovation (CSSI), have evolved unique formalisms for specific kinds of exchanges. Exchange of complete observational data is complicated by the fact that many sensors are not well-calibrated and the reluctance of observers to reveal the quality of their instruments as metadata. Being able to exchange observation data allows each recipient to develop orbits consistently with personally trusted techniques.

12.2.2.2 Orbit data

There are many forms of orbit data. All are developed from observations, and all are used to estimate the future trajectory of orbital objects. Diverse representations of orbit data are discussed in depth in Vallado's text [5], and summarized compactly in the AIAA publication, Astrodynamics, Best Practices and Test Cases [6]. It has been established as ISO Work Item 1123 for confirmation as an international standard.

The Orbit Data Message standard developed by the Consultative Committee for Space Data Standards (CCSDS) and ISO TC20/SC14 (Space Operations and Ground Support) is the most significant underpinning of space situational awareness and traffic control. This standard has established international consensus formats for elements of orbit data and for essential metadata. There are three types of orbit data messages: the Orbit Parameter Message, the Orbit Ephemeris Message, and the Orbit Mean Element Message. Each has a header section, a metadata section, and a data section. The header identifies the originator, creation date, and message version number. These messages are maintained consistently with the state of the art in astrodynamics. They will be modified as necessary and by consensus. Since content may vary from version to version, while versions may not be deployed synchronously, version number is an important element in data exchange.

The Orbit Parameter Message presents a state vector, at least the position and velocity of a satellite, as a precise time. This should be sufficient for the user to propagate the state into the future with any propagator of choice. If non-conservative forces and gravitational perturbations were included in developing the state, accelerations may also be required. However, the state is always imprecise for reasons stated previously. Even if the state is the outcome of a purely theoretical analysis, there is inevitable process noise. Covariance information is essential. These are the first standards to encompass covariances.

The Orbit Ephemeris Message includes time-ordered instances of the position, velocity and, if necessary, acceleration of the satellite. The ephemerides may extend into the future, the originator having already performed the necessary calculations. The user need only interpolate between the ephemerides to estimate the state of the satellite continuously over the prediction interval. However, the ephemerides are imprecise, as previously discussed. The satellite can be in the reported state or, with quantifiable probability, in some other nearby state. Therefore, covariance information is permitted in a prescribed format.

The Orbit Mean Element Message is completely new. It recognizes the fact that approximate, analytical, and semi-analytical models are used widely and are often the best choice for long-term propagation and for important applications. Mean element data sets often include parameters that allow the user to propagate states with mean element theories. Therefore, the metadata must include information about the theory that was employed, particularly the force models and perturbations that were included. Mean elements by definition have no covariances. They are often created by fitting a hypothesized trajectory with free parameters to observational data, minimizing the mean squared variance between the hypothesis and the data (least squares). There are only variances, and those variances cannot necessarily be associated with causes (such as imprecise physical models). Some satellite operators infer covariances with statistical analysis of mean elements as they vary from one update to the next. These are also "non-causal", giving the user no guidance for correcting the offending analysis elements. However, covariances are essential for probabilistic conjunction and collision assessment. Therefore, the Orbit Mean Element Message includes fields for synthesized covariances.

These are the first standards to include comprehensive metadata so that users can understand differences among originator processes and accommodate those differences. There are important common elements of metadata among all of these standards, such as the reference frame within which the data is instantiated, the coordinate system used within that reference frame, the near-space environment employed, the orientation of the Earth relative to a fixed inertial reference, and the gravitational model that was used. Without this understanding, no operator can use data on an orbital object properly. The user might employ different force models, arriving at a future state different than that which the originator of the data would predict. The reader can infer many more potential inconsistencies that will (not would) make collaborative operations, particularly collision avoidance or traffic management, impossible.

Each of these messages can be implemented in clear, simple language in a keyword (value format with units stated explicitly) or in a modern XML schema within a consensus namespace. A broad user community considers the Orbit Data Message development a milestone in space traffic management, establishing the environment for interoperability and collaboration.

12.2.3 Population of space objects

Space situational awareness and traffic management are conducted relative to a background of space activity. The space environment and trajectories of individual satellites are necessary information but not sufficient. Many experts confirm that the population of resident objects in Earth orbit is not maintained well enough [7]. Greater standardization will mitigate the deficiencies.

There are two aspects to maintaining the catalog of objects in orbit: observing satellite motion and estimating satellite trajectories between observations. Two observation sites 90° apart in latitude on the equator will be able to observe every satellite with an inclined orbit every day. Antipodal locations are redundant because they would see the same satellites descending as their antipodal partners would see ascending. Only geostationary satellites beyond the field of regard would escape them. This simple solution is infeasible because there are no suitable permanent and accessible locations so disposed on the equator. (Accessibility includes political and environmental considerations.) It is also insufficient because many satellite orbits change more frequently, requiring more than just two optimal sites to access them.

Space surveillance is a complex choreography of observations and propagation schemes balanced according to the orbit accuracy and precision that is desired, timeliness of position reports, the orbital regime, and the distribution of sensors. Sensor tasking, data acquisition, data distribution, and data application are not uniform across spacefaring nations. Non-spacefaring nations can contribute to the mission, since many enjoy good atmospheric or geographical accesses.

No nation, not even the USA, has enough sensors, let alone well-distributed sensors, to gather observations on all satellites of interest. The geostationary belt is the best example, since almost all nations lack sensors with reliable access to antipodal geostationary satellites. Repositioning geostationary satellites often puts them out of view of the owner's sensor systems, risking conjunction with other geostationary satellites. Space-based sensors might mitigate the problem, but they add even more objects to the satellite catalog. The Orbit Data Message standard would help if it were applied widely. An orbit observation standard would be better, since it would allow each user and operator to determine and propagate orbits consistent with their operational practice.

The European Space Agency (ESA) and Russia, among others, are evolving space surveillance capabilities that complement the United States Space Surveillance Network. In addition to uniformly useful and understandable tools for exchanging orbit and observation data, a collaborative network requires communication and data storage architectures. These, too, enjoy applicable standards developed with international consensus through the CCSDS and similar organizations. The CCSDS Reference Architecture for Space Data Systems is an excellent guide for developers [8]. There are also communication protocol standards across the Open Systems Interconnect (OSI) paradigm optimized for transmitting data among ground sites (Space Line Extensions, SLE) [9], and between satellites and ground stations (Space Communication Protocol Standards, SCPS) [10]. Data integrity and security are mandatory elements in these standards.

An effective tasking and collaborative space surveillance system should be based on quantified growth of orbit data accuracy and precision. "Covariance-based tasking" has been proposed independently many times but is not yet the accepted practice [11]. Revisits would be executed by the most accessible and capable sensors when orbits grew sufficiently inaccurate and imprecise for designated missions. The standards discussed above facilitate that concept, particularly because they allow covariance information.

12.2.4 Risks and threats

The universe of objects in Earth orbit is vulnerable to natural and man-made threats. It can also be a threat to terrestrial and airborne activity.

Risk is subjective. It depends on the probability that an untoward event will occur and what the consequences of that event might be. The consequences are in response to the susceptibility of the satellite to physical and other intrusions. Susceptibility, and a real threat that exploits the susceptibility, constitute the vulnerability of a satellite. These terms—risk, susceptibility, and vulnerability—are not standardized. We suggest this description as the basis for such a standard. Standard, unambiguous terminology is important in our multilingual community.

How one identifies threats and reacts to them may be a matter for individual operators to deal with, but there are mutual threats that require uniform, trusted interactions among those who might be threatened. Radio frequency interference (RFI) is one.

Locating and mitigating RFI is very demanding. It requires comprehensive information about the characteristics of the interference perceived by each affected party, and characteristics of the instruments and services that are affected. It also relies on very accurate and precise orbit information for stakeholder satellites. The process exploits space environment and orbit data standards already developed, and it needs standards for the other types of essential information. Those standards do not exist. The Satellite Users' Interference Reduction Group (SUIRG), a collaborative, ecumenical body of communication satellite owner/operators, is pursuing those diligently [12].

Collision risk and consequences dominate space situational awareness and space traffic management. All activity is motivated by assuring safe access to, and services from, space. Yet there is no uniform way to communicate threats in an actionable manner. Some organizations provide such "neighborhood watch" services for proactive operators. Formats have been developed by the CSSI that might serve as a point of departure for such standards [13]. Standards for communicating orbit data contribute greatly to this end, but threat characterization for successful avoidance is deficient.

CONCLUSIONS

Space situational awareness and space traffic management rely on an environment for perceiving, characterizing, and assessing activity in the space occupied by Earth-orbiting satellites. Standards create that environment. Standards are developed by consensus, based on well-stated need. Standards are applied voluntarily and can be tailored to specific circumstances. Many important standards toward this end have already been developed, notably the CCSDS/ISO Orbit Data Message Standard. Many more standards are required. Greater involvement from the broad space community would accelerate these developments. There are also diverse standards-related organizations pursuing common goals independently and diversely: the ISO, the European Cooperation for Space Standardization (ECSS), the American Institute of Aeronautics and Astronautics, and the United States Satellite Industry Association, among others. More effective coordination among them would also help. The ESA has developed a comprehensive space situational awareness development and implementation plan that, at this writing, has not been fully released to the broad community. ESA collaborators have developed a cogent depiction of opportunities and needs for standardization in space situational awareness and space traffic management. We leave the reader with this vision (Figure 12.2).

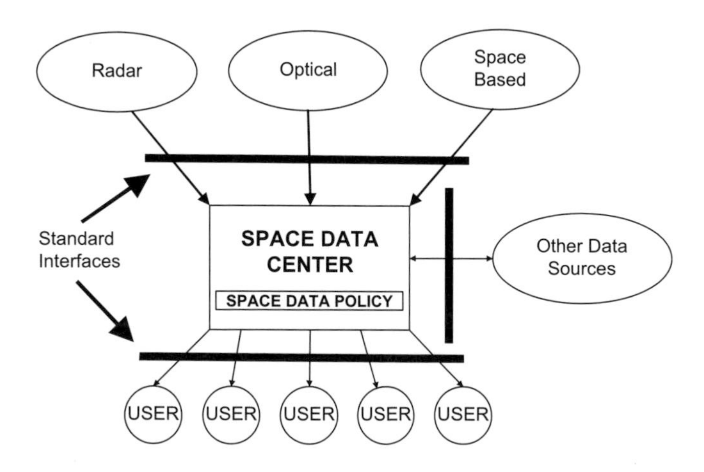

FIGURE 12.2

BNSC vision of opportunities for standardization in support of space situational awareness and traffic management.

Notes and references

[1] Vallado, D. (2008). *Fundamentals of Astrodynamics and Applications*, 3rd ed. Space Technology Library, Springer.

[2] COSPAR (2006). C4.1 2006-A-01661, New Solar Indices for Improved Thermospheric Densities.

[3] International Organization for Standardization (2008). ISO 22009, Space Systems—Space Environment (Natural and Artificial)—Model of the Earth's Magnetospheric Magnetic Field.

[4] International Organization for Standardization (2009). ISO 27852, Space Systems: Orbit Lifetime Determination.

[5] Vallado, D., *supra* note 1.

[6] AIAA Astrodynamics Standard (2009). Propagation Specifications, Test Cases, and Recommended Practices. American Institute of Aeronautics and Astronautics, Reston, VA.

[7] Butler, A. (2009). USAF Boosts Space Situational Awareness. *Aviation Week*, 3 July.

[8] CCSDS 311.0-R-1 (2008). Space Data System Standards, Reference Architecture for Space Data Systems. NASA Office of Space Communications (Code M-3), Washington, DC.

[9] This is a sequence of recommended practices beginning with CCSDS 913.1-R-1 (2006). Space Link Extension—Internet Protocol for Transfer Services.

[10] This is a sequence of standards beginning with CCSDS 714.0-B-2 (2006). Space Communications Protocol Specification (SCPS)—Transport Protocol.

[11] Vallado, D. and Alfano, S. (1999). A Future Look at Space Surveillance and Operations. AASIAIAA Space Flight Mechanics Meeting, Breckenridge, CO, 7—10 February, AAS 99-1 13.

[12] http://suirg.org/index.shtml

[13] Kelso, T. S. and Vallado, D. A. Center for Space Standards and Innovation; Chan, J. and Buckwalter, B. Intelsat Corporation; Improved Conjunction Analysis via Collaborative Space Situational Awareness, AMOS 2008, Maui, HI, USA, September.

PART

Regulating Commercial Space Flight

3

CHAPTER

Developing commercial human space-flight regulations

13

Kenneth Wong

US Federal Aviation Administration

CHAPTER OUTLINE

INTRODUCTION

The commercial human space-flight industry is emerging as several entrepreneurial companies begin developing launch vehicles to carry crew and space-flight participants. A regulatory framework governing commercial human space flight is now in place with the FAA's issuance of the human space-flight rule [1]. Significant events that influenced this rule include the enactment of legislation, formation of a Federal Aviation Administration (FAA) human space-flight team, establishment of sub-orbital definitions to clarify the regulatory regime for hybrid vehicles, and the licensing of sub-orbital reusable launch vehicle (RLV) missions with a pilot on board. This chapter discusses these events and

Space Safety Regulations and Standards. DOI: 10.1016/B978-1-85617-752-8.10013-3

their bearing on the development of the rule. Furthermore, the chapter addresses the comments received during the rule's development, as well as future issues that the FAA anticipates in the area of commercial human space flight.

13.1 LEGISLATIVE BACKGROUND

A review of the legislative history indicates that it took about two decades after passage of the Commercial Space Launch Act of 1984 (CSLA) before Congress enacted legislation for commercial human space flight. In 1984, Executive Order 12465 designated the Department of Transportation as the lead agency within the federal government for encouraging and facilitating commercial expendable launch vehicle activities by the US private sector. Licensing authority followed shortly with the CSLA, which authorized the Secretary of Transportation to license launches and operation of launch sites as carried out by US citizens or within the USA. The CSLA directed the Secretary to exercise this responsibility consistent with public health and safety, safety of property, and national security and foreign policy interests of the USA. The CSLA was drafted with commercial expendable launch vehicle operations in mind, such as launching of payloads and satellites, and did not contemplate commercial human space flight. The Secretary's licensing authority has been delegated to the Administrator of the FAA and further assigned to the Associate Administrator for Commercial Space Transportation (AST).

The CSLA was subsequently amended. The CSLA Amendments of 1988 revised insurance requirements and created a framework to govern third-party liability compensation when aggregate claims exceeded required financial responsibility amounts. In 1994, Congress recodified the CSLA at 49 USC Subtitle IX, chapter 701. The Commercial Space Act of 1998 extended licensing authority under the CSLA to re-entry vehicle operations and operation of re-entry sites by non-federal entities. The Commercial Space Launch Amendments Act of 2004 (CSLAA), which was signed into law on 23 December 2004, made the FAA responsible for regulating commercial human space flight and established an experimental permit regime for developmental reusable sub-orbital rockets. (Note: The CSLAA characterizes what is commonly referred to as a passenger as a space-flight participant, and defines this person to mean an individual, who is not crew, carried within a launch vehicle or re-entry vehicle.)

13.2 FAA HUMAN SPACE-FLIGHT TEAM DRAFT CREW GUIDELINES

Several years prior to the enactment of the CSLAA, the FAA's Office of Commercial Space Transportation (FAA-AST) started looking into the area of commercial human space flight. In response to the Commercial Space Act of 1998, the FAA in September 2000 issued the "Commercial Space Transportation Reusable Launch Vehicle and Re-entry Licensing Regulations". These specify the requirements for obtaining a license to launch and re-enter a reusable launch vehicle, re-enter a re-entry vehicle, and operate a re-entry site [2]. The September 2000 regulations focused on ensuring safety of the uninvolved public in the event of RLV launch and re-entry activities. At that time, the FAA acknowledged in the preamble to the rule that human space flight was an area that needed to be addressed in the future. Some commercial companies were already planning to develop RLVs to transport humans, both as crew and as passengers, to and from space.

In 2001, the FAA/AST examined the issue of crew and passengers on RLVs and prepared a paper that identified and discussed issues [3]. Subsequently, the FAA/AST formed a team of staff members to identify, research, and evaluate issues that might have a bearing on future FAA requirements associated with carrying humans on commercial RLVs. In April 2003, this team prepared an internal white paper that evaluated human flight safety issues and provided recommendations to resolve them. The white paper also identified the need to develop flight crew guidelines in anticipation of commercial human space flight in the future. Although Congressional hearings were held during this time, legislation did not exist yet on how the government should regulate commercial human space flight. However, the FAA had the authority to regulate flight crew safety when the flight crew was part of the flight safety system, and the team recommended that the FAA start developing guidelines for RLV flight crew.

To assist both the FAA and license applicants, in October 2003, the FAA developed draft guidelines for RLV flight crew. The guidelines focused on public safety and applied to crew on board a sub-orbital RLV when crew is part of the flight safety system. The FAA regarded the flight crew as part of the flight safety system, when the flight crew has control over the safety of the vehicle and the capability of averting a potential hazard to the uninvolved public on the ground and in the air. These guidelines covered crew qualification and training, environmental control and life support systems, fire detection and suppression, and human factors to ensure that the flight crew is able to protect the uninvolved public. These guidelines addressed what the FAA may expect to review and evaluate in a license application for a launch that has flight crew on board a sub-orbital RLV. Formation of the FAA/AST human space-flight team was extremely important and beneficial. Much of the work completed by this team, including the development of a white paper and draft guidelines, was later used to develop guidelines and regulations.

13.3 NEED FOR SUB-ORBITAL DEFINITIONS TO CLARIFY THE REGULATORY REGIME

In 2003, the FAA identified a need for definitions to differentiate between an aircraft and a sub-orbital rocket. The lack of definitions for the terms sub-orbital rocket and sub-orbital trajectory in the original CSLA led to some difficulty in identifying the appropriate regulatory regime for hybrid vehicles that had both aircraft and launch vehicle characteristics. Consequently, the FAA issued a Federal Register Notice in October 2003 to inform the public of FAA criteria used to differentiate civil aircraft subject to aircraft certification and operating standards for flight in airspace from a sub-orbital launch subject to licensing under 49 USC Subtitle IX, chapter 701 [4]. A modified version of the FAA's 2003 sub-orbital definitions appears in the Commercial Space Launch Amendments Act of 2004, as follows:

- **Sub-orbital rocket**—a vehicle, rocket propelled in whole or in part, intended for flight on a suborbital trajectory, and the thrust of which is greater than its lift for the majority of the rocket-powered portion of its ascent.
- **Sub-orbital trajectory**—the intentional flight path of a launch vehicle, re-entry vehicle, or any portion thereof, whose vacuum instantaneous impact point does not leave the surface of the Earth.

These definitions helped the suborbital RLV industry by clarifying the regulatory jurisdiction within the FAA for vehicle configurations and operations with both aircraft and launch vehicle characteristics.

13.4 LICENSING RLV MISSIONS WITH CREW ON BOARD

The FAA, in April 2004, issued two RLV mission-specific licenses: one for Scaled Composites and one for XCOR Aerospace in accordance with 14 CFR parts 431 [2] and 440. The FAA used its October 2003 draft flight crew guidelines to assist in these two license application evaluations because the CSLAA had not been enacted and human space-flight regulations did not yet exist. Scaled Composites won the X Prize on 4 October 2004, by being the first to finance privately, build, and launch a vehicle able to carry three people to an altitude of 100 km, return safely to Earth, and then repeat the trip within 2 weeks. Some of the lessons learned from licensing the *SpaceShipOne* launches were taken into consideration when the FAA later updated its crew guidelines and developed human space-flight regulations. (Note: The bill, HR 5382, Commercial Space Launch Amendments Act of 2004, which became Public Law 108-492 on 23 December 2004, was based on a previous bill, HR 3752.)

13.5 DRAFT SPACE-FLIGHT PARTICIPANT GUIDELINES

In June 2004, the FAA/AST human space-flight team reassessed the issues and recommendations proposed in its April 2003 white paper. Based on its reassessment, the FAA/AST human space-flight team determined that its next step should be the development of guidelines for space-flight participants to assist both the FAA/AST and potential license applicants. The FAA/AST human space-flight team felt that it should take the initiative to develop these guidelines and not wait until HR 37522 or another version of the bill became law; this is similar to the proactive approach that was taken by the team to develop the crew guidelines. Although HR 3752 had not yet been adopted and passed by the Senate, it provided insight on how some in Congress felt with regard to regulating space-flight participants. In addition, industry's views on the regulation of space-flight participants had been expressed at the biannual Commercial Space Transportation Advisory Committee (COMSTAC) RLV Working Group Meetings. COMSTAC provides information, advice, and recommendations to the FAA on matters relating to the US commercial space transportation industry.

As noted earlier, the RLV mission licensing regulations issued in 2000 contemplated crewed vehicles with passengers but the regulations did not address passengers. The team drafted space-flight participant guidelines with the intent that they would provide guidance on an interim basis and serve as inputs to future rulemaking. Specifically, these guidelines would assist both the FAA/AST and potential applicants in terms of providing guidance on what the FAA/AST expected to review and evaluate in a license application that proposed the carriage of space-flight participants.

13.6 RULEMAKING ACTIVITIES IN RESPONSE TO THE CSLAA

The CSLAA required the FAA to issue guidelines or advisory circulars for implementing the CSLAA as soon as practical; it proposed regulations relating to crew, space-flight participants, and permits for launch or re-entry of reusable sub-orbital rockets not later than 12 months after enactment of the CSLAA, and final regulations not later than 18 months after enactment of the CSLAA. The FAA's previous work in the area of human space flight made it possible to meet most of the aggressive timelines mandated by the CSLAA.

On 11 February 2005, less than 2 months after enactment of the CSLAA, the FAA issued "Draft Guidelines for Commercial Suborbital Reusable Launch Vehicle Operations with Flight Crew". Also on this date, the FAA issued "Draft Guidelines for Commercial Suborbital Reusable Launch Vehicle Operations with Space Flight Participants" and guidance for medical screening of commercial aerospace passengers. Expeditious issuance of these guidelines was possible because the FAA updated and modified the October 2003 crew guidelines to reflect the CSLAA and previous industry comments. In addition, the FAA had already begun developing draft space-flight participant guidelines, as mentioned earlier.

The FAA formed a rulemaking team consisting of the FAA/AST human space-flight team and representatives from the Office of the Chief Counsel and other FAA lines of business, such as Aerospace Medicine and Flight Standards, to develop the human space-flight regulations. The FAA issued the notice of proposed rulemaking (NPRM) for human space flight of crew and space-flight participants in December 2005, within 12 months of the enactment of the CSLAA [5]. The FAA applied a very aggressive and expedited rulemaking process to meet the congressional timeline for issuing the NPRM. After the NPRM underwent a public review and comment period, the FAA reviewed and dispositioned the comments received and issued a final rule in December 2006.

13.7 HUMAN SPACE-FLIGHT RULE

The following discussion addresses some of the key issues that arose during the development of the human space-flight rule. These are in the areas of crew qualification, training, and medical requirements. Considerations for medical requirements or guidance for space-flight participants are also discussed.

13.7.1 Pilot qualifications and training

The FAA requires a pilot of a launch or re-entry vehicle to possess and carry an FAA pilot certificate with an instrument rating. The FAA invited public comment on the proposed requirement and received differing views. Some agreed with the requirement for a pilot to have an instrument rating. Others recommended against requiring pilots of launch vehicles that do not have aircraft characteristics to possess an FAA pilot certificate with an instrument rating. Some commentators considered the requirement too lenient, and indicated that a pilot certificate might only partially address the knowledge, skills, and abilities necessary for safety.

The FAA requires a pilot certificate so that a pilot of a reusable launch vehicle has a basic level of aeronautical experience, an understanding of the National Airspace System (NAS), and an understanding of the regulatory requirements under which aircraft in the NAS operate, including cloud clearance requirements and airspace restrictions. This awareness enhances overall safety of the NAS, regardless of whether a vehicle has wings. An instrument rating ensures that pilots of launch and re-entry vehicles have the skills of scanning cockpit displays, correctly interpreting instruments, and responding with correct control inputs. There may be times when a pilot will have to rely on instrument skills and competency regardless of the kind of vehicle used. Having a pilot certificate and aeronautical experience provides evidence of a basic level of knowledge of and experience with the NAS, such as communications, navigation, airspace limitations, and other aircraft traffic avoidance, that will help promote public safety.

When the FAA crafted the regulations, it recognized that just possessing a pilot certificate may not be sufficient for piloting a launch vehicle that has operational and vehicle-specific characteristics different from those of an aircraft. Hence, the FAA also requires a pilot to receive vehicle- and mission-specific training that includes simulation training, training on a similar aircraft, flight testing, or another training method approved by the FAA. In addition, a pilot must train for nominal and non-nominal conditions, which must include abort scenarios and emergency operations.

13.7.2 Requirements for a pilot license

Guidance also was provided that the pilot of a craft should possess an FAA pilot certificate with an instrument rating. Several who offered comments considered that this requirement was too lenient, and indicated that they believed that a pilot certificate might only partially address the knowledge, skills, and abilities necessary for safety.

The FAA requires a pilot certificate on several bases. These are that a pilot of a reusable launch vehicle has a basic level of aeronautical experience, an understanding of the National Airspace System (NAS), and an understanding of the regulatory requirements under which an aircraft in the NAS operates, including cloud clearance requirements and airspace restrictions. This awareness enhances overall safety of the NAS, regardless of whether a vehicle has wings. An instrument rating ensures that pilots of launch and re-entry vehicles have the skills of scanning cockpit displays, correctly interpreting instruments, and responding with correct control inputs. There may be times when a pilot will have to rely on instrument skills and competency regardless of the kind of vehicle used. Having a pilot certificate and aeronautical experience provides evidence of a basic level of knowledge of and experience with the NAS, such as communications, navigation, airspace limitations, and other aircraft traffic avoidance procedures, that will help promote public safety.

When the FAA crafted the regulations it recognized that just possessing a pilot certificate may not be sufficient for piloting a launch vehicle that has operational and vehicle-specific characteristics different from those of an aircraft. Hence, the FAA also requires a pilot to receive vehicle- and mission-specific training that includes simulation training, training on a similar aircraft, flight testing, or another training method approved by the FAA. In addition, a pilot must train for nominal and non-nominal conditions, which must include abort scenarios and emergency operations.

13.7.3 FAA Consideration of Medical Standards

The FAA proposed in the NPRM that each member of the flight crew would be required to possess and carry a second-class airman medical certificate issued in accordance with 14 CFR part 67 [5]. In response to comments received, the FAA in the final rule changed this to require only that each crew member with a "safety-critical role" possess and carry an FAA second-class airman medical certificate issued in accordance with 14 CFR part 67.

Several of those offering comments generally concurred with the FAA that requiring a medical certification was appropriate. However, several suggested that it might not be necessary for all crew members; that is, crew without a safety critical function should not need to possess an FAA second class airman medical certificate. Some commented that a second-class medical certificate was appropriate for the pilot but not necessary for other members of the crew because not all flight crew will have duties that affect public safety. As reflected in the final rule, the FAA agrees that requiring second-class medical certification for crew members who do not perform safety-critical functions may be overly burdensome. The FAA recognizes that there may be missions when a flight attendant or flight test engineer has duties that would not affect public safety.

Others offering comments recommended that the FAA adopt more stringent medical standards, especially for orbital missions. For example, a commenter noted that that a second-class medical certificate is acceptable for sub-orbital flight, but more stringent physical standards should be applied to orbital missions. Recognizing that second-class medical certification is insufficient for space flight, the FAA established a performance standard that requires the flight crew to demonstrate an ability to withstand the stresses of space flight sufficiently so that the vehicle will not harm the public. The stresses experienced in space flight may include high acceleration or deceleration, microgravity, and vibration. The performance standard provides an additional level of safety beyond basic medical certification because flight crew members will have to demonstrate an ability to perform duties in the space flight environment where they plan to operate. The FAA recognizes that different standards may be required for orbital flights than for suborbital flights. The FAA will gather data for the development of those standards over time and such standards may be implemented through future rulemaking.

13.7.4 FAA Consideration of Guidelines for Spacecraft Participants

The CSLAA requires that a space-flight participant be informed of the risks of going on a launch or re-entry vehicle, and specifies that the FAA may issue regulations requiring space-flight participants to undergo an appropriate physical examination. As noted in the rulemaking, the FAA decided against prescribing specific medical requirements for space-flight participants at this time. Those who offered comments in general agreed with this decision. Instead of establishing space-flight participant medical requirements, the FAA issued guidelines recommending that space-flight participants obtain an evaluation of their medical history to determine whether a physical examination might be appropriate. The guidelines recommend that a space-flight participant provide a medical history to a physician experienced or trained in the concepts of aerospace medicine. The physician would determine whether the space-flight participant should undergo an appropriate physical examination before boarding a vehicle destined for space flight. Guidance for the FAA research and development projects included that there should be a human space-flight training study and crew rest analysis. The training study involves a survey of organizations that provide training applicable to the flight crew of a launch vehicle. Results of this study will assist in developing the aforementioned crew training advisory circular. The crew rest analysis includes researching what should be the duty and rest period requirements for a pilot of a launch vehicle.

One of the issues identified for future evaluation includes the FAA's lack of on-orbit authority. The FAA has authority to regulate launch and re-entry operations. This issue involves assessing what potential effect this may have on regulating orbital human space flight in the future and maintaining the FAA's responsibility to ensure public safety through its regulatory and licensing process. Medical assessment of space-flight participants is provided in a memorandum, "Guidance for Medical Screening of Commercial Aerospace Passengers" [6].

The Federal Air Surgeon of the FAA's Office of Aerospace Medicine and the Director of the FAA's Civil Aerospace Medical Institute provided this guidance to the Associate Administrator for Commercial Space Transportation. This guidance notes that medical conditions indicating an individual should not participate in a mission should be identified. Then, participation may be avoided where a space-flight participant's involvement in a mission could aggravate a pre-existing medical condition and put the flight crew or other space-flight participants at risk. As noted earlier, the FAA highly recommends that a space-flight participant seek such medical advice for an orbital mission. Orbital missions are longer in duration than sub-orbital missions and expose space-flight participants to flight conditions or environments such as microgravity and radiation for a longer period of time.

13.7.5 Other requirements that focus on safety of the uninvolved public but indirectly relate to crew and space-flight participant safety

The human space-flight regulations establish a regulatory framework in accordance with the FAA's authority to protect the uninvolved public—that is, those persons who are not involved in the launch or re-entry. Although the human space-flight regulations do not specifically ensure safety of those on board, *per se*, they do address their safety in some areas to protect the uninvolved public: environmental control and life-support systems (ECLSS), smoke detection and fire suppression, human factors, verification program, space-flight participant training, and security. With regard to ECLSS requirements, the FAA requires an operator to provide atmospheric conditions adequate to sustain life and consciousness for all inhabited areas within a vehicle. The operator or flight crew must monitor and control the atmospheric conditions in the inhabited areas.

Another regulation that indirectly affects space-flight participant safety is the FAA requirement that an operator successfully verify the integrated performance of a vehicle's hardware and any software in an operational flight environment before allowing any space-flight participant on board during a flight. Verification must include flight testing. An operator needs to establish a safety record to disclose to a space-flight participant as required by the CSLAA, and a space-flight participant should not be present during flight testing to avoid distracting the flight crew from its public safety mission. The FAA intends early, experimental flight testing to take place with the flight crew's entire attention dedicated to the vehicle, not to anyone else on board.

A further requirement in place for public safety reasons, but indirectly related to space-flight participant safety, pertains to training. The FAA requires an operator to train each space-flight participant before flight on how to respond to emergency situations, including loss of cabin pressure, fire, smoke, and emergency egress. Without this training, a space-flight participant might inadvertently interfere with the crew's ability to protect public safety.

The security requirements established for public safety reasons also indirectly relate to crew and space-flight participant safety. The FAA requires an operator to implement security requirements to prevent any space-flight participant from jeopardizing the safety of the flight crew or the public. Some of the restrictions prohibit a person carrying explosives, firearms, knives, or other weapons from boarding an airplane. These security restrictions contribute to the safety of the public by preventing a space-flight participant from potentially interfering with the flight crew's operation of the vehicle. Any such interference might jeopardize the flight crew's ability to protect the public. Regardless of whether there are specific regulations addressing crew and space-flight participant safety, industry is cognizant that safety of those on board is extremely important in order for the commercial human space-flight industry to survive.

13.8 WAY AHEAD—FUTURE ACTIVITIES AND ISSUES

13.8.1 A strategic roadmap

Since the human space-flight regulations were issued, the FAA/AST human space-flight team continues to be proactive by planning for the future and developing a strategic roadmap that identifies current, near-term, and future topics, issues, and activities the FAA should be addressing or pursuing. Areas or activities identified by the team include research and development, future rulemaking, and development of advisory circulars, guidance documents, and assessment of issues to ensure that the FAA keeps pace with the human space-flight industry as it evolves. Some of these activities are currently under way, including the development of advisory circulars in the areas of human space-flight crew training and sub-orbital environmental monitoring and control. These advisory circulars will provide guidance and acceptable means of meeting some of the human space-flight regulations pertaining to crew training and the ECLSS.

13.8.2 Safety indicators

Another area the FAA is addressing concerns safety indicators and events that may trigger additional human space-flight rulemaking. The CSLAA is specific as to when the FAA may issue regulations governing the design or operation of a launch vehicle to protect the health and safety of crew and space-flight participants. Issuance of such regulations must apply only to launches where a vehicle will carry a human being for compensation or hire, and be limited to restricting or prohibiting design features or operating practices that have:

- resulted in a serious or fatal injury to crew or space-flight participants; or
- contributed to an unplanned event during a licensed or permitted launch that posed a high risk of causing a serious or fatal injury to crew or space-flight participants.

Furthermore, the CSLAA specifies that, beginning on 23 December 2012, the FAA may propose regulations without regard to the above restrictions. Any such regulations shall consider the evolving standards of safety in the commercial space-flight industry. The FAA has held some preliminary internal discussions on the circumstances under which it would regulate crew and space-flight participant safety before 2012 in the event of a casualty or close call.

Some key safety indicators that the FAA plans to monitor include safety-related anomalies, safety-critical system failures, incidents, and accidents. The FAA plans to track these indicators for any precursors, trends, or lessons learned that would warrant additional FAA regulations affecting the commercial human space-flight industry at large. In general, a triggering event for rulemaking would be one that would affect more than one entity and not be unique to a specific company, launch vehicle or operation. Unique anomalies or failures resulting from a unique launch vehicle design or operation would be addressed under the terms and conditions of a license or permit. Launch vehicle operators are required to report mishaps and safety-related anomalies and failures to the FAA, and take appropriate corrective actions prior to the next launch.

13.8.3 Human space-flight study

The CSLAA required that the Secretary of Transportation, in consultation with the Administrator of the National Aeronautics and Space Administration (NASA), entered into an arrangement with a non-profit entity for a report analyzing safety issues related to launching human beings into space. This study was conducted by the Aerospace Corporation, George Washington University and MIT, and the final report was submitted to the Senate Committee on Commerce, Science, and Transportation and the House of Representatives Committee on Science by the specified deadline of December 2008.

Congress required that this report analyze and make recommendations about:

- the standards of safety and concepts of operation that should guide the regulation of human space flight;
- the effectiveness of the commercial licensing and permitting regime in ensuring the safety of the public and of crew and space-flight participants, and whether changes are needed;
- whether expendable and reusable launch and re-entry vehicles should be regulated differently, and whether either of those vehicle types should be regulated differently when carrying humans;
- whether the federal government should separate the promotion of human space flight from the regulation of such activity;
- how third parties could be used to evaluate the qualification and acceptance of new human space-flight vehicles prior to their operation;
- how non-government experts could participate more fully in setting standards and developing regulations concerning human space-flight safety; and
- whether the federal government should regulate the extent of foreign ownership or control of human space-flight companies operating or incorporated in the USA.

The FAA met with NASA to discuss the initial planning of this study and achieved coordination through discussions and meetings with NASA and the study team, as well as with COMSTAC, as the study progressed.

CONCLUSIONS

Development of commercial human space flight by entrepreneurs and privately funded companies as a new segment of the aerospace industry is under way. Although some challenges and issues still remain, there is now less regulatory uncertainty due to the FAA's establishment of a regulatory framework and issuance of human space-flight regulations. These regulations are expected to evolve to keep pace with industry as it develops beyond sub-orbital space tourism and moves toward point-to-point travel and orbital space flight. Challenges that lie ahead include how the industry should be regulated beyond informed consent, if crew and space-flight participant safety were to be regulated in the future. Furthermore, public, congressional, and industry reaction to an accident involving commercial human space flight is a topic that needs to be addressed with regard to the effect it would have on this nascent industry. Despite these challenges, the FAA and industry are both learning and working together in this exciting and dynamic environment created by the prospect of having human space flight available to the public.

References

[1] Department of Transportation, Federal Aviation Administration (2006). 14 CFR Parts 401, 415, 431, 435, 440 and 460, Human Space Flight Requirements for Crew and Space Flight Participants; Final Rule, 71 Fed. Reg. 75615 (15 December).

[2] Department of Transportation, Federal Aviation Administration (2000). 14 CFR Parts 400, 401, 404, et al. Commercial Space Transportation Reusable Launch Vehicle and Reentry Licensing Regulations; Final Rule, 65 Fed. Reg. 56617 (19 September).

[3] Wong, K. (2001). Safety Issues and Considerations Concerning Humans Aboard Commercial Reusable Launch Vehicles, AIAA-2001-4618.

[4] Department of Transportation, Federal Aviation Administration (2003). Commercial Space Transportation; Suborbital Rocket Launch, 68 Fed. Reg. 59977 (20 October) and corrected notice 68 Fed. Reg. 61241 (27 October).

[5] Department of Transportation, Federal Aviation Administration (2005). 14 CFR Parts 401, 415 et al. Human Space Flight Requirements for Crew and Space Flight Participants; Proposed Rule, 70 Fed. Reg. 77261 (29 December).

[6] FAA memorandum on "Guidance for Medical Screening of Commercial Aerospace Passengers", 31 March 2003; guidance later published as Report No. DOT/FAA/AM-06/1 (January 2006).

CHAPTER

Regulation and licensing of US commercial spaceports[1]

14

Michael Mineiro

Boeing Fellow in Air and Space Law, Institute of Air and Space Law, McGill University

CHAPTER OUTLINE

14.1 US FEDERAL LAW AND REGULATION GOVERNING COMMERCIAL SPACEPORT ACTIVITIES

US federal law and regulation governing spaceport activities derives from a system of federal governance established under the US Constitution. The Constitution establishes three branches of government, each with their own authority and obligations [1]. Congress enacts legislation, the Executive branch (i.e. President) implements the legislation, and the Judiciary ensures Congressional and Executive acts are within the bounds of law [2]. To that end, Congress has passed legislation to regulate commercial spaceports. This legislation is the Commercial Space Launch Act of 1984 (CSLA) and related amendments [3]. The CSLA and related amendments are codified in 49 USC §70101 ("the Act") [4]. Pursuant to this legislation, the Executive branch has issued regulations

[1] This chapter comprises excerpts from a paper that was originally printed and presented in the *Journal of Air Law and Commerce*, "Law and Regulation Governing US Commercial Spaceports: Licensing, Liability, and Legal Challenges" Vol. 73 (4), p. 759 (2008). Reprinted in amended form with permission from the *Journal of Air Law and Commerce* and Southern Methodist University.

Space Safety Regulations and Standards. DOI: 10.1016/B978-1-85617-752-8.10014-5

governing commercial spaceport activities ("Regulations"). In addition, the President has issued National Space Policy and Space Transportation Directives relevant to spaceport activities ("Directives") [5]. Together, the Act, Regulations, and Directives are the primary laws and regulations governing US commercial spaceport activity.

14.1.1 The principal framework for commercial space launches and spaceports

The CSLA is the principal law governing the licensing and regulation of commercial space transportation in the USA, including commercial spaceports [6]. The Act does not apply to spaceport operations or other space activities the US government carries out for the government itself [7]. As originally enacted in 1984, the CSLA was limited to the regulation of expendable launch vehicles (ELVs) and launch sites [8]. This regulatory authority was granted to the Department of Transportation (DOT) [9]. To implement this authority, the DOT established the Office of Commercial Space Transportation and, later, the Associate Administrator for Commercial Space Transportation (AACST) under the administration of the FAA [10]. In 1988, the CSLA was amended to provide a three-tier liability risk-sharing regime, including conditional indemnification for catastrophic accidents [11]. In 1998, the CSLA was amended to extend DOT licensing authority to re-entry licensing, allowing effective licensing of re-entry sites and reusable launch vehicles (RLVs) [12]. In 2004, Congress amended the CSLA "to promote the development of the emerging commercial human space-flight industry" [13], and granted the DOT the authority to implement regulatory standards to govern commercial human space flight [14].

With regard to commercial spaceports, the Act can be conveniently divided into six parts:

1. **Opening Provisions.** These consist of a statement of purposes, definitions, and a statement of general authority granted to the Secretary of Transportation (SOT) [15].
2. **Licensing Provisions.** These explain when a license is required, the conditions to receive a license, the scope of licenses, and under what conditions and to what extent a license can be modified, transferred, suspended, or revoked [16].
3. **Post-Licensing Provisions.** These establish SOT authority to monitor licensees and to enforce the Act and Regulations, assess penalties for violations of the Act and Regulations, and issue orders prohibiting, suspending, or ending a licensed activity [17].
4. **Financial Responsibility Provisions.** These require licensees to obtain insurance or to demonstrate the capacity to compensate for certain claims, and establish federal indemnification provisions for certain catastrophic losses [18].
5. **SOT Regulatory Authority.** These provisions establish and define the scope of SOT authority to issue regulations [19].
6. **Other Provisions.** The Act also contains provisions regarding inter-agency consultation, space advertising, pre-emption of scheduled launches/re-entries, acquisition of federal property and services, experimental sub-orbital rocket permits, human space flight-related provisions, administrative hearings/review, and the relationship of the Act to other executive agencies, law, and international obligations [20].

14.1.1.1 Opening Provisions: launch and re-entry sites

As discussed earlier, the term spaceport is commonly used to refer to launch and re-entry sites [21]. The Act defines "launch site" as "the location on Earth from which a launch takes place [as defined in a license the Secretary issues or transfers under this chapter] and necessary facilities at that location" [22]. "Re-entry site" is "the location on Earth to which a re-entry vehicle is intended to return [as defined in a license the Secretary issues or transfers under this chapter]" [23]. It is important to note that launch and re-entry sites, as defined in the Act, do not necessarily exhibit characteristics generally associated with spaceports. Launch and re-entry sites, in theory, may not have fixed launch infrastructure, launch service facilities, or other related buildings. Essentially, a launch site or re-entry site could be as simple as an open stretch of desert. Spaceports, while not defined in the Act, tend to bring to mind a launch or re-entry site with infrastructure to support operations. In this sense, launch and re-entry sites, as defined in the Act, encompass a range of sites greater than the common usage of the term spaceport.

14.1.1.2 Licensing Provisions: general scope and requirement to obtain

The SOT issues licenses to operate launch and re-entry sites (i.e. spaceport operator licenses) "in accordance with the representations contained in the licensee's application, with terms and conditions contained in any license" and subject to compliance with the Act [24]. The Act grants separate authorizations to licensees of launch sites and re-entry sites, essentially creating a dual licensing regime. Launch site licenses "authorize ... a licensee to offer its launch site to a launch operator for each launch point for the type and any weight class of launch vehicle identified in the licence" [25]. Re-entry site licenses "authorize ... the licensee to offer use of the site to support re-entry of a re-entry vehicle for which the three-sigma footprint [i.e. the area a re-entry vehicle will land within three standard deviations from the mean at the center] of the vehicle upon re-entry is wholly contained within the site" [26].

As a result, spaceport operators that want to support launch and re-entry activities must be licensed as both launch and re-entry sites [27]. This dual licensing regime is a product of the Act's historical development. Originally, the Act only granted the SOT authority over launch activities [28]. Authority over re-entry activities was not granted until the 1998 CSLA amendments [29]. Congress added the re-entry licensing language and failed to integrate launch and re-entry site activities into one all-encompassing general spaceport/site license. Furthermore, issuance of a license to operate a launch or re-entry site does not authorize a vehicle operator to use that site [30]. Before vehicle operators can utilize a spaceport, they have to demonstrate through Federal Aviation Administration Office of Commercial Space Transportation (FAA/AST) licensing procedures that the spaceport is suitable for the use proposed by the vehicle operator in accordance with the Regulations [31].

Whether or not a launch or re-entry site license (i.e. spaceport license) is required depends on the type of commercial space activity undertaken, who is performing the activity, where the activity is occurring, and whether or not the US government has any agreements with foreign countries to provide jurisdiction over the activity [32]. All individual citizens of the USA and entities organized or existing under the laws of the USA or a state of the USA (i.e. US corporations) are required to have a license or permit for the operation of a launch or re-entry site, regardless of the territory in which these activities take place [33]. This is consistent with an interpretation of Article VI of the Outer Space Treaty that obligates authorization and supervision of non-governmental activities undertaken by nationals, regardless of where the activity is taking place [34].

US citizens or corporations that operate spaceports in a foreign country will be required to comply with two legal licensing regimes: licensing as required under US law and licensing as required under the law of the foreign country where activities are undertaken [35]. It is unclear what impact, if any, this dual licensing regime will have. What is clear is that a host of possible issues may arise, including:

1. US regulatory standards may impose additional costs upon US spaceport operators as compared to foreign competitors (creating a cost disadvantage for US spaceport operators).
2. Other States party to the Outer Space Treaty may not obligate licensing for nationals outside of their territory, resulting in a non-uniform interpretation and application of Article VI Outer Space Treaty obligations.
3. US citizens or corporations may attempt to skirt US extraterritorial licensing requirements by operating spaceport ventures through foreign corporations.

The Act anticipates attempts by US citizens or corporations to skirt US extraterritorial licensing requirements and has created a long-arm statute requiring "an entity organized or existing under the laws of a foreign country if the controlling interest [as defined by the SOT] is held by an individual or entity" of the USA, to acquire DOT licenses to operate launch/re-entry sites (i.e. spaceports) [36]. This long-arm provision applies when activities are undertaken outside the territory of either the USA or the territory of the foreign country where the entity is organized or exists [37]. While some may criticize the USA for this exercise of extraterritorial authority, the Act does remove a lacuna under international law, effectively shutting down flags of convenience for commercial spaceport operators (at least with regard to entities of the country in which a US citizen or company maintains a controlling interest), where a foreign country does not exercise jurisdiction (and hence fails to authorize or supervise activities) outside of its territory (i.e. the high seas or outer space) [38]. This long-arm provision does not apply if there is an agreement between the US government and the government of the foreign country (where the entity is organized or exists), which provides that the government of the foreign country has jurisdiction over the launch, operator, or re-entry [39]. The Act does allow for the application of the long-arm provision in the territory of a foreign country if there is an agreement between the US government and the government of the foreign country (where the entity is organized or exists), which provides that the US government has jurisdiction over the launch, operator, or re-entry [40].

The Act grants the SOT significant discretionary authority to waive spaceport license requirements or even the need to obtain a spaceport license [41]. The SOT may exercise this authority if it "decides that the waiver is in the public interest and will not jeopardize the public health and safety, safety of property, and national security and foreign policy interests of the United States" [42].

Licensing Provisions: private exclusive-use launch sites

Normally, a license is required for an entity to operate a launch or re-entry site. One notable exception to this rule involves private exclusive-use launch sites. Licensed vehicle operators may conduct launches from a launch or re-entry site exclusive to the licensed vehicle's use without obtaining a separate launch site license [43]. Essentially, the FAA is writing into the launch vehicle license, operational parameters that allow the launching of the vehicle from an exclusive private site. Private exclusive-use launch and re-entry sites are still required to satisfy regulations governing safety and environmental issues [44]. Also, although not licensed, private exclusive-use launch sites are still launch sites for the purposes of the Act and Regulations [45].

Licensing Provisions: modification, transfer, suspension, or revocation

The SOT specifies the period for which a license issued or transferred is in effect [46]. The Regulations state that launch site licenses are valid for 5 years from the date of issuance and renewable upon application [47]. The Regulations do not prescribe a specific period of validity for re-entry licenses.

Licenses may be issued or transferred in accordance with the Act and Regulations. Currently, only the FAA can transfer a license [48]. Transfer applicants undergo an application process similar to licensees [49]. Transfers are granted when applicants have satisfied the bases for the issuance of the launch/re-entry site license to be transferred [50]. The FAA may modify licenses on application of a licensee or on its own initiative [51]. Licenses are modified through either the issuance of a license order or written approval to the licensee that adds, removes, or modifies a license term or condition [52]. Licensees are required to apply for modification if:

1. The licensee proposes to operate the launch site in a manner that is not authorized by the license; or
2. The licensee proposes to operate the launch site in a manner that would make any representation contained in the license application that is material to public health and safety or safety of property, no longer accurate and complete [53].

FAA authority to suspend or revoke licensees is established under the Act [54]. This authority may be exercised in three situations. First, the FAA may suspend or revoke a license if a "licensee has not complied substantially with a requirement of [the Act] or a regulation prescribed under [the Act]" [55]. Second, the FAA may suspend or revoke a license "to protect the public health and safety, the safety of property, or a national security or foreign policy interest of the United States" [56].

The last basis for suspension was established under the Commercial Space Launch Act Amendments of 2004 (CLSAA-2004). The CLSAA-2004 grants the SOT authority to suspend a license when a previous launch or re-entry under the license has resulted in a serious or fatal injury (as defined in 49 CFR 830, in effect on 10 November 2004) to crew or space-flight participants and the Secretary has determined that continued operations under the license are likely to cause additional serious or fatal injury (as defined in 49 CFR 830, in effect on 10 November 2004) to crew or space flight participants [57].

What is interesting is that the language of the CLSAA-2004 does not explicitly limit the authority to suspend, solely to launch vehicle operator licenses. Rather, the Act grants the SOT the authority to suspend "a license", without specifying a particular type [58]. Therefore, the Act grants the SOT the authority to suspend any license granted under the Act, including launch or re-entry site operator licenses. As a result, spaceport operators need to be advised that their operator license could be suspended if a launch or re-entry vehicle utilizing their spaceport has an accident that results in a serious or fatal injury, even if the spaceport operators have complied substantially with the Act and Regulations and their spaceport operations are not a threat to the public health or safety, safety of property, national security, or foreign policy.

Unless the FAA specifies otherwise, modifications, suspensions, and revocations take effect immediately and remain in effect during administrative review [59]. The Act creates an exception to this general rule, mandating that suspensions based on "serious or fatal injury … to crew or space-flight participants" [60] be as brief as possible and cease when "the licensee has taken sufficient steps to reduce the likelihood of a recurrence" or "have modified the license … to sufficiently reduce the likelihood of a recurrence" [61].

14.1.1.3 Post-Licensing Provisions: enforcement and penalty

Spaceport operators must allow federal officers or employees, or other individuals authorized under the FAA/AST, to observe any activity associated with the licensed operation of the spaceport [62]. This monitoring authority also extends to a spaceport operator's customers, contractors, or subcontractors, to the extent their activities are associated with the licensed operation of the spaceport [63].

The Act grants specific enforcement authorities that have been delegated to the FAA/AST. In carrying out the Act, the FAA/AST may conduct investigations and inquiries, administer oaths, take affidavits, and enter a spaceport to inspect an object to which the Act applies, or a record or report required to be made or kept [64]. The object, record, or report may be seized "when there is probable cause to believe the object, record, or report was used, is being used, or likely will be used in violation of [the Act]" [65]. Spaceport operators should be advised that violating the Act, Regulations, or launch/re-entry site license terms could result in civil penalties [66]. Within the purview of the Act, violations may result in civil penalties of not more than $100,000 per violation [67]. However, under the Act, "[a] separate violation occurs for each day the violation continues" [68].

Therefore, it is important that spaceport operators monitor spaceport operations closely and hire competent legal counsel to ensure spaceport operations are in conformity with the Act, Regulations, and license terms. Spaceport operators should also be aware that other federal, state, or municipal laws applicable to spaceport operations could potentially impose civil and criminal penalties.

Post-Licensing Provisions: prohibition, suspension, and end of spaceport operations

The FAA/AST has the authority to issue an order prohibiting, suspending, or ending spaceport operations without revoking the spaceport operator license if the FAA/AST decides that spaceport operations are "detrimental to the public health and safety, the safety of property, or a national security or foreign policy interest of the United States" [69]. This order is "effective immediately and remains in effect during a review" [70].

14.1.1.4 Financial Responsibility Provisions

The Act does not require spaceport operators (i.e. licensed launch or re-entry site operators) to obtain liability insurance or demonstrate financial responsibility [71]. Instead, Congress has mandated that vehicle licensees, not launch or re-entry site licensees, obtain coverage to compensate for the maximum probable loss of claims by a third party or the US government for death, injury, property damage, or loss resulting from an activity carried out under the vehicle operator license [72]. This discrepancy is rational if one assumes that catastrophic injury or loss will most likely arise from vehicle operations and not spaceport operations. The problem with this assumption is that the nature of commercial space activities is changing, and these changes undermine the financial responsibility provisions contained in the Act.

Traditionally, commercial spaceports serve ELV customers that launch commercial or government payloads. Such activities pose little risk to third parties. Most likely, the primary risk traditional commercial spaceport operations pose to third parties is potential environmental damage resulting from ground, air, or water contamination. Government compensation is not an issue, because this provision is designed to protect government spaceport facilities and infrastructure from vehicle damage [73].

Today, some commercial spaceports are supporting non-traditional operations, such as horizontal take-off and landing (HTOL) reusable launch vehicles (RLVs) and other types of human space flight operations. These operations challenge the presumptions of the financial responsibility provisions, because potential third-party liability and the possibility of catastrophic losses for spaceport operators increase once HTOL RLVs and human space-flight participants are spaceport customers. For example, regularly scheduled RLV flights that traverse national airspace may require navigation and communication services from spaceport operators. Providing this service will expand the scope of potential third-party liability for spaceport operators to parties injured as a result of negligent navigation and communication services rendered. Also, consider that spaceports will be open to space-flight participants, crew, or the general public, all of which may be injured on site, hence increasing the universe of potential third-party liability. Further, consider that airports in major cities may convert to aerospace ports supporting HTOL RLVs in order to serve the needs of major US metropolitan markets. In this event, spaceport operations will no longer be in remote locations, but rather in major population centers, resulting in an increased risk of third-party and catastrophic losses.

Whether or not spaceport operators should be required to maintain liability insurance or demonstrate financial responsibility is a policy decision. When the commercial spaceport and space transportation industry develops more fully, this may be an issue of importance. Arguments favoring mandatory liability insurance or financial responsibility include the positions that such a mandate assures compensation for innocent third parties, assures reimbursement to the USA in the event of liability established under corpus juris spatialis, and provides equitable treatment for vehicle and site licensees. Arguments against mandatory liability insurance or financial responsibility include the positions that spaceport operators should have the freedom to assume risk and decide whether or not to obtain insurance, that mandatory insurance is not necessary given the limited risk spaceport operations incur, and that the issue should be decided by state governments and not the federal government.

14.1.1.5 Secretary of Transportation Regulatory Authority

The SOT has general authority to carry out the Act [74] and shall issue regulations to carry out the Act [75]. The SOT is given a broad grant of regulatory authority as is "necessary to protect the public health and safety, safety of property, national security interests, and foreign policy interests of the United States" [76]. To that end, the SOT may prescribe regulations necessary to ensure compliance with the Act, including on-site verification [77]. The SOT is obligated to promulgate regulations within certain periods of time and to "establish procedures … that expedite review of [spaceport license applications] and reduce the regulatory burden for an applicant" [78].

The Act has granted the SOT specific regulatory authority to license the operation of spaceports [79]. The SOT is authorized to regulate spaceport license application procedure in the form and way it prescribes [80]. As a general rule, all requirements of the laws of the USA are applicable to the operation of a spaceport and are requirements for obtaining a license [81]. The SOT may prescribe by regulation that a requirement of a law of the USA not be a requirement for a license [82]. This is not to say that the law will not apply to the operation of spaceports. Rather, the Act grants discretion to the SOT with regard to licensing spaceport operations.

14.1.1.6 Other Provisions

The Act also contains provisions regarding inter-agency consultation, space advertising, pre-emption of scheduled launches/re-entries, acquisition of federal property and services, experimental sub-orbital rocket permits, human space flight-related provisions, administrative hearings/review, and the relationship of the Act to other executive agencies, laws, and international obligations [83]. Several of these provisions are directly relevant to spaceport operations.

Space advertising

Advertising is an alternative method for spaceport operators to generate revenue. Spaceports that serve human space-flight participants (SFPs) may be in a particularly strong position to generate advertising revenues. Spaceport operators and advertisers should be able to market products to SFPs during their training, orientation, space flight, and post-flight activities. Given the cost of human space flight, SFPs will mostly be wealthy persons, a definite advantage when marketing advertising for high-end products and services. In addition, spaceports serving SFPs may have a substantial amount of visitors to view space-flight launches or to inquire about spaceport and space-flight operations. These non-SFP persons may also be potential audiences for spaceport advertising.

In 1993, Space Marketing Inc., a US corporation, proposed to orbit a 1-mile-wide display satellite at an altitude of 180 miles that would be "legible to the naked eye" [84]. A public uproar ensued, and Congress passed a provision of the Act prohibiting licensees from launching payloads to be used for "obtrusive space advertising" [85]. "Obtrusive space advertising" is defined by the Act as "advertising in outer space that is capable of being recognized by a human being on the surface of the Earth without the aid of a telescope or other technological device" [86]. As a result, advertising in outer space has been limited to non-obtrusive advertising such as corporate sponsorship logos and product placement [87]. Spaceport operators should be aware of this provision and should not support vehicle launches that will violate this provision.

The Act specifically allows non-obtrusive space advertising, including advertising on spaceport launch and support facilities [88]. A careful reading of this provision reveals that obtrusive terrestrial advertising is not prohibited. As a result, terrestrial spaceports can advertise obtrusively on site, so long as their advertising does not violate any other federal, state, or local laws.

Acquisition of US government property and services

Purchasing launch or re-entry property from the USA may be a cost-efficient procurement method for commercial spaceports developing or expanding launch, re-entry, and support facilities. The Act provides for private sector and state government acquisition of excess US government launch or re-entry property [89]. Property can be acquired by sale or transaction at fair market value [90]. The price for property not acquired by sale or transaction "is an amount equal to the direct costs, including specific wear and tear and property damage, the government incurred because of acquisition of the property" [91].

Administrative hearings and judicial review

The SOT is obligated to provide a hearing to spaceport license applicants "for a decision [by the SOT] to issue or transfer a license with terms or deny the issuance or transfer of a license" [92]. In addition, for any modification, suspension, or revocation of a spaceport license, as well as the prohibition, suspension, or end of spaceport operations, the SOT must provide a hearing [93]. A final action by the SOT under the Act is subject to judicial review [94].

Relationship to other executive agencies and laws

In addition to the Act, commercial spaceports are subject to a range of federal laws. The Federal Communication Commission, Department of State, Environmental Protection Agency, Department of Defense, National Aeronautics and Space Administration, and other federal agencies all have been delegated regulatory authority over some aspect of commercial space activity. The FAA/AST has been deemed the coordinator for licensing commercial space activities, effectively communicating and coordinating on behalf of the license applicant, subject to the provisions of the Act.

Except as provided for in the Act, a person is not required to obtain a license to operate a spaceport from an executive agency [95]. In theory, this should result in lower licensing costs and improved licensing efficiency for the commercial spaceport industry. States, or political subdivisions of a state, may adopt or have in effect laws, regulations, standards, or orders that are in addition to, or more stringent than, a requirement of the Act or Regulations, so long as the state or local law and regulations are not inconsistent with the Act [96].

14.1.2 Commercial Space Transportation Regulations

The FAA/AST has promulgated regulations in accordance with authority delegated by the SOT established under the Act. The Regulations are listed as Commercial Space Transportation Regulations in Title 14 of the Code of Federal Regulations, §§400-1169. The Regulations are divided into three subchapters:

1. Subchapter A—General
2. Subchapter B—Procedure
3. Subchapter C—Licensing.

Legal counsel for commercial spaceport operators have to pay special attention to the regulations governing license application procedure, criteria and information requirements for obtaining a license, license terms and conditions, responsibilities of licensees, and investigation and enforcement.

14.1.2.1 Spaceport licensing process

The primary regulatory function of the FAA/AST is to license commercial space activities. To that end, this section will examine the spaceport licensing process established by the Regulations.

The spaceport licensing process is outlined in Figure 14.1 [97].

Pre-application consultation

Applicants are required to consult with the FAA/AST before submitting an application [98]. During this consultation, the FAA/AST discusses the application process, identifies possible regulatory issues and issues relevant to the FAA's licensing decision, and helps the applicant make any changes to the proposed application in an effort to prevent significant delay or costs to the applicant [99].

Policy review and approval

The FAA/AST conducts an inter-agency review of a license application to assess whether it presents any issues affecting national security, foreign policy interests, or international obligations of the USA [100]. For the policy review, applicants must submit information on the proposed launch site operator, launch site, foreign ownership interests, and launch site operations [101].

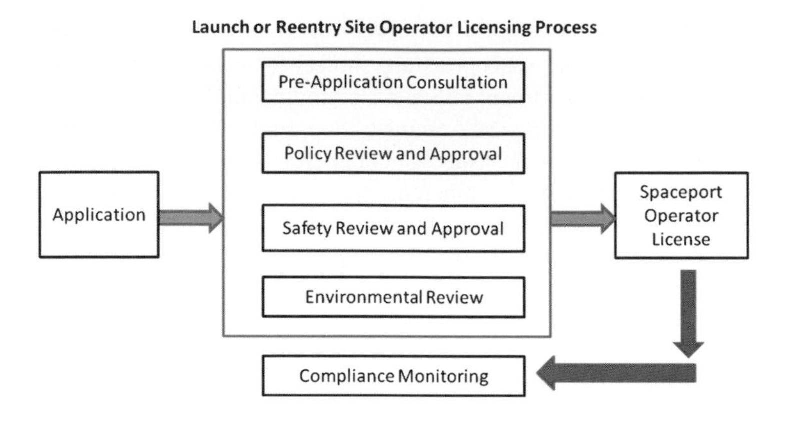

FIGURE 14.1

The spaceport licensing process.

Safety review and approval

The Regulations impose different safety review requirements for launch site and re-entry site applicants. The safety review for launch site licenses is much more detailed and stringent than for re-entry sites. Spaceports must be separately licensed to support both launch and re-entry operations. Launch site applicants must pass a safety review of the proposed launch site location and launch site operations.

To gain approval for a launch site location, an applicant must demonstrate that for each launch point proposed for the launch site, at least one ELV or RLV can be flown from the launch point safely [102]. If the applicant proposes more than one type of vehicle flown from a launch point, the applicant must demonstrate that every proposed vehicle can be flown safely from that launch point [103]. Also, the applicant must demonstrate that the heaviest weight class planned to be flown from the launch point can be flown safely [104]. Proposed launch site operations that are reviewed include control of public access, scheduling of site operations, notifications, FAA Air Traffic Controller (ATC) and Coast Guard coordination agreements, accident investigation plans, record keeping, handling of hazardous materials/explosives, handling of propellants, and lightning protection [105]. The main safety requirement for a re-entry site is that the site "support re-entry [of vehicles] for which the three-sigma footprint of the vehicle upon re-entry is wholly contained within the site" [106]. The FAA/AST reserves the right to issue a re-entry site license on a determination that operation of the site does not jeopardize public health or safety [107]. The FAA will most likely exercise this reserved authority in more regulatory detail when re-entry sites begin servicing the next-generation RLVs, such as single-stage take-off and landing (SSTL) and HTOL vehicles.

Environmental review

The National Environmental Policy Act (NEPA) requires the FAA to issue a detailed statement for every major federal action significantly affecting the quality of the human environment [108]. The decision to license commercial spaceports is a major federal action under the NEPA, and the FAA/AST is responsible for analyzing the environmental impacts of the proposed spaceport and complying with

NEPA requirements [109]. Applicants must provide information as requested by the FAA/AST for an analysis of the environmental impact associated with the proposed spaceport [110].

FAA Order 1050.1E implements FAA policy and procedures for compliance with the NEPA [111]. "NEPA analysis can be accomplished through various forms of environmental documentation depending on the size and type of proposed action. Such documentation can be a Categorical Exclusion (CATEX), an Environmental Assessment (EA), or an Environmental Impact Statement (EIS)" [112].

Compliance monitoring

The FAA monitors licensees to ensure compliance with the Act, Regulations, and license terms and conditions. To that end, spaceport operators must allow federal officers, employees, or other individuals authorized by the FAA/AST to observe any activity associated with the licensed operation of the spaceport [113]. In the event of non-compliance, the FAA/AST has the authority to suspend or revoke licenses [114], issue emergency orders [115], and impose civil penalties [116].

14.1.3 US National Space and Space Transportation Policy

National Space Policy and Space Transportation Directives are executive policy instruments "understood as a statement of goals or objectives which a President sets and pursues. Whether these directives have the force of law [is a matter of legal debate and] depends upon such factors as the President's authority to issue them, their conflict with constitutional or statutory provisions, and their promulgation in accordance with prescribed procedure" [117]. This article does not examine the legality of presidential directives. Instead, the directives are examined as statements of policy that are meant to be implemented by the Executive branch in accordance with the law. To that end, these directives provide context for the implementation of law and regulations governing commercial spaceport activities.

On 31 August 2006, President George W. Bush authorized a new US National Space Policy Directive ("National Directive") "that established overarching national policy ... governing the conduct of US space activities" [118]. While the National Directive covers a variety of space-related activities, our interest lies in the Commercial Space Guidelines [119]. The Commercial Space Guidelines require Executive departments and agencies to "maintain a timely and responsive regulatory environment for licensing commercial spaceports" [120], as well as to "ensure that United States government space activities, technology, and infrastructure are made available for private use on a reimbursable, non-interference basis to the maximum practical extent, consistent with national security" [121].

These guidelines have two practical impacts. First, the FAA/AST is compelled to process commercial spaceport licenses in a timely manner. This is important for commercial spaceport operators, because FAA/AST licensing activities impose both direct and indirect costs on commercial operators. Second, federal ranges are made available for private use, resulting in commercial spaceports co-locating on federal ranges.

The US Space Transportation Policy Directive ("Transportation Directive") establishes "national policy, guidelines, and implementation actions for [US] space transportation programs and activities to ensure the Nation's ability to maintain access to and use space for US national and homeland security, and civil, scientific, and commercial purposes" [122]. The Transportation Directive contains three provisions of special importance to commercial spaceport activities.

First, the Transportation Directive elucidates that access to federal space launch bases and ranges, as well as other government facilities and services, is to be provided for commercial purposes on a stable, predictable, and direct-cost basis (as defined in 49 USC §70101) [123]. As a result, commercial spaceports co-located on federal ranges, as well as launch and service providers, receive the benefit of federal space infrastructure and personnel on an at-cost basis.

Second, "private sector and state and local government investment and participation in the development and improvement of space infrastructure", including non-federal spaceports, are encouraged [124]. Given the relative strength of the federal government, as compared to state and local governments, simply implementing a policy of encouragement instead of dissuasion supports state and local participation. As is discussed later in this article, state and local governments are participating significantly in the development and improvement of non-federal spaceports, in part because of Congressional legislation and Executive policy [125].

Third, commercially available US space transportation products and services are to be purchased "to the maximum extent possible, consistent with mission requirements and applicable law" [126]. If spaceports, and the services they provide, are considered products and services, this provision appears to mandate that US government departments and agencies use commercial spaceports, instead of federal spaceports, when consistent with mission requirements and applicable law, including national security. In practice, it is not clear whether the government is utilizing commercial spaceports to the maximum extent possible. While the policy is in favor of government purchasing commercial spaceport products and services, if the commercial spaceport offers services and mission parameters comparable to those of federal spaceports most non-federal commercial spaceports cannot offer comparable service and mission parameters [127]. The commercial spaceport industry needs to further develop before this policy provision can fully take effect.

14.2 STATE LAW: COMMERCIAL SPACEPORT INITIATIVES

Several states have enacted, or are proposing, legislation that fosters the development of commercial spaceports. This type of legislation is known as spaceport initiatives. The ultimate goal of these initiatives "is to generate state economic growth and ... improve a state's revenue base" [128]. The Act does not prohibit this legislation, so long as the spaceport initiatives are not inconsistent with the Act [129].

California, Florida, New Mexico, Oklahoma, Texas, and Virginia have enacted legally binding spaceport initiatives. In addition, during the 2008 legislative session, the Hawaii legislature proposed an initiative to fund the costs associated with applying for a commercial space transportation license with the FAA/AST [130]. Each state's initiative is unique and contains various provisions designed to create, improve, and promote commercial spaceport infrastructure development and use. Sometimes these initiatives are passed as a series of laws over a period of months, or even years.

While each state's initiatives are unique, there are common strategies that the states have pursued in their spaceport initiatives. These include the establishment of spaceport authorities [131], tax incentives [132], state and local taxing and bonding authorization [133], military spaceport infrastructure conversion [134], trust funds [135], liability immunity [136], and spaceport infrastructure development [137]. Legal counsel for spaceport operators and service providers should be aware of these initiatives and advise their clients on the comparative advantages and disadvantages each state provides.

Notes and references

[1] US Constitution, Articles I, II, and III.
[2] *Ibid.*
[3] 14 CFR §§400-1199 (2008).
[4] 49 USC §70101 (2000, Suppl. 2004).
[5] See, e.g., Project of the Nuclear Age Peace Foundation, Presidential Directive on National Space Policy, http://nuclearfiles.org/menu/key-issues/space-weapons/issues/national-space-policy-presidential-directive.html (last visited 1 October 2008).
[6] Hughes, T. R. and Rosenberg, E. (2005). Space Travel Law (And Politics): The Evolution of the Commercial Space Launch Amendments Act of 2004. *J. Space Law,* 1: 11–12.
[7] 49 USC §70117(g) (2000).
[8] Hughes and Rosenberg, *supra* note 6, at 12.
[9] 49 USC §70103(a) (2000, Suppl. 2004); Hughes and Rosenberg, *supra* note 6, at 12.
[10] Hughes and Rosenberg, *supra* note 6, at 13 n.41.
[11] Id. at 16–17.
[12] Id. at 19–20.
[13] Commercial Space Launch Amendments Act (2004). Pmbl., Pub. L. No. 108-492, 118 Stat. 3974.
[14] Hughes and Rosenberg, *supra* note 6, at 48.
[15] 49 USC §§70101-70103 (2000, Suppl. 2004).
[16] Id. §§70104-70105, 70107.
[17] Id. §§70106, 70108, 70115.
[18] Id. §§70112-70113.
[19] See, e.g., id. §§70103, 70105, 70120(a).
[20] Id. §§70105a, 70109-70111, 70116-70117.
[21] See Part II, *supra* notes 3–8.
[22] 49 USC §70102(7) (Suppl. 2004).
[23] Id. §70102(15).
[24] 14 CFR §420.41(a) (2008).
[25] Id. §420.41(b).
[26] Id. §433.5. For a more detailed explanation of three-sigma footprints and re-entry launch sites, see Commercial Space Transportation Reusable Launch Vehicle and Re-entry Licensing Regulations, 64 Fed. Reg. 19,626 (21 April 1999). The three-sigma footprint describes the area where the vehicle will land with a 0.997 probability rate, assuming no major system failure. The statistical term "three-sigma" refers to three standard deviations from the mean, or average point, assuming a standard normal distribution. The area that is within three standard deviations from the mean point encompasses the area surrounding it with the mean at its center. An area within two or even one standard deviation of the mean point is a smaller, more precise measure; however, statistically there is less chance of an event falling within that range. The larger the area, the higher degree of confidence one has of an event falling within its boundary limits, assuming a normal distribution of events (quoting 64 Fed. Reg. 19626).
[27] See id. at 19,631.
[28] Id. at 19,630.
[29] Id.
[30] 14 CFR §433.3.
[31] Id.
[32] 49 USC §70104 (2000, Suppl. 2004).
[33] 49 USC §§70104(a), 70102(1) (2000).

[34] See Outer Space Treaty, *supra* note 6, Art. VI.
[35] See 49 USC §70104.
[36] Id. §70102. The Secretary of Transportation currently defines "controlling interest" as "ownership of an amount of equity in such entity sufficient to direct management of the entity or to void transactions entered into by management with ownership of at least fifty-one percent of the equity … creating a rebuttable presumption that such interest is controlling." 14 CFR §401.5.
[37] 49 USC §70104(a).
[38] See Id.
[39] Id. §70104(a)(3).
[40] Id. §70104(a)(4).
[41] 49 USC §70105(b)(3) (2000, Suppl. 2004).
[42] Id.
[43] Commercial Space Transportation Reusable Launch Vehicle and Re-entry Licensing Regulations, 65 Fed. Reg. 56,618, 56,648 (19 September 2000). See also 14 CFR §§415.101, 417.9, 417.111, 417.403 (2008); Commercial Space Transportation; Waiver of License Requirement for Blue Origin's Pre-flight Preparatory Activities Conducted at a US Launch Site, 71 Fed. Reg. 62,037 (20 October 2006).
[44] 65 Fed. Reg. at 56,648. "Safety and environmental issues associated with private use of a launch site by a launch or re-entry licensee, as well as an RLV mission licensee, would be addressed as part of the license to operate the vehicle." Id. See also 14 CFR §§415.101, 417.9, 417.111, 417.403.
[45] See 71 Fed. Reg. at 62,038 n.1. "Under current FAA policy, the FAA does not require Blue Origin to obtain a part 420 license for the operation of West Texas Launch Site. Nonetheless, although not licensed, West Texas Launch Site is still a launch site." Id.
[46] 49 USC §70107.
[47] 14 CFR §420.43.
[48] Id. §420.45(a).
[49] Id. §420.45(b). Transfer applicants submit an application in accordance with 14 CFR §413. Id.
[50] The Regulations require transfer applicants, in accordance with 14 CFR §413, to satisfy the requirements of 14 CFR §§420.15 and 420.17. Id.
[51] 49 USC §70107(b). Note that the Act grants authority to the SOT, who has delegated that authority to the FAA Administrator. Id.
[52] 14 CFR §420.47(e).
[53] Id. §420.47(b).
[54] 49 USC §70107(c) (2000).
[55] Id. §70107(c)(1).
[56] Id. §70107(c)(2).
[57] 49 USC §70107(d)(1) (Suppl. 2004).
[58] See 49 USC §§70107(c)-(d) (2000, Suppl. 2004).
[59] 49 USC §70107(e) (Suppl. 2004).
[60] Id. §70107(d)(1).
[61] Id. §70107(d)(2).
[62] See 49 USC §70106 (2000, Suppl. 2004); 14 CFR §405.1 (2008).
[63] See 49 USC §70106; 14 CFR §405.1.
[64] 49 USC §70115(b).
[65] Id.
[66] 49 USC §70115(c)(1) (2000).
[67] Id.
[68] Id.

[69] Id. §70108(a).
[70] Id. §70108(b).
[71] Id. §70112(a).
[72] Id.
[73] See id. §70112.
[74] 49 USC §70103(a) (2000, Suppl. 2004).
[75] 49 USC §70120(a) (2000).
[76] 49 USC §70105(b)(2)(B) (2000, Suppl. 2004).
[77] 49 USC §70105(b)(2)(A) (2000).
[78] 49 USC §70105(d) (Suppl. 2004).
[79] 49 USC §70105(a) (2000, Suppl. 2004).
[80] Id.
[81] Id. §70105(b)(1).
[82] Id. §70105(b)(2)(C). In order for the SOT to rule that a requirement of a law of the United States not be a requirement for a license, the SOT must consult "with the head of the appropriate executive agency [and decide] that the requirement is not necessary to protect the health and safety, safety of property, and national security and foreign policy interests of the United States." Id.
[83] Id. §§70105(a), 70109–70111, 70116–70117.
[84] Browne, M. W. (1993). City Lights and Space Ads May Blind Stargazers. *New York Times*, 4 May, at C1.
[85] 49 USC §70109a(b) (2000).
[86] 49 USC §70102(9) (Suppl. 2004).
[87] See Pizza Hut Becomes First Company in History to Deliver Pizza to Residents Living in Outer Space, Bus. Wire, 22 May 2001, http://findarticles.com/p/articles/mi_m0EIN/is_2001_May_22/ai_74847510. Pizza Hut has placed corporate logos on launch vehicles and even delivered the world's first space consumable pizza to the International Space Station. Id.
[88] 49 USC §70109a(c) (2000).
[89] Id. §70111(a)(1)(A).
[90] Id. §70111(b)(2)(A).
[91] Id. §70111(b)(2)(B).
[92] 49 USC §70110(a)(1) (2000, Suppl. 2004).
[93] 49 USC §§70110(a)(3)(A)–(B) (2000).
[94] Id. §70110(b).
[95] Id. §70117(a).
[96] Id. §§70117(c)(1)–(2).
[97] This chart was created by the author, Michael C. Mineiro, on 23 March 2008.
[98] 14 CFR §413.5 (2008).
[99] Id.
[100] 49 USC §§70116(a)–(b) (2000). See also Fed. Aviation Admin., Launch Site Policy Review and Approval, http://www.faa.gov/about/office org/headquarters offices/ast/licenses permits/launch site/policy/ (last visited 23 March 2008).
[101] 14 CFR §420.15.
[102] Id. §420.19(a).
[103] Id. §420.19(b).
[104] Id. §420.19(c).
[105] Id. §§420.51–420.71.
[106] Id. §433.5.
[107] Id. §433.3.

[108] National Environmental Policy Act of 1969, 42 USC §4332(C) (2000).
[109] 14 CFR §§415.201, 433.7. See also 40 CFR §1508.18 (2007). n194.14.
[110] 14 CFR §§415.201, 433.7. See also 40 CFR §1508.18.
[111] See generally Fed. Aviation Admin., Order 1050.1E, Environmental Impacts: Policies and Procedures (2004). Available at http://www.faa.gov/regulations policies/orders notices/media/ALL1050-1E.pdf.
[112] Assoc. Admin. For Commercial Space Transp., Fed. Aviation Admin., Guidelines For Compliance With the National Environmental Policy Act and Related Environmental Review Statutes for the Licensing of Commercial Launches and Launch Sites 6 (2001), available at http://www.faa.gov/about/office org/headquarters offices/ast/licenses permits/media/epa5dks. pdf.
[113] 49 USC §70106(a) (2000, Suppl. 2004); 14 CFR §405.1.
[114] 49 USC §70107(c)(1) (2000); 14 CFR §405.3.
[115] 14 CFR §405.5.
[116] 49 USC §70115(c)(1).
[117] Relyea, H. C. (2007). Cong. Res. Serv., Presidential Directives: Background and Overview 2, available at http://www.fas.org/sgp/crs/misc/98-611.pdf.
[118] Office of Sci. and Tech. Policy, US National Space Policy 1 (2006). Available at http://www.ostp.gov/galleries/default-file/unclassified%20National%20 Space%20Policy%20-%20FINAL.pdf.
[119] See id. at 6–7.
[120] Id. at 7 ("Maintain a timely and responsive regulatory environment for licensing commercial space activities and pursue commercial space objectives without the use of direct Federal subsidies, consistent with the regulatory and other authorities of the Secretaries of Commerce and Transportation and the Chairman of the Federal Communications Commission.").
[121] Id.
[122] Office of Sci. and Tech. Policy, US Space Transportation Policy 1 (2005). Available at http://www.ostp.gov/galleries/Issues/Space Transportation Policy05.pdf.
[123] Id. at 6.
[124] Id.
[125] See *infra* Part V.
[126] Office of Sci. and Tech. Policy, US Space Transportation Policy, *supra* note 120, at 6.
[127] See Handberg, R. and Johnson-Freese, J. (1998). State Spaceport Initiatives: Economic and Political Innovation in an Intergovernmental Context, 28 Publius. *J. Federalism*, 91:91.
[128] Id.
[129] 49 USC §70117(c) (2000).
[130] HR 2259, 24th Leg., Reg. Sess. (Haw. 2008).
[131] See, e.g., Cal. Govt. Code §§13,999–13,999.4 (West 2005); Wis. Stat. Ann. §§114.60–114.78 (West Suppl. 2007); HR 89, 47th Leg., 2d Sess. (NM 2006).
[132] See, e.g., HR 2259, supra note 214; S.B. 286, 2008 Leg., Reg. Sess. (Va. 2008).
[133] See, e.g., HR 473, 47th Leg., 2d Sess. (NM 2006).
[134] See, e.g., Okla. Stat. Ann. tit. 74, §§5201-5237 (West 2002).
[135] See, e.g., Tex. Govt. Code Ann. §481.0069 (Vernon 2004).
[136] See, e.g., Va. Code Ann. §8.01-227.9 (West 2007).
[137] See, e.g., HR 835, 47th Leg., 2d Sess. (NM 2006).

CHAPTER

Regulations and licensing at the European Spaceport

15

Bruno Lazare, Jean-Pierre Trinchero

CNES General Inspectorate and Quality Directorate, Safety Central Office

CHAPTER OUTLINE

INTRODUCTION

The prime objective of regulations and the "licensing" process at the European Spaceport is to protect people, properties, public health, and the environment. At the European Spaceport, these regulations issue from history, and the space safety requirements and processes are mainly justified from lessons learned. This last decade, the context has changed with the range opening to different launch vehicles,

Space Safety Regulations and Standards. DOI: 10.1016/B978-1-85617-752-8.10015-7

industrial companies' reorganization, worldwide competition, etc. The legal framework should be adapted to this major evolution. This chapter details the historical situation and framework, describes the lawmaking process, reminds of the international, European and national context that leads to this new framework, and then presents the French Act's scope and main features, including the interfaces with others' existing territory regulations. However, this chapter focuses only on space preparation and launching activities performed under French Governmental Responsibility or Liability from the European Spaceport.

In just a few words, the general purpose of the French Space Operation Act (FSOA) is to set up a coherent national regime of authorization and control of space operations under French jurisdiction, or for which the French government bears international liability either under UN Treaties' principles (namely the 1967 Outer Space Treaty, the 1972 Liability Convention, and the 1976 Registration Convention) or in accordance with its European commitments with the European Space Agency (ESA) and its Member States.

15.1 ORIGIN OF THE FSOA

15.1.1 International legal sources

France, like any State party of the 1967 Outer Space Treaty, is bound by obligations specified in articles VI and VIII to authorize and control national space activities carried out by its national non-governmental entities (private companies) and to exercise its full jurisdiction and control on its registered space objects. In addition, as a "Launching State" on the basis of the 1972 Liability Convention, France potentially bears, at the first level, i.e. toward victims, the full burden of indemnification for damage that could be caused to third parties:

- On ground or airspace, by a French private space launch service provider (Arianespace, Starsem); or
- By any launch service provider operating from the French territory or jurisdiction (especially from the Guyana Space Centre, the European Launch Base, under French territory); or
- By any French satellite operator that procures a launch service from a foreign country or to a foreign company (e.g. Eutelsat's satellite launched from Russia or by Sea Launch); the French government's liability can also be retained, without any time limit, for damage, in orbit or on earth re-entry, caused by a foreign satellite having being launched under French responsibility without having being registered by their appropriate state. All these UN commitments (the 1967 Outer Space Treaty, the 1972 Liability Convention, and the 1976 Registration Convention), as regularly signed, ratified, and published in the *Journal Officiel de la République Française*, have been directly enforceable under French jurisdiction, without any other formality. The issue arising, then, was not so much the lack of any domestic space legislation in France as the necessity to provide common and predictable implementing measures for any new activity.

15.1.2 The existing European and domestic technical and legal framework for preparation and launch operations from Guyana Space Centre since 1976

The Guyana Space Centre (GSC) is a French launch base, set up in the mid-1960s, to follow up national launcher programs. At the same time, the GSC was declared open to international and

European cooperation. In 1965, the Centre National d'Etudes Spatiales (CNES) proposed the GSC for use to the European Launcher Development Organization (ELDO), a former ESA organization, which was developing the "*Europa II* Launcher" program. This first European launcher program was finally stopped in 1972, after a series of launch failures that led to the dissolution of the ELDO. This decision was taken on 31 July 1973, during the VIth European Space Conference, together with the dissolution of the European Satellite Research Organization (ESRO), the creation of the ESA (effective in1975), and development of the Ariane program, whose management was specially delegated to the CNES (Ariane Agreement signed on 21 September 1973). In this context, the French government and the ESA signed an agreement relating to the GSC on 5 May 1976.

Under this framework agreement:

- The ESA recognizes the French government has delegated to CNES, the French Space Agency, its responsibilities over the general safety, security, and base management at the GSC.
- France and the ESA set up indemnification sharing rules for damages caused by launchers operated from the CSG, under which the ESA will be liable for damage caused by its own launcher programs (*Ariane 1* development, before its qualification flight) or by its defecting satellites, and France for other damage (in particular after 1980, for damage caused by an Ariane launch during its future production phase).
- France and the ESA share the financial burden of the launch base maintenance costs on a one-third/two-thirds basis.

Another intergovernmental treaty was signed on 14 January 1980 by the European government, to participate in the Ariane production phase (which comprises the manufacturing, the marketing, and the launching of the launcher, etc.), and to specify the liability regime of Ariane, to be operated by the private operator Arianespace. This agreement, called the "Ariane Production Declaration", reaffirms the full and exclusive liability of the French government as a Launching State under the UN 1972 Liability Convention.

By the end of the 1990s, the private operators of Arianespace had located their main offices in France and had begun to consider the possibility of other launch operations from the GSC beyond the Ariane launch vehicle. This led to studies of national legislation to authorize such additional launch operations at the GSC.

15.1.2.1 Ground range safety

As a French Territory, ground range safety is subject to four main applicable regulations. These regulations are applicable to the activities of others on the territory, and are not specific to space activities:

- Labor Acts dedicated to workers' occupational health and safety.
- Pyrotechnics safety regulations, dedicated to workers' occupational health and safety in pyrotechnics establishments. These regulations are mainly derived from pyrotechnics and the military. Industrial activity is, more or less, UN-standard compatible.
- Environmental protection acts are edited to control the pollution and accident hazards of installations involving dangerous substances and focus on protection of the environment and the public.
- "Seveso II" regulations set up to prevent major accidents for high-level industrial risk establishments.

This set of regulations is completed and detailed through the Range Space Safety Regulations, focusing on system safety (design, qualification, and operational rules including flight hardware). These requirements are supported by national and international space standard as a best practices base (e.g. BNAE, ECSS, ISO, MILSTD, etc.). These specific regulations are organized into the chapters listed below:

- general safety rules (applicable to all facilities and all contractors);
- information and coordination on hazardous activities issued from different facilities;
- payload and associated preparation facilities;
- launch vehicles and associated preparation facilities including the launch pad;
- general management of emergency plans.

15.1.2.2 Flight safety

These dispositions were implemented to control technical risks generated by a spaceship flight on the Earth's surface, aircraft and space objects in flight, or in atmospheric or extra-atmospheric space in order to protect people, properties, and the environment. To qualify for approval, a launch vehicle must go through a technical qualification review. Part of this process, flight safety, is conducted through the submission of iterative procedures from the design, qualification, and operational program phases. It includes: flight termination system approval; a calculation method for risk assessment and determination approval; trajectories approval, including compatibility with the flight safety officer; real-time procedures; and interfaces with the launch range means—localization, neutralization ground and flight hardware; TM data exploitation from range and down-range tracking station networks; and meteorological conditions applicable for the launch window.

These main features, relative to flight safety, are synthesized, as follows:

- Determinist approach in the close field area, by assessment and management of the flight termination system.
- End the flight, if necessary, from ground/onboard safety devices (fail operational) independent of functional devices (no matter if the failure mode is predictable or not).
- The flight termination decision must be based on at least one external (from the launch vehicle) localization system and the computerized data must be processed out of the launch vehicle environment.
- Best practices approach over ocean area without flight cross over inhabited areas.
- Flight termination decision made by launch vehicle internal or external means.
- Mitigation risk measure—end the flight in the case of non-orbiting mission before flight over inhabited areas.
- Inhibit the flight termination system before flight cross over inhabited areas to avoid large debris footprint in case of late launch vehicle failure.
- Probabilistic approach for flight crossover inhabited areas—trajectory optimization, relative to risk level mitigation.
- Space debris mitigation rules.
- Avoid orbiting debris on protected area.
- Stage passivation.

15.2 THE FRENCH SPACE OPERATION ACT LAWMAKING PROCESS

In 1999, the first discussions started regarding the opening of the GSC to launchers other than the *Ariane* family, in particular the *Soyuz* launch vehicle. This led to a special intergovernmental agreement signed on November 2003 between Russia and France.

In parallel, a first appraisal report "On the evolution of space law in France" was issued by the Ministry of Research (in charge of space affairs) in 2003. The study was based on work carried out over 18 months by four groups specializing in launch services, Earth observation by satellite, space radio communication and navigation by satellite, and ownership and securities on space objects. These four groups have mobilized the contributions of over 100 experts or representatives of the main ministries, or public and private organizations, involved in space activities, in France and overseas. The minister concluded that setting up a legal regime specific to space activities was necessary. A few months later, the prime minister officially seized the Council of State in order to conduct thorough consultations and legal studies in this field. The Council published a positive report appending a first draft of a Space Act project in 2006.

After a 3-month period for the governmental instruction process, the first official draft was issued at the end of April 2007. The text was first submitted to the French Senate, which discussed and adopted it, after a few amendments, on 16 January 2008. It was then discussed and amended by the National Assembly on 9 April 2008. The French Space Operation Act (FSOA) was finally adopted by the French Senate on 22 May 2008 after a thorough parliamentary consultation process. It was enacted by the French President on 3 June 2008 and published in the *Official Journal of the French Republic* on 4 June. The two implementing decrees, one specifying the authorization and control regime for space activities and the other, responsibilities of the CNES, were enacted by the government on 9 June 2009 and published in the *Official Journal of the French Republic* on 10 June.

15.3 FSOA: MAIN FEATURES

15.3.1 Concerned activities: definition parameters

Although the FSOA affects all space operations, this chapter is only dedicated to regulations and licensing relevant to the European Spaceport. On-orbit control and Earth return are also covered by this law, but are outside the scope of this chapter.

The FSOA contains a certain number of definitions that help to define the parameters of application of the Space Operation Act:

- Definition of "space operation" (FSOA Art. 1): "any activity consisting in launching or attempting to launch an object in outer space, or of ensuring the command of a space object during its journey in outer space, including the Moon and other celestial bodies, as well as during its return on Earth".
- Definition of the "operator "(Art. 1): "any entity carrying out, under its responsibility and in an independent way, a space operation. Not a subcontractor under operator's authority."
- Definition of "launching phase" and "command phase" in Arts 1.4 and 1.5. "Launching phase" is the period of time which, as part of a space operation, starts at the moment when the launching operations become irreversible and which, without prejudice to provisions contained, if necessary, in the authorization granted pursuant to the present act, ends when the object to be

put in outer space is separated from its launch vehicle. "Command phase" is the period of time starting as part of a space operation at the moment when the object to be put into outer space is separated from its launch vehicle and ending when the first of the following events occur:
- when the final maneuvers for de-orbiting and the passivation activities have been completed; when the operator has lost control over the space object;
- the return to Earth or the full disintegration of the space object into the atmosphere.

15.3.2 FSOA licensing/authorization and control regime

15.3.2.1 General principles

Authorizations are granted by the research ministry in charge of outer space affairs after completion of the following process:

- An administrative review by the ministry in charge of outer space affairs so that the ministry shall assess moral, financial, and professional guarantees of the operator.
- A technical review of the space system definition and procedures to be carried out by the applicant, in order to check compliance with technical regulations issued by the ministry in charge of outer space affairs. The technical assessment related to operations is delegated to the CNES (Art. 4). Exemptions of technical assessments (see Art. 4.4 for foreign operations) may be granted by the ministry.
- Operations carried out by the CNES in the scope of a "public mission" (governmental programs, science, development of space systems) are not subjected to this authorization process. Nonetheless, the CNES applies the technical regulations, and checks their application through internal independent procedures on a voluntary basis, as exemplary values.

15.3.2.2 Specific regime

Two specific regimes are covered by the FSOA.

Launch from a foreign country

This covers foreign launch services purchased by a French satellite operator (vs. application of domestic legislation or technical regulations, see Art. 4.4 SOA).

For an authorization application related to an operation to be carried out from the territory of a foreign State or from means or facilities falling under the jurisdiction of a foreign State, the ministry in charge of outer space affairs may exempt the applicant from all, or any part, of the compliance assessment with the technical regulations when the national (national operators contracts, liability ceilings) and international (UN treaties, bilateral agreements) commitments of that State, as well as its legislation and internal practices, provide sufficient guarantees regarding the safety of persons and property and the protection of public health and the environment, and regarding liability matters.

Development phase

Although the FSOA licensing/authorization regime is formally applicable to space operations, as regards space safety, a non-mandatory "consultation regime", which is prior to and independent of the

FSOA formal authorization procedure, is possible. It starts at the system level, from the previous design up to the real-time mission. This consultation enables the CNES to certify systems or subsystems under development at given milestones. Such certification issued by the CNES may be used by the operator as enforceable, to document the authorization procedure and to facilitate the granting of the overall authorization.

15.3.2.3 Authorization/licenses

Simple authorization is in place: every space operation is subject to authorization. In order to simplify the authorization procedures, several types of licenses have been introduced. Licenses certifying, for a determined time period, that a space operator satisfies moral, financial, and professional guarantees, may be granted by the administrative authority competent for issuing authorizations.

Licenses may also attest to the compliance of the systems and procedures referred to in the FSOA's technical regulations. These licenses will lighten the file associated with a dedicated launch operation, by treating the recurrent part of it, covering the definition of the system. The license, alone, will not authorize launch operations. A complementary justification file, covering all the aspects specific to the mission and the vehicle model used, is required for any launch operation.

15.3.2.4 "Prescriptions" associated with authorization: authorizations are granted on the basis of technical information available at the time of application

Technical key points will be assessed after the granting of authorization through specific "prescriptions".

The authorizations granted pursuant to the present act may include "prescriptions" set forth for the safety of persons and property, protection of public health and the environment, in particular in order to limit risks related to space debris.

Concerning the launch of a space object, the administrative authority, or the agents acting on its authority and empowered by it to this end, may at any time give instructions and require any measures they consider necessary for the safety of persons and property, and the protection of public health and the environment. This licensing/authorization regime is set up referring to technical safety regulations independently of the range for which the launch operation is issued.

15.3.2.5 Control

An "a posteriori" control regime based on specific prescriptions contained in the authorization/license is set up. The State's agents (including authorized CNES agents) are allowed to carry out the required controls in order to ensure that the operators comply with the obligations set forth in the aforementioned prescriptions:

- Visiting and inspecting the buildings, offices, and facilities from which the operations are undertaken, including the space object.
- Requirement of any useful document or file.
- Controls during the carrying out of the operation.
- Technical administrative investigation in the case of serious incident, or in the case of an accident.
- Measures considered necessary to guarantee the safety of people and property, the protection of public health and the environment.

15.3.3 Technical regulations

The technical regulations are set forth, in particular for the safety of persons and property, the protection of public health, and the environment. The CNES is entitled by the FSOA to assist the government in defining the technical regulations relating to space operations. As mentioned in section 15.1.2.2 of this paper, the technical regulations, as regards safety, are based on the current practices of the CNES, improved by decades of experience in space operations. Their scope is applicable to actual conducted operations: commercial or governmental unmanned space flight, and launch vehicle lift-off from the ground. These practices are very similar to the current best practices and standards among the international space community.

The main technical regulations include the following.

15.3.3.1 Technical files to be prepared by the applicant

- The general notification of compliance with the technical regulations.
- The internal standards and quality management provisions applicable to the space operation to be conducted.
- The danger analysis, including surveys of hazards and risk control plans, taken by the applicant to guarantee the safety of people and property, and to protect public health and the environment.
- The environmental impact studies and measures designed to avoid, reduce, or offset the harmful effects on the environment, including:
  - the risk prevention plan relating to risks caused by the fallback of the space object or fragments thereof;
  - the prevention plan relating to environmental damage, as defined in Article L.161-1 of the environment code;
  - the space debris limitation plan;
  - the collision prevention plan;
  - as applicable, the nuclear safety plan;
  - as applicable, the planetary protection plan;
  - the risk control measures planned during the performance of the space operation;
  - the emergency measures planned.

15.3.3.2 Organizational requirements to manage risks (e.g. safety and quality corporate management)

Specific technical requirements are relative to:

- Mission analysis, flight dynamics, mission robustness, etc.
- Onboard neutralization capacities
- Flight data record capacities
- General safety objectives
- Objectives concerning nominal stages of re-entry, including the non-creation of wreck
- Space debris mitigation rules
- Collision risk avoidance
- Technical compatibility with the involved launch site's technical means and procedures for localization, neutralization, and TM data transmission.

15.3.4 Specific regime for operations conducted from the European Spaceport (Guyana Space Centre)

The President of the CNES shall exercise, on behalf of the State, special police for the safe exploitation of the facilities of the Guyana Space Centre (GSC), within the range perimeter. This perimeter includes the physical territory of the range and the vicinity up to the end of neutralization ground segment capability (the so-called close field flight safety area). As such, it shall be in charge of a general mission of "safeguard" consisting of control of the technical risks related to the preparation and performance of the launches from the Guyana Space Centre, in order to ensure the safety of persons, property, public health, and the environment, on the ground and during the flight, and it shall set out to this end the specific regulations applicable within the limits of the perimeter defined above.

Under the authority of the government representative in the Département of Guyana, the President of the CNES shall coordinate the implementation, by companies and other entities settled in the GSC perimeter as described above, of measures taken in order to ensure the security of the facilities and of the activities undertaken therein, and shall verify that those companies and agencies fulfill their obligations in this respect. The President of the CNES may take, for any space operation, by delegation of the administrative authority mentioned in Article 8 of Act no. 2008-518 dated 3 June relating to space operations, the necessary measures provided for in the same article to ensure the safety of persons and property, as well as the protection of public health and the environment.

The president of the CNES shall be responsible for the special policing of the GSC. To that end, he shall formulate the safeguard actions applicable to the facilities located within the perimeter of the GSC, in particular as regards the activities of designing, preparing, producing, storing, and transporting space objects and their constitutive parts, as well as the tests and operations performed within the perimeter or out of the GSC.

The specific regulation applicable to the Guyana Spaceport covers, in particular, ground range safety as detailed in section 15.1.2.1 of this chapter, and specific flight safety rules imposed by the GSC, as technical interfaces compatible with the ground support equipment for the flight termination system managed by CNES flight safety officers. The specific regulation applicable to the GSC and the technical regulations are complementary regulations that must be considered together for a launch operation from the GSC.

CHAPTER

Accommodating sub-orbital flights into the EASA regulatory system[1] 16

Jean-Bruno Marciacq*, Yves Morier†, Filippo Tomasello‡, Zsuzsanna Erdélyi§, Michael Gerhard||

**Project Certification Manager, Certification Directorate, †Head of Product Safety, Rulemaking Directorate, ‡ATM Airport Safety Officer, Rulemaking Directorate, §Airport Rulemaking Officer, Rulemaking Directorate, ||Legal Adviser, Rulemaking Directorate; European Aviation Safety Agency (EASA), Cologne, Germany*

CHAPTER OUTLINE

[1] The views expressed in this chapter are those of the authors and do not commit the EASA to any regulatory position nor do they necessarily represent the Agency's view or the view of the Community Institutions. Some parts of the original paper have been modified for editorial purposes. Please refer to the Proceedings of the 3rd IAASS Conference in Rome, Italy, 21–23 October 2008 (ESA SP-662, January 2009) for more details [1].

Space Safety Regulations and Standards. DOI: 10.1016/B978-1-85617-752-8.10016-9

INTRODUCTION

The recent, rapid, and successful developments in the domain of commercial space flights have highlighted the need to develop corresponding regulations, in order to protect paying passengers and ensure that risks to people on the ground or in the air are appropriately mitigated.

As explained in further detail hereafter, the present regulatory responsibilities of the European Aviation Safety Agency (EASA) focus on aircraft, the definition of which excludes rockets and capsules. The scope of this chapter therefore meets the definition of space tourism as recently defined by the European Space Agency (ESA): "*sub-orbital flights [performed] by privately funded and/or privately operated vehicles*" [2,3], but is limited to winged aircraft, including rocket-powered aeroplanes, and excluding rockets.

Thus, the term "sub-orbital aeroplanes" has been chosen to address this field. This term encompasses both the operational pattern (sub-orbital, therefore requiring less speed/energy to climb and be spent on the return) and the type of vehicle, namely an aeroplane (airborne with wings) able to climb up to the upper limits of the atmosphere, which may also be considered as the lower limit of outer space, as discussed hereafter for the potential legal implications this may have. Although presenting specific design characteristics, these vehicles would be very similar to existing aeroplanes in operations for most of their flying pattern [4]. Therefore, the approach chosen by the EASA is to complement existing rules to capture the specific features of such sub-orbital aeroplanes, rather than developing new specifications from scratch. This approach has, for instance, also been followed by the EASA for unmanned aerial systems (UAS). This "small steps" approach, as defined and applied by the late Marcel Dassault in his projects, allows the accommodation of new technologies and operational ranges, while minimizing the effort, resources, and associated programmatic risk.

Since the Ansari X-Prize competition, dozens of sub-orbital space projects have begun to flourish throughout the world, including Europe, along with the actual beginning of the development of the associated required ground infrastructures called "spaceports", some of which are currently under construction [5].

In particular, the EASA has started to be approached by potential applicants, which reinforced the need for preparations to support the first application(s). With this goal in mind, the EASA Internal Safety Committee of 8 July 2008 agreed to proceed with the preliminary steps. These steps included writing and presenting a paper at the 3rd IAASS Conference in Rome. This also led to preliminary investigations on how to accommodate such projects in the existing EASA regulatory framework.

16.1 INTERNATIONAL LEGAL CONTEXT

This preliminary investigation was under the international legal perspective, provided by the Chicago Convention. It has also taken into consideration the aviation-related Regulations and Directives of the European Union [6]. For comparison, the US concept will be mentioned although the regulatory approach taken in the USA has been different in several perspectives, including a regulatory process defined under US law to include aeroplanes and rocket systems and operating under experimental permits with an emphasis on protection of people and property on the ground. Finally, as suborbital aeroplanes will also touch, if not enter outer space, international space law issues must also be considered.

16.1.1 Chicago Convention

Among the agencies of the United Nations or similar international intergovernmental bodies, the International Civil Aviation Organization (ICAO) can be considered a "success story". In fact, it has allowed the development of international civil aviation for more than 60 years, to the point that this mode of transportation has become a favorite for long range travel between different countries and overflying a number of other States or stretches of high seas. The ICAO was established by the Chicago Convention, which was signed on 7 December 1944 by 52 Contracting States, which today number almost 200 [7]. Article 1 of the Chicago Convention recognizes the complete and exclusive sovereignty of each Contracting State over the airspace above its territory (including territorial waters). As a consequence, international standards adopted and published by the ICAO are addressed to the said States and not directly to natural or legal persons. In other words, ICAO standards do not have direct force of law in the Contracting States. Article 12 of the Chicago Convention clarifies that each Contracting State undertakes to adopt measures to implement and enforce rules of the air in its territory. The same article requires that these regulations in force in the States should conform, to the greatest possible extent, to the standards adopted by the ICAO. Similarly, Article 38 of the Chicago Convention refers to regulations applicable in the Contracting States in order to actually implement the ICAO technical annexes to the Chicago Convention. However, Article 44 of the same Convention assigns to the ICAO objectives for the safe and orderly growth of aviation, in a larger geographical area (i.e. "throughout the world") than the territory and territorial waters of Contracting States. In other words, flights over water between Contracting States also come under the scope of the ICAO. Even in this case, however, the ICAO in practice does not establish law directly applicable to natural or legal persons. Article 17 of the Convention establishes that aircraft have the nationality of the State in which they are registered. Furthermore, the ICAO Council has divided the airspace over the entire world into Flight Information Regions (FIRs) spanning also the high seas. Responsibility to establish rules of the air and to provide air navigation services has then been delegated to the Contracting State to whom each FIR has been assigned.

This is, however, balanced by Article 33, which guarantees the mutual recognition of certificates issued by Contracting States. In conclusion, ICAO provisions are not directly applicable to aviation personnel or operators, unless transposed into national law by Contracting States. In practice, most States have done so, in order to remain part of the international aviation system, which brings them significant economic and social benefits.

16.1.2 Aviation law in the European Union

The principle of sovereignty implies that only States can adopt, publish, and enforce measures with force of law. The Chicago Convention is based upon this principle, which had been totally applied for international relations until 1957. However, in that year, six European States signed the Treaty establishing the European Economic Community on 25 March in Rome, Italy. Such a Treaty, while maintaining independence and sovereignty for each of the six signatory States, established institutions (e.g. the Council and the European Commission) with delegated powers to adopt and publish legally binding measures, which are directly applicable to natural and legal persons acting in the territory of the Community. In other words, some attributes of the sovereignty had been delegated to a supranational entity without the need to create a larger state.

The Treaty of Rome contained Articles 74–84, giving the Community the ability to establish a common policy and common rules for transport. Article 84(2) required unanimity in the Council, in order to establish common provisions for air transport. This requirement has greatly delayed the development of binding common rules for aviation in Europe. Nevertheless, technical cooperation has existed. However, this led to the harmonization of several technical rules under the Joint Aviation Authorities, on a "best endeavors" principle. In 1986, the Single European Act was signed by the 12 States belonging to the European Community at that time. The Single European Act introduced the principle of qualified majority voting in order to adopt common policies, in particular (Article 16.5 therein) for air transport. It entered into force on 1 July 1987 and was widely used by the European Commission as led by Jacques Delors (1985–94) in order to establish Community law in a number of fields. Since then, the number and scope of the basic EU legal instruments applicable to aviation safety have constantly expanded. Some major ones are summarized in Table 16.1.

Table 16.1 EASA Main Milestones and Associated Regulations

Year	Act	Topic
1991	Directive 670	Mutual recognition of aeronautical licenses
1991	Regulation 3922	Harmonization of technical aeronautical rules
1994	Directive 56	"Independent investigators"
2002	Regulation 1592	Establishment of the EASA
2003	Directive 42	Safety occurrence reporting
2004	Directive 36	Safety Assessment of Foreign Aircraft (SAFA)
2004	Regulations 549, 550, 551, 552	"Package" of four regulations on the "Single European Sky" (SES)
2005	Regulation 2111	"Black list"
2006	Regulation 1899	EU-OPS for commercial air operators
2008	Regulation 216	First extension of EASA to operations and flight crew licensing
2008	COM final 388, 389, 390	Proposals for SES reform and for extending the EASA to aerodromes and ATM (Air Traffic Management)/ANS (Air Navigation Systems)

Source: F. Tomasello, EASA

The Member States of the Community currently number 27. Today the Community is also known as the European Union, as established by the Maastricht Treaty [8]. These Member States have progressively discharged their obligation to transpose the ICAO standards into law applicable in their territory, not individually but collectively. This is not contradicting any of the Articles of the Chicago Convention. Conversely, EU law does not contain the "minimum" requirements, but "the" requirements. In fact, minimum requirements are necessary to ensure safety, but additional requirements adopted nationally may distort the internal market. EU Member States are therefore not allowed to establish additional requirements in the fields where community competence has been established by the legislature.

16.1.3 Role and procedures of the EASA

The EASA was established by Regulation (EC) No 1592/2002 [9], since repealed and replaced by Regulation (EC) No 216/2008 [10] (hereafter referred to as the "Basic Regulation"). Initially, its competence was limited to airworthiness and environmental compatibility of aeronautical products (i.e. aircraft, engines, or propellers). The Basic Regulation does not explicitly define the term aircraft. Therefore, the ICAO definition contained in Annex 8 to the Chicago Convention applies: "*an aircraft is any machine that can derive support in the atmosphere from the reactions of the air other than the reactions of the air against the Earth's surface*". For example, "hovercraft" are therefore not considered as aircraft and are out of the EASA's competence. On the other hand, civilian aeroplanes and helicopters, and also sailplanes and aerostats (i.e. either balloons or airships), are subject to EASA common rules for airworthiness, based on Article 5 of the Basic Regulation. It has to be noted that some aircraft are not subject to the Basic Regulation (see Art. 4(4) and Annex II of this Regulation). Examples are some historic aircraft, some aircraft specifically designed for research, experimental, or scientific purposes, aircraft in service in military, or police forces.

In this context, the EU legislature has decided that:

- Legally binding implementing rules for airworthiness, in compliance with the Basic Regulation, can only be adopted by the European Commission having received an opinion from the EASA for that purpose.
- Airworthiness codes, i.e. certification specifications (CS) and applicable means of compliance (AMCs) applicable to specific products (e.g. large aeroplanes), can be adopted and published by the EASA, but they are not legally binding.
- A "certification basis" has to be defined for each product subject to certification, based on the airworthiness codes, but adapting them for each specific case through "special conditions" (SCs).

Only the EASA is competent to issue type certificates in the EU, Norway, Iceland, Lichtenstein, and Switzerland. The type certificate attests that the design of a product complies with its individual certification basis.

Sub-orbital aeroplanes, deriving support from the atmosphere for the largest part of their flight, are considered as aircraft by the EASA. Being considered as aircraft, the legal framework of the EASA also applies to that specific product. The EASA is therefore ready to be consulted by interested designers, and even to receive applications for airworthiness approvals.

If and when necessary, depending on developments and proposals by industry, the EASA might even issue a "policy", i.e. a document offering guidance in order to develop the certification basis for sub-orbital aeroplanes. Should the need arise, even a specific CS could be issued for such products, once sufficient experience has been acquired. But, as explained above, the absence of a CS at the moment does not prevent the airworthiness approval of the design of a sub-orbital aeroplane. Any new policy will, in any case, be developed through the rulemaking procedure, which means public consultation on the content of the rules, through the Notice of Proposed Amendment (NPA) process [11].

16.1.4 The US concept

In 1984, the US Space Launch Act established the basis for licensing and promoting commercial space flights in the USA, which led to the creation of the Office of Commercial Space Transportation under the Clinton administration, which was then relocated to the Federal Aviation Administration Office of Commercial Space Transportation (FAA/AST) [12–15].

The US National Space Policy [16], authorized by George W. Bush in 2006, further aimed to "*encourage an innovative commercial space sector, including the use of prize competitions*" such as the successful $10 million Ansari X-Prize, won by Burt Rutan of Scaled Composites and their backer Paul Allen with the *SpaceShipOne/WhiteKnight* carrier two-stage-to-space (TSTS) system on 4 October 2004 [17,18].

Having faced this launch of the first commercial sub-orbital spaceflight (*SpaceShipOne*), the USA needed to describe the legal basis applicable to those activities. The solution found was carried by the fact that the same authority (namely the FAA) was responsible to issue certificates/approvals regarding aviation safety, as well as to license launches or to allow experimental permits for launches into outer space. This solution was codified when the US Congress adopted the Commercial Space Launch Amendment Act, which was signed into law by the US President on 23 December 2004. The Commercial Space Launch Act is now the legal basis for the FAA to regulate commercial human space flight. Under this legislation, the aim is to protect the safety of people on the ground, the uninvolved public in terms of launch and re-entry, including air- and space-worthiness, protection of health and safety of flight crew and flight participants, training, and medical check of flight crew and participants.

In December 2006, the FAA/AST published its Final Rulemaking "Human Space Flight Requirements for Crew and Space Flight Participants", the purpose of which is to establish minimum standards and specific requirements for providing experimental permits for space launches. The significant difference between issuance of experimental permits, licensing, and certification must be underlined at this point, since it bears substantial consequences for the approach chosen. In the instance of experimental permits or licensing, the operator bears the full responsibility of its operations, whereas in the latter the certificating authority takes a part of the responsibility. There are advantages and drawbacks to both methodologies, which may have to be adopted in order to best fit into the existing regulatory framework of each country.

16.1.5 Applicability of international space law

In addition to the international and European air law context described above, international space law has to be considered. Sub-orbital aeroplanes may also enter extra-atmospheric (i.e. outer) space. Despite not knowing where airspace ends and outer space starts, this can almost be taken for granted, since a reason for offering such flights is to bring people into outer space, i.e. to, at least, "touch" outer space. However, although a clear and commonly accepted legal delimitation does not exist [19], it is at least possible to legally draw a line where outer space is attained: every flight which goes beyond 100/110 km above sea level must legally be considered as having entered outer space. This 110 km delimitation is used by many outer space lawyers and goes back to a Russian proposal in the United Nations Legal Subcommittee to the Committee on the Peaceful Uses of Outer Space on the Agenda Item of “Definition and delimitation of outer space”. So far only one State has seen the necessity to delimit airspace and outer space in a national legislation and has therefore also chosen the 100 km opinion (see Section 8 of the Australian Space Activities Act 1998). Those who are not supporting the (arbitrarily set) 100 km (or 110 km) delimitation take a physical approach: airspace ends where the air cannot support the machine any more and outer space begins where an object can (at least briefly) maintain an orbit.

This opinion was first brought up by Theodore von Kármán, which is why the line between airspace and outer space is sometimes named the Kármán line. Calculations of that line differ, and this is why some people see it at 53 miles (c. 84 km) and others at 60 miles (almost 100 km).

Almost all provisions on outer space activities in international law are applicable whenever the line between air and space law is crossed [20]. Only few provisions require that an orbit is taken. This is why also “sub-orbital” activities, including sub-orbital aeroplanes, are potentially subject to most of the space law provisions. If a sub-orbital aeroplane passes that line and enters outer space, it is subject to a set of outer space rules, which are, very often, quite different from the rules provided for in (international) air law. While air law originates from national rules set up under the sovereignty of States about their airspace and eventually harmonized on an international level through treaties and agreements, space law has its origins in international law. No State can claim sovereignty over any part of outer space. Legal provisions, therefore, can only be set up by the international agreements of the Community of States (i.e. on the UN level). Only those issues explicitly mentioned in these agreements as being a national issue can be regulated by States. In the case of the EU, a number of sovereign States exercise together their respective sovereignties in the fields where common action has been agreed.

The legal issues highlighted in the following section aim to identify different concepts in (international) air and space law that need to be solved, if one considers that both air and space law apply to activities of sub-orbital space flights.

16.2 AIRWORTHINESS AND CERTIFICATION

16.2.1 General

The conventional way to certify a complex aeroplane is to issue a type certificate (TC). The type certificate is issued by the EASA after a process by which the Agency notifies the type certification basis and, after a technical investigation, finds that the type design is compliant with the certification basis. The designer is responsible for showing compliance to this certification basis and must declare that compliance has been shown. The type certification basis comprises the applicable airworthiness code complemented by special conditions if necessary to address:

1. Unusual features, or
2. Unusual operations, or
3. Features for which experience in service on similar design has shown that an unsafe condition may develop.

The type certification basis must ensure compliance with the essential requirements for airworthiness included in Annex 1 of the Basic Regulation. As their name already indicates, the essential requirements are the conditions to be fulfilled by a product, a person, or an organization to ensure, as much as possible, that the public is not unduly affected by their operations or activities. They address, therefore, the means by which risks associated with a specific activity can be eliminated or reduced to an acceptable level. To achieve this goal, hazards and associated risks must be identified and analyzed to determine the requirements that are essential to mitigate the unacceptable risks. As far as mitigating measures are concerned, it is also important to insist that they must be proportionate to the safety objective. This means that they must not go beyond that which is necessary to achieve the expected safety benefit without creating undue restrictions that are not justified by that objective. The airworthiness codes are the standard means to show compliance of products, parts, and appliances with the essential requirements. These airworthiness codes contain specifications relative to performance, handling qualities, structures, design and construction, power-plant installation, systems and equipment, operating limitations, and technical limitations.

In addition, the designer must demonstrate its capability to design by obtaining a Design Organization Approval, also issued by the Agency. Individual aircraft receive a certificate of airworthiness issued by the Member States when they comply with the type design and are in condition for safe operations. To be pragmatic, this general approach could be used in the case of sub-orbital aeroplanes. The EASA is legally competent for aeroplanes, irrespective of the fact whether they enter outer space or not. There are, of course, no airworthiness codes for such machines yet, but the idea would be to take an existing one, adapt it, and complement it by the necessary special conditions, to address the points that are listed in this regard as noted above. The issue will be to define which airworthiness code is used as a starting basis. Looking at the aeroplanes that are under design now, CS-23 [21] or CS-25 [22] are obvious candidates. This approach takes into account that below the rocket ignition/ballistic phase, these aeroplanes would behave as classical aeroplanes. However a "type certificate" may not be issued if it is judged that conformance with essential requirements is not met (i.e. during the rocket/ballistic phase). In such a situation, the possibility of issuing a restricted type certificate (RTC) exists. RTCs may be issued when the "type" certificate is inappropriate and the aircraft is being designed for a special purpose. This is provided when the Agency agrees that this special purpose justifies deviations from the established "essential requirements". The general approach is the same as for the issuance of a "type certificate", but it allows including in the basis for the certification a list of paragraphs of the applicable airworthiness codes that the Agency finds inappropriate for the special purpose for which the aircraft is to be used. Restricted type certificates would include any additional limitations for the use related to the special purpose. Thus, RTCs seem the most realistic avenue for sub-orbital aeroplanes. TCs and RTCs assume that the aeroplane will be produced in large numbers. In the case where there would be only a very limited series of aeroplanes produced, a Restricted Certificate of Airworthiness (RCofA) based on Specific Airworthiness Specifications (SAS) would also be a possibility. The Restricted Certificate of Airworthiness, in this case, would be issued by the Member States but the approval of the design would stay with the Agency.

However, in the case of sub-orbital aeroplanes, this approach is not favored by the Agency for continuing airworthiness reasons. Another type of airworthiness certificate may be considered in this context: the Permit to Fly (PtF). However, permanent PtFs cannot be used in the context of commercial operations and complex aircraft, but would be issued pending the delivery of the "type certificate" or more likely the restricted type certificate for the sole purpose to allow flight testing. Permits to Fly are issued by the Member States, and the associated flight conditions are approved by the Agency. Permits to Fly and associated flight conditions may also be issued by an appropriately approved design organization.

Airworthiness does not stop at certification. The continuing airworthiness of the aeroplane must be ensured, as this is a fundamental point for safety. Modifications and repairs must be approved. Occurrences in service must be analyzed and corrective action taken, if necessary. The designer has a key role to play here. This is the reason why the Agency supports the "type certificate" or most likely the "restricted type certificate" approach. This would mean that the holder of such a certificate must demonstrate its capability for design.

The issue of production must also be considered. Manufacturers of sub-orbital aeroplanes should obtain a production organization approval. The issue of such an approval attests to the ability to produce. Organization approvals are a key feature of the approach that is adopted by the legislator for airworthiness. These approvals contain the following elements:

- Means necessary for the scope of work (e.g. facilities, personnel, procedures)
- Management system
- Arrangements with other organizations as are needed
- Occurrence reporting and handling system.

Finally, the EASA plans to propose to the Commission an amendment to the Annex of Regulation (EC) No. 1702/200314 (Part 21) [23] requiring TC and RTC holders to obtain an operational suitability certificate. This certificate will facilitate the entry into service of aircraft by defining among others a minimum syllabus for pilots and maintenance, certifying staff, type ratings, and a master minimum equipment list.

16.2.2 Technical issues

Besides existing certification requirements for standard aeroplanes, called CS-23 for small and CS-25 for large aeroplanes in the EASA regulatory system, additional requirements would have to be developed to address the specific characteristics and operations of sub-orbital aeroplanes. The aim would be to ensure an equivalent level of safety as currently pertains to existing aeroplanes, as far as possible, considering the inherent risks linked to such endeavors at the outer limit of the atmosphere, and the novelty of this domain.

The regulatory mechanisms used to cover such cases are Equivalent Safety Findings (ESF) and special conditions (SCs). These additional requirements, being publicly available, are covered by Certification Review Items (CRIs) in the EASA system. Each CRI provides a link to additional or different requirements and explains the need for these further specifications. For example, some CRIs corresponding to high-altitude/high-speed characteristics have already been developed for small high-performance aircraft (>25.000 ft, Mach > 0.6), such as the very light jets (VLJs). Thus the CRIs would complement the CS-23 and reach an equivalent level of safety (ELoS) as required from the large aircraft as per the CS-25. This avoids imposing directly all CS-25 requirements on smaller aircraft that would otherwise have been inapplicable. The CRIs are in fact proprietary exchange letters between the applicant and the authority, calling out a variety of public requirements or guidelines as necessary to cover the required cases: (i) special conditions (SCs); (ii) acceptable means of compliance (AMC); (iii) interpretative/guidance material (IM/GM), etc. CRIs are confidential, since they contain industrially sensitive information about the resolution of a specific certification issue. However, all the above listed documents, once established, may be used by all projects, when applicable.

In order to facilitate the indicative reference to existing requirements, we have chosen to quote them in square brackets [] hereafter, with the origin of the reference (e.g. [FAA 460.11]). In the case of a link to EASA existing requirements, i.e. CS-23 and CS-25, for simplification purposes, only the reference of the paragraph is indicated (e.g. [EASA 831] links to both CS-23.831 and CS-25.831). Moreover, the following items should not be taken as a comprehensive list of requirements as such, but rather as elements for discussion. Therefore, because of the very nature of sub-orbital aeroplane projects, i.e. diverse and innovative, the main technical issues (and others) would have to be re-discussed thoroughly for each project on a case-by-case basis. Last but not least, should design requirements not be completely fulfilled by the suggested design, because of technical and/or verification impossibilities, the corresponding hazards would have to be mitigated by Operational Hazard Controls, such as labels, placards, procedures, and/or training of the crew and passengers.

16.2.2.1 Basic requirements

Since sub-orbital aeroplanes are very similar to conventional aircraft in their design and operations besides the rocket-propelled and ballistic part of their flight, all basic requirements shall be fully applicable for the aerial phase of the flight. In particular, Regulations (EC) Nos. 216/2008 and 1702/2003, including the Annex to Regulation (EC) No. 1702/2003 (Part 21) and the Certification Specifications based on this Annex (CS-23 [13] and CS-25 [14]), are deemed fully applicable for the ground/air phase of the flight, at the exclusion of the rocket-powered and ballistic sub-orbital phases of the flight. Nevertheless, a trade-off study shall be carried out for each application, in order not to apply too stringent requirements to aeroplanes simply based on their weight. The purpose is to avoid increasing the weight of a project because of more stringent requirements. This could result in having to increase subsequently the fuel capacity to reach the same altitude, thus making them eventually potentially less safe in the end than aeroplanes with less stringent requirements. This shall be discussed and demonstrated on a case-by-case basis.

16.2.2.2 Complement to existing requirements

When not included in existing airworthiness codes, additional requirements would have to be created for specific systems (the corresponding guidelines from FAA 14 CFR Part 460 Human Space Flight—Application Checklist [24] are mentioned in brackets for information).

Environmental Control and Life-Support Systems (ECLSS) [FAA 460.11]

Because of the necessity of a closed-loop system in the stratosphere, and although the basic requirements shall apply, a *toxic gases detection system* shall be implemented in order, for example, to ensure that CO_2 levels do not reach a threshold value which would incapacitate the crew [FAA 460.11 (a)], so that, for example, CO levels stay below 1/20,000 ppm [EASA 831(a)] and CO_2 levels do not exceed 0.5% in volume reported against a Standardized Atmosphere as defined by the ISO (25°C, 1013.2 hPa at sea level) [EASA 831(b)]. It would therefore be recommended to use non-offgassing materials in the construction and outfitting of the vehicle, where possible, and, if not otherwise possible, to use the NASA SMAC system to determine the potential maximum concentrations and effects of hazardous substances in the cabin, and to mitigate them by appropriate means (manufacturing processes, containment, etc.) [NASA 1700.7b, IAASS-S-1700] [16,25].

Ventilation [EASA 831]

Another side-effect of microgravity is the lack of air circulation, because of the closed-loop system and the absence of convection in microgravity. Therefore, sufficient ventilation must be provided to crew and passengers throughout the stratospheric part of the flight. Also, because the pressurization system needs to be modified for a closed-loop system, redundant means of avoiding an untimely depressurization must be provided, along with the means to prevent crew incapacitation in case of *depressurization*. The corresponding indications must also be provided to the crew [EASA 841]. To prevent adverse effects of an untimely depressurization, the utilization of pressurized suits may be considered, but then these suits would have to undergo a comprehensive qualification process according to Military Standards, since this type of equipment has not yet been used for civilian purposes. On top of the applicable requirements for *oxygen supply* [EASA 1441–1449], this system shall be fully redundant [FAA 460.11(b)] and keep the PPO_2 in the cabin below 24.5% in all phases of the flight to prevent potential ignition. If the Oxygen Supply System is chemical, existing requirements would apply [EASA 1450]. A further requirement is defined for the *carbon dioxide removal* system, especially to prevent overheating and subsequent fire. The *temperature* control system must be able to maintain not only acceptable temperatures for the cabin atmosphere (e.g. in the range of 18–25°C), but also be able to prevent touch temperatures above 49°C and below −18°C to be accessible to the crew and passengers during the whole mission [NASA 1700.7B] [25].

Smoke detection and fire suppression [EASA 851–865, FAA 460.13]

On top of the existing requirements, ventilation shall be provided for the weightless part of the flight, in order to be able to detect smoke throughout the whole flight envelope. Also, means of fire suppression must be compatible with a closed cabin, and depressurization can only be used as a last resort and without impairing the crew or harming passengers, as per the above-mentioned requirement to prevent cabin depressurization [FAA 460.11I].

Human factors [FAA 460.15]

There are a number of specifications that relate to human factors as follows:

- **Personnel and cargo accommodation.** On top of existing requirement [EASA 771], in case of a free-floating phase for the passengers, the Pilot compartment and Cabin is to be padded, with no sharp edges (minimum radius >5 mm) or pinching points accessible at any time [NASA 50005] [26]. Also, all parts of the equipment must not create any shatterable material release or sharp edges when broken (e.g. diodes shall be made of plastic, no glass parts exposed) [NASA JSC 28354] [27]. Furthermore, the *pilot compartment view* must be sufficient for the pilot to safely perform any required maneuvers throughout the flight envelope [EASA 773]. If not otherwise possible, this requirement could be alleviated by the use of optical indirect viewing systems and the assistance of robust avionics. Besides the existing requirements for *cockpit controls* [EASA 777], those would have to be designed for zero-*g* free-floating operations (protected against untimely contact). Also, the design of the controls for the Reaction Control System (RCS) should allow the aircraft to be maneuvered throughout its flight envelope while preventing any inadvertent maneuver. *Door(s)*: [EASA 783] the door(s) are to be designed to prevent untimely opening by crew and passengers during all phases of flight (no handling point). Conversely, they must be quickly openable by one person in case of an emergency. An emergency pressure equalization system must be put in place for this purpose.

- **Seats, berths, litters, safety belts, and shoulder harnesses [EASA 785].** The seats shall be adjusted in order for the crew and passengers to best support all potential accelerations during all phases of flight, especially at pull-up with rocket engine on, or during the resource upon re-entry (proposed requirement: 6*g* constant during 30 seconds). Apart from the zero-*g* free-floating phase, the crew and passengers shall be restrained to their seats by the means of harnesses.
- Specific **personal protective quipment (PPE)** may be considered for certain phases or all flight (pressurized suits, helmets, visors against sun glare, etc.), but this is to be considered on a case-by-case basis. Beyond being impact resistant (inside/outside) as per [EASA 775], *windshield and windows* would have to be non-shatterable and non-scratchable, in order not to cause loose debris to float inside the cabin. On top of the existing required *pax information signs* [EASA 791], specific "0*g*" (e.g. Blue) signs might be considered, as well as return to seat/fasten seat belts (Amber + Gong) annunciations, similar to the ones used in larger aircraft for parabolic flights, in order for all passengers to be seated and strapped before the re-entry begins. Also, a countdown to return to gravity should be displayed in the cabin.
- **Emergency evacuation [EASA 803] and emergency exit [EASA 805-813].** On top of existing requirements for emergency evacuation on the ground, additional requirements would have to be adapted on a case-by-case basis to the specific projects (e.g. rounded shape vs. square shape to better accommodate pressurization and weight constraints); the provision for an in-flight bail out must be considered, along with the use of parachute rescuing systems. As per EASA 813, the emergency exit shall not be obstructed by obstacles; in other words the internal cabin layout must not prevent evacuation. However, if instrument panels or equipment are part of the emergency detachable part [18], this will be reviewed on a case-by-case basis.
- **Emergency equipment [EASA 1411-1415].** Depending on mission profile, the corresponding emergency equipment shall be considered.
- **Onboard recorders [EASA 1457-1459].** Because of the novel design characteristics of such aeroplanes and their mission profile, flight data recorders and cockpit voice recorders shall be provided, in addition to telemetry, if available.

16.2.2.3 Specific systems and operations (vehicle)

At the level of the vehicle, additional requirements could be introduced, as follows:

- **Exposure to high/cold temperatures.** For temperature, a sufficient air-conditioning and/or heat sink system shall be provided, should any part of the vehicle become extremely warm or cold during a part of the flight such as it would jeopardize the safety of the vehicle or its occupants.
- **Exposure to radiations** shall be counteracted by design, should the exposure time and levels be above commonly acceptable norms. If deemed critical, radiation measuring devices may be integrated inside the vehicle and/or provided to each member of the crew.

- **High skin temperatures/kinetic heating.** All parts of the aircraft subject to kinetic heating during acceleration and/or re-entry shall be designed to accommodate the heat generated in order not to jeopardize the integrity of the structure and the proper functioning systems contained inside. Potential deformation of structures due to thermal expansion/retraction or degradation shall be accounted for, so that it would not jeopardize the handling qualities, proper functioning, and/or structural integrity of the aircraft. Temperature sensors and/or indicators may be used to monitor critical parameters.
- **Rocket boosters/engines.** Rocket engines shall be designed to minimize burst hazards and prevent inadvertent firing. In particular, for liquid or hybrid rocket engines, pressure relief valves and fuel dump systems must be provided, as well as a fuel cut-off in order to be able to interrupt the burn at any time, should a contingency situation occur. For solid rocket engines, a jettisoning system shall be implemented, in order to allow an aborted climb and safe return of the vehicle when already ignited. In such cases, the booster(s) should have their own guidance, signalization (transponder) and recovery/flight termination system, in order not to endanger the safety of the public on the ground or other aircraft. All rocket engines shall be fully tested successfully several times on the ground in their flight configuration and for the full duration of the burn before being used for actual flights. In the case vectored thrust is used to control a critical phase of the flight, it shall be two-fault tolerant.
- **Attitude/reaction control systems.** The same principle applies to Reaction Control Systems, which shall be at least redundant and should jeopardize neither the safety of ground personnel nor that of the carrier aircraft (if the case of a two-stage-to-space configuration), they thus being secured both against inadvertent firing and runaway.
- **Propellant(s).** Toxic/explosive fuels shall be contained at all times to prevent exposure of ground personnel and crew, inadvertent spillage and non-explosive burst.
- **Containment/crashworthiness.** In order to cope with an aborted mission and premature return, as well as with unused fuel after a successful mission, the crashworthiness of tanks and solid rocket engines must be ensured such as to prevent any spillage and ignition. In the case of liquid/gaseous propellants, a fuel jettisoning system must be implemented, provided environmental requirements are observed.
- **Operations in weightlessness (0*g*).** All fluids shall be contained throughout the parabolic part of the flight, i.e. in weightlessness (fuel, oil, etc.), in order not to jeopardize the safety of the flight or the restart after the maximum expected 0*g* period.
- **Engines "safing" for restart.** In the particular case of hybrid/single-stage-to-space (SSTS) vehicles, the restart of the turbofan engines shall be ensured after the rocket-propelled and ballistic part of the flight. To this end, comprehensive shutdown and "safing" procedures shall be established in order to minimize hazards and ensure a safe restart in due time. In the case where the engines will not restart, the vehicle shall be able to be flyable as a glider in order to return safely to the ground, on to main or alternate landing sites. In particular, electrical power shall be available to essential systems until safe return to the ground.
- **Instruments/avionics.** In addition to, and/or in combination with, conventional aeroplane instruments, specific space instruments shall be provided to the pilot(s) in order to maximize their situation and orientation awareness in all phases of flight, especially at high altitudes and unconventional flight attitudes. For example, a spheric-type Earth-like horizon, radar altimeter and efficient guidance and navigation system (e.g. redundant and efficient GPS) must be made

available. Because of the degradation of GPS reactivity at high speeds, alternate/independent means of guidance shall also be considered for redundancy. Flight directors/pilot assisting devices shall be customized to provide the pilot(s) with the best understandable information in a synthetic form, in order for them to optimize the trajectory in all phases of flight and recover from abnormal situations. Cabin pressure and gas monitoring devices must be provided to the crew, with the associated cautions and warning visual and/or aural annunciations.

16.2.2.4 Intravehicular activities (IVA) requirements

For the parabolic part of the flight in weightlessness, standard intravehicular activities requirements would apply, as per [NASA NSTS 1700.7b] [28]:

- **Specific equipment (payload) requirements.** In particular, NASA Form 1230 could be taken into reference for payloads and equipment installed on board the vehicle, and which shall remain secured throughout the dynamic parts of the flight. In short, this checklist addresses the structural failure of sealed/vented containers, sharp edges (min. radius 5 mm in IVA), pinch points, shatterable material release (no glass), flammability, offgasing, batteries, touch temperatures (−18/+49°C), rotating devices and mechanisms, electrical power (bounding [EASA 867], mating/demating), and contingency return/rapid "safing". Should one of the requirements not be met, a unique hazard report would have to be developed according to existing standards [NASA 13380] [29].

16.2.2.5 Ground support equipment

A ground safety data package would have to be established in order to address all potential hazards on the ground, especially with respect to fueling operations, untimely ignition of engines and/or activation of reaction control systems, as well as protection of (ground) crews, passengers, and public during all ground operations: assembly, testing, parking/stowage, fueling, ground checks, boarding, taxiing, take-off, and (premature) return and landing [NASA KHB 1700.7] [30].

16.2.2.6 Environmental requirements

Potential "showstoppers", such as the disruption of the ozone layer, atmospheric contamination because of smoke residues or release of toxic propellants, would need to be addressed in accordance with, and in complement to, existing environmental requirements [5].

16.2.2.7 Crew/PAX qualification and training

The crew is considered an integral part of the safety of the system; therefore it must be trained and qualified accordingly [FAA 460.5], taking into account all potential off-nominals and contingencies, abort/emergency, and human behavior and performance (HBP) aspects in all phases of flight, including microgravity [31], in order to minimize hazard to the public and passengers. Crew training shall be performed using representative hardware and applying standards for training records and CQRM [FAA 460.7] [32]. Space-flight participants shall also comply with training requirements [FAA 460.51] for emergency cases and be informed of the risks [FAA 460.45]. In order to define and implement such training, it would be useful to consult existing entities that are already providing this type of training to astronauts, such as the European Astronaut Center (EAC) located in Cologne, Germany.

16.2.2.8 Verification program [FAA 460.17]

An integrated verification program would have to be conducted successfully before allowing passengers on board, including flight testing.

16.2.3 Legal implications of space-worthiness when entering outer space

While the EASA is competent to certify the type of sub-orbital aeroplane as being an aircraft, the jurisdiction of the EU (and all States) ends where outer space begins. States have agreed on a different legal concept for activities carried out in outer space. As a part of that concept, the Community of States has established a national responsibility for States for national activities in outer space (Art. VI of the Outer Space Treaty) [33]. States have to authorize and supervise such activities. This responsibility is not transferred to the EU as part of the aviation competences. The EASA cannot deal with that (very short) outer space part of the sub-orbital flight, unless it agrees with the States to enforce this responsibility on their behalf.

This responsibility indirectly includes aspects of space-worthiness as described above (in addition to the airworthiness). The space-worthiness of a sub-orbital aeroplane is not directly mentioned within the international treaties (or in Community law). However, States can be held internationally liable (according to Art. VII of the Outer Space Treaty as well as according to the Convention on International Liability for damages caused by space objects) for damages caused by a space object. Some States [34] therefore have included the approval of the space-worthiness of a space object within their national legislation about space activities, requiring the operator to carry out the activity in a safe manner and without causing damage to persons or property. Such national legislation would apply in addition to what will be established by the EASA as the certification basis for the air flight part.

Unlike the approval of airworthiness under the EASA legal framework, only individual objects (not the type of product) are subject to this space-worthiness approval.

As stated above, the approval of space-worthiness is not subject to the competences of the European Union, including the EASA. If a wider scope is required in the future, in order to bring such activities on the crossroads of air and space under one single jurisdiction, specific action must be decided by the (European) legislature. However, appropriate agreements could already be established today between the EASA and the responsible State for the activity in outer space (typically the State of the operator), in order to avoid double processes that would be a burden for industry. The EASA is ready to explore this possibility with the administrations involved.

16.3 FLIGHT CREW LICENSING

There are a number of regulatory issues that must be addressed with regard to the various aspects of flight crew licensing.

16.3.1 Crew training and qualification

It is obvious that the flight crew of a spacecraft has to fulfill certain requirements for initial training, proficiency, testing, and medical fitness [35]. Sufficient scientific, operational, and managerial experience for this exist, mainly in public space organizations such as NASA or the ESA. However, when looking at commercial space tourism operations carried out by private operators, the issue is not only whether crews possess sufficient knowledge; it is also necessary to ensure the following:

- Proper rules exist in order to clearly establish responsibilities and privileges for natural and legal persons.
- Such rules are accompanied by acceptable means of compliance (AMCs) and published (e.g. training syllabi).
- Mechanisms exist to oversee and enforce the application of the rules (e.g. issuing, suspending, or revoking pilot licenses).

Article 7 of the Basic Regulation has extended the mandate of the EASA to propose legally binding implementing rules and to issue AMCs in relation to flight crew, their medical fitness, and their training organizations, as well as to the person entitled to check medical fitness, to train the pilots, or to test their skill. On this legal basis, the Agency has issued its Notice of Proposed Amendment (NPA) with regard to pilots (commercial or private) of aeroplanes, airships, balloons, helicopters, and sailplanes [36]. According to the rulemaking procedure, the EASA will issue a Comment Response Document (CRD) summarizing the comments received to the NPA, followed by an opinion addressed to the European Commission. It is expected that such common rules for the "traditional" pilots will be published in 2010.

However, new demands are also emerging for pilots "on the ground", taking responsibility for the flight of unmanned aerial systems (UAS). Since both sub-orbital aeroplanes and UAS are aircraft, and the EASA has legal competence to establish rules for their pilots, it is not excluded that the EASA might undertake specific additional rulemaking tasks, aiming at establishing legal certainty also for these categories of pilots across the EU (and Norway, Iceland, Liechtenstein, and Switzerland). According to the EASA's rulemaking procedure, such a rulemaking task can only be initiated if it is included in the rulemaking program. Such a program is decided in cooperation with stakeholders. The first step is for one organization to complete the appropriate Rulemaking Proposal Form and send it to the EASA [37].

Should the above possible task(s) be undertaken by the EASA, of course, the medical requirements applicable to the mentioned new pilot categories will need to be defined. One could say that UAS pilots and sub-orbital aeroplane pilots are at two opposite extremes. The former, in fact, operate on the ground in normal room conditions (i.e. for them the medical requirements could be similar to those for air traffic controllers). By contrast, sub-orbital aeroplane pilots operate in an environment that is even more severe than the cockpit of a traditional aircraft, whose operations are limited to the dense atmosphere.

16.3.2 Passenger safety

Furthermore, in the case of space tourism, paying passengers will inhabit the sub-orbital aeroplane cabin. They would also be in a potentially severe environment under abnormal conditions. Even during normal operations they will experience intense accelerations and the absence of gravity. It seems obvious that they should be medically fit for such experiences. However, the issue here for the EASA is not whether sufficient scientific knowledge exists on this matter, but which legally binding rules have to be established. Currently, no rules have been established for the medical fitness of paying passengers on board commercial air transport aircraft. This approach (i.e. "do nothing") could be continued even for sub-orbital flights. Some might, however, argue that the peculiarities of sub-orbital flights do require the establishment of rules to protect potential passengers from a medical standpoint, and also in order to avoid jeopardizing the safety of a flight because of passenger sickness or loss of consciousness. But even if such a principle is accepted, many alternatives do exist, such as obliging operators to simply publish medical guidelines before selling tickets, or obliging passengers to acquire a medical certificate, or to require operators to check the medical fitness of the passengers before accepting the reservation, or, perhaps, to do so before confirming the flight.

Should it come to proposals, the EASA will consult stakeholders, according to the rulemaking procedure mentioned above. In particular, several entities in Europe that are already delivering medical certifications to astronauts, such as the MEDES/IMPS in Toulouse, France or the DLR-Flight Clinic in Cologne, Germany, or to parabolic flight participants, such as the CEV in France, may be consulted in order to benefit from their expertise.

16.3.3 Legal implications of flight crew licensing when entering outer space

The international space law treaties do not contain any rules on how to deal with flight crew certification or passenger training. However, another issue comes up for flight crew and passengers, being slightly different from what is ruled by Art. 25 of the Chicago Convention, namely that each Contracting State undertakes to provide such measures of assistance to aircraft in distress in its territory as it may find practicable, and to permit, subject to control by its own authorities, the owners of the aircraft or authorities of the State in which the aircraft is registered to provide such measures of assistance as may be necessitated by the circumstances.

The Rescue Agreement [38] establishes additional rights that apply to the crew (and maybe the passengers) of a sub-orbital aeroplane. Unlike the title, the text of the Agreement refers to "personnel of the spacecraft". The term "spacecraft" is not defined in the Agreement. Based on the principles of the ICAO taxonomy, a vehicle may be considered as a spacecraft if it is designed to fly in the environment of outer space, especially if it is not, or no longer, deriving support from the reactions of the air. This applies to sub-orbital aeroplanes. Hence, the personnel of such aeroplanes enjoy the rights and privileges of that Agreement, i.e. rescue support and assistance from any State in which the spacecraft has landed in the case of accident, distress, emergency, or unintended landing (Art. 2), as well as from the States that are in a position to do so in a case where the landing took place on international territory (Art. 3), and safe and prompt return to the representatives of the launching authority (Art. 4). From the wording, it is questionable whether passengers can be considered as "personnel of the spacecraft", but such consideration might be necessary bearing in mind the fundamental reasons that led to the Agreement.

16.4 AIR OPERATIONS

Air operations include a number of technical, administrative licensing, and training aspects.

16.4.1 Technical considerations

On 27 March 1977, two "jumbo jets" filled with passengers collided at Los Rodeos airport on the island of Tenerife. To this day, it is still the worst aviation accident ever to have occurred in terms of the number of fatalities. At that time, it was common among ICAO Contracting States, when trying to understand the causal factors of an accident, to consider the "machine", the human, and the environment [39]. However, a few years before (1972), Professor Elwyn Edwards of the University of Birmingham had acknowledged that modern machines, and especially aeroplanes, have more and more software embedded in them. This development led to the conception of the "SHEL model" to analyze safety events: software + hardware + environment + "live ware" (= human operator). Captain Frank Hawkins, involved in the investigation of the Tenerife accident, then proposed in 1979 to turn the SHEL model licensing for aviation into SHELL, i.e. repeating the "L" twice, to mean that not only the individual (e.g. the pilot) had to be considered in respect of aviation safety, but also the "organization" (i.e. the air operator). The SHELL model was accepted by the ICAO in 1989 [40]. Immediately after (1990), the ICAO adopted Amendment 19 to its Annex 6, requiring States to establish rules in order to issue a specific "air operator certificate" (AOC) to organizations offering commercial air transport to citizens. Consequently, in the EU the legal obligation for the AOC was established in 1992 [41].

Besides their development, the main activity regarding sub-orbital aeroplanes would be considered commercial air transportation. Therefore, in the present regulatory framework, EU-OPS [42] would apply, with the need of exemptions in order to adapt to the very specifics of sub-orbital flight operations. The corresponding air operator certificate would have to be issued by the Member States for operators operating from their territory.

Moreover, compliance with EU-OPS also includes compliance with the EASA maintenance requirements, which would require the operators of sub-orbital aeroplanes to obtain a Continuous Airworthiness Management Organization Agreement (CAMO). Also, sub-orbital aeroplanes would have to be maintained and refurbished between flights in Part 145 organizations, in full compliance with existing maintenance organization requirements, providing the necessary adjustments to capture the very specific characteristics of sub-orbital aeroplanes (e.g. handling of hazardous/toxic fluids/explosives). Last but not least, for the sake of convenience and consistency between aircraft and spacecraft maintenance organizations and systems, it may also be worth developing a common Space Transportation Association (STA) nomenclature, similar to the well-known ATA-100 Chapters used worldwide.

Finally, Article 8 of the Basic Regulation gives the European Commission the responsibility to establish detailed implementing rules applicable to air operators. A specific Notice of Proposed Amendment (NPA), containing the EASA's opinion for such an implementing regulation concerning operations of conventional aircraft, was planned to be published in 2009.

It is, however, obvious that specific rules for operators of UAS and of sub-orbital aeroplanes are also required, in terms of mission planning, filing a flight plan, briefing passengers on safety and so on. In the case of sub-orbital aeroplanes, it is likely that the safety briefing may be quite extensive and necessary to be delivered before boarding the cabin, for instance, through on-ground training. Even in this case, a specific rulemaking task could be planned and initiated, depending on the requests by stakeholders, following the rulemaking procedure.

16.4.2 Legal implications of air operations when entering outer space

As already explained, at present the EASA is legally not competent to deal with the very short period of the sub-orbital flight that would take place in outer space. For this very specific part of the flight, the operator would have to take additional operational requirements into consideration.

In principle, the jurisdiction (and control) over an object launched into outer space is held by one of the Launching States, which launched the object, procured the launching, from whose territory the object was launched, or from whose facility the object was launched. The State of Registry is one of those States, as jointly determined by the Launching States. As the Treaties clearly define which States can become Launching States (and therefore State of Registry), it will not in all cases necessarily be the same State that has jurisdiction of the vehicle as an aircraft (see Art. 17 of the Chicago Convention), although it is also referred to as the State of Registry.

Nevertheless, this does not cause any problems for sub-orbital aeroplanes. Although such aeroplanes are considered space objects, they are not subject to registration as space objects. This is because space objects only need to be registered if they are launched into an Earth orbit or beyond an orbit, as per Art. II of the Registration Convention. A sub-orbital aeroplane is launched into outer space, but not into an orbit or beyond an orbit. Sub-orbital aeroplanes therefore do not need to be registered as space objects. The purpose of registration is to keep track of objects which remain in outer space on a certain orbital position. This is not the case for sub-orbital aeroplanes. No State therefore holds jurisdiction and control over a sub-orbital aeroplane as a space object according to Art. VIII of the Outer Space Treaty.

Theoretically, this means that there is no jurisdiction over this specific object while in outer space (other than the jurisdiction over the sub-orbital aeroplane as an aircraft while in airspace). However, in practice, operational provisions are applied by the State responsible for the activity according to Art. VI of the Outer Space Treaty. This is a lawful application, as Art. VI of the Outer Space Treaty applies to all activities in outer space and not to objects launched into an orbit (and therefore subject to registration as a space object).

These operational provisions, set by the State responsible for the space activity, are a few basic requirements for the space flight. They bind private space activities through requirements in the national legislation, mostly within the general authorization allowing the applicant to undertake space activities. While in outer space, the sub-orbital aeroplane has to respect international law, promote international cooperation, and avoid harmful contamination of the outer space as well as adverse changes to the Earth's environment. National legislation, however, might need to establish additional operational requirements.

16.5 TRAFFIC MANAGEMENT

Traffic management is one of the most controversial issues when it comes to commercial space operations. Some of the issues have become more heated since the collision of the *Iridium 33* satellite and the retired Russian *Cosmos 2251* satellite on 10 February 2009.

16.5.1 Collision avoidance with other aviation traffic

Community competence on Air Traffic Management (ATM) was already established in 2004 on the basis of the "Single European Sky" (SES) package of four regulations [43].

In turn, ATM comprises Air Traffic Services (ATS) and the latter also Air Traffic Control (ATC) in certain classes of airspace, as defined by the ICAO. All the upper airspace over the EU belongs to "Class C", which means that, therein, the flights operating under instrument flight rules (IFR) are "separated" by an air traffic controller from the ground. It is assumed, herein, that spacecraft will indeed operate under IFR and they will cross the upper layers of the atmosphere during climb and descent. In fact, most Upper

Air Traffic Control Centers (UACCs), at least in theory, are declared to have vertical competence from a defined flight level (e.g. FL 195 = 19,500 feet) up to "unlimited", although today, in practice, they do not execute any task in order to control space flights. Nevertheless, when developing the implementing rules for ATM, the upper limit of its competence could be set around FL 3300–3600 (i.e. 330,000–360,000 ft, around 100–110 km), or even higher (e.g. FL 4000, around 120 km) to cover the highest sub-orbital flights and overlap with space controlled areas, although specific rules should be established for this ballistic part of the flight.

From an ATM point of view, the operation of a sub-orbital aeroplane can be split into different phases as follows:

- Filing a flight plan at the conclusion of the mission planning—no major differences in this respect are foreseen in relation to conventional aviation.
- Receiving instructions for "flow management" (e.g. delays before departure, because of high traffic density foreseen at a certain time in defined volumes of airspace)—this could be minimized by selecting an appropriate departure aerodrome and hours for the mission, avoiding areas and times of high traffic congestion (e.g. central Europe in the early morning).
- Operating at the departure aerodrome, which, depending on the characteristics of the sub-orbital aeroplane and on the traffic therein, may impose restrictions.
- Operating from an aerodrome with less other traffic, which may alleviate interaction issues.
- Climbing through the atmosphere (up to FL 450 = 45,000 ft = around 14 km from Earth's surface). If the performance of the sub-orbital aeroplane is similar to that of conventional aviation, this phase should not pose problems in relation to separating these flights from other traffic. If the performances are significantly different in terms of rate of climb or speed (either very low, e.g. in the case of aerostatic lift, or very high), special ATC procedures may apply. However, the basic procedures to cater to special needs (e.g. air shows, large military exercises, climb of weather observation balloons) are already in place today and could easily be adapted to the needs of the sub-orbital aeroplane. Of course, the likelihood of "flow management" measures (i.e. delayed departure) will greatly depend on the time and geographical area chosen by the operator and include the fact that some sub-orbital aeroplanes will return to land by gliding, requiring traffic priority.

The same considerations apply to the descent (below FL 450) and landing phase, during which the sub-orbital aeroplane will operate sustained by atmospheric lift. The segment of the atmospheric flight above FL 450 during climb is even easier, since during that phase the sub-orbital aeroplane will be fully controllable by the crew, while those layers of the atmosphere are not populated by significant conventional aviation traffic.

Conversely, some aspects of the sub-orbital mission require further study in relation to ATM, including the following:

- The rocket-propelled and ballistic portions of the flight, during which the ATC service may not even be required (e.g. having properly planned the mission in advance and having avoided the simultaneous presence of two sub-orbital aeroplanes in the same space volume).
- The initial descent phase, which may involve flight at high Mach number.
- The procedures to exit from ATC procedures and then to resume them, which is a new topic not necessary to be analyzed for conventional aviation.

For the latter, the predictability of the ballistic trajectory could be very relevant, also because the future ATM system (currently being developed for Europe by the SESAR Joint Undertaking) is expected to be largely based on predicted trajectories, more than on an intervention by the Air Traffic Controller in real time.

16.5.2 Radio frequencies

As soon as the sub-orbital aeroplane leaves airspace and enters outer space, the allocation of frequencies (e.g. for communication with ground control) would need to be assigned on an international level. However, the ICAO has already standardized the aeronautical mobile (communication) satellite services, which have now been operational for almost 10 years and are compliant with the ITU Radio Regulations. This service and its foreseen developments could possibly also be used for that portion of the flight. The State of the operator has to oversee that appropriate communication has been established between the operator, ATC, and the vehicle.

16.5.3 Legal implications for traffic management in outer space

No legal requirements exist yet for Space Traffic Management (STM). There have been some research projects on such requirements, stressing the need to establish such services [44]. This does not only include the traffic management between air flights and space flights, but also the traffic management whilst being in outer space.

Rules intending to achieve safe access to outer space, safe operations of space activities, and collision avoidance, as well as the prevention of pollution can be found within national space acts and other national regulations (e.g. licensing regimes). But one has to take into account that these rules originally were not meant for dealing with traffic management. One may differentiate between rules applicable to all space operations and those applicable only to a certain phase of space operations (launch/in-orbit/re-entry).

Specific attention should be paid to the mitigation of the risk of collision of sub-orbital aeroplanes with space debris. Here again, some States have set up national rules as part of their space legislation, which is not affected by any competence of the European Union established for Air Traffic Management. Those rules are part of the initial license for undertaking space activities [45].

16.6 LIABILITY ISSUES IN CONTINGENCY CASES

The operator of a sub-orbital aeroplane faces several liability risks. Two of these risks need a deeper evaluation with regard to their specificities of air and space activity.

16.6.1 Passenger liability

Air carriers might be held liable by passengers (or his/her legal successors) for damages caused to passengers, baggage and goods, and also for damage caused by delay. This liability is based on the "Warsaw System". The Warsaw System is a set of rules based on the Warsaw Convention [46], which have been modified and amended several times throughout the last century because of the growth of air traffic [47]. The latest module to the Warsaw System was set in 1999, when the Montreal Convention was signed [48].

The applicability of the Warsaw System to sub-orbital space flights might already be questionable, as the carriage has to be an international one. That means, broadly speaking, that the point of departure and the point of destination, or an agreed stopping place as agreed by the parties, have to be in different States. If, nevertheless, assuming the applicability of the Warsaw System, the carrier can be held liable for injury or death, it can also be held liable for damage to baggage (payloads) and goods.

This legal system as described for air carriage seems to contradict the legal system established for space carriage: international space law never looked at passenger rights, other than crew. Therefore, no rules yet exist, in principle. However, there is one article upon which the States agreed when dealing with third-party liability rights. Although it was probably not intended, this article excludes the possibility for passengers to claim for liability. Third-party liability is excluded for any damage caused to a national of the Launching State, as well as to foreign nationals during that time that they participate in the operation of that space object. However, it has to be noted that the Liability Convention deals with the Liability of States (i.e. Launching States). This liability is established in addition to the liability of the operator of an aircraft (according to national tort law).

The Warsaw System—only applicable to international flights—is therefore not contradicting the liability concept established by international space law.

16.6.2 Third-party liability

In the case where a sub-orbital aeroplane causes damage to a third party (anybody to whom no contractual relationship was established), the legal systems of international air and space law vary.

A protection of damages caused by air traffic to third parties on the ground was established in the Rome Conventions [49]. Only a few States have signed and ratified these Conventions; hence they have a very narrow application. Nevertheless, the principles described below have been adopted in the legislations of many States and are, therefore, becoming important. The Rome Conventions establish an absolute liability to the operator of the aircraft (Art. 2 Rome Convention 1952) if the damage was caused by an aircraft in flight. The operator may be exonerated if it proves that the damage was caused solely through the negligence or other wrongful act or omission of the person who suffered the damage. The liability of the operator is also limited. The exact amount depends upon the weight of the aircraft (Art. 11). No limitations exist if the person who suffered the damage proves that the damage was caused by a deliberate act (or omission) of the operator (Art. 12). Collisions of aircraft are hardly covered at all by the Rome Conventions. Only Art. 7 states that in such cases both operators shall be liable (to a third party).

Third-party liability in outer space law is dealt with differently. In principle, liability is established for damages caused by space objects. A space object is any object that is in outer space (or intended to be launched into outer space). As soon as a sub-orbital aeroplane is in outer space at one moment of its flight, it is a space object to which the Liability Convention applies. Unlike the Rome Conventions, it is not the operator of a space object that is held liable according to international space law. The operator of a space object might be liable only according to national (tort) law. These national (tort) laws applicable to space objects are not harmonized by an international Convention. Rather than holding the operator liable, international space law establishes an additional State liability. The liability is established to the Launching States (the State that launches or procures the launching of a space object, as well as the State from whose territory or facility a space object is launched) according to Art. VII of the Outer Space Treaty and the Liability Convention. This liability is also established as an absolute liability in the case where the damage occurred on the surface of the Earth or to an aircraft in flight, but it is a fault-based liability if the damage is caused to (another) space object. Other than in the Rome Conventions, these liabilities are unlimited.

The relationship and appropriateness of both legal systems to apply to sub-orbital aeroplanes must still be clarified [50]. Thus, for example, it could be up to determination as to whether the State liability, as established by Art. VII of the Outer Space Treaty and the Liability Convention, is acceptable and appropriate for such sorts of commercial sub-orbital space flight. Similar discussions have already taken place in order to determine the applicable law in the case where an aircraft collides with a ship on the high seas [51]. As the Treaties apply, however, third parties can rely on both liability concepts.

CONCLUSIONS

If we limit ourselves to the case of sub-orbital aeroplanes as defined in this chapter, the EASA has the regulatory framework and the procedures to consider certifying them as aircraft. The certification would most likely take the form of a restricted type certificate (RTC). The challenge will be in adapting existing airworthiness codes and the development of special conditions necessary to cover this category of aircraft to ensure the appropriate level of safety. The Agency would need to complement its existing expertise in aircraft structures, systems, flight, power-plant, etc. by accessing the expertise specific to rocket/ballistic flight.

With the same assumption, operational rules (EU-OPS today, EASA Part-OPS in the future) and maintenance regulations [52] provide a basis for operations and maintenance respectively, but would need exemptions to cover the case of sub-orbital aeroplanes. The challenge will be in the identification, development, and agreement of such exemptions.

While waiting to develop a policy, and in order to follow closely the novel design and techniques used by such aeroplanes, the EASA may also offer a cooperative research framework to applicants, in order to jointly prepare the terrain optimally for their applications. Also, in order to pave the way for future long-lasting rulemaking activities, especially concerning operations and licensing, potential applicants are encouraged to route their request for proposed rulemaking as soon as possible via their representatives in EASA consultative forums, such as the AeroSpace and Defence (ASD) group at the Safety Standards Consultative Committee (SSCC). Due to the distribution of responsibilities between the different actors in the Community system as well as to the distribution of responsibility for the air flight parts and the space flight part, a close cooperation between the Agency, the Commission, and the Member States would be necessary. Last but not least, cooperation with the ESA, FAA, and ICAO is deemed essential as their expertise and experience would help the Agency considerably in the first phase of our proposed approach, which is to adapt aviation requirements to the novelties introduced by sub-orbital aeroplanes. In the future, should the scope be extended to space planes, those beyond the outer limits of the atmosphere, more global cooperation with all parties would be necessary to explore the technical, operational, and legal aspects of this fascinating and challenging endeavor.

References

[1] Marciacq, J.-B., Morier, Y., Tomasello, F., Erdelyi, Z. and Gerhard, M. (2008). Accommodating Sub-Orbital Flights into the EASA Regulatory System. *Proceedings of 3rd IAASS Conference—Building a Safer Space Together*, Rome, Italy, 21–23 October (ESA SP-662, January 2009).

[2] "ESA's Position on Privately-Funded Suborbital Space Flight", 10 April 2008. Online, available from http://www.esa.int/esaCP/SEM49X0YUFF_index_0.html (cited 22 May 2008).

[3] Galvez, A. and Naja, G. (2008). Space Tourism. *ESA Bulletin 135*, August.

[4] Grayson, C. (2007). International Association for the Advancement of Space Safety (IAASS) Independent Space Safety Board. *Proceedings of the 2nd IAASS Conference*, 14–16 May, Chicago, IL (ESA SP-645, July), US National Space Policy, 31 August 2006, Para 4.1, Suborbital vehicles, "aero-spacecraft".

[5] Pelton, J. N. *Space Planes and Space Tourism: the Industry and the Regulation of its Safety*, a research study of the Space and Advanced Communications Research Institute (SACRI), George Washington University.

[6] Von Wogau, K. (2008). *Report on Space and Security*. European Parliament 2004–2009 Session Document, 10 June, Ref. A6-0250/2008.

[7] Convention on International Civil Aviation (1944). As amended 2006 (ICAO Doc 7300/9).

[8] Treaty on European Union (1992). *Official Journal*, C 191, 29 July.

[9] Regulation (EC) No. 1592/2002 of 15 July 2002 on "Common rules in the field of civil aviation and establishing a European Aviation Safety Agency". *Official Journal*, L 240, 7 September.

[10] Regulation (EC) No 216/2006 of 20 February 2008 on "Common rules in the field of civil aviation and establishing a European Aviation Safety Agency, and repealing Council Directive 91/670/EEC, Regulation (EC) No 1592/2002 and Directive 2004/36/EC". *Official Journal*, L 79, 19 March.

[11] http://www.easa.europa.eu/ws_prod/g/doc/About_EASA/Manag_Board/2007/MB%20Decision%2008-2007%20amending%20rulemaking%20procedure9ba9.pdf?page=3

[12] Wong, K. (2007). Developing Commercial Human Space Flight Regulations. *Proceedings of the 2nd IAASS Conference*, 14–16 May, Chicago, IL (ESA SP-645, July).

[13] Repcheck, R. (2005). FAA's Implementation of the Commercial Space Launch Amendments Act of 2004—The Experimental Permit. *Proceedings of the 1st IAASS Conference*, 25–27 October, Nice, France (ESA SP-599, December).

[14] US Executive Order 12465.

[15] Title of the United States Code, Subtitle IX, Chapter 701.

[16] US National Space Policy, 31 August 2006.

[17] Pelton, J. N. (2007). Space Planes and Space Tourism: Developing New Safety Standards and Regulations. *Proceedings of the 2nd IAASS Conference*, 14–16 May, Chicago, IL (ESA SP-645, July).

[18] Linehan, D. (2008). *SpaceShipOne—An Illustrated History*. Ian Allan.

[19] UN Doc A/AC.105/769; Oduntan, G. (2003). The Never Ending Dispute: Legal Theories on the Spatial Demarcation Boundary Plane Between Airspace and Outer Space. *Hertfordshire Law Journal*, p. 64 et seq.; UN Doc A/AC.105/769, para 3.

[20] Convention on Registration of Objects Launched into Outer Space, Art. 2(1), 1976 (1023 UNTS 15).

[21] EASA Certification Specifications for Normal, Utility, Aerobatic and Commuter Category Aeroplanes, CS-23 Initial Iss., 14 November 2003.

[22] EASA Certification Specifications for Large Aeroplanes, CS-25 Amendment 7, 21 October 2009.

[23] Commission Regulation (EC) No. 1702/2003 of 24 September 2003 laying down implementing rules for the airworthiness and environmental certification of aircraft and related products, parts and appliances, as well as for the certification of design and production organizations. *Official Journal*, L 243, 27 September.

[24] FAA 14 CFR Part 460, Commercial Space Transportation—Human Space Flight Checklist, Version 2.0 1/4/2008.
[25] IAASS Independent Space Safety Board. Space Safety Standard—Commercial Manned Spacecraft, IAASS-ISSB-S-1700 Draft, August 2006.
[26] NASA-STD-3000: Anthropometrics Requirements, SSP 50005 Revision A.
[27] Jenkins, M. (1998). Human-Rating Requirements JSC-28354, Office of the Director, NASA Johnson Space Center, Houston TX, June.
[28] NASA Safety Policy and Requirements for Payloads Using the Space Transportation System, NSTS 1700.7B, January 1989.
[29] NASA Payload Safety Review and Data Submittal Requirements for Payloads using the Space Shuttle and ISS, NSTS/ISS 13830 Revision C, July 1998.
[30] NASA/USAF (1999). *Space Shuttle Payload Ground Safety Handbook*, KHB 1700.7 Rev.C, 45 SW HB S-100, 19 August.
[31] Marciacq, J.-B. and Bessone, L. (2007). Training the Crew for Safety/Safety of Crew Training, an Integrated Process. *Proceedings of the 2nd IAASS Conference*, 14–16 May, Chicago, IL (ESA SP-645, July).
[32] Marciacq, J.-B. and Bessone, L.(2009). In *Safety Design for Space Systems—Chapter 25: Crew Training Safety* (Musgrave, G., Larsen. A. and Sgobba, T., eds). International Association for the Advancement of Space Safety. Elsevier.
[33] Treaty on Principles Governing the Activities of States in the Exploration and Use of Outer Space, Including the Moon and Other Celestial Bodies, 1967 (610 UNTS 205).
[34] On the European level, namely: Norway (Act on Launching Objects from Norwegian Territory etc. into Outer Space, 1969); Sweden (Act on Space Activities, 1982, Decree on Space Activities, 1982); UK (Outer Space Act, 1986); Belgium (Law on the activities of launching, flight operations or guidance of space objects, 2005); the Netherlands (Rules Concerning Space Activities and the Establishment of a Registry of Space Objects—Space Activities Act, 2006); and France (Bill No. 2008-518 relating to spatial operations, 2008).
[35] NASA Space Flight Health Requirements, JSC 26882, January 1996.
[36] http://www.easa.europa.eu/ws_prod/r/doc/NPA/NPA%202008-17a.pdf
[37] http://www.easa.europa.eu/ws_prod/r/doc/rp/Rulemaking%20Proposal%20Form.doc
[38] Agreement on the Rescue of Astronauts, the Return of Astronauts and the Return of Objects Launched into Outer Space, 1968 (672 UNTS 119).
[39] ICAO Circular 18-AN/15 published in 1951.
[40] ICAO Human Factors Digest No. 1 (Circular 216-AN/13).
[41] Article 9 of Council Regulation (EEC) No. 2407/92 of 23 July 1992 on licensing of air carriers. *Official Journal*, L 240, 24 August 1992.
[42] Annex III to Council Regulation (EEC) No. 3922/91 of 16 December 1991 on the harmonization of technical requirements and administrative procedures in the field of civil aviation. *Official Journal*, 373 L 4, 31 December 1991, as last amended by Regulation (EC) No. 859/2008. *Official Journal*, L 254, 20 September 2008.
[43] Regulation (EC) No. 549/2004 of the European Parliament and of the Council of 10 March 2004 laying down the framework for the creation of the single European sky; Regulation (EC) No. 550/2004 of the European Parliament and of the Council of 10 March 2004 on the provision of air navigation services in the single European sky; Regulation (EC) No 551/2004 of the European Parliament and of the Council of 10 March 2004 on the organization and use of the airspace in the single European sky; Regulation (EC) No 552/2004 of the European Parliament and of the Council of 10 March 2004 on the interoperability of the European Air Traffic Management network. *Official Journal*, L 96, 31 March 2004.

[44] Contant-Jorgenson, C., Lala, P. and Schrogl, K.-U. (eds) (2006). *IAA Cosmic Study on Space Traffic Management*. IAA.

[45] Sect. 5 (2)(g) UK Outer Space Act; 14 CFR Sect. 415.39, Sect. 417.07, Sect. 417.129 and Appendix A to Sect. 440 of the US Commercial Space Transportation Licensing Regulation.

[46] Convention for the Unification of Certain Rules Relating to International Carriage by Air, 1929.

[47] Listed, e.g., in Diederiks-Verschoor, I. H. (2006). *An Introduction to Air Law*, 8th revised ed. Kluwer Law International p. 102 et seq.

[48] Convention for the Unification of Certain Rules for International Carriage by Air, 1999.

[49] Convention on Damage caused by Foreign Aircraft to Third Parties on the Surface, 1952, including the Protocol to Amend the Convention on Damage caused by Foreign Aircraft to Third Parties on the Surface, Montreal, 1978; International Convention for the Unification of Certain Rules Relating to Damage Caused by Aircraft to Third Parties on the Surface, Warsaw, 1929.

[50] Haanappel, P. C. (2006). Envisaging Future Aerospace Applications—Passenger and Third Party Liability in Aerospace Transport. In *Project 2001 Plus—Global Challenges for Air and Space Law at the Edge of the 21st Century* (Hobe, S., Schmidt-Tedd, B. and Schrogl, K.-U., eds), p. 231 et seq. Cologne.

[51] Diederiks-Verschoor, I. H. (2006). *An Introduction to Air Law*, 8th revised ed. Kluwer Law International p. 241.

[52] Commission Regulation (EC) No 2042/2003 of 20 November 2003 on the continuing airworthiness of aircraft and aeronautical products, parts and appliances, and on the approval of organisations and personnel involved in these tasks. *Official Journal*, L 315, 28 November 2003.

PART

International Regulatory and Treaty Issues

4

CHAPTER

International regulation of emerging modes of space transportation

17

Ram S. Jakhu, Yaw Otu M. Nyampong
Institute and Centre for Research of Air and Space Law, McGill University

CHAPTER OUTLINE

Space Safety Regulations and Standards. DOI: 10.1016/B978-1-85617-752-8.10014-5

INTRODUCTION

Sub-orbital–orbital space flight has become a reality in our world today. As compared to conventional air transportation, it presents the potential for transporting people and freight across large distances on the surface of the Earth through space in remarkably shorter periods of time and also provides for transport to orbit.

A recent American study titled "The Space Transportation Annex to the Next Generation Air Transportation System (NextGen) Concept of Operations" [1] envisions three phases for the development of Future Space Transportation Systems (FSTS). Although this vision is based on a series of incremental development steps, it is expected that, eventually, FSTS will be comprehensive and fully integrated. The third stage of the FSTS, called "mass public space transportation", which is predicted to start in or about 2030, is "envisioned to begin when the economics and technology of space travel align with the demands of a mass market characterized by safe, routine, [and] affordable commercial space travel" [2]. At that time, launch safety and reliability, according to the study, will be at the same level as that of air transport today. From all indications, it is obvious that as the technology matures, sub-orbital–orbital space flight will spawn a new era of human space flight that will significantly influence the way we live our daily lives on Earth.

Although, at present, space tourism appears to be the major driver behind the emerging modes of space transportation, it should be kept in mind that the new modes of space transportation will not be used solely for space tourism. In fact, they will significantly enhance the orbiting of satellites, particularly small and/or nano satellites in low Earth orbit for military, scientific, and commercial purposes. In addition, as current observations suggest, there is a strong possibility of using sub-orbital and orbital flights as a means of providing point-to-point travel on Earth via space [3] for time-sensitive rich and business people, as well as for precious cargo, especially small packages and letters handled by various courier service operators. Therefore, the design, construction, and operation of ground facilities will have to be appropriately suited for the provision of safe and efficient services to all such users.

But, unlike what pertains in the realm of civil air transportation, there are no international legal regimes in existence today that specifically address the safety, responsibility, and liability issues that will inevitably flow from space transportation. The only international legal regimes in existence that come close to that which exists in the field of civil air transportation are contained in the basic legal principles laid down by the existing international space treaties and conventions adopted starting from the 1960s. These treaties and conventions generally govern space exploration and use, as well as the responsibilities and liabilities attached thereto. But these provisions are couched in very broad terms and as such are limited in scope, content, and effect. Thus, there is an urgent need for new and/or additional international legal regimes to be negotiated and adopted in order to provide effective safety regulation as well as responsibility and liability mechanisms applicable to space transportation.

Historically, the discovery of a new (and potentially dangerous) means of human flight has always provided justification for the negotiation and adoption of new and/or additional international regulatory regimes. For example, facing the prospect of increased commercial civil air transportation following the First World War, the international community negotiated and adopted the Convention for the Regulation of Aerial Navigation in 1919 (Paris Convention) [4], which set out the basic legal framework for international civil aviation. This major milestone was shortly followed by the adoption of the Convention for the Unification of Certain Rules Relating to International Carriage by Air of 1929 (Warsaw Convention) [5]. This important accord laid out an international liability regime applicable to civil air transport. Just as with the manufacturing of aircraft, only a few countries presently have the capability to build aerospace vehicles. By constructing and operating appropriate ground facilities, all other nations can participate in this newest mode of transportation, and thereby benefit from the commercialization of space. Aerospace vehicles will soon be "flying" not only domestically, but also on international routes (i.e. taking off and landing on various runways (spaceports) located in different countries around the world). Thus, it is essential at this point in civilization that, as sub-orbital–orbital space flight develops into a viable means of space transportation, efforts are put in place to establish effective international regulatory regimes to address those important issues that will inevitably become associated therewith, such as safety, responsibility, and liability. Before delving into an analysis and assessment of the existing international legal regimes that may have some application to space transportation, the next section of this chapter provides an overview of recent developments in space transportation.

17.1 RECENT DEVELOPMENTS IN SPACE TRANSPORTATION: THE ADVENT OF AEROSPACE VEHICLES, SPACEPORTS, AND SPACE TOURISM

17.1.1 Aerospace vehicles

From the beginning of the space age, space transportation systems have almost exclusively consisted of single-use expendable rockets launched vertically, with the notable exception of the American space shuttle, which is partially reusable. After a space mission, the shuttle returns to the Earth as a glider (i.e. an aircraft) that lands on a specifically designed and designated runway. The Russian *Soyuz* personnel/cargo supply spacecraft is another exception, but its descent capsule returns to the Earth and lands in an uninhabited remote and safe site by deploying parachutes. Developments occurring in recent years suggest, however, that the era of exclusive reliance upon single-use expendable space vehicles is gradually being phased out.

Following the success of the *SpaceShipOne* flight to the edge of space in 2004, a number of space vehicles are being developed and tested for routine trips to and from space. The most promising one seems to be *SpaceShipTwo*, which is being built by California-based Scaled Composites for Sir Richard Branson's space travel company, Virgin Galactic [6]. Though some of the new space transportation systems will be designed to travel to and from space vertically, starting with *SpaceShipOne*, serious efforts are being made to design, build, test and operate fully reusable vehicles that will take off and land horizontally, just like aircraft. These include *SpaceShipOne*, *SpaceShipTwo*, *WhiteKnightTwo*, etc. The main incentive for such efforts is to significantly reduce the cost of space transportation, which has been the single most expensive element of space operations. It is generally believed that the advent of reusable aircraft-like vehicles will, in turn, boost space commercialization by expanding space activities. Any such vehicles that will "fly" through the airspace and outer space should be called "aerospace" vehicles or aerospace craft [7].

17.1.2 Spaceports

In the last few years, while aerospace vehicles are being developed and perfected, there already seems to have been a "rush", or "competition" as some call it, in the establishment of spaceports, not only in the USA, but also in several other countries around the world [8]. Below is a list of well-established or actively planned spaceports designed and/or constructed for a variety of new space transportation operations:

US spaceports

- Spaceport America, New Mexico
- Mojave Spaceport, California
- Oklahoma Spaceport, BurnsFlat, Oklahoma
- Mid-Atlantic Regional Spaceport (MARS), Virginia
- Kodiak Launch Complex, Alaska
- Corn Ranch Spaceport, Texas.

Non-US spaceports

- Baikonur Spaceport, Kazakhstan
- United Arab Emirates Spaceport, Dubai
- Spaceport Sweden, Kiruna Sweden
- Spaceport Singapore, nr. Changi Airport Singapore.

Trends so far observed from the development of new spaceports suggest that most of them will have the same characteristics as airports, and aerospace vehicles will be operating from there just like aircraft do from airports. Writing in 2000, Kevin Brown [9] suggested the following evolutionary development of five generations of spaceports; namely:

- **First-generation spaceport:** today, Eastern and Western Ranges, based on missile test legacy and governed by EWR 127-1—responsible only for ascent phase of space flight.
- **Second-generation spaceport:** 2005–2010, 10× cost improvement, in transition from military to civil control. Spaceports operate under Federal Aviation Administration (FAA) license, but some facilities, systems, and procedures are similar to first generation, and many safety regulations are legacy 127-1; manages all phases of space flight.
- **Third-generation spaceport:** 2015–2025, 100× cost improvement, full civil control, similar to today's small airports. These new spaceports operate under FAA license with little resemblance to 1990s era launch sites.
- **Fourth-generation spaceport:** 2030–2040, Earth surface and earth orbit nodes; i.e. an orbital space station can serve as an integrated "spaceport" within the spaceline routing network; "orbit traffic control" is an integral part of the system, monitoring and controlling all objects in Earth orbit.
- **Fifth-generation spaceport:** 2040–2060, a spaceport on the moon or other planet and the notion of interplanetary traffic control are introduced.

It is interesting to note that Kevin Brown predicted that during the third generation, which will start some time in 2015, spaceports will be "similar to today's small airports".

The USA is undoubtedly the world leader in establishing spaceports, not only in terms of their physical construction, but also with regard to their regulation, including the establishment of technical safety standards. Spaceport regulation in the USA is carried out pursuant to a relatively well-developed legal regime under the Space Commercial Launch Act of 1984, as amended [10], and implemented by the FAA's Associate Administrator for Commercial Space Transportation (FAA/AST) [11]. Each spaceport in the USA must procure a license before it can start operations. Prior to issuing any such license, the FAA/AST carries out a thorough safety review of all the facilities of the planned spaceport.

17.1.3 Space tourism

Presently, some commercial space tourism operators are already offering sub-orbital space flights as a means of delivering the experience of real weightlessness in space to those members of the general public who are willing and able to pay the premium. On 11 October 2008, Richard Garriott, an American computer game millionaire, became the sixth paying space tourist when he boarded a Russian *Soyuz* spacecraft for a 10-day trip to the International Space Station [12]. Almost a year later, in September–October 2009, Guy Laliberté, a Canadian billionaire and owner of the famous *Cirque du Soleil*, also traveled on board a Russian *Soyuz TMA 16* spacecraft to the International Space Station, where he spent 11 days in orbit [13]. In so doing, he became the seventh private space-flight client of Space Adventures Ltd. Sir Richard Branson's Virgin Galactic has been taking reservations for regular sub-orbital trips for ordinary people, starting sometime in 2010 at a ticket price of about US $200,000. There is a growing interest in space travel as numerous individuals are making deposits to reserve their "flights"; revenues from commercial space-flight activities are increasing, and capital investment from private sources is expanding. According to a recent study carried out by the Tauri Group of Alexandria, Virginia, USA, "deposits and revenue for direct commercial human space-flight services, such as flights of private citizens to the International Space Station and deposits on sub-orbital commercial human space flights, rose to $50.0m in 2008, compared to $38.8m in 2007 … [and] investment[s of up to] of $1.46 billion ha[ve] been committed to the industry since January 2008" [14].

These are small but important pieces of evidence pointing to a growing interest in space tourism, an area that is soon expected to become a multimillion dollar industry. At present, global travel and tourism is the largest service sector industry in the world, of which adventure travel is worth about $150 billion in the USA alone [15]. Space tourism will be a part of adventure tourism, at least in the beginning, but it will continue to expand as it increasingly becomes economically affordable. However, the realization of this projected enormous expansion will heavily depend upon the safety of space transportation systems.

In sum, recent developments in space transportation point to the fact that new space vehicles akin to aircraft are being developed along with the ground facilities from which they will be operated. The success of any new space transportation systems will depend not only upon their operational safety, but also on the safety of the associated ground facilities. These ground facilities are presently referred to as ranges, launch sites, or spaceports. In this regard, it should be noted that if a facility handles aircraft (as currently defined) for transportation to and from two or more points on the surface of Earth, it is an airport. On the other hand, the term "spaceport" generally refers to a facility that handles machines or craft for transportation to and from space. Since technological developments are making it possible to conduct transportation to, from, and via space, using machines or craft that are essentially the same as aircraft (when considered separately or in combination), the ground handling faculties should be called "aerospace ports" [16]. In this chapter, the term "spaceport" is therefore used essentially in the sense of "aerospace port".

At present, the major driver behind these developments appears to be the space tourism industry, although there is immense potential for these emerging modes of space transportation to be used to further advance the conduct of other—more traditional—space activities. As noted above, different types of aerospace vehicles are being developed and spaceports have begun spreading around the world in places both within and outside of the USA. However, there is no detailed and uniform international regulatory system (including technical standards) in existence for ensuring their operational safety and efficiency, and for guaranteeing uniformity in their operations. The operation of aerospace vehicles and spaceports will undoubtedly raise old and new questions concerning responsibility and liability between States, entities, and individuals. Again, it remains to be seen whether these issues are sufficiently dealt with by the existing international legal regimes. The next section of this chapter therefore outlines and assesses the extent to which existing international legal regimes governing the exploration and use of outer space address these matters.

17.2 OVERVIEW AND ASSESSMENT OF EXISTING INTERNATIONAL LEGAL REGIMES CONCERNING OUTER SPACE EXPLORATION AND USE

17.2.1 Safety regulation in space transportation

From the dawn of the space age in 1957, the international regulatory regime governing space activities has been developed primarily through the Committee on Peaceful Uses of Outer Space (COPUOS) of the United Nations General Assembly [17]. The Committee, as created in 1958, has been mandated to address legal problems arising from the exploration and use of outer space, a function it performs essentially through its Legal Subcommittee, while its Scientific and Technical Subcommittee deals with more technical space-related issues. So far, the Committee has drafted five international treaties [18] and five declarations [19] that cumulatively have created the fundamental principles of the international regulatory regime governing space activities.

The Treaty on Principles Governing the Activities of States in the Exploration and Use of Outer Space, including the Moon and Other Celestial Bodies (the "Outer Space Treaty"), which is considered to be the constitution of outer space, lays down several foundational and general legal norms. Some of them specify that: (a) the exploration and use of outer space must be carried out for the benefit and in the interests of all mankind; (b) outer space and celestial bodies are free for exploration and use by all States on the basis of equality and in accordance with international law; (c) outer space and celestial bodies are not subject to national appropriation by any means; (d) States party to the Treaty are obliged not to place in orbit around the Earth any objects carrying nuclear weapons or any other kind of weapon of mass destruction; and (e) in the exploration and use of outer space, States are to be guided by the principle of cooperation and mutual assistance and must conduct all their space activities with due regard for the corresponding interests of other States. In addition, Article VIII of the Treaty entitles the State of Registration of a space object to retain jurisdiction and control over such object, and over any personnel thereof. Ownership of space objects, and of their component parts, is not otherwise affected by their presence in outer space. Such objects or component parts, if found beyond the limits of the State of Registration, are required to be returned to that State.

The provisions of Article VIII of the Outer Space Treaty have been further expanded upon by the 1968 Agreement on the Rescue of Astronauts, the Return of Astronauts and the Return of Objects

Launched into Outer Space (the "Rescue Agreement") and the1975 Convention on Registration of Objects Launched into Outer Space (the "Registration Convention"). In the absence of an international treaty to specifically regulate space safety, the provisions of Article VIII of the Outer Space Treaty and of the Registration Convention can be considered, at least by implication, to create a legal link or basis for assigning responsibility and possibly liability for the unsafe conduct of space transportation to the Launching State. The 1979 Agreement Governing the Activities of States on the Moon and Other Celestial Bodies (the "Moon Agreement") does not contain any additional pertinent provisions related to space safety and has also been ratified only by a small number of States, none of which is a major spacefaring nation.

The 1992 Principles Relevant to the Use of Nuclear Power Sources in Outer Space (the NPS Principles), adopted as a UN Resolution that was drafted by the Scientific and Technical Subcommittee of the COPUOS, contains the first set of principles and guidelines aimed at ensuring the safe use of nuclear power sources in outer space, particularly for the generation of electric power on board space objects for non-propulsive purposes. According to this Resolution, the use of nuclear power sources in outer space should "be based on a thorough safety assessment, including probabilistic risk analysis" for "reducing the risk of accidental exposure of the public to harmful radiation or radioactive material" [20]. States launching [21] space objects with nuclear power sources on board are obliged to protect individuals, populations, and the biosphere against radiological hazards. Nuclear reactors may be operated (i) on interplanetary missions, (ii) in sufficiently high orbits [22], and (iii) in low Earth orbits if they are stored in sufficiently high orbits after the operational part of their mission. Nuclear reactors may use only highly enriched uranium 235 as fuel. A Launching State is obliged to ensure that a thorough and comprehensive safety assessment is conducted. The results of any such assessments must be made publicly available prior to each launch [23]. After re-entry into the Earth's atmosphere of a space object containing a nuclear power source on board, or its components, the Launching States are obliged to promptly offer and, if requested by the affected State, promptly provide the necessary assistance to eliminate actual and possible harmful effects. These principles seem to have been consistently complied with. For example, the USA notified the UN about the launch of *Cassini*—a spacecraft powered by 33 kg of plutonium [24]. *Cassini*, a joint endeavor of NASA, the European Space Agency and the Italian Space Agency, was launched to study the planet Saturn and its magnetic and radiation environment.

It is well known that the current space debris environment poses a risk to spacecraft in Earth orbit and space travel. As the population of debris continues to grow, the probability of collisions that could lead to potential damage will consequently increase. Therefore, after several years of deliberations, the Scientific and Technical Subcommittee of the COPUOS adopted the Space Debris Mitigation Guidelines in February 2007 [25]. These Guidelines are to be applied for mission planning, design, manufacture, and the operational phases (i.e. launch, mission, and disposal) of spacecraft and space transportation systems. According to the Guidelines, *inter alia*, space systems should be designed so as not to release debris during normal operations. In addition, the intentional destruction of any spacecraft and space transportation systems should be avoided. States and international organizations are to take measures, through national mechanisms or through their own applicable mechanisms, to ensure that these Guidelines are implemented, to the greatest extent feasible, through space debris mitigation practices and procedures.

17.2.2 Responsibility and liability in space transportation

A general principle under the existing space law treaties is that State responsibility attaches to all national space activities, including space transportation. From the beginning of the space age, it was realized that since space activities require huge amounts of money, they would most likely be carried out with public funds by States as opposed to private entities. Secondly, the ultra-hazardous nature of space activities (e.g. the magnitude of loss of human lives and property damage likely to result from an accident in a space operation involving nuclear energy) provided a basis for the requirement of close State involvement (responsibility) for purposes of ensuring public safety. Consequently, all space activities are classified either as national activities carried out by States or their entities, or international activities undertaken by intergovernmental organizations. The Outer Space Treaty obliges States to bear international responsibility for national activities in outer space, whether they are carried out by their governmental agencies or by private entities, and to provide the necessary authorization and continuous supervision of such activities in order to ensure that national activities are carried out in conformity with the principles set forth in the Treaty [26]. When space activities are carried out by an international organization, responsibility for compliance with the Outer Space Treaty is to be borne both by the international organization and by the States party to the Treaty participating in such organization. This principle is an exception to a rule of general international law under which State responsibility can be incurred only where a "genuine link" exists between the State and the concerned activity/person [27].

The Outer Space Treaty, under Article VII, specifies that each Launching State (i.e. the State that launches or procures the launching of an object into outer space, or from whose territory or facility an object is launched) is internationally liable for damage to a foreign State or to its natural or juridical persons by such object or its component parts. This principle of international liability has been further elaborated upon in the 1972 Convention on International Liability for Damage Caused by Space Objects (the "Liability Convention"). Under the Convention, a Launching State is absolutely liable to pay compensation for damage caused by its space object on the surface of the Earth or to aircraft in flight. However, if the damage is caused in outer space, the Launching State can be held liable only if the damage is due to its fault or the fault of persons for whom it is responsible.

Whenever two or more States jointly launch a space object, they can be held jointly and severally liable for any damage thereby caused. If a space object belonging to an intergovernmental organization causes damage, the intergovernmental organization can be held liable alongside the Member States of the organization, who would also be jointly and severally liable. The term "damage" means loss of life, personal injury or other impairment of health, or loss of or damage to property of States or of persons, natural or juridical, or property of international intergovernmental organizations. This liability regime is victim-oriented as it calls for speedy payment of compensation, and there is no limitation on the amount of liability exigible. Legal principles related to international responsibility and possible liability, as incorporated in the Outer Space Treaty and the Liability Convention, can be deemed to create confidence in the safety of space transportation.

17.2.3 Analysis and assessment of the adequacy of the existing international regimes governing space activities

It is clear from the above description of the international space regulatory regime that no international space treaty specifically and effectively regulates space safety, particularly in relation to emerging trends in space transportation. However, the international space treaties assign to States, with respect to all space transportation operations, whether carried out by their public or private entities or by intergovernmental organizations in which they participate, (a) national jurisdiction and control over such activities, and (b) national obligation for international responsibility and liability. Therefore, these treaties seem to create a legal obligation for the concerned State to adopt *national* space safety standards and procedures as well as *national* means for implementing them in order to ensure the safe conduct of space transportation. In other words, the treaties do not include provisions dealing with *international* space safety standards and procedures or, for that matter, their *international* implementation.

There seems to be a small exception. The Scientific and Technical Subcommittee of the COPUOS has initiated the first set of *international* principles and guidelines to ensure space safety, in two documents, namely: (a) the 1992 Principles Relevant to the Use of Nuclear Power Sources in Outer Space that applies to the use of nuclear power sources in outer space, and (b) the 2007 Space Debris Mitigation Guidelines that apply to mission planning, design, manufacture, and operational phases of spacecraft and space transportation systems. Unfortunately, the 1992 Principles are limited in their scope as they only regulate nuclear power sources used for the generation of electric power on board space objects for non-propulsive purposes. It is therefore clear that the current international space regulatory regime is seriously limited in scope and character, and is thus inadequate to effectively regulate space safety, particularly pertaining to sub-orbital and orbital space transportation, which would mostly be of an international nature.

Secondly, the decision-making process, both in the COPUOS and its Subcommittees, is extremely slow, cumbersome, and tedious, as they work on the basis of an informal rule of consensus [28]. The consensus rule worked relatively well in the past, as several treaties and resolutions on major space law issues were successfully drafted and adopted. However, for the last two decades, the rule has become controversial and the decision-making process in the COPUOS is considered to have reached a stage of serious crisis. A small minority of States seems to exercise a kind of veto power as they have taken over the decision-making process, using the requirement of consensus. Since the adoption of the Moon Agreement in 1979, not a single new space law treaty has been drafted by the COPUOS. Lately, the Committee seems to have chosen to draft non-binding resolutions and guidelines that States may choose to follow or ignore. For example, there have been cases where some States did not hesitate to adopt national regulations or to take other actions that were contrary to the provisions of such resolutions [29].

As regards the responsibility and liability provisions of the existing space law treaties, although they appear to be very broadly couched and therefore in need of further detailing, they nonetheless serve the essential purpose of ensuring that States provide authorization and continued supervision and control over emerging modes of space transportation and also remain primarily liable to make reparations in cases where damage is caused as a result of such activities. At this nascent stage of their development, emerging modes of space transportation are well served by the existing provisions on responsibility and liability. As the sector matures, however, there will be the need to establish more comprehensive regimes of State responsibility and liability building upon the broad framework established by the existing regimes.

Therefore, based on the above analysis, it is our considered opinion that the current international space treaties are insufficient for current and future space utilization needs. Likewise, it would appear that COPUOS, as a functioning and effective international organization, has insufficient authority and capability to regulate space safety effectively and efficiently. What is ultimately required is an international regulatory regime with authority that is binding in nature and international in scope, and which is updated regularly by a technically competent, permanent, international organization. Thus, if this logic is accepted, there would appear to be a need for both a new regulatory regime and a new organization for space safety. As proposed in this chapter, the Chicago Convention and its Annexes, and the International Civil Aviation Organization (ICAO), after some necessary changes and adaptations, might serve that purpose quite well. It will therefore be useful to take a look at the universally accepted international regime of Standards and Recommended Practices (SARPs) adopted by the ICAO, which ensure the safe, efficient, and orderly growth of international civil aviation. Of course, this is just one of several possible alternatives. Others may take the form of negotiating entirely new international treaties and creating new entities for the same purpose. In the next section, we will outline and assess the relevant provisions of the Chicago Convention and its Annexes, examine how the SARPs are adopted and implemented, and then proceed to explore the extent to which such an aviation-specific model can be adapted and made applicable to safety regulation of emerging modes of space transportation. We will also consider how issues of responsibility and liability have been addressed in air transportation, from the days when it was considered to be an ultra-hazardous activity, to contemporary times, when it is deemed to be the safest mode of transportation.

17.3 OVERVIEW AND ASSESSMENT OF EXISTING INTERNATIONAL LEGAL REGIMES GOVERNING AIR TRANSPORTATION

17.3.1 The International Civil Aviation Organization

It is a well-recognized fact that civil aviation has become a powerful force driving progress in our modern global society. Twenty-four hours a day, 365 days a year, an aeroplane takes off or lands every few seconds somewhere on the face of the Earth. Each one of these flights is handled in the same uniform manner, whether by air traffic control, airport authorities, or pilots at the controls of their aircraft. This clockwork precision in procedures and systems is made possible by the existence of universally accepted standards known as international Standards and Recommended Practices (SARPs). SARPs cover all technical and operational aspects of international civil aviation, such as safety, personnel licensing, operation of aircraft, aerodromes, air traffic services, accident investigation, and the environment. Without SARPs, the global aviation system would be, at best, chaotic and, at worst, unsafe.

Creating and modernizing SARPs is the responsibility of the International Civil Aviation Organization (ICAO), the specialized agency of the United Nations whose mandate is to ensure the safe and efficient operation as well as orderly evolution of international civil aviation. The charter of the ICAO is the *Convention on International Civil Aviation* (the Chicago Convention), drawn up in Chicago in December 1944, and to which each ICAO Contracting State is a party. From its beginning in 1944, the ICAO has grown to an organization with about 192 Contracting States. It provides the forum where requirements and procedures in need of uniform aviation-related standards may be introduced, studied, adopted, and implemented [30]. According to the Chicago Convention, the ICAO consists of an Assembly, a Council, and a Secretariat. The chief officers are the President of the Council and the Secretary General. The Assembly, composed of representatives from all Contracting States, is the sovereign body of the ICAO. It meets every 3 years, reviewing in detail the work of the Organization, setting policy for the coming years, and establishing a triennial budget. The Assembly elects the Council, the governing body, for a 3-year term [31].

The Council is composed of members from 36 States who maintain their permanent offices and conduct their regular business at the ICAO Headquarters in Montreal, Canada. It is in the Council that Standards and Recommended Practices are adopted and incorporated as Annexes to the Convention on International Civil Aviation. With regard to the development of standards, the Council is assisted by the Air Navigation Commission (ANC) in technical matters, the Air Transport Committee in economic matters, and the Committee on Unlawful Interference in aviation security matters. The bulk of the SARPs are technical in nature; therefore, the ANC plays a predominant role in the development of SARPs. The ANC is composed of 15 experts with appropriate qualifications and experience in various fields of aviation [32]. Its members are nominated by Contracting States and are appointed by the Council. They are expected to function as independent experts and not as representatives of their States. The Commission is assisted in its work by the technical personnel of the Air Navigation Bureau, which is a part of the Secretariat.

17.3.2 Safety regulation in civil aviation

Among the numerous objectives of the ICAO, it is required under the Chicago Convention to "... develop the principles and techniques of international air navigation and to foster the planning and development of international air transport so as to ... promote safety of flight in international air navigation" [33]. Thus, safety is a fundamental goal of the organization. The safety aspects of international civil aviation are addressed in part by the substantive provisions of the Chicago Convention and in part by the SARPs contained in the Annexes. Three substantive provisions of the Chicago Convention are particularly relevant for purposes of safety regulation. Article 12—titled Rules of the Air—obliges Contracting States to regulate the flight and maneuver of aircraft within their respective territories by enacting appropriate rules and regulations and ensuring compliance with them. With respect to airspace beyond national jurisdiction (i.e. airspace over the high seas), the Convention mandates the ICAO to establish the applicable rules and regulations.

The Convention further obliges Contracting States to keep their own national regulations uniform, to the greatest extent possible, with those established by the ICAO under the Convention. Article 31 provides that each aircraft engaged in international air navigation shall be provided with a certificate of airworthiness issued or rendered valid by the State of Registry of the aircraft. Article 32, on the other hand, requires that every pilot or member of the operating crew of an aircraft engaged in international navigation shall be provided with certificates of competency and licenses granted or rendered valid by the State of Registry of the aircraft. The Convention empowers the ICAO to establish minimum standards for the issuance of the foregoing licenses and certificates by national authorities, and Contracting States are obliged to accept licenses and certificates issued by the appropriate authorities of other Contracting States, provided that they are based at least on the minimum standards set by the ICAO.

To enable the ICAO perform its functions, referred to in the preceding paragraphs, the Chicago Convention entrusts the ICAO with the power to make Standards and Recommended Practices (SARPs) that, for convenience, are designated as Annexes to the Convention. A *Standard* is defined as any specification whose uniform application is recognized as *necessary* for the safety or regularity of international air navigation and to which Contracting States will conform in accordance with the Convention. A *Recommended Practice*, on the other hand, is any specification whose uniform application is recognized as *desirable* for the safety, regularity, or efficiency of international air navigation [34].

17.3.2.1 Formulation of new or revised SARPs

SARPs are created through international cooperation, consensus, compliance, and commitment (i.e. the four "Cs" of aviation, namely: Cooperation in the formulation of SARPs; Consensus in their approval; Compliance in their application; and Commitment to adhere to this ongoing process). ICAO standards and other provisions are developed in the following forms:

- Standards [35] and Recommended Practices [36]—collectively referred to as SARPs
- Procedures for Air Navigation Services [37]—called PANS
- Regional Supplementary Procedures [38]—referred to as SUPPs
- Guidance Material in several formats [39].

The formulation of new or revised SARPs begins with a proposal for action from the ICAO itself, or from any of its Contracting States. Proposals also may be submitted by international organizations. For technical SARPs, proposals are analyzed first by the Air Navigation Commission (ANC). It is through a variety of meetings that most of the work is finalized and the necessary consensus reached. In their development, a number of consultative mechanisms are used. All Contracting States are invited to participate in these meetings with equal voice. Interested international organizations may be invited to participate as observers. Technical issues dealing with a specific subject and requiring detailed examination are normally referred by the ANC to a panel of experts. Less complex issues may be assigned to the Secretariat for further examination, perhaps with the assistance of an air navigation study group [40].

17.3.2.2 Review of draft SARPs

The ANC prepares a technical proposal for the new SARPs or amendments and circulates same to Contracting States and appropriate international organizations, for their comments. The comments of States and organizations are analyzed by the Secretariat and a working paper, detailing the comments and the Secretariat proposals for action, is prepared. Standards developed by other recognized international organizations can also be referenced, provided they have been subject to adequate verification and validation. Before submitting its recommendations to the Council, the ANC undertakes a final review and establishes the final texts of the proposed new SARPs or amendments to SARPs, PANS, and associated attachments [41].

17.3.2.3 Adoption/publication of amendments to Annexes

The Council reviews the proposals of the Air Navigation Commission and adopts the new SARP or amendment to an Annex if two-thirds of its members are in favor. After the adoption of a new SARP or an amendment by the Council, it is again dispatched to States, this time seeking their approval. The new SARP or amendment becomes effective on the notified Effective Date provided a majority of States have not registered their disapproval with it. On or before the Notification Date, which is 1 month prior to the Applicability Date, States must notify the Secretariat of any differences that will exist between their national regulations and the provision of the new SARP or amendment. The differences reported by States are then published in supplements to Annexes. On the Applicability Date, States must implement the amendments unless, of course, they have previously notified differences. The result of this adoption procedure is that the new or amended Standards and Recommended Practices become part of the relevant Annex [42].

To date, 18 Annexes, containing thousands of SARPs, have been adopted by the ICAO Council on the following matters:

- Annex 1—Personnel Licensing
- Annex 2—Rules of the Air
- Annex 3—Meteorological Service for International Air Navigation
- Annex 4—Aeronautical Charts
- Annex 5—Units of Measurement to be Used in Air and Ground Operations
- Annex 6—Operation of Aircraft
- Annex 7—Aircraft Nationality and Registration Marks
- Annex 8—Airworthiness of Aircraft
- Annex 9—Facilitation
- Annex 10—Aeronautical Telecommunications
- Annex 11—Air Traffic Services
- Annex 12—Search and Rescue
- Annex 13—Aircraft Accident and Incident Investigation
- Annex 14—Aerodromes
- Annex 15—Aeronautical Information Services
- Annex 16—Environmental Protection
- Annex 17—Security: Safeguarding International Civil Aviation Against Acts of Unlawful Interference
- Annex 18—The Safe Transport of Dangerous Goods by Air.

As demonstrated by the list of Annexes, the SARPs adopted by the ICAO address mainly operational and several important safety issues. On average, the adoption of a new or amended SARP takes up to 2 years from the Preliminary Review by the ANC to the Applicability Date. Although this process may seem lengthy at first glance, it provides for repeated consultation and extensive participation of States and international organizations in producing a consensus based on logic and experience. International cooperation, consensus, compliance, and an unfailing commitment to the ongoing implementation of SARPs have made it possible to create a global aviation system that has evolved into the safest mode of mass transportation ever conceived. The flight crew of an aircraft can count on standardized aviation infrastructure wherever they fly in the world. This internationally cooperative regulatory system has thus provided worldwide air transport industry with the vital infrastructure for achieving safety, reliability, and efficiency.

17.3.2.4 Implementation of SARPs

Under the Chicago Convention, the implementation of SARPs essentially lies with Contracting States. To assist them in the area of safety, the ICAO established a Universal Safety Oversight Audit Programme (USOAP) in 1999. The Program consists of regular, mandatory, systematic, and harmonized safety audits carried out by the ICAO in all Contracting States. The objective is to promote global aviation safety by determining the status of implementation of relevant SARPs, associated procedures, and safety-related practices. The audits are conducted within the context of critical elements of a State's safety oversight system. Under this program, the ICAO sends audit teams, comprising trained and certified personnel, on missions to Contracting States to evaluate their domestic implementation of the safety-related standards contained in Annexes 1, 6, 8, 11, 13, and 14. The missions produce formal reports of their findings to the ICAO, which in turn uses them solely for safety-related purposes. Since its inception, the Programme has proved effective in identifying safety concerns in the safety-related fields under its scope, while providing recommendations for their resolution. The Programme is being gradually expanded to include aerodromes, air traffic services, aircraft accident and incident investigation, and other safety-related fields.

As an offshoot of USOAP, the ICAO has established an International Financial Facility for Aviation Safety (IFFAS), a mechanism to provide financial support to Contracting States for the implementation of those necessary measures identified by the audit process (i.e. safety-related projects for which Contracting States cannot otherwise provide or obtain the necessary financial resources). Although there has been some debate regarding the legal basis for the ICAO's implementation of USOAP, the program has nevertheless been running smoothly over the years, and it continues to play a significant role in safety regulation of international civil aviation.

17.3.3 Responsibility and liability in civil aviation

With the exception of a handful of cases [43], the concept of State responsibility has not been relied upon in international civil aviation as much as it has been used in the treaties governing space exploration and use. Indeed, there are no international treaties in existence today that ascribe State responsibility to Contracting States for acts and/or omissions of their national air carriers emanating from commercial civil aviation.

With respect to the concept of liability, however, there are a number of international treaties in existence that have established international legal regimes dealing with different aspects of the subject. Until recently, liability for passengers, baggage, and cargo in international civil aviation was exclusively governed by the Warsaw System [44]. A new treaty, known as the Montreal Convention of 1999 [45], has been adopted as a replacement for the Warsaw System; it came into force in 2003. At present, however, both the Warsaw System and Montreal Convention continue to govern international air carrier liability for passengers, baggage, and cargo simultaneously as the new Convention strives to attain universal ratification by all the Contracting States of the old system.

Both the Warsaw System and the Montreal Convention establish a uniform international legal regime of air carrier liability. They ascribe to the air carrier (as opposed to its State of Registration) liability for damage sustained in the event of death, bodily injury, or wounding of a passenger if the accident that caused the said damage took place on board an aircraft operated by the carrier, or in the course of any of the operations of embarking or disembarking [46]. The carrier is also liable under the regime for damage sustained in the event of destruction, loss of, or damage to, registered luggage or cargo carried by air, if the occurrence that caused the damage so sustained took place during the carriage by air. Finally, air carriers are liable for damage occasioned by delay in the carriage of passengers, luggage, or goods.

A significant aspect of the uniform international passenger liability regimes established by the above-mentioned treaties is that they prescribe standard monetary limits on the liability of the air carrier, irrespective of where the damage occurs. Further, although the quantum of the liability limits differs across the two regimes, a claimant is generally not required to prove the fault of the carrier in order to succeed on a claim for damages within those limits. Stated differently, both regimes implement a trade-off by limiting the liability of the air carrier in exchange for strict liability (i.e. no requirement on the part of the claimant to prove fault). Within the limits of liability, a carrier may not avoid or reduce its liability by relying on any legal defenses except those specifically prescribed under the treaties. Beyond the limits, however, a claimant may have to prove fault on the part of the carrier in order to succeed, and the carrier may also rely upon any number of legal defenses.

In essence, the Warsaw System and its successor, the Montreal Convention, are widely perceived to be very successful attempts at comprehensively unifying numerous disparate rules of private law across national boundaries. In the majority of countries around the world where either of the two regimes apply, it provides the sole and exclusive remedy for the recovery of damages arising from bodily injury, death, or wounding to a passenger, or from destruction, loss of, or damage to, luggage or cargo, or delay in international air transportation.

On the other hand, liability for damage caused to third parties on the surface of the Earth by aircraft engaged in international civil aviation is governed by the 1952 Convention on Damage Caused by Foreign Aircraft to Third Parties on the Surface (Rome Convention) for the handful of States that are party to it [47]. Under the Convention, a person who suffers damage on the surface is entitled to compensation if he can prove that the damage was caused by an aircraft in flight. The third-party claimant is not required to prove the fault of the operator of the aircraft. The liability for damages attaches to the operator of the aircraft, defined as the person who was making use of the aircraft or who was the registered owner thereof at the time of the event that caused the damage. As in the case of the passenger liability regimes, the Rome Convention limits the amount of damages that can be recovered against the operator by any claimant, on the basis of the weight of the aircraft concerned. It also prescribes insurance requirements for air carriers, as well as rules of procedure to be followed in the event that a claim of this nature is to be made. Although it was opened for signature as far back as 1952, the Rome Convention has failed to attain universal acceptance by the international community, primarily due to its low limits of liability. To date, there are only 47 Contracting States who remain party to it.

As a result of the high level of international dissatisfaction with the regime of third-party liability established by the Rome Convention, the ICAO commenced efforts in 1998 to modernize the regime. These efforts were given a boost by the terrorist events of 11 September 2001, and they eventually culminated in the adoption of two international treaties on third-party liability: one dealing with third-party liability arising from general risks inherent in civil aviation (e.g. aviation accidents) [48], and the other dealing with liability arising from acts of unlawful interference with aircraft (e.g. terrorist attacks or hijacking) [49]. Both conventions provide for strict liability of the operator for third-party damages up to a maximum of 700 million System Definition Reviews (SDRs) per event, calibrated on the basis of the maximum certificated take-off mass of the aircraft involved. In addition, the Unlawful Interference Convention establishes an International Civil Aviation Compensation Fund to pay compensation for third-party damages of up to 3 billion SDRs in addition to the first tier of 700 million SDRs per event. The fund is to be financed from mandatory contributions levied against departing international passengers and cargo shippers. For damages in excess of the $3.7 billion provided in the first two tiers, third-party claimants under the Unlawful Interference Convention may bring claims against the operator of the aircraft and must prove the fault of the operator in order to succeed. At the time of writing, the General Risks and the Unlawful Interference Compensation Conventions had attained nine and seven signatories respectively and no ratifications at all. The next section contains an assessment of the above-described international treaty regimes governing air transportation.

17.3.4 Analysis and assessment of the existing international regimes governing civil aviation

The existing international regimes for safety regulation and the establishment of liability in relation to international civil aviation are fairly well established and advanced. Although the Chicago Convention is founded on the basic principle of State sovereignty over national airspace, it nevertheless attempts to establish a uniform and effective international regulatory regime for civil aviation. In conformity with the basic principle, the Convention recognizes the right of its Contracting States to make appropriate rules and regulations concerning air navigation within their national airspace. However, the authority to make rules applicable to airspace not subject to national jurisdiction has been given to the ICAO, and Contracting States are enjoined to ensure that their own national regulations conform to those established by the ICAO.

Mechanisms have also been established to enable the ICAO to monitor, evaluate, and correct State implementation of the relevant international rules. As noted, the Convention also entrusts safety regulation of international civil aviation to the ICAO, an international institution properly equipped for, and well suited to, the task. This has provided legal basis for the establishment of uniform international safety standards and also for their uniform implementation all over the world. In the opinion of one commentator, "ICAO has successfully managed to separate political from technical aspects of civil aviation, and the regulatory mechanisms contained in the Chicago Convention allow for a flexible adaptation of technical regulations to new developments" [50]. Although the Annexes to the Chicago Convention are so designated solely for the sake of convenience, and are by no means an integral part of the treaty, they nevertheless have the binding force of law or at least are so regarded by Contracting States.

Over the years, they have ensured the continued safety of civil aviation as a means of transport. The existing liability regimes applicable to international civil aviation have consistently ascribed liability arising from international air transportation to private air carriers instead of the States whose nationalities the airlines carry. At the commencement of commercial aviation, the activity was viewed as being very risky. Accordingly, in recognition of the need to safeguard and enhance the development of the industry, the liability regimes have (until recently) emphasized the use of limited liability mechanisms in order to shield private air carriers from the debilitating effects of unlimited liability. At the same time, these regimes have ensured that a uniform system of liability governs, irrespective of where an accident occurs or which carrier is involved. As the air transport industry matured, the focus of the liability regimes has shifted from one of protecting the industry and enhancing its development to one of ensuring that victims are awarded levels of compensation commensurate with the injuries they suffer. As such, besides raising the liability limits to comparatively higher levels, the Montreal Convention is also more victim-oriented than its predecessor, the Warsaw System. The same can also be said about the two new third-party liability conventions discussed above.

As a result of the application of these international liability regimes, most commercial air carriers are able, with reasonable certainty, to assess their liability exposure from their operations on a worst-case scenario basis. This, in turn, has enhanced the insurability of the risks associated with civil aviation; carriers are thus able to easily secure adequate insurance on the commercial markets to cover their liability exposure. Thus, irrespective of the drawbacks inherent in the international air carrier liability regimes (the biggest of which appears to be the use of liability limits to suppress the amount of compensation that may be awarded against air carriers), they have provided comprehensive and, to a very large extent, uniform international legal frameworks that govern liability issues in international air transportation. These, and other significant benefits, could also be realized in space transportation if comparable international legal regimes were to be established to regulate safety and establish responsibilities and liability with particular regard to emerging, and potentially risky, trends such as sub-orbital–orbital space flight.

17.4 PROPOSALS FOR NEW INTERNATIONAL LEGAL REGIMES RELATED TO SPACE

17.4.1 Proposal for safety regulation of space transportation

It is evident from the foregoing analyses that inasmuch as the existing international space treaties do not sufficiently regulate the safety aspects of space transportation, they also do not prohibit the development of new and/or additional international regimes on the issue. It is also clear that in the realm of international civil aviation, the Chicago Convention and its Annexes (as the basic legal framework) and the ICAO (as the institutional mechanism for its implementation at the international level) have been very effective in ensuring the safety of international air navigation.

Accordingly, one solution is to undertake all necessary and practically feasible adaptations of the Chicago Convention and its Annexes so as to provide a practical framework for the regulation of the safety aspects of space transportation. Numerous legal hurdles will be encountered and would need to be addressed in the implementation of this proposal, the most prominent of which will be the amendment of the Chicago Convention so as to broaden its scope to include space transportation. However, this might not be necessary. Although the Chicago Convention generally applies to international air navigation, it neither defines nor delimits what exactly constitutes the "airspace". Also, the Convention does not define what an "aircraft" is. The definition of "aircraft" commonly used by the ICAO is contained in one of the Annexes to the Convention and, as such, does not form an integral part of the treaty. Thus, technically, using the mechanism of the Annexes, the scope of application of the Convention could be widened to include "aerospace vehicles" flight instrumentalities that have all the attributes of aircraft in addition to the capability to operate in outer space—under the Convention. All that would be required is the development and adoption of a new and/or additional set of SARPs or Annexes to the Convention specifically designed to cater for the peculiar characteristics of such aerospace vehicles [51].

As noted above, the Annexes to the Chicago Convention are so designated solely for the sake of convenience. Although the Annexes provide an innovative and important means of implementing the provisions of the Convention, the SARPs contained therein do not have the same force and legal effect as the substantive provisions of the Convention. The significance of the Annexes, however, lies in their flexibility. Since Annexes are adopted by the Council of the ICAO, they are relatively easier to make and/or change as compared to the substantive provisions of the Convention. Thus, a widening of the scope of application of the Convention so as to regulate the technical and/or safety-related aspects of aerospace vehicles could be achieved by means of an ICAO Council decision to that effect. As a starting point, a working group could be established by the ICAO Council to study the feasibility of such an approach. The ICAO's own precedents justify this proposal. In 1998, the ICAO Assembly adopted a resolution at its 32nd session expressing concern about a missile test carried out by North Korea, and urging all Contracting States to comply strictly with the provisions of the Chicago Convention and its Annexes in order to prevent the recurrence of such potentially hazardous activities [52].

In that instance, the launch of a rocket-propelled missile in the vicinity of an international airway was considered to be a threat to the safety of civil aviation severe enough to justify the ICAO's intervention. Emerging trends in space transportation pose an even greater risk to the safety of civil aviation since aerospace vehicles will use the same airspace as conventional aircraft (at least for certain portions of their outbound flights to space and their inbound flights to Earth). Therefore, any efforts on the part of the ICAO to take the lead in the safety regulation of the same would be justified. Moreover, the ICAO Council has expressed the view that: "... the Chicago Convention applies to international air navigation but current commercial activities envisage sub-orbital flights departing from and landing at the same place, which may not entail the crossing of foreign airspaces. Should, however, foreign airspace(s) be traversed, and should it be eventually determined that sub-orbital flights would be subject to international air law, pertinent Annexes to the Chicago Convention would in principle be amenable to their regulation" [53]. Thus, the foregoing proposal seems to fall in line with the ICAO's own expectations of the future.

In sum, the existing regime of safety regulation of space activities consists primarily of non-binding international principles and guidelines intended to be implemented by national authorities. The proposed inclusion of space transportation under the rubric of the Chicago Convention and its Annexes will ensure that binding international standards and recommended practices are established for space transportation and that there is an effective mechanism for their implementation at the national level.

17.4.2 The starting point: expansion of the scope of Annex 14 (Aerodromes) to include spaceports

As a starting point, the experience gained in the aviation and airports field could be used for the international regulation of spaceports. Almost all the matters currently dealt with in Annex 14 are relevant and necessary for new (aero) spaceports. It is believed that segregated and inconsistent regulatory regimes and safety standards for aviation and space transportation systems spread across different countries could result in unnecessary duplication of efforts, wastage, confusion, and disaster for both systems. Therefore, the Member States of the ICAO are hereby encouraged to ask the ICAO Council to amend and expand the scope of Annex 14 in order to cover (aero) spaceports. As noted above, the ICAO possesses the legal authority to undertake such an expansion: Article 37 of the Chicago Convention expressly permits the ICAO Council to address and act upon "such other matters concerned with the safety, regularity, and efficiency of air navigation as may from time to time appear appropriate" in adopting SARPs [54]. However, there will no doubt be differences in requirements for aircraft and aerospace vehicles, and their respective "passengers", such as air and space traffic considerations, passenger processing and checking in, medical facilities and certification, variety of fuel to be used, sizes and dimensions of runways, navigational patterns and aids, variety of cargo and baggage handling, communication means and data. Thus, in revising and expanding Annex 14, the special needs of aerospace operations must be taken into consideration and appropriate standards must be adopted.

National implementation and application of the revised and expanded Annex 14—as proposed—might also require the adoption of, or the introduction of, changes to existing national legal regimes. Ultimately, international jurisdiction for all relevant matters such as aerospace vehicle/airspace worthiness, pilot/astronaut certification, and navigation services, should be placed within the province of the ICAO. It is recommended that an evolutionary approach should be followed by the ICAO and its Member States (i.e. solutions to space transportation issues should be sought as and when the problems arise). At present, the proposed revision and expansion of Annex 14 will not require an amendment of the Chicago Convention. Eventually, once air and space transportation systems become fully integrated, there will no doubt be the need for appropriate amendments to the Chicago Convention, which could be retitled the Convention on International Civil Aerospace Transportation, and the current ICAO could also be renamed as the International Civil Aerospace Organization.

17.4.3 Proposals for the establishment of responsibility and liability in space transportation

As noted earlier, the existing responsibility and liability regimes governing space activities in general, and space transportation in particular, impute responsibility and liability to the State(s) that carry out, or whose entities undertake, the space activities in question. At the other end of the spectrum, the liability regimes applicable to international air navigation ascribe liability to the private air carriers, although responsibility for ensuring that their operations are carried out safely ultimately rests with the State of Registration of the aircraft. Space transportation is presently at the infant stage and it continues to remain an ultra-hazardous activity. Accordingly, it is hereby proposed that the existing responsibility and liability regimes applicable to space activities generally should be retained with respect to emerging trends in space transportation.

This approach has a number of advantages. As it will ensure that States remain ultimately responsible for the space activities of their entities, States will be more inclined to see to it that proper regulations are enacted and enforced at the national level in order to guarantee the safety of those activities. Also, since States traditionally have deeper pockets than their subjects, retention of the existing liability regimes would likely enhance the probability that victims would be properly and adequately compensated. Moreover, the States would share the liability risks of space transportation companies, especially during the nascent years of their operations. However, with time, as space transportation proliferates, there will be the need to reconsider the relevance and utility of the existing liability regime, and to gradually initiate the shift from State-based responsibility and liability towards private carrier/operator-based liability.

CONCLUSIONS

Existing international space regimes on safety, responsibility, and liability (treaty provisions as well as non-binding principles and guidelines) are not sufficient to regulate emerging trends in space transportation. It has also been demonstrated that existing comparative regimes in international civil aviation have been particularly effective, especially in the area of safety regulation. It is against this background that this chapter provides food for thought by suggesting that long-standing international legal regimes governing safety, responsibility, and liability in international civil aviation may, to the extent that this is feasible, be adapted as a practical means of establishing effective legal regimes to govern similar aspects of space transportation. With respect to responsibility and liability, however, it has been proposed that the existing international regimes governing space be retained at least until such time that expanded development of space transportation systems justifies the creation of new or expanded legal regimes.

Notes and references

[1] The Space Transportation Annex to the Next Generation Air Transportation System (NextGen) Concept of Operations (ConOps), Draft Version 2.1, 2004, prepared by representatives from the Federal Aviation Administration (FAA), National Aeronautics and Space Administration (NASA) and the Department of Defense (DoD—Air Force Space Command). It is interesting to note that the "purpose of this ConOps is to outline a common national vision for conducting space-flight operations in the future. Space-flight operations includes the planning, scheduling, coordination, and management of space transportation activities, including the shared use of spaceport and range support elements worldwide, to accommodate multiple simultaneous flights of different types of vehicles to, through, and from space using a variety of control centers tied together through a distributed network-centric architecture to coordinate ground and flight operations and support."

[2] *Ibid.*

[3] It is, however, recognized that there are currently several technological and economic factors that make point-to-point travel challenging [see Webber, D. (2008). Point-to-Point Sub-orbital Space Tourism: Some Initial Considerations. *1st IAA Symposium on Private Manned Access to Space*, Arcachon, 28—30 May], but eventually this sort of travel is undoubtedly expected to become a reality primarily because of the entry into this industry of the private sector with available cash and entrepreneurial drive.

[4] Convention for the Regulation of Aerial Navigation, signed at Paris, on 13 October 1919 ("Paris Convention") 11 LNTS 173. This convention, which was subsequently superseded by the Chicago Convention of 1944, established elaborate rules relating *inter alia* to: the registration and nationality of aircraft; their fitness for flight; the sufficiency of their equipment; the competence of aircrews; and the navigation and landing of aircraft.

[5] Convention for the Unification of Certain Rules Relating to International Carriage by Air, signed at Warsaw, on 12 October 1929 ("Warsaw Convention") 137 LNTS 11.

[6] Boyle, A. (2008). First Look at *SpaceShipTwo*, 23 January, http://cosmiclog.msnbc.msn.com/archive/2008/01/23/601315.aspx (accessed 15 October 2008).

[7] The term "spacecraft" is used for vehicles which essentially travel to, through, and from outer space. On the other hand, the term "aircraft" is used for a machine that flies in airspace. According to an official definition, as included in several Annexes to the 1944 Convention on International Civil Aviation, an aircraft is "any machine that can derive support in the atmosphere from the reactions of the air other than the reactions of the air against the Earth's surface".

[8] For details, see: "Next Step for Spaceport America", 24 April 2008, http://www.personalspaceflight.info/category/spaceports/ (accessed 31 July 2008); "Spaceports: Blasting Off Around the World", 29 January 2008, http://www.travelburner.com/2008/01/29/spaceports-blasting-off-around-the-world/ (accessed 31 July 2008); Foust, J. (2007). Spaceports Still Taxiing Towards Takeoff. *The Space Review*, 17 December, http://www.thespacereview.com/article/1023/1 (accessed 31 July 2008); Brown, M. (2007). Spaceports Spring Up All Over. *Nature*, 30 January, http://www.nature.com/news/2007/070130/full/news070129-6.html (accessed 31 July 2008); Boyle, A. (2006). 'Competition' Reigns Among Spaceports: Rivals Work

Together on Broader Issues Surrounding Future Space Travel. *MSNBC*, 18 October, http://www.msnbc.msn.com/id/15320942/ (accessed 31 July 2008); Boyle, A. (2004). Spaceports Compete in Race for Business: What Will It Take to Attract Well-off Space Passengers? *MSNBC*, 7 October, http://msnbc.msn.com/id/6191567/ (accessed 12 April 2007).

[9] Brown, K. R. (2001). The Next Generation Space Launch Range. *Space Technology and Applications International Forum*, AIP Conf. Proc. 2 February 2001, Volume 552, pp. 680–685.

[10] Commercial Space Launch Activities, 49 USC §701 et seq. and Commercial Space Transportation Regulations, 14 CFR §400 et seq.

[11] For details, see Mineiro, M. C. (2008). *Law and Regulation Governing US Commercial Spaceports: Licensing, Liability, and Legal Challenges*, August.

[12] "Latest Paying Space Tourist Blasts off from Baikonur", 12 October 2008, http://www.novinite.com/view_news.php?id=97812 (accessed 15 October 2008); "Russian Spacecraft Docks with Orbital Station", 14 October 2008, http://www.cbc.ca/technology/story/2008/10/14/tech-spacestation.html (accessed 15 October 2008).

[13] "Orbital Spaceflight", http://www.spaceadventures.com/index.cfm?fuseaction=orbital.Guy_Laliberte (accessed 2 November 2009).

[14] Isakowitz, M. (2009). "Investment in Commercial Spaceflight Grows," *Spacemart*, Washington, DC, 11 November, http://www.spacemart.com/reports/Investment_In_Commercial_Spaceflight_Grows_999.html (accessed 11 November 2009). It is also interesting to note that some knowledgeable people, like Australian astronaut Andy Thomas, are already envisioning the commercial trips by tourists to the moon by 2020 with NASA's latest *Ares I-X* rocket. See Crawford, C. (2009). "Tourists May Land on the Moon by 2020", 27 October, http://www.news.com.au/couriermail/story/0,26265864-17102,00.html (accessed 3 November 2009).

[15] Peeters, W. and Crouch, G. (2008). *Space Tourism = Space + Tourism*. International Space University.

[16] It is interesting to note that point-to-point travel on Earth via space and aerospace ports may not only be on the surface of the Earth but also in space (i.e. orbiting aerospace ports or hubs), where several types of aerospace vehicles from various nations may dock for embarking and disembarking space travelers (passengers), delivery or picking up cargo, and servicing their crafts purposes. In fact, this has already been done, though to a limited extent, as astronauts from the Russian Federation and the USA traveled from and to their respective countries using Russian *Soyuz* launch vehicle and American space shuttle to and from Mir Space Station and the International Space Station (ISS). In other words, the places of their embarkation and disembarkation on the surface of the Earth have been different; thus in some sense, both Mir and ISS have been used as aerospace ports in space.

[17] The membership of the COPUOS has periodically been expanded to the current 67 States. See United Nations General Assembly Resolution on International Cooperation in the Peaceful Uses of Outer Space (A/RES/59/116) that was adopted without a vote on 10 December 2004.

[18] The Treaty on Principles Governing the Activities of States in the Exploration and Use of Outer Space, including the Moon and Other Celestial Bodies (the "Outer Space Treaty", adopted by the General Assembly in its resolution 2222 (XXI)), opened for signature on 27 January 1967, and entered into force on 10 October 1967 [98 ratifications and 27 signatures (as of 1 January 2006)]; the Agreement on the Rescue of Astronauts, the Return of Astronauts and the Return of Objects Launched into Outer Space (the "Rescue Agreement", adopted by the General Assembly in its resolution 2345 (XXII)), opened for signature on 22 April 1968, entered into force on 3 December 1968 [88 ratifications, 25 signatures, and one acceptance of rights and obligations (as of 1 January 2006)]; the Convention on International Liability for Damage Caused by Space Objects (the "Liability Convention", adopted by the General Assembly in its resolution 2777 (XXVI)), opened for signature on 29 March 1972, entered into force on 1 September 1972, 83 ratifications [25 signatures, and three acceptances of rights and obligations (as of 1 January 2006)]; The

Convention on Registration of Objects Launched into Outer Space (the "Registration Convention", adopted by the General Assembly in its resolution 3235 (XXIX)), opened for signature on 14 January 1975, entered into force on 15 September 1976 [46 ratifications, four signatures, and two acceptances of rights and obligations (as of 1 January 2006)]; and the Agreement Governing the Activities of States on the Moon and Other Celestial Bodies (the "Moon Agreement", adopted by the General Assembly in its resolution 34/68), opened for signature on 18 December 1979, entered into force on 11 July 1984 [12 ratifications and four signatures (as of 1 January 2006)].

[19] The Declaration of Legal Principles Governing the Activities of States in the Exploration and Uses of Outer Space (UN General Assembly resolution 1962 (XVIII) of 13 December 1963); the Principles Governing the Use by States of Artificial Earth Satellites for International Direct Television Broadcasting (UN General Assembly resolution 37/92 of 10 December 1982); the Principles Relating to Remote Sensing of the Earth from Outer Space (UN General Assembly resolution 41/65 of 3 December 1986); the Principles Relevant to the Use of Nuclear Power Sources in Outer Space (UN General Assembly resolution 47/68 of 14 December 1992); and the Declaration on International Cooperation in the Exploration and Use of Outer Space for the Benefit and in the Interest of All States, Taking into Particular Account the Needs of Developing Countries (UN General Assembly resolution 51/122 of 13 December 1996).

[20] The Principles Relevant to the Use of Nuclear Power Sources in Outer Space (UN General Assembly resolution 47/68 of 14 December 1992), Preamble.

[21] *Ibid.* Principle 2(1)1 defines the terms "State launching" and "Launching State" as "the State which exercises jurisdiction and control over a space object with nuclear power sources on board at a given point in time relevant to the principle concerned".

[22] *Ibid.* Principle 1(2)(b) defines the term "sufficiently high orbit" as the "one in which the orbital lifetime is long enough to allow for a sufficient decay of the fission products to approximately the activity of the actinides. The sufficiently high orbit must be such that the risks to existing and future outer space missions and of collision with other space objects are kept to a minimum. The necessity for the parts of a destroyed reactor also to attain the required decay time before re-entering the Earth's atmosphere shall be considered in determining the sufficiently high orbit altitude."

[23] *Ibid.* Principle 4: Safety assessment.

[24] Note verbale dated 2 June 1997 from the Permanent Mission of the United States of America to the United Nations (Vienna) addressed to the Secretary-General, UN Document A/AC.105/677 of 4 June 1997. Also see "*Cassini* Skirts Earth with 33 kgs of Plutonium", available at: http://www.spacedaily.com/news/cassini-99c.html (accessed 4 May 2007).

[25] Committee on the Peaceful Uses of Outer Space, "Report of the Scientific and Technical Subcommittee on its forty-fourth session, held in Vienna from 12 to 23 February 2007", United Nations General Assembly, Document no. A/AC.105/890 of 6 March 2007, paragraph 99 and Annex IV on Space Debris Mitigation Guidelines of the Scientific and Technical Subcommittee of the Committee on the Peaceful Uses of Outer Space.

[26] The Treaty on Principles Governing the Activities of States in the Exploration and Use of Outer Space, including the Moon and Other Celestial Bodies, Article VI.

[27] For details see, Oude Elferink, A. G. (1999). *The Genuine Link Concept; Time for a Post Mortem?* Netherlands Institute for the Law of the Sea (NILOS), Utrecht University, Utrecht, the Netherlands, final version, March 1999, available at http://www.uu.nl/content/genuine%20link.pdf (accessed 4 May 2007).

[28] According to N. Jasentuliyana, former head of the UN OOSA: "The process of drafting [international agreements] is necessarily detailed, laborious, and time-consuming, involving formal statements of position, general discussions, detailed negotiations, editorial review, and most important, numerous informal consultations which allow delegations to make compromises without having to formally depart

from stated positions." Jasentuliyana, N. (ed.) (1992). The Lawmaking Process in the United Nations. *Space Law: Development and Scope*, p. 33.

[29] Jakhu, R. (2006). Legal Issues Relating to the Global Public Interest in Outer Space. *Journal of Space Law,* 31, 32.

[30] International Civil Aviation Organization (2008). *Making an ICAO Standard*, online at ICAO website: http://www.icao.int/icao/en/anb/mais/ (accessed 22 October 2008).

[31] *Ibid.*

[32] *Ibid.*

[33] Convention on International Civil Aviation, signed at Chicago, on 7 December 1944 (Chicago Convention) ICAO Doc. 7300/9, art. 44(h).

[34] See Memorandum on ICAO, ICAO website, online: www.icao.int/icao/en/pub/memo.pdf (accessed 3 May 2007).

[35] A Standard is defined as any specification for physical characteristics, configuration, material, performance, personnel or procedure, the uniform application of which is recognized as necessary for the safety or regularity of international air navigation and to which Contracting States will conform in accordance with the Convention; in the event of impossibility of compliance, notification to the Council is compulsory under Article 38 of the Convention.

[36] A Recommended Practice is any specification for physical characteristics, configuration, material, performance, personnel or procedure, the uniform application of which is recognized as desirable in the interest of safety, regularity or efficiency of international air navigation, and to which Contracting States will endeavor to conform in accordance with the Convention. States are invited to inform the Council of non-compliance. SARPs are formulated in broad terms and restricted to essential requirements. The differences to SARPs notified by States are published in Supplements to Annexes.

[37] Procedures for Air Navigation Services comprise operating practices and material too detailed for Standards or Recommended Practices—they often amplify the basic principles in the corresponding Standards and Recommended Practices. To qualify for PANS status, the material should be suitable for application on a worldwide basis.

[38] Regional Supplementary Procedures have application in the respective ICAO regions. Although the material in Regional Supplementary Procedures is similar to that in the Procedures for Air Navigation Services, SUPPs do not have the worldwide applicability of PANS.

[39] Guidance Material is produced to supplement the SARPs and PANS and to facilitate their implementation. Guidance Material is issued as attachments to Annexes or in separate documents such as manuals, circulars, and lists of designators/addresses. Usually it is approved at the same time as the related SARPs are adopted. Manuals provide information to supplement and/or amplify the SARPs and PANS. They are specifically designed to facilitate implementation and are amended periodically to ensure their contents reflect current practices and procedures. Circulars make available specialized information of interest to Contracting States. Unlike manuals, circulars are not normally updated.

[40] International Civil Aviation Organization (2008). *Making an ICAO Standard*, online at ICAO website: http://www.icao.int/icao/en/anb/mais/ (accessed 22 October 2008).

[41] *Ibid.*

[42] *Ibid.*

[43] Notable cases in which the concept of State responsibility has been relied upon in relation to civil aviation accidents have primarily involved State-sponsored terrorism (as in the bombing of Pan Am flight 103 over Lockerbie, Scotland by Libyan operatives in 1988), or the military shooting down of commercial aircraft (as in the Russian shooting down of Korean Air flight KE 007 in 1983, and the US shooting down of Iran Air flight 655 in 1988).

[44] The Warsaw System refers to the Warsaw Convention of 1929 as subsequently amended and/or modified by the Hague Protocol of 1955, the Guadalajara Convention of 1961, the Guatemala City Protocol of 1971, and Additional Protocol Numbers 1, 2, 3, and 4, all 1975.
[45] Convention for the Unification of Certain Rules for International Carriage by Air, 1999 (Montreal Convention), signed at Montreal, on 28 May 1999, ICAO Doc. 9740.
[46] For a detailed analysis of the provisions of the Warsaw System and those of the Montreal Convention, see Dempsey, P. S. and Milde, M. (2005). *International Air Carrier Liability: The Montreal Convention of 1999*. (Centre for Research in Air and Space Law, McGill University.
[47] Convention on Damage Caused by Foreign Aircraft to Third Parties on the Surface (Rome Convention), signed at Rome, on 7 October 1952, ICAO Doc. 7364.
[48] Convention on Compensation for Damage Caused by Aircraft to Third Parties, 2 May 2009, ICAO Doc. DCCD No. 42 ("General Risks Convention").
[49] Convention on Compensation for Damage to Third Parties, Resulting from Acts of Unlawful Interference Involving Aircraft, 2 May 2009, ICAO Doc. DCCD No. 43 ("Unlawful Interference Compensation Convention").
[50] Hobe, S. (2004). Aerospace Vehicles: Questions of Registration, Liability and Institutions—A European Perspective. *Annals of Air and Space Law*, XXIX:377, at 387.
[51] For details, see Dempsey, P. and Mineiro, M. (2008). ICAO's Role in Regulating Safety and Navigation in Suborbital Aerospace Transportation. Paper presented at the 2008 IAASS Conference in Rome, Italy.
[52] See ICAO Assembly Resolution A 32-6: Safety of Navigation. In *Resolutions Adopted at the 32nd Session of the Assembly*, available online at ICAO website: http://www.icao.int/icao/en/assembl/a32/resolutions.pdf. See also announcement of the ICAO Assembly Resolution on Japanese Ministry of Foreign Affairs website, online at: http://www.mofa.go.jp/announce/announce/1998/10/1003.html (accessed 6 May 2007).
[53] See Concept of Sub-orbital Flights, ICAO Doc. CDEC 175/15 of 17/6/05 (based on C-WP/12436 of 30 May 2005).
[54] Convention on International Civil Aviation, at Chicago, 7 December 1944, ICAO Doc. 7300/9.

CHAPTER

An international Code of Conduct for responsible spacefaring nations

18

Michael Krepon*, **Samuel Black**†

Co-founder, Henry L. Stimson Center, † Research Associate, Henry L. Stimson Center

CHAPTER OUTLINE

INTRODUCTION

Political scientists write about an anarchic system of states, but national leaders are usually not drawn from among the ranks of anarchists. Our world is interconnected in profound ways, and unless states can find ways to cooperate amidst their competition, public safety can be placed at great risk, even for major powers. One way that states have found to cooperate is through codes of conduct. Interstate and international travel would be chaotic without agreed "rules of the road". Military establishments have also agreed upon codes of conduct for managing close proximity operations of ground, sea, and air forces during peacetime. In recent years, codes of conduct have gained prominence in reducing the likelihood of proliferation. These codes of conduct can take the form of political compacts, legally binding multilateral agreements, or treaties. Some rules of the road have also been established for outer space, but many aspects of space operations remain completely discretionary. We argue, in this chapter, for a more encompassing set of rules of the road for space, including a provision for no harmful interference against man-made space objects.

In our view, a Code of Conduct for responsible spacefaring nations could provide valuable near-term results in further mitigating space debris and in reducing the likelihood of further anti-satellite (ASAT) weapon tests in space. A Code of Conduct clarifying responsible spacefaring activities might also usefully include provisions regarding space traffic management, information sharing, and consultation arrangements. The Code of Conduct could take many forms, including bilateral or multilateral arrangements between or among spacefaring nations. It could be drawn up by major

Space Safety Regulations and Standards. DOI: 10.1016/B978-1-85617-752-8.10017-0

spacefaring nations in an ad hoc body created especially for this purpose. Key elements of a Code of Conduct might be developed in the Conference on Disarmament (CD) or the Committee on the Peaceful Uses of Outer Space (COPUOS). However, we would argue for a greater sense of urgency in proceeding, rather than for a particular way to proceed. If there is sufficient support for a Code of Conduct, then interested parties can devise an appropriate work plan to make this concept a reality.

18.1 THE FUNDAMENTAL DILEMMA—SATELLITES ARE ESSENTIAL BUT VULNERABLE

A Code of Conduct for space would help address the fundamental dilemma of space operations—that satellites are both essential and vulnerable. The vulnerability of satellites cannot be "fixed" by technical or military means, since marginal improvements in satellite survivability can be trumped by the growth of space debris caused by ASAT weapons, collisions in space, or other factors. The use of destructive methods for military purposes in space can be especially problematic, since escalation control will be difficult to establish in conflicts between major space powers. If asymmetric warfare can be waged in space, as on the ground, and if the weaker party still has the means to disrupt or disable the satellites of a more powerful foe, then the initiation of warfare in space is likely to be a lose–lose proposition. Space warfare can also be waged by attacking ground stations, or electronic and cyber links to satellite operations. Asymmetrical warfare can also occur if the weaker space power resorts to the use of debris-causing weapons or nuclear detonations that would produce long-lasting, indiscriminate effects.

18.1.1 Responses to the dilemma

Spacefaring nations are well aware of the difficulties of escalation control in the event of warfare in space, and are rightly wary of relying on war plans that are predicated on the "fallacy of the last move", wherein one side's initiative works well—until the other side makes a painful counter-move. Multiple responses to the central dilemma of relying on essential but vulnerable satellites are required. Most are non-controversial, such as: improving space "situational awareness"; having spare satellites available when these are affordable; moving over time to distribute the services provided by singular, expensive space assets to reduce the reliance on very few high-value targets; supplementing space-based assets with terrestrially based systems; and improving satellite survivability when it is cost-effective to do so. Since the dilemma of satellite vulnerability can be partially managed but not solved, the USA also relies on terrestrial power projection capabilities that might dissuade or deter other governments from engaging in actions against US satellites.

Advanced spacefaring nations are also pursuing hedging strategies in the event of warfare in space. Hedging strategies can take the form of researching, developing, and flight-testing multipurpose technologies that could be used for peaceful or for war-fighting purposes in space. Hedging strategies can also manifest themselves in the flight-testing in an ASAT mode of military systems or capabilities designed for other purposes. Tests of "dedicated" ASAT weapons have been relatively infrequent phenomena during the space age, at least compared to the testing of nuclear weapons or ballistic missiles. One reason for the infrequent testing of ASAT weapons is the negative political and public repercussions that result. Another is that advanced spacefaring nations have considerable "latent" or "residual" means to damage satellites, and thus have little need to repeatedly demonstrate dedicated ASAT capabilities.

If a nation has medium- or longer-range ballistic missiles, space-tracking capabilities, and the means to insert a satellite into a precise orbital slot, it also has the means to attack satellites. If a nation lacks space-tracking and precise orbital insertion capabilities, but possesses medium-range ballistic missiles and nuclear warheads, it too can severely damage satellites operating in low-Earth orbit. National capabilities to do great and indiscriminate harm to satellites are becoming more distributed over time, as nuclear and missile proliferation occurs and as more nations expand their satellite operations.

Some have argued that, because of the inherent value of space for national and economic security, warfare against space objects is inevitable. But if this were true, then such warfare would have already occurred during the Cold War. Since the "inevitable" hasn't already happened, this suggests that the principles of deterrence between major space powers may apply to space wars as well as to nuclear wars. Satellites are, after all, intimately connected to nuclear forces, which rely on space assets for targeting, early warning, intelligence, command and control, communication, and forecasting information. To attack the satellites of a major space power, therefore, runs the risk of prompting nuclear exchanges.

Deterrence is by no means assured; it could break down because of conscious choices, accidents, or uncontrolled crises. Nonetheless, the history of the space age to date suggests that warfare in space between major powers is by no means inevitable. Prior restraint may be due, in part, to the ability of major spacefaring nations to respond with latent ASAT capabilities and other hedges to damage each other's satellites. Mutual vulnerability in space could serve, in part, to keep future warfare in space from occurring.

18.2 THE ROLE OF DIPLOMACY

Diplomacy is also essential in dealing with the central dilemma of satellite dependence and vulnerability. US space policy during the George W. Bush administration placed significant constraints on diplomacy. Specifically, the Bush administration's space policy dictated that: "Proposed arms control agreements or restrictions must not impair the rights of the United States to conduct research, development, testing, and operations or other activities in space." The Bush administration's allergy to space diplomacy continued even after the Chinese test of a kinetic energy ASAT weapon in January 2007. This test was the worst producer of lethal debris in the history of the space age to date. The Bush administration, under whose auspices the US Air Force embraced a new doctrine of power projection in, through and from space, was hard-pressed to object to similar efforts by potential adversaries. To avoid being drawn into negotiations that might reduce its military freedom of action in space, the Bush administration confined its public response to the January 2007 Chinese test to the rather lame observation that such practices would make Beijing's desire for space cooperation more remote.

In February 2008, the Bush administration also tested a destructive ASAT weapon, ostensibly to prevent a chemical spill from an inactive, de-orbiting US intelligence satellite. While this test was announced in advance and created minimal debris, all of which has since de-orbited, it compounded the disquieting effects of the Chinese ASAT tests. The acceleration of national hedging strategies is likely in Russia, India, France, Israel and, perhaps, elsewhere. Additional ASAT tests may occur.

18.3 ARGUMENTS AGAINST NEW MULTILATERAL AGREEMENTS

ASAT tests have clarified the weakness of arguments that no new rules of the road are needed for space. The reasoning against new multilateral agreements for space boils down to five arguments. The first is that because there is no likelihood of an arms race in space, there is no need for new multilateral arrangements. It is true that an arms race is unlikely, since arms racing has now been replaced by asymmetric warfare. But an arms race isn't needed to do lasting damage to space, as the Chinese ASAT test demonstrated. It was surely evident from this test, if it was not clear beforehand, that very few kinetic kill tests or uses of ASAT weapons are needed to result in long-lasting damage to low-Earth orbit. New diplomatic initiatives are required precisely because an arms race isn't necessary to prevent the peaceful uses of outer space.

The second argument advanced by opponents of multilateral agreements for space is that arms control is a vestige of the Cold War and not terribly relevant to contemporary security concerns. There is partial truth in this argument, because "classic" arms control arrangements dealt with a superpower competition that ended with the demise of the Soviet Union. What used to be known as arms control has now morphed into cooperative threat reduction agreements, including rules of the road that clarify responsible behavior and facilitate corrective steps against those acting irresponsibly. Semantic arguments aside, the Bush administration itself championed multilateral agreements in the form of Codes of Conduct to prevent proliferation, such as The Hague Code of Conduct, as well as the Proliferation Security Initiative. It is unnecessary to argue over whether these codes of conduct constitute arms control to conclude that such creative initiatives can be sensible.

A Code of Conduct for space would also be quite useful in ending the practice of kinetic-kill ASAT tests. If Codes of Conduct relating to missiles and exports make sense for preventing proliferation, then surely a Code of Conduct also makes sense for activities in space. After all, troubling activities in space could also prompt proliferation on the ground.

The third argument advanced by the Bush administration against new diplomatic initiatives for space is that there could be no agreed definition of what constitutes "space weapons". Moreover, verification of agreed limitations in space would be extremely problematic. Consequently, some have argued that no multilateral agreements should be negotiated barring such weapons. Past experience has indicated that the difficulties in defining and verifying space weapons are formidable at best. A Code of Conduct would, however, focus on verifiable activities, not on definitions of what constitutes a space weapon. For example, one key element of a Code of Conduct would surely be that responsible spacefaring nations do not engage in activities that deliberately produce persistent space debris, such as the Chinese ASAT test. This key element of a Code of Conduct would obviate the need to define this particular category of space weapons, since actions, not definitions, lie at the core of a rules of the road approach. Detection of purposeful, persistent, debris-causing events would be more straightforward, since it is hard to hide deliberate acts directed against space objects.

A fourth argument against new diplomatic initiatives for space is that the USA must preserve its right to self-defense—including the right to defend space assets. This argument is certainly valid, but it, like its predecessors, does not justify rejecting a Code of Conduct. With such a Code, the USA would still possess more capabilities than ever before to deter and, if necessary, punish states that take actions against US satellites. A Code of Conduct does not nullify the right of self-defense. But without rules, there are no rule-breakers. A Code of Conduct would clarify rules and rule-breakers, making actions against the latter more likely to garner support.

The final argument employed against space diplomacy by the Bush administration was that US military freedom of action in space must not be constrained. By this standard, the Nuclear Non-proliferation Treaty, the Outer Space Treaty, President Ronald Reagan's Intermediate Nuclear Forces Treaty, and President George H. W. Bush's Strategic Arms Reduction treaties were all dreadful errors in judgement, since every one of these agreements limited the US military's freedom of action in some key respects. Indeed, using this reasoning, the Geneva Conventions were also unwise, as were codes of conduct long in place for the United States Army, Navy, and gravity-bound Air Force.

The weakness of this argument can also be measured by the growth of space debris during the Bush administration. The growth of this indiscriminate hazard to space operations caused by ASAT testing and other means has curtailed US freedom of action in space. Unless more concerted actions are taken to address the creation of space debris and to establish a space traffic management system—two critical elements of our proposed Code of Conduct—debris will continue to grow significantly, which will further curtail US freedom of action in space.

18.4 RUSSIA AND CHINA SUBMIT A DRAFT TREATY BANNING WEAPONS IN SPACE

While the Bush administration opposed space diplomacy initiatives that would curtail freedom of action in space, China and Russia submitted an ambitious draft treaty to the CD that would ban the deployment of weapons in space. This draft treaty was noticeably reticent on the subject of testing and producing terrestrial ASAT systems. This draft treaty suffered from a number of other weaknesses, including its vague definition of what constitutes "space weapons". An extended effort would likely be required to reach consensus on how to define, monitor, and enforce agreed prohibitions under this draft treaty. It is very unlikely that the USA would endorse such an ambitious effort, or that the US Senate would consent to the ratification of a poorly defined, unverifiable treaty.

A narrow treaty that focuses on destructive tests against man-made space objects might be viewed in a different light as it addresses verifiable actions that contribute to the pressing problem of space debris. A treaty of this kind would avoid many of the problems of scope, definitions, and verification that have bedeviled ambitious proposals to tackle the "weaponization" of space. Since many military technologies and weapon systems that can be used against satellites also serve other essential military and peaceful purposes, the treaty definition of a "space weapon" is either likely to be too narrowly drawn, and thus ineffective, or too broadly defined, thereby capturing capabilities primarily designed for other essential purposes. Undertaking a treaty of this kind could entail lengthy and fruitless negotiations. Those seeking a near-term alternative might well turn to a Code of Conduct to preserve and advance the peaceful uses of outer space.

A Code of Conduct approach could avoid difficult dilemmas associated with drafting a treaty banning the use of force in, from, or through space. It would focus on responsible and irresponsible actions, rather than on what might constitute a space weapon. Additional rules of the road are needed because existing standards for responsible activities in space, while extremely valuable, include many loopholes. The use of space is expanding, and the potential for friction is growing. Adding new rules of the road does not mean that they will be adhered to in all cases and for all time. The addition of new rules could, however, lessen the likelihood that rules will be broken, while increasing the probability that rule-breakers will be isolated and penalized in some fashion. In contrast, the absence of new rules means the continued absence of standards for responsible behavior.

18.5 THE STIMSON CENTER DRAFT CODE OF CONDUCT

The Stimson Center, in partnership with NGOs from other spacefaring nations, has released a draft Code of Conduct for Responsible Spacefaring Nations [1]. This code includes a number of rights and responsibilities that might help solidify and internationalize existing best practices. Among the rights included are: the right to access to space, to interference-free space operations, to self-defense, and the right to consultations on matters pertaining to the Code. The responsibilities include protecting the rights of other spacefarers, enforcing appropriate regulations, developing and abiding by safe operating procedures, and consulting with others when requested.

In December 2008, the European Union issued its own draft Code of Conduct [2]. This draft, like that of the Stimson Center, places great importance on the need for responsible spacefaring nations to avoid harmful interference against space objects [3]. One such act of harmful interference would be destructive ASAT tests. The other key elements of a Code of Conduct, including debris mitigation, space traffic management, and radio frequency and orbital slot allocation and coordination, would be undermined if acts of harmful interference were carried out, especially destructive ASAT tests that could endanger satellite operations for many decades. The same logic applies to the need for a traffic management regime in space. The collision of a functioning US commercial satellite and a dead Russian satellite in February 2009 has further clarified the need to further improve space situational awareness and promote timely exchanges of information.

CONCLUSIONS

This analysis suggests the utility of a work program to identify and flesh out the key elements of a Code of Conduct for responsible spacefaring nations. This would not be a simple undertaking, but it would be far easier and less time-consuming than negotiating a treaty that seeks to define and ban the use of weapons in space. Rules of the road to prevent dangerous military activities exist for navies, ground forces, and air forces. The codification of rules of the road for responsible spacefaring nations can also make a significant contribution to national, regional, and international security.

References

[1] The Stimson Center's Model Code of Conduct for Responsible Spacefaring Nations is available at http://www.stimson.org/pub.cfm?ID=575

[2] The European Union's draft code of conduct for outer space activities is available at http://www.stimson.org/space/pdf/EU_Code_of_Conduct.pdf

[3] See Black, S. (2008). No Harmful Interference with Space Objects: The Key to Confidence-Building. Stimson Center Report No. 69, July, http://www.stimson.org/pub.cfm?ID=646

CHAPTER

19 The ICAO's legal authority to regulate aerospace vehicles

Paul Stephen Dempsey*, Michael Mineiro†

* *Director, Institute of Air and Space Law, McGill University,* † *Boeing Fellow in Air and Space Law, Institute of Air and Space Law, McGill University*

CHAPTER OUTLINE

INTRODUCTION

The Chicago Convention does not place restrictions on the International Civil Aviation Organization (ICAO)'s authority to regulate civil aircraft simply because the aircraft traverses the upper reaches of Earth's atmosphere. In principle, civil aircraft engaged in international flight are subject to ICAO jurisdiction, whether or not the aircraft traverses the upper reaches of Earth's atmosphere. So the question arises: Does the Chicago Convention grant the ICAO the legal authority to adopt Standards and Recommended Practices (SARPs) applicable to civilian sub-orbital aerospace vehicles engaged in international flight?

19.1 THE TERM "AIRCRAFT"—ARTICLE 3(A) OF THE CHICAGO CONVENTION

A threshold question regarding the ICAO's authority to adopt SARPs regulating sub-orbital aerospace transportation is whether the Chicago Convention Article 3(a) term "aircraft" includes sub-orbital aerospace vehicles. Article 3(a) limits application of the Convention "to civil aircraft, and shall not be applicable to state aircraft" [1]. The Convention does not define either "civil aircraft", "state aircraft" or simply "aircraft". The Convention does provide a *presumption iuris tantum* (rebuttable

Space Safety Regulations and Standards. DOI: 10.1016/B978-1-85617-752-8.10019-4

presumption) that aircraft used in military, customs, and police services shall be deemed to be state aircraft [2]. "Aircraft" is defined in the Annexes of the Chicago Convention as "any machine that can derive support in the atmosphere from the reactions of the air other than the reactions of the air against the Earth's surface" [3].

A critical distinction exists between the term "aircraft" in Article 3(a) and the term "aircraft" adopted in Chicago Convention Annexes. This distinction is as follows: the term "aircraft" in Article 3 (a) may apply to vehicles, such as sub-orbital aerospace transportation vehicles (SATVs), even if the definition of "aircraft" adopted in the Annexes does not. If it is established that the term "civil aircraft" (as used in Article 3(a)) does include SATVs, then the Convention, and hence the ICAO's scope of authority, extends to SATVs, subject to other provisions of the Convention. Once established, the ICAO needs only to exercise its authority by defining "aircraft" within the appropriate Annexes to include SATVs.

In 2005, the ICAO Council considered the concept of sub-orbital flights in relation to the Chicago Convention [4]. The main question explored by the Council was whether sub-orbital flights would fall within the scope of the Chicago Convention and therefore the ICAO's mandate [5]. The Council's working paper appropriately reasoned that "should sub-orbital vehicles be considered (primarily) as aircraft, when engaged in international air navigation, consequences would follow under the Chicago Convention, mainly in terms of registration, airworthiness certification, pilot licensing and operation requirements (unless otherwise classified as State aircraft under Article 3 of the Convention)" [6].

An important distinction exists between whether the Chicago Convention *does* apply to SATVs engaged in international air navigation, or whether the Chicago Convention *should* apply. The question of whether the Convention *does* apply to SATVs is primarily a question of law. The question of whether the Convention *should* apply is primarily a question of policy, subject to a variety of considerations including the legal scope of the Convention's application. The Council failed to conclude whether or not the Chicago Convention *does* apply. The Council did reason that *if* it is determined that sub-orbital vehicles are considered "aircraft" (in accordance with the term used in Article 3(a) of the Chicago Convention), *then* the Chicago Convention does apply and sub-orbital vehicles will be subject to applicable SARPs.

Determining whether or not the Chicago Convention *does* apply requires an interpretation of the Convention and her relevant articles, in particular Article 3(a). The interpretation of the Convention is guided by general customary international law principles of treaty interpretation considered as reflected in the Vienna Convention of the Law of Treaties [7]. Treaties are to be interpreted in good faith, in accordance with the ordinary meaning to be given to the terms of the treaty in their context, and in light of its objective and purpose [8]. Also, in interpreting treaties, there shall be taken into account, together with the context, any subsequent agreement between the parties regarding the interpretation of the treaty or the application of its provisions, any subsequent practice in the application of the treaty that establishes the agreement of the parties regarding its interpretation, and any relevant rules of international law applicable in the relations between the parties [9].

A strong argument exists in favor of interpreting the term "aircraft" as including sub-orbital aerospace vehicles. According to its Preamble, the Chicago Convention serves two primary purposes. The first purpose is to facilitate the safe and orderly development of civil aviation [10]. The second is to establish international air transport services on an equal and economically sound basis [11]. Throughout the Articles of the Convention, one sees that a principal object and purpose of the Chicago Convention was to create a unified and harmonious regime of safety and navigation of airspace. Even where the Convention decreed that certain "state aircraft" were beyond its reach, it nevertheless required that Contracting States issue safety regulations for such aircraft having "due regard for the safety of navigation of civil aircraft" [12]. The fundamental purpose is clear—to ensure that safety shall prevail in air transportation.

At the time the Chicago Convention was drafted, sub-orbital aerospace vehicles were not considered (due in part to the state of technology and concepts of international air transportation at that time). Nonetheless, the drafters of the Convention did consider the need for standardizing air navigation procedures. To this end, the Convention created the ICAO with the aims and objectives of developing the techniques of international air navigation and to foster the planning and development of international air transportation [13]. Specific objectives of the ICAO include insuring the safe and orderly growth of international civil aviation throughout the world [14], to promote safety of flight in international air navigation [15], and to promote generally the development of all aspects of the international civil aeronautics [16]. The purpose of the Convention is *not* to regulate a specific type of vehicle, but rather to ensure international civil aviation is safe and orderly. It seems absurd to conclude the treaty was meant to be frozen in time, only regulating vehicles that fit within the concept of aircraft at the time of the drafting of the Convention. Were that the intention, the ICAO could have difficulty regulating jets, since no commercial jet existed in 1944. The largest commercial aircraft then operating was the DC-3, a mere gnat alongside today's behemoth Boeing 747 or the Airbus 380 airliners. Such an interpretation would conflict with the stated purpose of the Convention: to ensure the safe and orderly development of international civil aviation.

For example, if one interprets the term "aircraft" in Article 3(a) to exclude SATVs engaged in international air transportation, then the following result occurs:

> *An airplane (that is registered in a State party to the Convention) is engaged in the international carriage of passengers, traveling at 20,000 feet. This airplane and her State of registry are subject to the rules and obligations established in the Chicago Convention and promulgated by ICAO. At the same time, a SATV (that is registered in a State party to the Convention) is engaged in the international carriage of passengers, traveling at 20,000 feet. The SATV, and her State of registry, is not subject to the rules and obligations established in the Chicago Convention and promulgated by ICAO. Therefore, the aircraft and aerospace vehicle are operating in the same airspace without standardized rules of collision avoidance, navigation, and communication.*

The operation of SATVs traversing international air space without being subject to the Chicago Convention immediately undermines the fundamental purpose of the Convention.

19.2 THE ICAO AND SUBSEQUENT APPLICATION OF THE CHICAGO CONVENTION AS IT RELATES TO OUTER SPACE

Under customary international law, the subsequent practice in the application of a treaty that establishes the agreement of the parties regarding its interpretation is to be taken into account when interpreting a treaty [17]. Since the Chicago Convention entered into force, the ICAO General Assembly has adopted several resolutions relating to international civil aviation and outer space. These resolutions are subsequent practices that support an interpretation of the Convention that places outer space activities that affect international civil aviation within the purview of the ICAO.

The General Assembly has stated that events "relating to the exploration and use of outer space are of great interest to the ICAO, since many of these activities affect matters falling within the Organization's competence under the term of the Chicago Convention" [18]. In Resolution A15-1(1965), the General Assembly recognized that "although the Convention does not specifically define how the term 'outer space' should be interpreted, the space used or usable for international civil aviation is also used by space vehicles" and that "the use of the same medium by different fields of activity necessarily requires adequate coordination to achieve the normal and efficient functioning of both these fields" [19]. In Resolution A29-11 (1992) the General Assembly resolved "that the ICAO be responsible for stating the position of international civil aviation on all related outer space matters" [20]. The General Assembly has also directed the Council "to carry out a study of those technical aspects of space activities that affect international navigation and that, in its view, call for special measures, and report the results" [21] and for the Secretary General to "ensure that the international civil aviation positions and requirements are made known to all organizations dealing with relevant space activities and to continue to arrange for the Organization to be represented at appropriate conferences and meetings connected with or affecting the particular interests of international civil aviation in this field" [22].

These resolutions reveal the following: (1) the ICAO and its Member States believe the Chicago Convention has granted the ICAO the authority, responsibility, and obligation to ensure international civil aviation develops in a safe and orderly manner; (2) any outer space-related activities that affect international civil aviation, to the extent they do, are subject to the authority of the ICAO; (3) while the terms "airspace" and "outer space" are not clearly defined, any activity whether "space" related or "air" related that occurs or affects international civil aviation, in particular when the activity occurs in the medium of "airspace" that is traditionally utilized by civil aviation, requires coordination by the ICAO; (4) physical three-dimensional space used by "aircraft" is also used by "outer space vehicles". On these bases, the ICAO's current regulatory role could be expanded through the promulgation of SARPs to include SATVs because these vehicles pass through physical three-dimensional space used by "aircraft", affect the safe and orderly development of civil aviation, and require coordination.

19.3 PRECEDENT FOR A BROADER JURISDICTIONAL INTERPRETATION: SARPs ADDRESSING SECURITY AND THE ENVIRONMENT

Among the general rules for treaty interpretation established by the Vienna Convention on the Law of Treaties is consideration of "any subsequent practice in the application of the treaty which establishes the agreement of the parties regarding its interpretation" [23]. If the ICAO adopts SARPs to regulate SATVs on the bases of objectives listed under Article 44 and authority granted under Article 37, it would not be the first time the ICAO promulgated SARPs on these bases over an area related to international civil aviation, but not considered a traditional area for the ICAO to exercise authority. It must be remembered that although Article 37 lists 11 specific areas in which SARPs may be promulgated, it also provides that the ICAO may adopt SARPs addressing "such other matters concerned with the safety, regularity, and efficiency of air navigation as may from time to time appear appropriate" [24]. Two major examples of SARPs addressing matters not explicitly incorporated into the Chicago Convention in 1944 are SARPs addressing the environment and security—Annexes 16 and 17 to the Convention respectively.

Since the 1970s, the ICAO has taken a leading role in the regulation and implementation of policies relating to international civil aviation and the environment. In 1971, Annex 16 to the Chicago Convention was adopted to address concerns about air quality near airports [25]. This Annex has been expanded to address aircraft engine emission, aircraft noise, and aircraft engine certifications [26]. A careful reading of the Chicago Convention reveals no language that explicitly grants the ICAO the authority to regulate aircraft emissions, aircraft noise, or other environment-related matters. Nonetheless, in 1972 the General Assembly adopted Resolution A18-11, providing that "ICAO is conscious of the adverse environmental impact that may be related to aircraft activity and that of its Member States to achieve maximum compatibility between the safe and orderly development of civil aviation and the quality of the human environment" [27].

The language of A18-11 directly references the objectives of the ICAO listed under Article 44 of the Chicago Convention. Today, Resolution 35-5 (2004) is the statement of ICAO policies and practices related to environmental protection that directs the Council to promulgate SARPs relating to international civil aviation and the environment [28]. Resolution 35-5 relies upon an interpretation of the Convention preamble and Article 44 as granting the ICAO the legal authority to regulate the environment "to achieve maximum compatibility between safe and orderly development of civil aviation and the quality of the human environment."

Annex 17 of the Chicago Convention—entitled "Safeguarding International Civil Aviation Against Acts of Unlawful Interference"—addresses aviation security [29]. First promulgated as a SARP in 1974, it has since been expanded and updated many times [30]. It addresses preventive measures for aircraft, airports, passengers, baggage, cargo, and mail, as well as standards and qualifications for security personnel and responsive measures to acts of unlawful interference.

Annex 17 requires that each Member State "have as its primary objective the safety of passengers, crew, ground personnel and the general public in all matters related to safeguarding against acts of unlawful interference with civil aviation" [31]. It binds them to establish a national civil aviation security program [32] and to create a governmental institution, dedicated to aviation security, which would develop and implement regulations to safeguard aviation [33]. Contracting States also must develop a security training program [34], share aviation threat information [35], and otherwise cooperate with other States on their national security programs [36].

Contracting States must take action to prevent weapons, explosives, or other dangerous devices that might be used to commit an act of unlawful interference from being introduced on to aircraft [37]. Unauthorized personnel must not be allowed to enter the cockpit [38], and aircraft, passengers, baggage, cargo, and mail must all be checked and screened [39]. Each State also must ensure that originating hold [40] (as opposed to carry-on) baggage is screened prior to being loaded on to an international aircraft [41]. Furthermore, airports must establish security-restricted areas [42], and persons performing the security function must be subjected to background checks and selection procedures [43], hold appropriate qualifications for the position, and be adequately trained [44].

The drafters of the convention, as implied in the language of Article 44 and Article 37, acknowledge that challenges relating to international civil aviation unforeseen at the time of the Convention drafting would eventually arise. Therefore, the ICAO has been granted the authority to adapt to these challenges, subject to the limitations established under the Convention. To be certain, matters not related to the objectives of Article 44 are outside the purview of the ICAO's authority and the Convention would need to be amended if the ICAO would want to act on such matters. Nonetheless, "ensuring the safe and orderly growth of international civil aviation" is an expansive objective and can easily be read to include the regulation of SATVs, so long as SATVs affect the safe and orderly growth of international civil aviation [45]. Article 37 contains an equally wide catch-all provision for the ICAO to adopt and amend SARPs to address "any such matter concerned with the safety, regularity, and efficiency of air navigation" [46], including those not explicitly referenced in the Chicago Convention when it was drafted in 1944.

19.4 THE ICAO AND THE ITU

Other international organizations have faced legal challenges similar to the challenges SATVs present the ICAO. The International Telecommunications Union (ITU) is the leading UN agency for telecommunications. The ITU is responsible for the allocation, assignment, and registration of bands of the radio-frequency spectrum. The development of space technology in the 1950s forced the ITU to consider expanding the traditional terrestrial regulation of radio-frequencies to outer space activities. All satellites utilize radio-frequency bandwidth and have the capacity to cause harmful interference with terrestrial signals. In 1957, *Sputnik* produced the first conflict between earth and space transmissions [47]. In 1959, the ITU made allocations of frequency bands for space and earth-space radio communication services [48] and passed several resolutions applicable to telecommunication with and between space vehicles [49]. This authority was implicitly exercised through an interpretation of the ITU Convention in force [50]. The ITU Convention (in force at the time) contained no provision directly applicable to the regulation of space radio communications. The ITU convention did provide for the ITU to "effect allocation of the radio frequency spectrum and registration of radio frequency assignments in order to avoid harmful interference between radio stations of different countries" [51]. On this general authority, the ITU initially exercised authority over radio communications with space vehicles. The ITU would later amend the ITU Convention to include a provision for the regulation of satellite radio communication and geostationary orbital positions [52].

The ITU is an example of new space-related technology catalyzing the expansion of an international organization's traditional terrestrial regulatory functions to outer space-related activities. The lessons learned from this example illuminate how the ICAO could interpret the Chicago Convention to grant the initial exercise of authority over SATVs. As mentioned earlier, the ICAO's objective is "to develop principles and techniques of international air navigation and to foster the planning and development of international so as to … insure the safe and orderly growth of international civil aviation … meet the needs of the world for safe, regular, efficient and economical air transportation … [and] to promote safety of flight in international air navigation" [53]. To that end, Article 37 of the Convention grants the ICAO the authority to promulgate SARPs over "such matters concerned with the safety, regulation, and efficiency of air navigation as may from time to time appear appropriate" [54]. This broad language, coupled with the ICAO's stated mission, can be reasonably interpreted as granting the ICAO the authority to promulgate SARPs that regulate SATVs in order to ensure air navigation remains safe and efficient.

19.5 CONCLUSIONS AS TO THE ICAO's AUTHORITY TO REGULATE SATVs

The aims and objectives of the ICAO are to "develop principles and techniques of international air navigation and to foster the planning and development of international air transportation" [55].

Contracting States to the Chicago Convention have granted the ICAO the legal authority to achieve these goals, authority that includes the regulation of SATVs through the promulgation of SARPs. While the ambiguity of the law may present lacunae and conflicts, it does not prevent the ICAO from exercising its legal authority. This is true even if the exercise of this authority results in conflicting legal regimes with respect to either a region of space or a particular transportation vehicle.

Whether or not the ICAO promulgates SARPs will not be determined by the language of the Convention, for it clearly can be interpreted to grant the ICAO the legal authority. Ultimately, the decisions will be political, dependent on the interest of ICAO Member States, largely influenced by nations like the USA that have a direct interest in sub-orbital aerospace transportation.

19.6 THE NEED FOR A UNIFIED LEGAL REGIME

Future transportation systems will be highly influenced by the legal regime in which they are developed. Commercial development of space would be much enhanced by clarity, stability, and predictability of law [56]. Lack of uniformity of law, and conflicting and overlapping laws, will impair the market's interest in investment in space transportation, and the insurance industry's ability to assess and price risk [57].

Commercial investment in space transportation systems is expensive, depends on as yet unproven technology, and is fraught with risk. Clear legal rules can help define the degree, or consequences, of risk, and reduce uncertainty, providing the predictability necessary to support commercial investment. Conversely, legal uncertainty can increase risk and dampen enthusiasm for investment.

Many commentators have urged that legal rules be refined to take account of commercial needs in space [58]. Some have suggested that the emerging legal regime should be one of Air Law [59]. Others prefer the regime of Space Law [60]. Still others have urged immunity from liability for commercial activities in space for a developmental period [61].

It seems there are four alternatives: (1) sub-orbital vehicles remain unregulated internationally; (2) regulation occurs on a case-by-case basis through bilateral or regional agreements; (3) a new international organization is created to regulate them; or (4) the ICAO amends its Annexes to regulate them.

Probably the simplest, and most sensible, initial effort would be for the ICAO to amend its Annexes to redefine aircraft to include aerospace vehicles, so that when they fly in airspace used by civil aircraft, the rules of safety and navigation are the same [62]. It could do so by amending the definition of an aircraft to include aerospace vehicles. It created the definition of aircraft, and amended it to clarify that air cushion vehicles were not within the Chicago Convention; the ICAO could amend its Annexes again to clarify that sub-orbital vehicles fall within the definition of "aircraft".

One potential definition that might be used as a model was that promulgated by the US Congress in the Air Commerce Act of 1926: "any contrivance now known or hereafter invented, used or designed for navigation or flight in the air" [63]. The Canadian Parliament has defined an aircraft as "any machine capable of deriving support in the atmosphere from reactions of the air, and includes a rocket" [64]. Another source recommends that sub-orbital vehicles be included in the Air Law regime, and orbital vehicles be placed within the Space Law regime [65]. Alternatively, the ICAO could promulgate a new Annex 19 on "Space Standards". There is precedent for this as well. Article 37 of the Chicago Convention vests in the ICAO the authority to promulgate SARPs as Annexes to the Convention. As noted above, Article 37 lists therein 11 specific areas to which the ICAO is instructed to devote itself, mostly focusing on safety and navigation. Yet, since its creation, as air transport has grown and evolved, the ICAO has focused on other areas not explicitly listed in Article 37, including, for example, the promulgation of wholly new Annexes addressing environmental and security issues. Article 37 is sufficiently broad to permit such jurisdictional assertions, as it provides that the ICAO may promulgate SARPs addressing "such other matters concerned with the safety, regularity, and efficiency of air navigation as may from time to time appear appropriate" [66]. To avoid collisions, some international regulatory body is needed to provide uniform standards for national certification of space launch systems and vehicles, and their navigation through airspace.

The ICAO might also define the limits of airspace by amending an Annex, though some may argue that such a change would require a new Protocol amending the Chicago Convention itself, or perhaps an entirely new multilateral convention. This is by no means a new proposal. As early as 1956, Professor John Cobb Cooper urged that the definition of airspace should be determined by the United Nations and that pertinent regulations should be promulgated by the ICAO [67].

Others may argue that a separate space traffic management system, under a new international space management organization, should be established. As early as 1960, one source insisted: "It has been questioned whether the ICAO should amend its Annexes and widen the scope of its definition of aircraft so as to include rockets and missiles and even satellites. In view of the specific character of outer space law and inasmuch as those contrivances are mostly used in outer space, it is suggested that spacecraft are different from the contraptions regulated by the air law conventions and should be dealt with in separate international instruments" [68]. Yet, these observations were written at a time when the only space activities were those launching satellites into orbit. Today, we confront the issue of sub-orbital vehicles, which are very similar to "contraptions regulated by the air law conventions" [69]. Moreover, it would be difficult to justify replication of the able and detailed work already done by the ICAO on issues such as safety, navigation, security and liability, at least with respect to flights in the Earth's atmosphere.

Dr Nandasiri Jasentuliyana has called for the Committee on Peaceful Uses of Outer Space (COPUOS) to promulgate "Space Standards" similar to the ICAO's SARPs and to draft a convention creating an international framework for space vehicles [70]. Yet, for three decades, COPUOS has been unable to promulgate any multilateral legal instrument for ratification by States [71]. If COPUOS is able to break its deadlock, so much the better. If not, as the United Nation's arm for air transportation, the ICAO should provide clarification on the issues of what is contemplated by aircraft, and what is contemplated by airspace, and then set about providing standards of harmonization as SARPs, which Member States would be obliged to follow. Under the Chicago Convention, a Member State is obliged "to collaborate in securing the highest practicable degree of uniformity" on such issues [72], and to "keep its own regulations … uniform, to the greatest possible extent" with SARPs [73].

It is important to note that interpreting the definition of "aircraft" in the Chicago Convention, an instrument of international public law, to include SATVs may have consequences, both intended and unintended, with regard to the interpretation and application of other public and private international air law instruments. No international air law treaties, including the Warsaw Convention, Montreal '99, or Rome Convention, define the term "aircraft". As a result, these treaties implicitly defer to a generally accepted definition that derives in usage from the ICAO annexes. If the ICAO interprets or defines the term "aircraft" to include SATVs, questions will be raised as to the applicability of other international instruments to SATVs. This issue should be identified, researched, and discussed. The

ICAO, as the organization that drafted the Montreal Convention of 1999 addressing air carrier liability, and the several aviation security conventions, may be the appropriate organization to lead the process of clarifying whether aerospace vehicles fall under these other treaties as well.

Formal clarification of what law applies is needed. The conflicts and inconsistencies embedded in the respective legal regimes of Air Law and Space Law threaten to unravel the uniformity of law that the Conventions seek to attain, inhibit investment, and retard the development of commercial space transportation.

The time has come for the international community to address the issue of harmonizing Air and Space Law with an eye to facilitating—and indeed, promoting—commercial activity in space. The laws governing commercial space transportation need to be integrated with the prevailing rules of safety, navigation, security, and liability applicable under Air Law. The public's safety demands no less.

Notes and references

[1] Although Article 3(a) of the Chicago Convention states "This Convention shall be applicable only to civil aircraft, and shall not be applicable to state aircraft," Article 3(c) states "No state aircraft of a Contracting State shall fly over the territory of another State or land thereon without authorization by special agreement or otherwise, and in accordance with the terms therein." Therefore, in some instances, the Convention does apply to state aircraft, regardless of the Article 3(a) prohibition.

[2] Milde, M. (2000). Status of Military Aircraft in International Law. In *Air and Space Law in the 21st Century* (Benko, M. and Kroll, W., eds), p. 161. Carl Heymanns Verlag.

[3] Annexes 2, 7, and 11 of the Chicago Convention all define aircraft as "any machine that can derive support in the atmosphere from the reactions of the air other than the reactions of the air against the Earth's surface". In 1967, the ICAO modified the definition to its present form for the purpose of excluding hovercraft from the definition.

[4] Summary Page of the Concept of Suborbital FlightsWorking Paper, ICAO Council 175 Session, 30 May 2005, C-WP/12436.

[5] Section 1.1 of the Concept of Suborbital Flights Working Paper, ICAO Council 175 Session, 30 May 2005, C-WP/12436.

[6] Section 2.3 of the Concept of Suborbital Flights Working Paper, ICAO Council 175 Session, 30 May 2005, C-WP/12436.

[7] The International Court of Justice considers the principles of general treaty interpretation embodied in Articles 31 and 32 of the Vienna Convention on the Law of Treaty (1969) to reflect customary international law. Paraphrasing Aust, A. (2007). *Modern Treaty Law and Practice*, p. 232. Cambridge University Press.

[8] Article 31(1) of the Vienna Convention on the Law of Treaties (1969). Done at Vienna on 23 May 1969, entered into force on 27 January 1980. UN *Treaty Series*, vol. 1155, p. 331.

[9] Article 31(3)of the Vienna Convention on the Law of Treaties (1969).

[10] Preamble to the Chicago Convention.

[11] Preamble to the Chicago Convention.

[12] Article 3(d), Chicago Convention.

[13] Article 44, Chicago Convention.

[14] Article 44(a), Chicago Convention.

[15] Article 44(f), Chicago Convention.

[16] Article 44(i), Chicago Convention.

[17] Article 31(3)(b), Vienna Convention on the Law of Treaties (1969).

[18] ICAO General Assembly Resolution A29-11: Use of Space Technology in the field of Air Navigation (1992). Doc. 9600, A29-RES.
[19] ICAO General Assembly Resolution A15-1: Participation by ICAO in Programmes for the Exploration and Use of Outer Space (1965). Doc. 8528, A15-P/6.
[20] ICAO General Assembly Resolution A29-11: Use of Space Technology in the field of Air Navigation (1992). Doc. 9600, A29-RES.
[21] ICAO General Assembly Resolution A15-1: Participation by ICAO in Programmes for the Exploration and Use of Outer Space (1965). Doc. 8528, A15-P/6.
[22] ICAO General Assembly Resolution A29-11: Use of Space Technology in the field of Air Navigation (1992). Doc. 9600, A29-RES.
[23] Article 31(3)(b), Vienna Convention.
[24] Article 37(k), Chicago Convention.
[25] Dempsey, P. S. (2007). *Coursepack: Public International Law*. Faculty of Law, McGill University, at 163; referencing Goh, J. (1995). Problems of Transnational Regulation: A Case Study of Aircraft Noise Regulation in the EU. *Transp. Law J.*, 23:277, 284.
[26] *Ibid.*
[27] ICAO General Assembly Resolution A18-11: ICAO Position at the International Conference on the Problems of the Human Environment (Stockholm, 1972).
[28] ICAO General Assembly Resolution A35-5: Consolidated Statement of Continuing ICAO Policies and Practices Related to Environmental Protection (2004). Doc. 9848, A35-5.
[29] Chicago Convention, Annex 17.
[30] Amendment 10 of Annex 17 was adopted on 7 December 2001. It was based on proposals emerging from the 10th and 11th Aviation Security Panel meetings held in April 2000 and April 2001. Annex 17 is also supplemented by the ICAO Security Manual for Safeguarding Civil Aviation Against Acts of Unlawful Interference (Doc. 8973—Restricted), 6th ed., 2002, first published in 1971, and its Strategic Action Plan. Abeyratne, R. I. R. (1998). Some Recommendations for a New Legal and Regulatory Structure for the Management of the Offense of Unlawful Interference with Civil Aviation. *Transp. Law J.*, 25:115, 121–130.
[31] Chicago Convention, Annex 17, §2.1.1.
[32] Id. §3.1. Airports and aircraft operators must also establish security programs. Id. §§3.2.1, 3.3.1.
[33] Id. §§2.1.2, 3.1.2–3. States must also establish a national aviation security committee that coordinates security activities between various governmental institutions. Id. §3.1.6.
[34] Each Contracting State must establish a security training program. Id. § 3.1.7. They are also obliged to cooperate with other States in the development and exchange of training program information . Id. §2.3.3.
[35] Id. §2.3.4.
[36] Id. §2.3.2.
[37] Id. §4. Weapons brought on board by law enforcement and other authorized persons must have special authorization in accordance with the State's domestic law. Id. §§4.6.4–6.
[38] Id. §4.2.3.
[39] Id. §§4.2–4.6.
[40] In other words, goods transported in the "belly" of the aircraft.
[41] Id. §4.4.8. Until then, States are recommended to conduct hold baggage screening. Id. §4.4.9.
[42] Id. §4.7.1.
[43] Id. §3.4.1. Those granted unescorted access to security restricted areas of airports are subject to similar background checks. Id. §4.7.2. Security checks should be reapplied on a regular basis. Id. §4.7.6.
[44] Id. §3.4.2.
[45] Article 44, Chicago Convention.

[46] Article 37, Chicago Convention.
[47] Estep, S. D. and Kearse, A. L. (1962). Space Communications and the Law: Adequate International Control after 1963? *Michigan Law Review*, 60(7):873–904, at 896 footnote 86.
[48] *Ibid.* at 874.
[49] CCIR Documents of the IXth Plenary Assembly (Los Angeles, 1959). Published by the ITU (1960), Geneva; Resolution 40 (Influence of Troposphere on frequencies used for telecommunication with and between space vehicles), Resolution 47 (Effects of the ionosphere on radio waves used for telecommunication with and between space vehicles beyond the lower atmosphere).
[50] In 1959, the International Telecommunication Convention (1952) of Buenos Aires was in force.
[51] *Ibid.* ITU Convention (Buenos Aires, 1952), Article 4(2)(a).
[52] Article 1(2), Constitution of the International Telecommunication Union (done at Geneva, 22 December 1992, as amended by the Plenipotentiary Conferences (of Kyoto 1994, Minneapolis 1998, and Marrakesh 2002).
[53] Article 44, Chicago Convention.
[54] Article 37(k), Chicago Convention.
[55] Article 44, Convention on International Civil Aviation (also known as the "Chicago Convention").
[56] See, generally, Hobe, S. (2004). Aerospace Vehicles: Questions of Registration, Liability and Institutions. *Annals of Air and Space Law*, XXIX:377 (hereafter Hobe).
[57] What is needed is a "secure framework of regulations and legal responsibility ... [to] encourage increased activities in the future". Nesgos, CP. (1991). Commercial Space Transportaton: A New Industry Emerges. *Annals of Air and Space Law*, XVI:393, 412.
[58] Wassenbergh, H. (1999). Access of Private Entities to Airspace and Outer Space. *Annals of Air and Space Law*, XXIV:311, 325; Wassenbergh, H. (1998). The Art of Regulating International Air & Space Transportation. *Annals of Air and Space Law*, XXIII:201; Stockfish, B. (1992). Space Transportation and the Need for New International Legal and Institutional regime. *Annals of Air and Space Law*, XVII-II:323.
[59] Ryabinkin, C. (2004). Let There Be Flight: It's Time to Reform the Regulation of Commercial Space Travel. *J. Air Law Com.*, 69:101.
[60] Freeland, S. (2005). Up, Up and ... Back: The Emergence of Space Tourism and Its Impact on the International Law of Outer Space. *Chin. J. Int. Law*, 6:1. Blending functionalist and spatialist principles, Prof. Freeland argues: "the most appropriate approach seems to be the application of space law ... to the entire journey on the basis of the proposed function of the spacecraft carrying tourists—that is, the intention that it involves flight in outer space. The alternate 'exclusive' approach—to apply air law to the entire space tourism activity—appears unworkable given the lack of sovereignty that exists in outer space." Id. at 9. Prof. Hobe makes a similar argument: "the provisions of the Chicago Convention are based on the principle of sovereignty in national airspace and are therefore generally not applicable to activities which take place in outer space." Hobe, *supra* at 382. Similarly, Prof. Zhao argues: "The air transportation regime, characterized by State sovereignty over air space, substantially differs from the space travel regime ... This fundamental difference justifies the necessity of developing a distinct legal regime for space travel." Zhao, Y. (2005). *Developing a Legal Regime for Space Tourism: Pioneering a Legal Framework for Space Commercialization.* Am. Institute of Aeronautics and Astronautics. It is unclear why it is unworkable to have an Air Law regime apply to non-territorial outer space, inasmuch as a sophisticated body of both Public and Private International Air Law has developed involving intercontinental flights over the high seas, where no State has sovereignty. Over the high seas, which comprise more than 70% of the planet, the rules of the air are those established by the ICAO. See Article 12, Chicago Convention.
[61] Trepczynski, S. (2006). The Benefits of Granting Immunity to Private Companies Involved in Commercial Space Ventures. *Annals of Air and Space Law*, XXXI:381, 403.
[62] Jakhu et al., *supra.*

[63] Cooper, J. C. (1962). The Chicago Convention and Outer Space. Address before the American Rocket Society Conference on Space Flight, New York, 24 April.
[64] Aeronautics Act, RSC §3(1) (1985).
[65] Vissepo, V. (2005). Legal Aspects of Reusable Launch Vehicles. *J. Air Law Com.*, 31:165, 214.
[66] Article 37, Chicago Convention.
[67] Cooper, J. C. (1956). Legal Problems of Upper Space. Address before the American Society of International Law, Washington, DC, 26 April, quoted in Haley, A. (1958). The Law of Space—Scientific and Technical Considerations, 4th NYL Forum, 266. In 1962, Professor Cooper wrote that the ICAO should interpret what is contemplated by "airspace" under Article I of the Chicago Convention. Cooper, J. C. (1962). The Chicago Convention and Outer Space. Address before the American Rocket Society Conference, New York, 24 April.
[68] Verplaetse, J. (1960). *International Law in Vertical Space*, p. 157. Rothman.
[69] *Ibid.*
[70] Jasentuliyana, N. (1999). *International Space Law and the United Nations*, pp. 379–382. Kluwer.
[71] COPUOS has drafted guidelines and principles, but since the ill-fated Moon Agreement, has failed to achieve the consensus necessary to advance a treaty.
[72] Article 37, Chicago Convention.
[73] Article 12, Chicago Convention.

CHAPTER

International launch and re-entry safety standards

20

Jerold Haber, Paul Wilde

CHAPTER OUTLINE

20.1 LAUNCH AND RE-ENTRY RISK MANAGEMENT OBJECTIVES

As an increasing number of nations of the world join the spacefaring community, issues of computing risks, tolerable risk levels, and risk management have emerged as important topics for government and private organizations involved with national security, regulations, diplomacy, emergency response, and safety associated with both launch ranges and space lift programs. Diverse factors contribute to the concerns and objectives of these parties. Development and implementation of appropriate risk characterization and risk management approaches require an understanding of these diverse concerns [1]. The ultimate objectives of risk management are to mitigate risks to the largest extent feasible and to ensure that the risks tolerated are within prescribed limits.

The launch industry operates under a different paradigm from the aviation industry. For example, States exercise a supervisory role (responsibility) with respect to planes (and ships) owned by the private sector, without accepting financial liability for the actions of these assets. By contrast, Launching States have both supervisory and financial responsibility because the United Nations space treaties make countries both "responsible" and "liable" for the space activities of their nationals (person, companies, etc.). Furthermore, aviation safety approvals typically come in the form of certification that an unlimited number of flights are authorized, but launches are typically approved on a mission-by-mission basis. Even though much of the fundamental groundwork may be established for a launch program planned to include many launches over a period of time of the same space booster with varying payloads, the timing of individual launches is highly uncertain. Planned launch dates typically slip days to months prior to actual launch. Program hardware and supporting processes and instrumentation frequently evolve over the life of the program. Furthermore, launch vehicles often evolve substantially over time to incorporate improvements made in response to failures, improve performance, etc., such that the reliability of a launch vehicle typically changes substantially [2]. Consequently, the nature of today's launch vehicles and today's technology effectively constrains

Space Safety Regulations and Standards. DOI: 10.1016/B978-1-85617-752-8.10020-0

approvals to be granted on a launch-by-launch basis. Any viable risk acceptability regime must allow the acceptability of a launch mission to be evaluated on its own merits without examining the effects of other space lift traffic from other ranges or even the launching range.

Safety offices at launch ranges are responsible for approving a proposed flight plan and for providing the launch authority with the assurance that, from a safety perspective, all conditions are acceptable to proceed with a launch. Personnel in these offices are, typically, technically sophisticated enough to address the nuances of the different types of risk and risk-mitigating measures.

Regulatory agencies typically ensure that a launch does not jeopardize public safety due to planned events such as jettisoned stages, or unplanned events such as explosions during flight or pre-launch operations. Regulatory agencies typically require a launch operator to submit evidence to demonstrate that the launch satisfies a set of requirements, and issue approvals for a single launch or a series of similar launches. Regulatory approval of orbital launches typically involves a three-pronged licensing process that includes a quantitative risk analysis, a system safety program, and operational controls. In the USA, Congress mandated a streamlined regime to permit an unlimited number of launches and re-entries for a particular sub-orbital rocket designed for the purposes of research and development, showing compliance with requirements as part of the process of obtaining a launch license, or for crew training prior to obtaining a license.

National security agencies (such as the US Department of State) must establish policies regarding deliberate and accidental intrusions into national airspace. It must also identify tolerable risks to its aircraft, to its shipping lanes, to its ground-based populations, to its national assets, and to levels of environmental degradation. Further it must establish national policy toward participating in international responses to emergencies generated by failed spacecraft and uncontrolled re-entry of spacecraft. This is, typically, a policy-oriented, non-technical group. They must be able to place risks in the context of risks with which they are familiar, understand what is technically feasible, precedents for tolerable actions, communications protocols for avoiding and managing risks, and viable communications to international, national and local emergency responders. As a group, they generally respond to a mandate similar to the one articulated by the International Association for the Advancement of Space Safety (IAASS) Working Group on "An ICAO for Space?" [3]:

> *Ensure that citizens of all nations are equally protected from "unreasonable levels" of risk from over flight by missiles, launch vehicles and returning spacecraft.*

In developing practical approaches to define and mitigate "unreasonable levels" of risk, the US launch safety community has endorsed a policy that the general public should not be exposed to a risk level greater than the background risk in comparable involuntary activities, and the risk of a catastrophic mishap must be mitigated. In this context, the US launch safety community considers "comparable involuntary activities" as those where the risk arises from man-made activities that:

1. Are subject to government regulations or are otherwise controlled by a government agency, and
2. Are of vital interest to the USA, and
3. Impose involuntary risk of serious injury or worse on the public.

Emergency response agencies are scenario-oriented. For any particular mission, they need to know the expected outcomes and what might happen if the booster fails catastrophically. Relevant risk measures must characterize the individual launch, assist the agency in understanding the relative risk of the particular launch in comparison to other launches, and allow the agency to prepare contingency plans.

20.2 RISK CHARACTERIZATION

A risk metric must account for both probability and severity. There are five common risk metrics for space lift operations, three of which address risk to people and two of which address risk to assets:

1. Individual risk—the probability that the maximally exposed individual suffers a given level of harm.
2. Societal or collective risk—the expected number of persons who will suffer a given level of harm.
3. Catastrophic risk—the chance that a large number of people will suffer a given level of harm.
4. Asset risk—the likelihood of damage to national or commercial assets.
5. Environmental risk—the potential for significant impacts to critical habitats, wildlife, or environments.

Complete risk characterizations must also specify the severity of harm to be measured (i.e. minor injuries, serious injuries, fatalities), the timeline of consequence (prompt injury/damage, delayed effects), and the time span associated with the risk exposure (annual or some other time-frame or mission by mission). Attention has often been focused on protecting against fatalities. The trend has, however, been toward a more comprehensive evaluation addressing lesser levels of injury, property damage, and environmental consequences. Because responsibilities for these different adverse effects may span several offices or agencies, they pose procedural challenges in addition to the technical challenges. In the USA, consensus standards for risk acceptability have been published [4].

In the broader safety community, individual risk in the form of *annual fatality* probability is one of the most common risk statistics published. (Recently, individual risk statistics from disease and accidents have also been published in the form of annual non-fatal injury rates and morbidity, or frequency of disease.) Launch ranges, typically, set limits on the probability that the maximally exposed individual will be seriously injured or killed.

Societal risk is less frequently considered than individual risk. Regulators typically use $F-N$ curves depicting the annual frequency of incidents resulting in N or more fatal injuries when addressing societal risk. Agencies planning for emergency preparedness and assessing the value of alternative risk mitigations frequently develop "planning scenarios" characterizing various accident or natural disasters. Key components of such characterizations are the expected number of injuries requiring medical attention, the expected number of people who will require hospitalization, the expected number of prompt fatalities, and the expected number of delayed fatalities. The US launch safety community characterized societal risk by the expected number of casualties (serious injuries or worse, including fatalities) or fatalities per launch. In other countries, a variety of measures are used, including the probability that the launch will result in one or more fatalities.

Catastrophic risk is commonly interpreted to mean the chance of a single event resulting in large numbers of injuries, fatalities, or extensive property damage. $F-N$ curves are used by many industries to characterize such events. The launch safety community has not formally defined catastrophe, but in the USA the OSHA promulgated a formal definition of catastrophe in 29 CFR 1960.2: "An accident resulting in five or more agency and/or non-agency people being hospitalized for inpatient care." More fundamentally, catastrophe has also been interpreted to refer to the chance of an event that would have sufficient adverse outcomes so as to preclude further launches for an extended period of time.

Asset risk characterizes property damage or loss of the ability to use property for its intended function. Emergency management agencies typically employ a combination of annual expected losses from particular hazards and expected losses from various planning scenarios. In the USA, measures of expected annual property damage from various hazards are fundamental to organizations involved with insurance. Emergency first-responders need detailed characterizations of the types of damage anticipated from a "planning scenario". Important elements include the potential for interfering with communications, transportation, and the water supply. It is also important to know the region's capacity to treat the injured and related emergency needs. When inherently hazardous activities are regulated, such as launch operations, the regulator is particularly concerned about assuring that third-party losses are highly unlikely and, if a catastrophic accident does occur, that the organization responsible for the hazardous activity provides the resources to compensate for any losses. Thus, for example, the US Federal Aviation Administration Office of Commercial Space Transportation (FAA/AST) requires a commercial launch operator to carry insurance sufficient to cover the "maximum probable loss (MPL)". The MPL is defined as "the greatest dollar amount of loss for bodily injury or property damage that is reasonably expected to result from licensed launch activities; losses to third parties … that are reasonably expected to result from licensed launch activities are those having a probability of occurrence on the order of no less than one in ten million; losses to government property and government personnel involved in licensed launch activities that are reasonably expected to result from licensed launch activities are those having a probability of occurrence on the order of no less than one in one hundred thousand." The MPL includes losses associated with casualties, but most US ranges limit their quantification of risk to property to estimates of the likelihood of an event that may damage an asset.

In the USA, the potential for environmental damage for planned activities is, typically, considered as part of the environmental impact assessment process. Environmental damage concerns have also driven air emissions regulations and groundwater contamination concerns. Although there have been cases of severe environmental damage, such as toxic contamination in Kazakhstan, management of these risks for individual launches continues to be the exception rather than the rule.

Why are risks most frequently characterized on an annual basis? Most regulated events occur on a frequent or continuous basis. Automobile traffic is on the road every day of the week, every hour of the day. Manufacturing facilities, and chemical and nuclear facilities operate throughout the year. Moreover, decisions to allow operating conditions to continue or to require changes to mitigate risks are made infrequently. Exceptions to mitigate high-risk, easily defined situations tend to follow simple defined rules, such as closing an airport for adverse weather conditions.

Launch and re-entry operations differ in several important ways. They occur infrequently; a large launch complex may be responsible for as many as 50 launches in a year. They are of short duration. Powered flight during ascent may last a few minutes; re-entry timelines are somewhat longer, of the order of a few hours. These operations individually represent a tiny fraction of a year. Moreover, attempts to quantify mission risks prospectively (for the year ahead) are further complicated by the difficulty in forecasting when any particular launch will occur. In contrast to commercial aircraft operations, where take-offs and landings typically occur within an hour of scheduled time, launch operations commonly slip from hours to months from the original schedule, making accounting for the specific launch operations that will occur within a specific year highly uncertain. Furthermore, a single launch accident commonly results in the entire fleet of similar launch vehicles being grounded during the investigation to determine the cause and corrective actions, while certified aircraft typically continue to operate during an accident investigation.

Moreover, even for a single launch complex, there may be substantial variations of key characteristics of the launch (exact ascent vehicle design, payload, trajectory azimuth and shape, and weather conditions). Consequently, decisions to allow a launch to occur or to proceed are made on a case-by-case basis, considering the conditions and risks seen to flow from the alternative decision options. Similarly, training scenarios for emergency planning are well-defined hypothetical fact patterns. Risks must be characterized for a particular launch. Moreover, beyond a baseline case, planning scenarios will typically require characterization of the outcomes of hypothetical, but credible, failure scenarios. Baseline cases assist planners in assessing the "degree of readiness" required and to prioritize resources as to the regions most likely to suffer adverse consequences. Particular failure scenarios allow "desktop" or real-world exercises to prepare first-responders for the actions that may be required of them.

As noted above, the risk characterization needs of launch agencies differ significantly from those of most other interested parties. When the launch site is within the territorial boundaries of the launching nation and there is a sufficient geographical separation from the territories of other nations, these differences are less significant to the international community. When, however, the launch site or the immediately surrounding region lies within the political boundaries of another country (e.g. Kazakhstan), these differences may be significant. Moreover, when the nationality of the launch complex operator, the host nation, and the nations desiring to launch from the complex are different, such as occurs for the Guyana Spaceport, resolving these differences may become complex.

In the launch area, risks tend to be dominated by launches from a single launch facility. The populations being subjected to the risk or their representatives should be informed that the risk levels are managed to levels consistent with other tolerated risks and that they are not being forced to assume disproportionate risk levels. To our knowledge, this guideline has been applied so far only to a few launch facilities [5]. Three alternative approaches have arisen throughout the world:

1. **Might makes right.** When the operator and launching nation has sufficient power to impose its will without challenge, the informed risk principle is ignored. Launches occur at the convenience of the launching nation.
2. **Everything is negotiable.** Notwithstanding the theoretical position stated above, it is possible and, indeed, it has become the practice in some locations, to communicate the risks on a per-launch basis to the host nation. Within an appropriate economic and diplomatic framework, this may produce results satisfying the needs of the host nation.
3. ***De minimis* risk.** The third approach is for the launching agency to manage launch risks on a per-event basis and to constrain the tolerated risk to the host nation to such low levels that the annual risk levels are of no concern because they are vanishingly small.

While each of these cases could create a need for emergency preparedness, in the first case the launching nation is unlikely to take such actions, and in the third case it is desirable but less critical. Accidents within the extended launch area may result in large fragments, or even entire stages, falling back to the Earth. Explosions on impact and toxic spills or releases of toxic gases are potential outcomes with catastrophic consequences. While the time after the accident for debris to impact is short, there will frequently be time to warn affected populations before toxic clouds reach them. Planning must address exclusion areas on land, at sea, and in the air. Emergency first-responders should be advised of the areas where they may have to control toxic spills or fires resulting from the accident. Reverse 911 systems may be deployed to warn and protect populations from toxic gases.

As a launch vehicle moves downrange, it depletes fuel and jettisons spent stages. The combination of vehicle altitude and velocity result in significantly longer time intervals from the time a failure could occur and the time vehicle debris could pose a threat. In addition, when downrange failures occur, the hazards are typically spread over large regions. The extent of the regions that may be hazarded precludes alerting first-responders prior to a launch. By contrast, the increased time from failure to danger provides an opportunity to divert aircraft and ships from the hazarded region and to notify downrange nations to deploy first-responders. For example, the aircraft that were most significantly at risk from the *Columbia* accident flew through the debris field between 8 and 20 minutes after the accident [6]. A recent study on the potential to warn aircraft about debris from unplanned re-entries of spacecraft or discarded upper stages found that threatening fragments could reach the airspace anywhere from 13 to 19 minutes after an observation made when the object break-up was near the characteristic altitude at approximately 78 km (42 nmi) [7]. Other studies have shown that similar amounts of time may be available between an orbital launch failure and when planes flying in areas at least 100 km from the launch point would be at significant risk from debris. As shown in Figure 20.1, the shortest dimension of typical re-entry debris fields are such that a typical jet aircraft would need only a few minutes to clear the threatened area if properly directed. The shortest dimensions of launch accident debris fields are predicted to be somewhat wider, but prohibitively so.

In the USA, the Federal Aviation Administration (FAA) is investigating procedures and tools to facilitate appropriate mitigations following such launch or re-entry accidents [8]. The development of internationally accepted standards and protocols related to the protection of ships and aircraft would clearly facilitate and promote launch safety. For example, there are currently no internationally recognized standards on (1) the level of protection aircraft should be afforded from planned or unplanned launch hazards, (2) the vulnerability of various types of aircraft to launch hazards, or (3) the use of rapid communication technologies to issue real-time warnings in response to launch or re-entry hazards. Voluntary consensus standards on the first two subjects have been developed in the USA by

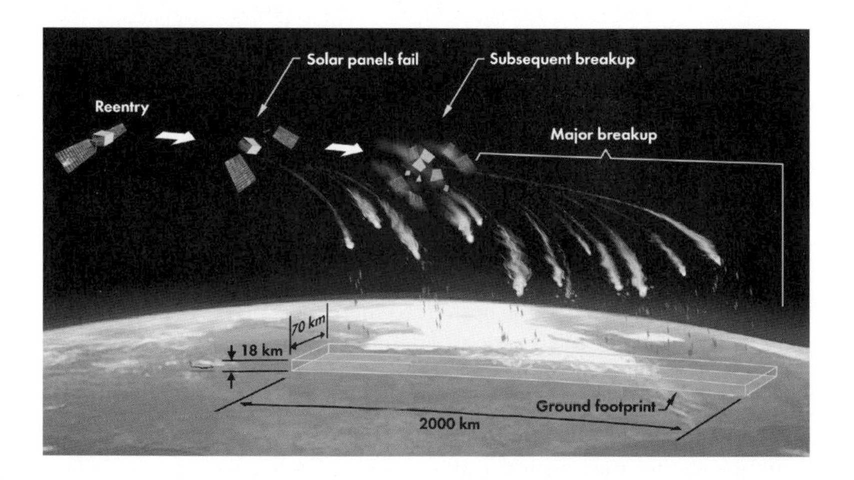

FIGURE 20.1

Dimensions of airspace affected by a spacecraft re-entry event.

Courtesy of the Aerospace Corporation

safety experts in 2007 [9], but 2 years later these had not yet been implemented by many of the agencies involved.

Controlled re-entries are a viable means to remove expired satellites and launch vehicle upper stages, as well as to return individuals from space. Hazard areas for controlled re-entries are computed in advance so that warnings to ships and aircraft can be issued, whether or not these vehicles are designed to survive re-entry. However, uncontrolled re-entry as a means to "dispose" of orbiting objects presents some unique risk management challenges. Approximately 100 large, man-made objects re-enter the Earth's atmosphere randomly every year. These randomly re-entering objects are not designed to survive re-entry, and re-entry heating and loads disintegrate each object into a number of fragments that are spread over a long, narrow footprint, as illustrated in Figure 20.1. About 10–40% of the dry mass of such objects prior to re-entry typically survives to impact the Earth's surface and the fragments are large enough to be a hazard to people and property. There are no warnings currently available on where debris from these events is likely to fall.

The US Air Force maintains a catalog of objects in Earth orbit that can be used to estimate when an object will re-enter (defined as when the object will intersect the entry interface, the top of the sensible atmosphere, generally defined to be at an altitude of 120 km or 400,000 ft) based on periodic radar and optical observations. Variability in the conditions of the extreme upper atmosphere and dynamics of the re-entering body mean that, as a general rule, re-entry predictions using this tracking data have an error of approximately 10% in time: if re-entry is predicted to occur in 1 hour after an observation of the body state, there is a ± 6 minute error in that prediction. Given the orbital speed of the object (4 nmi/s), this error translates to an uncertainty in the re-entry point of approximately ± 1500 nmi. Without special tasking, good estimates of final orbits are generally not computed within 1 hour of re-entry, and thus estimates of the location of re-entry debris are typically too uncertain to be used for issuing safety warnings of any kind.

A recent study suggested that the risk to an aircraft *exposed* to the debris field from a typical uncontrolled re-entry is currently above the long-term acceptable risk, but below the short-term acceptable risk based on risk acceptability guidelines used by the FAA for other types of threats. The currently viable mitigations include equipping the vehicle with the capability to perform a controlled re-entry, using low-melting-point materials and other design features that promote demise during re-entry heating, limiting the size of re-entry objects, and limiting the orbital inclination during re-entry to overfly only sparsely populated regions. A more robust observation network would be valuable to facilitate re-entry warnings and mitigate potentially catastrophic collisions with large ships and aircraft at least.

20.3 INTERNATIONAL RISK COMMUNICATION, EMERGENCY PLANNING, AND EMERGENCY RESPONSE

As noted earlier, while the highest risk areas are typically in the vicinity of the launch site, downrange locations may be hazarded at times by launches from multiple launch complexes located in several countries. Figure 20.2 depicts the overlapping regions potentially hazarded by launches from the US NASA Wallops Flight Facility and the Eastern Range (Kennedy Space Center). Near term, from a risk management perspective, the most important issues are the creation of a communication network for

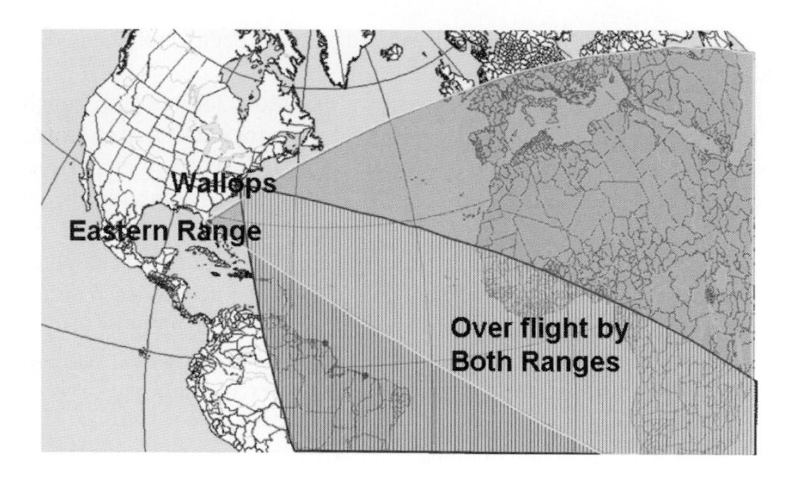

FIGURE 20.2

Overflight risks accrue from multiple launch sites.

alerting downrange nations should a failure occur late in flight, and establishing emergency response protocols. However, from the perspective of security agencies and government agencies, there is a growing need to demonstrate awareness of risk levels and to be able to show their constituencies that these risks are known and accepted, much as the nations of the world have established protocols to tolerate the risk of overflight of commercial aircraft. What risk measures should be employed for this purpose and how should they be developed? The following outlines one possible process for accumulating this information and disseminating it to appropriate agencies.

The process must provide emergency planners and political representatives with information that allows them to affirm that they are aware of the annual risk to which their citizens are being subjected and that it is both reasonable and consistent with the risk other regions are bearing. This may be developed by creating a database showing the annual individual risk from ascending trajectory overflights and re-entries of spacecraft on a grid with spacing sufficient to resolve significant changes in individual risk levels (e.g. spacing of the order of 5 minutes latitude by 5 minutes longitude). Figure 20.3 illustrates such a grid, together with hypothetical individual risk contours from launches from two different launch complexes. Summing the individual risks within a grid over all launches occurring during the year produces the annual individual risk for the grid cell. It may be anticipated that downrange overflight individual risks will be lower than historically accepted annual individual risks for a single range; annual fatality probabilities are expected to be less than one in 1 million. The annual individual risks for a grid cell can easily be combined with population data for that grid cell by the local authorities to compute the societal risk and assure levels are acceptable.

Most launching agencies compute individual and societal risk in the extended launch area. It is a fairly easy extension to require that these agencies also compute the downrange overflight individual risks. The IAASS as a professional society or an international regulatory body (ICAO for Space?) could design and maintain a worldwide database with this information. As a launching agency finalizes its risk analyses in support of flight plan approval, it would post to the database the probability of

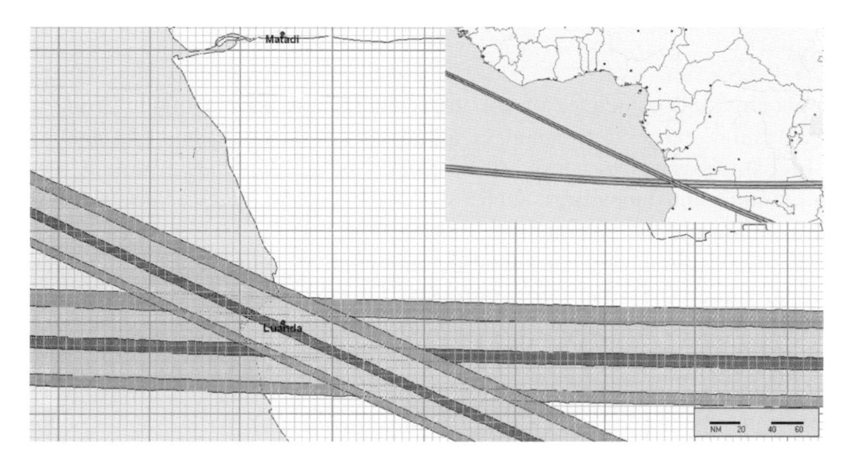

FIGURE 20.3

Individual risks from two launch sites.

serious injury or fatality per person for each grid cell affected by a mission. The agency would also record its identification and the identification of the launch.

Supporting database software would compute for each grid cell the annual individual risk using a 1-year retrospective set of analyses. Perhaps the most significant challenge in establishing such a system will be enlisting the cooperation of all launchers.

To become meaningful, a sufficient amount of data must be accumulated. Until the database is adequately populated and the results achieve some level of stability access, it should be limited to the organizers and their researchers. When the database achieves a stable status it can provide the basis for the risk communication to the affected communities. Perhaps even more significantly, the database can provide important information to regional emergency preparedness planners. Higher risk zones can be identified. Correlation of risk zones with launch locations and mission types can be established. If, as anticipated, space launch traffic increases at a significantly higher rate than booster reliability improvements, measures may be required to assure that annual risk to the public is managed. The US consensus standard on launch risk acceptability already calls for ranges to "periodically conduct a formal review to ensure that its activities and its mission risk acceptability policy are consistent with the annual risk acceptability criteria". Total annual traffic is significant for locations in the launch area; an alternative concept is needed for downrange overflight and re-entry risks. An approach such as the one described above, providing the ability to evaluate the aggregate risks to receptors and to map the sources of the risks, can become an important tool for negotiating risk allocations to different launch sites and programs.

Since planned debris jettisons, as well as launch failures, often occur over the oceans, a similar database to communicate the risks to individuals on ships would be a logical extension. The limitations associated with ship traffic data may justify the simplifying assumption that an individual is present on the deck of a ship at each grid point, such that the reported risks would be conditional on the presence of such an exposed individual.

SUMMARY

The diverse concerns and objectives of the government and private organizations developing an interest in launch and re-entry safety cannot be addressed with a single set of measures and criteria. A complete set of risk measures to address these concerns must include planning scenarios, launch specific measures, and annual measures. Such a set of risk measures must incorporate protection of the individual, the community, and community assets.

Communication is essential as a part of public policymaking and the political processes, as well as emergency preparedness and emergency response. Limited progress has been made in defining and implementing adequate structures to define what risk characterization needs to be communicated and establishing the channels for such information accumulation and dissemination. Addressing and implementing methods for multinational development of risk data and methods for dissemination will take time and development of trust. The opportunity for creating such systems without a crisis is greatest before a catastrophic accident precipitates an international incident and before the number of spacefaring nations grows to generate overwhelming communication and coordination challenges. Despite the enduring differences between the aviation and space industries, the channels of communication and international cooperation established to facilitate aviation safety may provide an important building block for the development of international launch and re-entry safety standards.

Notes and references

[1] Haber, J. and Lamoreaux, R. (2008). Launch Risk Acceptability: The Public Speaks. *Proceedings of the Third IAASS Conference*, Rome, Italy, October.

[2] Guikema, S. D. and Pate-Cornell, M. E. (2005). Probability of Infancy Problems for Space Launch Vehicles. *Reliability Engineering and Safety Systems*, No. 87, 303–314.

[3] Bahr, N., Sgobba, T., Jahku, R., Schrogl, K.-U., Haber, J., Wilde, P., Kirkpatrick, P., Trinchero, J. P. and Baccini, H. (2007). "An ICAO for Space?", an IAASS White Paper.

[4] Cather, C. and Haber, J. M. (2007). Common Risk Criteria for US National Ranges—An Update. *Proceedings of the Second IAASS Conference*, Chicago, USA, May.

[5] The US FAA holds public hearings before granting a launch site license. The US Eastern Range holds annual coordination meetings with the Brevard County of Emergency Preparedness. In addition, during the countdown the County Emergency Preparedness staff are provided real-time information to allow them to promptly respond to an accident.

[6] Larson, E. W. F., Wilde, P. and Linn, A. (2005). Determination of Risk to Aircraft from Space Vehicle Debris. *Proceedings of the First IAASS Conference*, Nice, France, October.

[7] Ailor, W. and Wilde, P. (2008). Requirements for Warning Aircraft of Reentering Debris. *Proceedings of the Third IAASS Conference*, Rome, Italy, October.

[8] Wilde, P. et al. (2005). Public Safety Standards for Launch and Reentry of Spacecraft. *Proceedings of the First IAASS Conference*, Nice, France, October.

[9] Range Commanders Council (2007). *Common Risk Criteria for National Test Ranges: Supplement*, RCC 321-07. White Sands Missile Range, New Mexico.

CHAPTER

21 The United Nations and its efforts to develop treaties, conventions, or guidelines to address key space issues including the de-weaponization of space and orbital debris[1]

Theresa Hitchens

Director, United Nations Institute for Disarmament Research (UNIDIR)

CHAPTER OUTLINE

INTRODUCTION

Following what seems to have emerged as an annual pattern, 2009 once again has proven a dramatic year for outer space. On 10 February 2009, the first known collision of two intact satellites—a working US Iridium communications satellite (*Iridium 33*) and a defunct Russian Cosmos (*Cosmos 2251*)—created a massive debris field in one the most used orbits (c. 790 km in altitude), once again raising widespread fears around the globe about the ability to sustain safe space operations [1].

On 29 May 2009, the Geneva-based Conference on Disarmament (CD)—for the first time in 13 years—agreed to a program of work that includes start-up of a formal working ground on the "Prevention of an Arms Race in Outer Space (PAROS)" empowered "to discuss substantively, without limitation, all issues related to space weaponization" [2]. This decision opened the door to possible multilateral efforts (potentially via treaty) to restrain or prevent dangerous military activities in space for the first time since the passage of the 1967 Outer Space Treaty.

At its 52nd Session in Vienna (3–12 June), the Committee on the Peaceful Uses of Outer Space (COPUOS) approved a new agenda item for its 2010 work to discuss "Long-Term Sustainability in

[1] The views presented in this paper are solely those of the author in her personal capacity, and do not represent the position of the United Nations or UNIDIR.

Space Safety Regulations and Standards. DOI: 10.1016/B978-1-85617-752-8.10021-2

Outer Space" [3], which will be based on the efforts during the past 2 years of an informal working group of government and space industry representatives. That informal working group has been preparing a report on the issue that includes: debris mitigation/remediation; improving the safety of space operations; managing the finite electromagnetic spectrum; space weather; and a review of international mechanisms that could support improvements in long-term space sustainability. The final report of the working group, led by French space scientist and former COPUOS head, Gérard Brachet, is expected to be released by the end of 2009 [4]. While directed at space activities during peacetime, it is almost inevitable that any development of "rules of road" for space operations will, *ipso facto*, also affect military and dual-use satellite operators. Due to the nature of the space environment, there is no realistic way to separate military from civil operations—the laws of physics apply equally to both.

Thus, in mid-2009, for the first time in several decades, multilateral efforts under the auspices of the United Nations were ongoing to address both the civil/peacetime and military aspects of protecting space for future generations. Indeed, with awareness of issues such as debris and orbital crowding/ spectrum interference, recognition was dawning at the two key UN agencies that the traditional separation of space safety and security issues into "civil" and "military" categories is not sustainable—given the physics of the space environment, the imminent threat to both civil and military satellites from the increasing debris population, and the realities of dual-use technology. Indeed, tentative efforts for outreach and even possible cooperation between the CD and COPUOS, long considered politically taboo, were beginning to be seen in informal forums such as the 15–16 June annual space conference at the United Nations Institute for Disarmament Research (UNIDIR) in Geneva [5].

However, the welcome momentum for progress by the international community in addressing the myriad and complex threats to the space environment came to a screeching halt on 10 August 2009. After months of haggling about procedures under an implementation plan for the agreed CD work program, Pakistan metaphorically firebombed the process by making demands for changes to the proposed implementation plan that in effect sought to reopen the delicate compromise on the work program itself [6]. While Pakistan's objections have nothing to do with the space issue—but are apparently centered on reservations about entering into negotiations on a treaty to stop production of/ reduce stockpiles of (the question of scope of the treaty is yet to be determined) fissile materials used to create nuclear weapons—they have had the effect of also blocking further CD work on PAROS. As of this writing, it is highly unclear whether the impasse with Islamabad can be breached.

21.1 ACTIVITIES IN THE CD

The PAROS issue has been on the agenda of the CD since 1985, when an ad hoc committee was founded to examine the problem. After 9 years of little substantive result, the committee was disbanded in 1994 and since that time only informal discussions of space-related weapons and warfare have been undertaken. The lack of progress can primarily be attributed to the Cold War tensions between the USA and the USSR/Russia on all issues under the CD purview during the 1980s, and since 1999, and to similar tensions (although primarily revolving around US plans for missile defense) between the USA and China.

China and Russia stepped up efforts to push PAROS on to the CD's formal agenda in 2002, following the abrogation of the Anti-Ballistic Missile Treaty by the administration of US President George W. Bush. On 27 June 2002, Beijing and Moscow jointly submitted (along with a handful of co-sponsors) a working paper to the CD, "Possible Elements of a Future International Legal Agreement on the Prevention of the Deployment of Weapons in Outer Space, the Threat or Use of Force Against Outer Space Objects" [7]. Despite the Bush administration's rejection of the PAROS issue as neither viable or necessary [8], China and Russia continued to pepper the CD with a variety of so-called non-papers on dimensions of a possible space weapons ban treaty between 2002 and late 2007, and on 12 February 2008, formally introduced a draft treaty, "Treaty on the Prevention of the Placement of Weapons in Outer Space, the Threat or Use of Force Against Outer Space Objects" [9], or PPWT for short.

Somewhat ironically, the introduction of the PPWT came just more than a year after Beijing had shocked the world with its successful flight test in January 2007 of a ground-based, kinetic-energy (hit-to-kill) anti-satellite (ASAT) weapon against an aging Chinese weather satellite (*Fenguyn-1C*)—breaking the *de facto* moratorium on ASAT testing that had held for 25 years [10]. Besides raising fears about a resumption of an arms race in outer space, the test also heightened international concerns about the increasing levels of debris in many useful orbital bands [11].

Predictably, especially in the light of the Chinese ASAT test, the Bush administration continued to flatly refuse to consider negotiations of the PPWT in the CD. However, Washington was active in negotiating the "substantive discussion" language of CD Resolution 1840—ultimately with success. In particular, Bush administration officials stressed that a key problem is the draft treaty's lack of applicability or coverage of the development and deployment of direct-ascent, kinetic-energy ASATs (KE-ASATs), which are one of the easiest technologies for attacking satellites to attain. Indeed, any nation operating mid-range ballistic missiles (at least 12 currently do so) and a reasonably well-developed space program could likely develop a similar weapon. Unfortunately, as witnessed by the Chinese test, this kind of ASAT is one of the most hazardous to the future safety of space because of the large debris fields that would occur from testing and/or use.

The incoming Obama administration in the USA has up to now said little on the PPWT, or directly on the issue of space weaponization. While a candidate for the presidency, Barack Obama stated that he was opposed to space weaponization and, shortly after his inauguration, language appeared on the White House website that pledged the administration to "seeking a worldwide ban on weapons that interfere with military and commercial satellites". That language, however, was almost immediately disavowed by administration officials, and shortly thereafter was removed from the website and the replacement language spoke only of cooperation with US allies and the private sector to protect US and allied satellites—breaking no new ground from previous American policy [12].

As noted above, on 29 May 2009, the CD approved a program of work that included setting up a formal working group on space (Working Group 3). The group was mandated to "discuss substantively, without limitation, all issues related to space weaponization". This language is important, in that it forges a compromise between the standing American position against any PAROS negotiations, and the erstwhile insistence of Russia and China that such negotiations be launched as part of any CD work program. Obviously, however, Moscow and Beijing have continued to stress that their draft PPWT be the centerpiece of those discussions—which, unless another nation formally tables an alternate proposal, will be the default situation.

There has been considerable speculation about the possibility that Canada might introduce an alternate treaty text—following several presentations made by Ottawa this year in different forums. On 26 March 2009, Canadian Ambassador to the CD Marius Grinius introduced a working paper, "The Merits of Certain Draft Transparency and Confidence Building Measures and Treaty Proposals for Space Security" [13]. The paper argues that transparency and confidence building measures

(TCBMs) could pave the way toward a treaty, and argues that the CD should "consider security guarantees, such as a declaration of legal principles, a code of conduct or a treaty that would: (a) ban the placement of weapons in space, (b) prohibit the test and use of weapons on satellites so as to damage or destroy them, and (c) prohibit the use of satellites themselves as weapons." At the 2009 space security conference sponsored by UNIDIR [14] on 15–16 June 2009, in Geneva, Phillip Baines, deputy director, space security and conventional weapons non-proliferation and disarmament, Canadian Department of Foreign Affairs, expanded on the rationale for the suggested Canadian approach, detailing threats that the package of proposals would address and detailing how each of the three concepts above could work [15]. However, it remains unclear if Ottawa will seek to formally table a proposal based on the working paper.

Meanwhile, the Obama administration is undertaking a sweeping space policy review that is not expected to be completed until early in 2010. Thus, Washington is unlikely to have a firm position for discussion within the CD working group on space if, indeed, that working group is set up this year—which now looks rather unlikely. According to Washington insiders, there is an ongoing debate among space experts in political and military circles, pitting those who would like to see the administration move in the direction of a multilateral treaty designed to prevent space weaponization (at a minimum, an instrument to bar testing and use of KE-ASATs) and those who would rather pursue solely voluntary measures (thus keeping US options for future ASAT capabilities open). That said, given the tenor of the discourse in the USA and the powerful concerns of the US military (particularly of the US Air Force), it would be safe to assume that the new administration is highly unlikely to do a 180° turn from the previous Bush and Clinton administration policies and support the PPWT. It is also highly unlikely that the USA would be willing to accept a treaty or an accord that prevented the use of so-called temporary and reversible means of disabling a satellite during wartime. Although one can make a strong argument that it would be in the self-interest of the USA—as the nation most dependent on space assets for its military operations—to set the normative bar against satellite interference as high as possible, the strong desire of the US Air Force to retain the option of satellite attack, supported by a large number of members of Congress, represents a domestic political roadblock. It remains a question whether the Obama administration will be willing to expend much of its political energy on achieving a multilateral instrument on space security, when the White House faces a number of other domestic battles of higher priority (such as healthcare reform, economic stimulus).

What is clear is that when the CD restarts again in January 2010, there will be every effort to relaunch the program of work approved this May, approve an implementation plan, and launch the PAROS discussions. Whether Pakistan will be ready to deal on the nuclear issues at that moment, however, is not clear. Nor is it clear as to whether, at that time, they might be joined by other nations who had initial qualms about the May consensus. At the moment, all bets are off.

21.2 ACTIVITIES IN COPUOS

Although the CD has essentially been moribund for the past decade or more, work to frame normative and legal concepts regarding the use of space has trundled along in the 67-member COPUOS over that time. While progress has been achingly slow—and on some issues, such as agreeing to a legal definition of "outer space," nonexistent—there nonetheless has been progress on a number of issues. COPUOS was established by the UN General Assembly in 1959 to encourage international cooperation in the peaceful uses of outer space, as well as to study legal problems associated with the exploitation of the space domain. Its work falls under two subcommittees: the Science and Technical Subcommittee and the Legal Subcommittee.

The most recent success story is the establishment of voluntary guidelines to mitigate the creation of space debris, which were endorsed by the UN General Assembly in December 2007 [16]. The guidelines were the result of 7 years of technical study, led by the Inter-Agency Debris Coordinating Committee (IADC), which comprises the civil space agencies of China, France, Germany, India, Italy, Japan, Russia, Ukraine, and the USA, plus the European Space Agency. While the technical recommendations of the IADC provided to COPUOS in 2002 were watered down during deliberations by the Science and Technical Subcommittee and the full COPUOS in 2007 [17], the guidelines eventually adopted by the General Assembly in December 2007 are nonetheless significant. In particular, Article 4, which pledges nations not to create long-lived debris, is of importance due to the fact that it does serve as a political constraint on further ASAT tests of the type conducted by China. At the same time, the guidelines' voluntary nature and the "escape clauses" inherent in the language—for example, for reasons of national security—raise questions about whether the guidelines will be fully implemented by nations.

The success of the debris guidelines effort led to a push, begun in mid-2007, for a wider approach by COPUOS to issues surrounding safety of operations in space through the creation of "rules of the road" or "best practices". Gérard Brachet, COPUOS chairman at the time, put forward in June 2007 a working paper recommending that the committee take up deliberations on the "long-term sustainability of space operations" using a similar "bottom-up" approach to that of the debris guidelines development process [18]. The government of France, in early 2008, proposed the creation of an informal working group of spacefaring nations and commercial operators on the subject—and signaled to COPUOS that it intended to make an initiative to place the idea on the committee's formal agenda. The group, comprising 20 nations and three multinational commercial operators (Intelsat, Inmarsat, and SES Global), met in the margins of the 11–20 June 2008 COPUOS meeting in Vienna, and again on the sidelines of the 59th International Astronautical Congress in Scotland held from 29 September to 3 October 2008. With additional inputs from a number of international organizations with responsibilities in the space arena—including the International Telecommunications Union, the World Meteorological Organization, the International Space Environmental Service, and the International Association for the Advancement of Space Safety—the group is working on a report addressing a number of elements that should be addressed in any space sustainability regime. These include:

- Space debris mitigation and remediation
- Improving the safety of space operations
- Managing the electromagnetic spectrum
- The impact of space weather and other natural causes of interference
- Review of existing international mechanisms to improve the safety and sustainability of space activities [19].

France successfully proposed COPUOS uptake of the sustainability issue at the 52nd Session, held 3–12 June 2009, with the result that the matter will be put on the agenda of the Science and Technical Subcommittee in 2010. The subcommittee's initial deliberations will be based on the informal working group report, which is currently expected to be completed by the end of 2009.

While the COPUOS effort specifically avoids the issue of space weapons, it does increasingly have convergence with the work of the CD to consider confidence-building measures. Obviously, any new regime—formal or informal—to ensure safe and secure space operations will require nations and operators to share more data about their orbital assets. Space object data-sharing is also a critical component in building confidence in the security arena. Further, any "rules of the road" agreed to by nations are by their very nature going to have to be applied to the operating of military, as well as civil and dual-use, satellites. One simply cannot imagine that agreed safety procedures, governing satellite maneuvers for example, would be applied only to satellites with no military applications—this would make little practical sense and would negate the purpose of the exercise in the first place.

21.3 ACTIVITIES IN THE GENERAL ASSEMBLY

The other traditional arena for UN activity regarding space has been the General Assembly (GA) in New York, which began considering space issues at the dawn of the space age—issuing its first resolution on the peaceful uses of outer space in 1958. Questions related to PAROS are the purview of the First Committee, whereas questions related to peaceful uses fall under the Fourth Committee. While GA resolutions are not binding, they do serve to set international norms, highlight areas of concern in the international community, and lay open policy divisions among nations. The more successful resolutions can lay, and often have done, foundations for future UN action or action in other multilateral forums.

The first GA resolution supporting negotiations of a treaty on PAROS was passed in 1981, and similar resolutions have been passed nearly every year since [20]. Until 2004, these resolutions were passed without any "no" votes—although several nations, including the USA, abstained. In 2005, in a change of policy implemented by the administration of President George W. Bush, the USA began to cast a nay vote (sometimes supported by Israel). The most recent resolution, adopted in December 2008, was supported by 177 nations, with the USA the lone "no" vote and Israel abstaining [21].

Aside from PAROS, Russia in 2005 began spearheading the introduction of a resolution seeking to establish a set of transparency and confidence-building measures for space. The December 2005 resolution called upon Member States to provide the UN Secretary General with their views on the advisability of further developing international outer space transparency and confidence-building measures in the interest of maintaining international peace and security and promoting international cooperation and the prevention of an arms race in outer space [22]. The 2006 resolution asked for "concrete proposals" delineating potential measures, as did the 2007 and 2008 versions. All four of the resolutions implicitly linked TCBMs to future negotiations on PAROS. Traditionally, the USA alone has voted against the resolutions, with Israel abstaining. The debate in 2008, however, introduced a more nuanced US position. According to Bush administration officials at the time, Washington initiated discussions with Moscow to try to negotiate away the linkage between TCBMs and PAROS in order for the USA to reverse its position. That effort failed. Indeed, the 2008 resolution arguably tightens the link by directly mentioning the Russian–Chinese PPWT [23].

It is certain that both the PAROS resolution and the Russian TCBM resolution will be re-tabled during the First Committee meeting in October 2008. It will bear watching as to whether the compromise language reached in the CD regarding "discussions" rather than "negotiations" of PAROS will be transmitted into the language of either resolution (such a change is more likely regarding the TCBM resolution) and whether that would allow the Obama administration to change the US voting pattern.

CONCLUSIONS

The burst of more substantive activity in United Nations forums in recent years regarding the future security of outer space can be attributed to a number of factors. Most fundamentally, the increasing number of space "players" has raised awareness around the world of the importance of the space realm to human security and development. Some 50 nations now own and/or operate spacecraft, with about 900 active satellites in orbit. Satellites provide vital services for nation states: from telecommunications and internet connections to enabling banking transactions to telemedicine and tele-education to support of military operations. The growing use of space has, in turn, resulted in an increase in concerns about potential satellite interference, orbital crowding and, most importantly, the threat of an ever-more polluted space environment to the continued functioning of satellites and spacecraft. The improved access to more sophisticated space technologies has also resulted in increasing application of space assets for military purposes, raising tensions among major spacefaring nations (most importantly Russia, China, and the USA). This problem was most recently highlighted by the January 2007 test of an anti-satellite weapon by China, and the February 2008 move by the USA to "shoot down" an ailing spy satellite using a missile-defense-equipped Aegis cruiser—a decision Washington justified on the grounds of public safety—but that was widely seen as a *de facto* ASAT test.

While, as detailed above, few of these efforts have resulted in substantial steps forward in protecting future space security—with the exception of the GA adoption of the COPUOS voluntary guidelines on space debris mitigation—the trajectory of activity is in the right direction. In particular, hope can be taken from the fledging convergence of thinking between the COPUOS and the CD on basic measures to improve cooperative management of space operations and build confidence among states regarding each other's activities. Perhaps what is most critical in the near future is to maintain and bolster the sensibilities of nations to the fact that space is truly a global commons, an environment that can either be responsibly exploited or, instead, abused and possibly destroyed. In no other domain of human activity is it so important to remember that the actions of any one nation or commercial entity will have effects, whether negative or positive, on all others. In other words, there will be no safety and security in space in the absence of collective action. Given the importance of space operations to the well-being of humankind, it would be catastrophic for the world if space is allowed to become an arena dominated by national competition and even warfare, rather than a sphere of international cooperation.

Notes and references

[1] According to the private satellite monitoring organization, Celestrak, the collision created more than 1000 pieces of debris bigger than 10 cm in diameter and thus highly dangerous to spacecraft. See Kelso, T. S. (2009). Iridium 33/Cosmos 2251 Collision. *Celestrak*, updated 13 May, http://celestrak.com/events/collision.asp

[2] Decision for the establishment of a Programme of Work for the 2009 session, CD/1864, 29 May 2009, Conference on Disarmament, Geneva.

[3] The formal report is not yet available.

[4] Brachet, G. (2009). How Does the Set of Best Practices Interact with the EU Proposed Code of Conduct. IFRI-SWF Workshop, 18–19 June, Paris, http://www.ifri.org/files/Espace/GBrachet.pdf

[5] See UNIDIR (2009). "Space Security 2009: Moving Toward a Safer Space Environment", 15–16 June, http://www.unidir.org/bdd/fiche-activite.php?ref_activite=455

[6] Acheson, R. (2009). More Delays and Rotting Fruit. *CD Report*, 10 August, Reaching Critical Will, http://www.reachingcriticalwill.org/political/cd/speeches09/reports.html#10august
[7] Found at http://www.reachingcriticalwill.org/political/cd/speeches02/chiruswp_062703cd.html
[8] See statement of House, K. E. (2008). United States Public Delegate to the 62nd Session of the United Nations General Assembly, delivered in the Debate on Outer Space (Disarmament Aspects) of the General Assembly's First Committee, 20 October, *Arms Control Update*, US Delegation to the Conference on Disarmament, Geneva, http://geneva.usmission.gov/CD/updates/1020OuterSpace.html
[9] Found at http://www.reachingcriticalwill.org/political/cd/papers08/1session/Feb12%20Draft%20PPWT.pdf
[10] While the USA and USSR tested ASATs during the Cold War, the last Soviet test was in 1985 (involving a missile interceptor known as the Co-orbital ASAT that exploded a conventional payload near the target satellite) and the last declared US test was in 1985 (involving a kinetic-energy missile launched from an F-15 aircraft). However, many nations and outside experts consider the February 2008 'shoot-down' of an ailing US satellite (*USA 193*) by an Aegis cruiser modified for missile defense purposes to comprise a *de facto* ASAT test. See Grego, L. "A History of Anti-satellite (ASAT) Programs", Union of Concerned Scientists website, http://www.ucsusa.org/nuclear_weapons_and_global_security/space_weapons/technical_issues/a-history-of-anti-satellite.html
[11] For an analysis of the test's debris legacy, see Kelso, T. S. Chinese ASAT Test. *Celestrak*, http://celestrack.com/events/asat/asp
[12] See http://www.whitehouse.gov/issues/defense
[13] Found at http://reachingcriticalwill.org/political/cd/speeches09/
[14] In partnership with the Secure World Foundation, and with the support of the governments of Canada, China, and Russia and the Simons Foundation.
[15] Baines, P. J. and Côté, A. (2009). Promising Confidence- and Security-Building Measures for Space Security. Presentation to *Space Security 2009: Moving Toward A Safer Space Environment*. United Nations Institute for Disarmament Research (UNIDIR), http://www.unidir.org/pdf/activites/pdf2-act455.pdf
[16] United Nations General Assembly, Resolution A/Res/62/217, 10 January 2008, http://www.oosa.unvienna.org/pdf/gares/ARES_62_217E.pdf
[17] The Science and Technical Subcommittee adopted a revised set of guidelines based on the IADC recommendations in February 2007. See UN Office for Outer Space Affairs, "Report of the Scientific and Technical Subcommittee on its Forty-Forth Session, Held in Vienna from 12 to 23 February 2007", A/AC.105/890, 6 March 2007, http://www.oosa.unvienna.org/pdf/reports/ac/105/AC105_890E.pdf; COPUOS adopted those same recommendations in June 2007, see United Nations General Assembly, Report of the Committee on the Peaceful Uses of Outer Space, a/62/20, supplement no. 20, 26 June 2007, http://oosa.unvienna.org/pdf/gadocs/A_61_20E.pdf
[18] Brachet, op. cit.
[19] *Ibid.*
[20] See http://www.oosa.unvienna.org/pdf/gares/ARES_36_97E.pdf
[21] Blount, P. J. (2008). UN Passes Resolutions on PAROS and TCBMs. *Res Communis*, The University of Mississippi School of Law, 3 December, http://rescommunis.wordpress.com/2008/12/03/un-passes-resolutions-on-paros-and-tcbms/
[22] United Nations General Assembly (2006). "Resolution adopted by the General Assembly: Transparency and Confidence-Building Measures in Outer Space Activities," Sixtieth session, Agenda item No. 7, A/RES/60/66, 6 January, http://www.oosa.unvienna.org/pdf/gares/ARES_60_066E.pdf
[23] Blount, P. J. (2008). Draft Resolution on Transparency and Confidence-Building Measures in Outer Space Activities Approved the First Committee. *Res Communis*, The University of Mississippi School of Law, 2 November, http://rescommunis.wordpress.com/2008/11/page/8/

CHAPTER

Toward an International Space Station Safety Authority

22

Tommaso Sgobba

European Space Agency, Directorate of Manned Spaceflight and Microgravity, Noordwijk

CHAPTER OUTLINE

INTRODUCTION

At the 40th IAF Congress in China 13 years ago, the incumbent NASA Associated Administrator for Safety and Mission Assurance, George Rodney, concluded his paper on Space Station safety with the following remarks: "*NASA single-point oversight of SRM&QA is thus important to the shared safety of all SSF crew members. Over the long run the safety of all human beings in the global commons of space is a responsibility that must be shared by all spacefaring powers*". This is a remarkable statement if we consider that it came at the time of the Space Station Freedom (SSF) program in which the USA had no match among the partners, in terms of manned space-flight experience, investment, and transportation capabilities.

The involvement of Russia in the space station program brought a new perspective to the NASA–partners relationship, due to the Russians' outstanding historical achievements, current technical capabilities, and essential contributions to the new International Space Station (ISS) configuration. During the first part of the cooperation, the Shuttle–Mir program, also called ISS phase I,

Space Safety Regulations and Standards. DOI: 10.1016/B978-1-85617-752-8.10022-4

the safety responsibilities were basically kept separate, with NASA having only limited insight into the Mir systems. Safety interface issues were resolved through a co-chaired coordination group called the Joint American–Russian Safety Working Group (JARSWG). The overall approach received harsh criticism following the series of problems on board Mir in 1997, in particular a fire first (a European Space Agency astronaut was on board) and a collision later of a Progress resupply vehicle with the Spektr module. The US Congress became so concerned about the safety of American astronauts operating on board Mir such as to state, as part of the appropriation bill for the years 1998/1999, that:

> *NASA shall not place another United States astronaut on board the Mir space station, without the space shuttle attached to Mir, until the administration certifies to Congress that the Mir space station meets or exceeds United States safety standards. Such certification shall be based on an independent review of the safety of the Mir space station.*

Is time mature for the establishment of an international safety authority for the ISS? Why would such an authority be more suitable than the current arrangements? This chapter attempts to answer these questions by demonstrating that the international structure of the program, the upcoming transition to the operational phase, and the promotion of commercialization, all require adjustments and rationalization of the current ISS safety. As a matter of fact, some modifications are already taking place, not as part of a grand strategy but to solve practical problems.

22.1 OVERVIEW OF LEGAL ISSUES RELATED TO ISS SAFETY

22.1.1 United Nations Space Treaties

The ISS partners are among the nations that had signed and ratified the following four UN international space treaties and agreements:

- Treaty on Principles Governing the Activities of States in the exploration and Use of Outer Space, including the Moon and Other Celestial Bodies (the Outer Space Treaty, 1967)
- Agreement on the Rescue of Astronauts, the Return of Astronauts and the Return of Objects Launched into Outer Space (the Astronaut Treaty, 1968)
- Convention on International Liability for Damage Caused by Space Objects (the Liability Convention, 1972)
- Convention on Registration of Objects Launched into Outer Space (the Registration Convention, 1976).

It is useful to examine some fundamental principles that have relevance to safety responsibilities and roles, in particular with reference to private or commercial activities in space. During the drafting of the Outer Space Treaty, there was some initial disagreement regarding the legal status of private sector space activities. The USA wanted to leave the door open to private sector involvement in future space exploitation. The USSR opposed this idea, up to the point that the draft they proposed included the following statement: "*All activities of any kind pertaining to the exploration of outer space shall be carried out solely and exclusively by the State …*". The USA then proposed a compromise solution, accepted by the Russians, according to which each country should bear the responsibility for the activities of its nationals in space. The compromise was incorporated in article VI of the Outer Space Treaty as follows: "*States … shall bear international responsibility for national activities in outer space ... whether such activities are carried on by government agencies or by non-governmental entities, and for assuring that national activities are carried out in conformity with ... [this] Treaty. The activities of non-governmental entities in outer space ... shall require authorization and continuing supervision by the appropriate State party to the Treaty*".

Different from both maritime and air law, the UN space treaties make countries both "responsible" and "liable" for the space activities of their nationals (persons, companies, etc.). States exercise a supervisory role (responsibility) with respect to ships and planes owned by the private sector but do not accept the financial risk (liability) for the actions of these assets. In contrast, in space, governments have both supervisory and financial responsibility.

According to some interpretations, Article VI would prohibit strictly private, unregulated activity in space. The terms "authorization" and "continuing supervision", therefore, appear to require a certain minimum of licensing and enforced adherence to government-imposed regulations. Furthermore, in the case of activities that could "cause potentially harmful interference with activities of other States", a State, under article IX of the Outer Space Treaty, must "undertake appropriate international consultation before proceeding with any such activity".

Finally, the Registration Convention establishes the principle that "A State ... on whose registry an object launched into space is carried shall retain jurisdiction and control over such object and over any personnel thereof, while in outer space ...". For elements of the ISS developed by space agencies, the State jurisdiction will be formalized by registration of each element on the relevant national registry and notification to the UN Secretary General, in accordance with the Registration Convention.

In the case of a privately owned element of the ISS, the owner may decide to register the element on the registry of any ISS Partner State (similarly to the practice for ships). This will affect the legal status of the element (e.g. applicable intellectual property law) but will not modify the responsibility and liability of the State of which the element's owner (person or company) is a national. The rule in aviation and the latest trend in shipping (to fight terrorism) are to ensure a genuine link between "ship and flag".

Application of the Outer Space Treaty to the case, for example, of an ISS commercial payload, developed by a German company to be flown on the Russian segment, appears to require the primary involvement of the European Space Agency (ESA) in the safety certification process. The ESA is the representative space agency (also) of the German government in the ISS program. In accordance with the Outer Space Treaty, the ESA's safety certification would need some form of concurrence by the Russians, and most probably by the other ISS partners' agencies too.

22.1.2 ISS agreements

Legally, the ISS is not a single national spaceship, although the USA has a leading role, and it is also not truly an international one. The ISS partners had four possible legal options: (1) *a national space station*, under the jurisdiction and control of one country; (2) *a multinational space station,* under the joint jurisdiction and control of several nations; (3) *a multinational space station*, the individual modules of which are under the jurisdiction and control of separate nations; or (4) *an international space station,* under the jurisdiction and control of an international governmental organization similar to Intelsat. The third option was the one eventually selected for the ISS, thus recognizing the "fundamental objective" of each partner to retain "*responsibility for design, development, exploitation and evolution of ... identifiable elements of the space station together with the responsibility for their management ...*" (ESA/C-M/LXVII/Res.2, 31 January 1985).

The ISS Inter-Governmental Agreement (IGA) states, in fact, in Article 5 that: "each partner shall retain jurisdiction and control over the elements it registers ... and over personnel in or on the space station who are its nationals". A major exception to the multinational principle is the area of safety. Safety is defined in the ISS IGA and Memorandums of Understanding (MoUs) as a shared responsibility that is discharged by each partner under the lead role and overall certification responsibility of NASA. NASA has the management responsibility to establish, with the support of the partners, the overall safety requirements and plans, while each partner is responsible for "... management of their own programs, including their utilization activities; system engineering ... development and implementation of detailed safety requirements and plans ..." (Article 7). The ISS MoUs further establish responsibilities for safety certification. Each partner has the responsibility to review and certify the safety compliance of the elements and payloads it provides, while NASA has the overall responsibility to review and certify the space station as a whole, as well as its own elements and payloads.

Unfortunately, such approach does not apply to all ISS elements. The transportation vehicles not owned by NASA (i.e. Soyuz and Progress vehicles of Russia, Autonomous Transfer Vehicle (of ESA) and H-1 Transfer Vehicle of Japan) are defined as elements of the International Space Station (ISS) in the International Governmental Agreement (IGA), but they are not and cannot be legally under NASA's overall safety responsibility. They are under the complete and sole safety responsibility of the "Launch State", as long as they are not in proximity of the ISS or docked to the ISS. Such lack of uniformity and single overall safety responsibility complicates the flight certification process, in particular of ISS payloads, as we will further discuss later on.

The ISS commander issue demonstrates the fact that, to ensure system safety, special legal instruments are sometimes required to overcome the limitations of the ISS's multinational nature. As we have seen, the ISS is legally an "assembly of ships" that navigate under several flags, yet a single commander's authority and responsibility to enforce safety procedures and crew rescue procedures is mandatory. To give a single ISS commander the authority over an international crew, in September 2000 the ISS partners formally approved the "Code of Conduct" for the International Space Station Crews, which included such role definition. The Code has been legalized in parallel by all Partner States to overcome the jurisdiction/control conflict due to the multinational legal nature of the Station. The commander's authority is defined as strictly limited to safety and security and does not extend, for example, to the disposal of payloads and elements. Finally, it is interesting to note that the ISS IGA includes a cross-waiver of liability that extends to any "users and customers" of a Partner State.

22.1.3 Lack of specific safety rules for ISS commercial activities

What are ISS commercial activities? Here, we intend neither to assess progress and timeline of ISS commercialization activities nor to present their status, but only to identify the general classification of such activities for the purpose of identifying the kind of safety regulations that should be in place.

The end of the Cold War opened the door to changes in US space policy toward space commercialization and wider international cooperation. In 1988, President Reagan approved a revised national space policy that, among other goals and principles, encouraged private sector investment and the commercial use and exploitation of space technologies. 1988 saw also the amendment of the Commercial Space Launch of 1984, setting the stage for the current Federal Aviation Administration (FAA) involvement as the regulatory body for licensing and safety certification of US commercial launchers/spaceports.

With reference specifically to the ISS, the US International Space Station Authorization Act of 1995 included in section 6 the following policy statement about commercialization of the space station: "*The Congress declares that a priority goal of constructing the International Space Station is the economic development of Earth orbital space. The Congress further declares that the use of free market principles in operating, allocating the use of, and adding capabilities to the space station, and the resulting fullest possible engagement of commercial providers and participation of commercial users, will reduce space station operational costs for all partners and the Federal Government's share of the United States burden to fund operations.*"

Although not much is known about formal policy statements from the Russian side on commercialization, the facts speak loud and clear. According to the press, in recent years Russia's manned-space program has been financed by almost 50% through selling orbital access to other governments' astronauts and payloads, and, most recently, to ISS space tourists. Russian designers have also been involved in the conceptual design of a possible privately owned module called "Enterprise" to be attached to the space station. Initially conceived as an orbiting broadcast studio and private research park, it was later re-proposed as sleeping quarters to be rented out to NASA and the ESA.

The ESA has launched a wide initiative aimed at promoting specific commercial opportunities to companies interested in research, development, and technology. It will also pave the way for innovative space station-based activities in entertainment, advertising, and sponsorship by businesses that would not normally think about space. At the same time, the ESA is considering streamlining its direct management involvement in the day-to-day running of the station by contracting much of the operational-level responsibility to an industry consortium.

Of all aspects of ISS commercialization, space tourism has attracted the most public attention and some harsh criticism of perceived weaknesses in the international agreements governing the project. At the time of the Tito debate, a bill was even introduced in the US Congress to ban tourists from visiting US parts of the space station or traveling to the ISS on US government launch vehicles. At the same time, the bill included provision for $2 billion in federal loan guarantees for firms developing space tourist habitation and transportation systems and to assign space tourism regulatory responsibility to the US FAA.

The Tito episode was the catalyst for the preparation of a set of rules, approved by all ISS partners at the end of 2001, which will govern the selection, training, and flight certification of "commercial crew members". For the rest, it can be said that, although the policies of the partners' governments encourage ISS commercialization, and the ISS IGA does not include any article that conflicts with commercial usage of the space station, all lower-level implementation rules have been written almost exclusively with government operations in mind.

In view of defining later tasks and function of the proposed Safety Authority, we can identify the following typologies of ISS commercial or private activities:

1. Partner providing a commercial service—(a) commercial payloads transportation/operation; or (b) transportation/accommodation of paying passengers (so-called space tourists).
2. Corporate operating on behalf of a partner—operation of systems or equipment owned by a partner government.
3. Corporate providing a commercial service—transportation/operation of payloads (institutional and commercial) by means of privately owned systems, equipment, and personnel (civil astronaut).

It should be noted that not only privately owned payloads should be understood here to be "commercial", but also those owned by a Partner State government organization that has not been developed under the provisions of the relevant MoU. This would be the case, for example, for a European payload from a national space agency that does not fly on the ISS under ESA sponsorship.

Currently, there are no specific safety rules for commercial activities on the ISS except the guidelines issued regarding space tourism. Specific rules are needed to fulfill three different needs. One, the most obvious, is to prevent an increase of the safety risk for the ISS and its crew. The second is that of maintaining a climate of trust and fair competition among ISS partners. The third is the need to promote the commercial use of the ISS by making related processes clear and by providing legal certainties to the users. In fact, a successful promotion of the space station will eventually very much depend on removing uncertainties concerning the application of certain laws such as product liability, export law, and intellectual property. Whether a company decides to conduct space research or to market a space product will depend in part on the administrative complexity and cost of the overall process, a significant part of which is compliance with safety regulations. In order to assess these variables, a company must know which State's law and safety regulations would apply, and what the likely outcome of a controversy would be.

The activities under point 1 above have the potential of creating some (legitimate) competition among the partners. In this respect, it should be noted that clear safety standards and certification rules, as well as their uniform interpretation and rigorous application, are fundamental to any free movement of goods across international borders. To prevent unfair practices and to ensure safety, mandatory and uniform hardware certification processes should be agreed upon by the partners. Here, the focus is on the partners' and practices for design certification, product conformance certification, user quality assurance system certification, and personnel certification. Furthermore, commercial payload developers that are not space companies may require substantial support from a partner in designing and testing their hardware to comply with safety requirements. To ensure independence of the control function, the partner organization and personnel in charge of safety reviews and certification should be different from those involved in such detailed engineering support.

With reference to point 2, it would be important to identify rules for delegating institutional functions and tasks to a corporate entity. In particular, it is necessary to identify those functions and tasks that, under no circumstances, should be transferred to a corporate entity, and those that may be delegated to qualified individuals of a corporate organization under their personal legal accountability. The most complex scenario is the one under point 3, in which privately owned and operated elements would be docked to the ISS or operated in close proximity. This scenario would require an extensive rewriting and adaptation of all current rules and standards relating to ISS safety.

22.2 CURRENT REQUIREMENTS, ORGANIZATION, AND PROCESSES

22.2.1 Shuttle payloads safety process as model

During the course of the Shuttle program, two different process models were established to manage the safety risk. One, for the vehicle and related systems, made use of risk management techniques, and the other, for payloads, was based on a risk avoidance approach. The two safety organizations were also much different, the first being enmeshed in the shuttle's development, while the second remained completely independent of payload developers.

The shuttle systems safety process model and related organization came under severe scrutiny during the *Challenger* disaster investigation. One criticism was that the assessment of safety risks, identified through reliability analyses (Failure Modes and Effects Analysis (FMEA)/Critical Items List (CIL)) and hazard analyses for management approval, was subjective, with little in the way of formal and consistent criteria for approval or rejection. The second criticism was the lack of independence in risk evaluation and in the verification and certification of critical hardware and software.

The process applied to shuttle payloads and related safety reviews organization has worked well and remained unaltered for 20 years, except for some fine-tuning. The process is based on an independent review, by the Payload Safety Review Panel (PSRP), of the results of safety analyses (including FMEA) performed to demonstrate compliance with well-defined safety technical requirements. The relevant document, NSTS 1700.7b, has been supplemented over the years by a consistent collection of "interpretation letters", which captured lessons learnt and acceptable means of compliance.

Of the two shuttle safety processes, the one selected for ISS elements is basically that used for shuttle payloads, enhanced with additional features traceable to recommendations issued during the *Challenger* disaster investigations. In particular, quantitative risk management techniques are used in support of safety requirements validation and to assess the overall safety risk. The elements of the ISS technical safety requirements are baselined in the SSP 50021 document, which is largely traceable to NSTS 1700.7b. The board in charge of ISS elements safety reviews is called the Safety Review Panel (SRP) and operates very much along PSRP lines. The PSRP chairman also co-chairs the SRP.

22.2.2 The problem of multiple certifications

The ISS MoU identifies two safety certification levels for the ISS: the International Partner that owns the element/payload and NASA, as overall ISS integrator, therefore responsible for final certification. In addition, safety certification must be obtained from the "Launcher State" safety authority for both flight safety and ground processing at the launch range. Table 22.1 gives an example showing organizations and panels sharing final certification responsibility for a typical ATV mission. The Safety Review Panel and the Payload Safety Review Panel in Table 22.1 are the NASA-level boards responsible for certifying ISS elements and payloads respectively. The Safety and Mission Assurance

Table 22.1 ATV Final Safety Certification Responsibilities

Safety Organization	Ground Safety	Flight Safety		
		Launch Phase	Re-entry Phase	On-Orbit Phase
ESA-ATV Project			X	
CNES-CSG	X	X		
PSRP[1]				X
SRP/(RSA)				X
SMART[2]		X		

[1] *Transported payloads.*
[2] *Non-hazardous cargo items.*

Review Team (SMART) is a less formal review team belonging to the Safety and Mission Assurance organization, which has delegated responsibility for certifying non-hazardous cargo items.

It should be noted that, in addition to the above certifications, one of the ISS partners has required fulfillment of its own safety certification requirements and processes whenever an item owned by another partner is expected to cross the "border" of an element under its jurisdiction and control, even if only for stowage purposes. In this respect the basic rule that safety requirements of a partner must "meet or exceed" the overall ISS requirements may create further complications if the hosting partner applies more stringent requirements. It may happen, in fact, that an item fully compliant with ISS requirements, for example offgassing, is certified as safe for operations inside one module but forbidden to enter another one, although the common air (in which the offgassed products will end up) circulates through both modules.

There are various issues that arise from the above multiple certifications:

1. Proliferation of safety reviews and data submittal requirements
2. Lack of harmonization of safety technical requirements
3. Possible conflicting assessment of means to achieve compliance
4. Parallel safety authorities without possibility of arbitration.

NASA has avoided the above problems, for shuttle transported payloads, by issuing the ISS payload safety requirements as an extension of those already existing for shuttle payloads (ISS Addendum to NSTS 1700.7b), and by unifying the safety review process for the two phases (i.e. transportation and on-orbit operations) under a single safety panel, the PSRP, and in a single review cycle. Furthermore, the participation as PSRP member of the Chairman of the Kennedy Space Center Ground Safety Review Panel (GSRP) prevents a duplication of effort with reference to flight certification. This is a very cost-effective approach that should be generalized and extended to all ISS transportation vehicles.

22.2.3 Need for an integrated and international safety organization

There are two reasons why the ISS safety organiszation should become more integrated and international. One is to streamline the overall safety certification process to make it more efficient, and the second, also very important, is to extend certain independent control functions that exist only at NASA and are located outside the ISS project organization.

Let us start from the latter point. There are three safety groups at NASA that are charged with independent safety audit, review or assessment: the Aerospace Safety Advisory Panel (ASAP), the Space-flight Safety Panel, and the Independent Assessment Office. The NASA ASAP was established by the US Congress as part of the NASA Authorization Act of 1968 as an independent safety review body charged with advising the NASA Administrator and Congress on safety systems and operational systems. Its establishment was precipitated by the *Apollo 1* spacecraft fire in January 1967, in which three astronauts lost their lives during a ground test. Currently, some believe that the panel does not have adequate time or breadth of expertise to thoroughly study the agenda it undertakes, and that the panel's value would be enhanced through refinement of its focus from detailed engineering questions to safety management and process issues.

The NASA Space Flight Safety Panel is chaired by a member of the Astronaut Corps and comprises six members representing Centers Directors and Mission Managers. The panel reports to the Associate

Administrator for Safety and Mission Assurance and is tasked with independently assessing the NASA Space Flight Safety Program. It also conducts reviews of selected issues or concerns. The ISS Independent Assessment (IA) Office reports directly to the Office of Safety and Mission Assurance at NASA Headquarters. Both offices (IA and S&MA Headquarters) were established following the recommendations of the "Report of the Presidential Commission on the Space Shuttle *Challenger* Accident". A primary stated objective of IA is to assist the ISS Program in risk management and mitigation. The IA Office includes an independent verification and validation (IV&V) group responsible for assessing mission-critical software. A representative of the Independent Assessment Team is a voting member of the ISS SRP.

The point here is that the ASAP, the Space Flight Safety Panel, and the IA Office have to limit *de facto* their independent surveillance activity to the US-managed part of the ISS program, due to the multinational definition of the ISS. As a consequence, a gap exists with reference to the parts of the program managed by the International Partners, which, as far as the authors know, do not include any equivalent "check and balance" mechanism.

The other reason to modify the current organization is that the integration of NASA and International Partners safety review panels, in a single system, is a necessary step to streamline the safety certification process and avoid multiple certifications. In this respect, some good progress is being made by the PSRP. An agreement was signed in 1999 between NASA and the Russian Space Agency, granting wide autonomy in performing payload safety reviews of Russian-owned payloads operating on the Russian segment of the space station. A second agreement, between NASA and the ESA, is being concluded that will establish a sort of "franchise" PSRP in Europe. The scope of the European PSRP will include final safety certification of ESA payloads to be operated on US and ESA segments of the ISS, as well as their transportation on board the shuttle. Conversely, NASA will autonomously certify its own payloads for transportation on board ESA-ATVs. The ESA-PSRP will operate in close coordination with and under NASA-PSRP oversight, and will follow exactly the same review process as documented in NSTS/ISS 13830C. The ESA-PSRP membership will be established on the basis of delegation agreements directly negotiated between each ESA function and represented in the panel and its counterpart in the NASA PSRP. The ESA-PSRP operations will be periodically audited by NASA and the ESA. Finally, a multilateral agreement is also in preparation that will allow each partner to autonomously certify their non-hazardous cargo for transportation to the ISS on any other partner's vehicle.

22.3 EXTENDING THE CIVIL AVIATION MODEL TO SPACE

The expansion of commercial aviation has progressed at the same pace as its safety record. No cost improvements would have been generated and millions of people attracted to fly daily without the astonishing improvements in safety. Commercial aviation success and safety records are intimately connected.

Looking to civil aviation as a model for space safety is not a new idea; it dawned in the post-*Challenger* era and focused on two particular aspects: safety technical requirements standardization and separation of certification authority from project management. Studies in these areas were performed by the ESA in the frame of the Hermes project. At the conference on "Spacecraft Rendevous and Docking", held in July 1990 at NASA-JSC at the direction of the US Congress, the need was addressed for developing (in the long term) a set of international design and operational standards that "*... would have a similar role of that of the international civil aviation certification standards which permit civil aircraft designed and manufactured in one nation to operate within the airspace, and to land at the airports, of other nations ...*". Also, Japanese engineers have been active in studying civil aviation requirements and certification processes for the possible application to their space tourism system concepts. The civil aviation model is currently being extended in the USA to commercial launchers, but on a national basis.

22.3.1 The International Civil Aviation Organization

In 1910, when aviation was in its infancy, on the invitation of France, the first important conference on an international air law code was convened in Paris. The treatment of aviation matters was a subject at the Paris Peace Conference of 1919, and it was entrusted to a special Aeronautical Commission. Later, on the basis of a proposal by France to the other principal Allied powers, an International Air Convention was established that created the International Commission for Air Navigation (ICAN). In 1922, a small permanent secretariat was located in Paris to assist the Commission in its tasks of monitoring the developments in civil aviation and to propose measures to States to keep abreast of developments.

In consideration of the great advancement being made in the technical and operational possibilities of air transport during World War II, the USA initiated in 1943 studies of post-war civil aviation. The studies confirmed, once more, the belief that civil aviation had to be organized on an international scale, or it would not be possible to use it as one of the principal elements in the economic development of the world. The consequence of those studies and subsequent consultations between the Allies was that the US government extended an invitation to 55 States to attend, in November 1944, an International Civil Aviation Conference in Chicago. The Convention on International Civil Aviation (also known as Chicago Convention) was signed on 7 December 1944 by 52 States and came into being on 4 April 1947.

The 96 articles of the Chicago Convention establish the privileges and restrictions of all Contracting States and provide for the adoption of International Standards and Recommended Practices (SARPs, also known as convention annexes) to secure the highest possible degree of uniformity in regulations and standards, procedures, and organization regarding civil aviation matters. The design and airworthiness certification standards are stated in ICAO Annex 8 of the ICAO Convention and in the Airworthiness Technical Manual. The Chicago Convention established the permanent International Civil Aviation Organization (ICAO). In October 1947, the ICAO became a specialized agency of the United Nations. The work of the ICAO covers two major activities:

1. Establishment and maintenance of rules and regulations concerning:
   - training and licensing of air and ground personnel;
   - communication systems and procedures;
   - rules for air traffic control systems;
   - airworthiness requirements;
   - registration and identification of aircraft meteorology, maps and charts.
2. Practical application of air navigation services and facilities by States and their coordinated implementation.

The first session of the ICAO Assembly in 1947 established a permanent Legal Committee to provide advice on related matters. The work of the Legal Committee has led to the establishment of important international law instruments such as international recognition of rights in aircraft, strict liability of the aircraft carrier in respect of the international carriage of passenger and baggage, and State of Registration of an aircraft as competent over offences and acts committed on board.

The ICAO secretariat is headed by a Secretary General. The secretariat is made up of staff members selected for technical competence in their respective fields, which supplies technical and administrative aid to the governmental representatives who make up the ICAO Council, Committee, and Divisions. The representative bodies of the ICAO are: the *Assembly*, composed of all Contracting States; the *Council*, composed of 33 Contracting States elected by the assembly; and a number of *Committees* (Air Navigation, Air Transport, Legal, etc.).

The implementation of the ICAO rules and regulations is carried out by national safety authorities like the Federal Aviation Administration in the USA or the Joint Aviation Authorities (JAA) in Europe. It should be noted that only in 1997 did the ICAO launch the first fully fledged mandatory audit program of national authorities, superseding a previous assessment program, on a voluntary basis, started in 1995.

22.3.2 Federal Aviation Administration and commercial launchers

Most of the existing laws, regulations, and agreements governing space activities were written to make it easier for governments to function in space. There has been a growing awareness that it is time to make it easier for the private sector to undertake the development of commercial space activities, by reforming the existing legal/regulatory framework that is primarily rooted in Cold War period realities and frame of mind. The first example in this direction is the licensing of commercial launchers and spaceports in the USA.

Although the ICAO is not currently involved in space, the role of the US civil aviation regulatory body, the FAA, has been extended in recent times to also cover commercial launchers licensing and safety certification. The commercial Space Launch Act, 49 USC Subtitle IX (as amended in 1998), makes the US Department of Transportation responsible for licensing and regulating non-government launch activities conducted in the USA (or anywhere in the world if a US corporation controls the launch) and for re-entry of reusable launch vehicles. The act assigned to the FAA the responsibility of issuing safety approvals for launch vehicles, safety systems, processes, services, and personnel.

In 1999, the American Institute of Aeronautics and Astronautics (AIAA) sponsored the issue of the American National Standard "ANSI/AIAA S-061-1998 Commercial Launch Safety". This standard is a sort of performance-based version of the military standard EWR 127-1, issued in 1995 as the first common standard for range safety at the Western and Eastern Ranges (i.e. Vandenberg Air Force Base and Cape Canaveral Air Station). The safety rules developed by the FAA for commercial launches, published in August 2002, have drawn on past (military) experience but injected certain criteria of rigor and consistency that are typical of civil aviation (single safety responsibility, elimination of quantitative targets and waivers, accountability, etc.). An important consequence of this approach, of establishing the regulatory framework for commercial space activities as an extension of aviation, is in the field of responsibilities of both civil penalties and criminal law.

Act 49 USC subtitle IX provides for the Department of Transportation to impose civil penalties if a person is found to have violated a requirement of the Act, a regulation issued under the Act, or any term or condition of a license issued or transferred under the Act. All authority under the Act has been delegated to the Administrator of the FAA. In January 2001 the FAA issued detailed rules (14 CFR Parts 405 and 406) on civil penalty actions in commercial space transportation that have been modeled on three current aviation rules: 14 CFR 13.16, which the FAA uses to assess civil penalties in certain aviation cases; 14 CFR 13.19, which the FAA uses to suspend and revoke aviation certificates such as pilot and air carrier operating certificates; and 14 CFR 13.29, which provides for civil penalty procedures for certain security violations.

With reference to criminal law, in April 2000, US President Clinton signed into law the "Wendell H. Ford Aviation Investment and Reform Act for the 21st Century". Section 506 of the Act, titled "Prevention of Frauds Involving Aircraft or Space Vehicle Parts in Interstate or Foreign Commerce" substantially increased the criminal penalties for false statements related to quality of parts (up to a life sentence in case of accident). The law was meant to better fight the growing market of counterfeit aircraft parts, but the reference to space vehicle parts and its generic definition may lead to wider interpretations, and application to practically any case of false statement of conformity related to space commercialization.

22.4 SETTING UP THE ISS SAFETY AUTHORITY

It is proposed that the MoUs between NASA and each partner should be supplemented by a multi-lateral safety agreement establishing: (a) basic principles regarding system safety in a mixed government/private environment, and (b) tasks and organization of the ISS Safety Authority.

The Safety Authority would be organized in two main layers. The first would be located at the safety organization "headquarters", perhaps in Europe, and staffed by top-notch engineers and legal experts, seconded from ISS partners' agencies in proportion to their share in the program. The second layer would be made of "local safety organizations", each co-located with the ISS organization of each partner and composed of personnel from their mission assurance, engineering and other functions, but independent of design, manufacturing, and operations functions.

Apart from the international composition envisioned here for the ISS Safety Authority, its functions and objectives would not be far from those of the so-called System Safety Engineering (SEE) described by Recommendation N.11 of the "Committee on Shuttle Criticality Review and Hazard Analysis Audit". In particular the "HQ organization" would be responsible for:

1. Design and certification standards
2. Rules for operations and mission control centers
3. Rules for training and licensing of space and ground personnel
4. Regulation and oversight of "local safety organizations"
5. Problem and mishap investigations
6. Probabilistic risk assessment
7. Approval of safety non-compliances and waivers.

The "local safety organization" would be instead responsible for:

1. Performance of safety reviews
2. Hardware, software, and facilities acceptance
3. Licensing of space and ground personnel

4. Problem and mishap reporting
5. Companies' certification and surveillance.

The NASA-nominated chairman of the ISS Safety Authority would report directly to the highest multilateral board in the ISS program. The ISS Safety Authority would also support the ISS program management in the case of higher-level inquiries.

CONCLUSIONS

Commercial activities involving safety risk usually require licensing and enforced adherence to government-imposed regulations. The International Space Station will probably evolve over time into a hybrid government/private system. Current rules and processes, written mainly with a multinational government type of operation in mind, need further standardization to properly address both cases. New rules may need to be added for certification of commercial products, services, personnel and facilities, which were never considered in previous space projects. The case of space tourism is an example. Rules would also be needed to prevent possible compromise of the overall independence, consistency, and effectiveness of the safety process by the delegation of some safety-related responsibilities to industry. It is also necessary to ensure that rules are applied evenly by all parties to avoid possible distortion of the commercial competition.

ISS commercialization and the international structure of the program would very much benefit, both in efficiency and effectiveness, by the establishment of a single independent Safety Authority to which ISS partners' space agencies participate and are represented, in respect of the roles and responsibilities assigned by the IGA and MoUs. The ISS Safety Authority should be empowered with effective means to achieve its mission. The methods and procedures of such an authority could be similar to those of commercial aviation airworthiness authorities, which have worked successfully since the founding Chicago Convention in 1944.

Bibliography

Bulletin Technique du Bureau Veritas No. 1 (1991). "The Certification of Manned Spaceflight." Jean-Louis de Montlivault, Robert Record.

Committee on Shuttle Criticality Review and Hazard Analysis Audit (1988). Post-Challenger Evaluation of Space Shuttle Risk Assessment and Management. National Academy Press, January.

ESA Internal Memo QS/90/213/60 to ESTEC Director, "International Spacecraft Rendezvous and Docking Conference—Working Group 4: Safety and Reliability", 9–12 July 1990.

ESA/PB-Ariane (98)49, Paris, (1998). "Ariane Safety: Responsibility and Roles", 3 June.

Farand, A. (2001). ESA Bulletin 105, "The Code of Conduct for International Space Station Crews", February.

GAO/T-NSIAD-00-128 (2000). "Space Station, Russian Compliance with Safety Requirements", 16 March, US GAO.

ISS Multilateral Crew Operations Panel (2001). Rev. A, "Principles Regarding Processes and Criteria for Selection, Assignment, Training and Certification of ISS (Expedition a Visiting) Crewmembers", November.

Jones, R. I. (1987). "Application of Airworthiness Requirements to Hermes Space Vehicle." College of Aeronautics, Cranfield, April.

NASA Office of Inspector General Report (1997). "NASA Aersospace Safety Advisory Panel Inspection", Case Number G-96-005, 11 March.

NASA Procedures and Guidelines (2001). NPG 1000.3—6.21, 1 March 2001, NASA Space Flight Safety Panel.

Report DNVID-51-000 (1989). ESTEC/Contract 7394/87/NL/IW "Study of Safety and Reliability of Terrestrial Systems", July.

Report by the Committee on Shuttle Criticality Review and Hazard Analysis Audit (1988). "Post-Challenger Evaluation of Space Shuttle Risk Assessment and Management". National Academy Press.

Rodney, G. A. (1989). The Space Station Freedom: International Cooperation and Innovation in Space Safety. 40th IAF Congress, Beijing, China, October, IAA-89-615.

Rogers, T. F. (1998). The Prospect for Space Tourism, 29 January. The Sophron Foundation and the Space Transportation Association.

US Congress (1986). Office of Technology Assessment, Space Station and the Law: "Selected Legal Issues—Background Paper", OTA-BP-ISC-41. US Government Printing Office, Washington, DC, August.

US Congress (1995). "International Space Station." Report 104-210, House of Representatives, July.

US Congress HR 1275 (1997). "Civilian Space authorisation Act, Report No. 105-65", 21 April.

US Congress HR 1000 (2000). "Wendell H. Ford Aviation Investment and reform Act for the 21st Century", signed into law 5 April.

US Department of Transportation—Federal Aviation Administration (2000). 14 CFR Parts 413, 415, and 417, 25 October, "Licensing and Safety Requirements for Launch; Notice of Proposed Rulemaking; Proposed Rule".

US Department of Transportation—Federal Aviation Administration (2001). 14 CFR Parts 405 and 406, 10 January, "Civil Penalty Actions in Commercial Space Transportation".

White House Office of Science and Technology Policy (1994). Fact Sheet, National Space Transportation Policy, August.

CHAPTER

The international challenges of regulation of commercial space flight

23

Joseph N. Pelton

Former Dean, International Space University

CHAPTER OUTLINE

INTRODUCTION

Although most of the commercial ventures embarked on the newly emerging space tourism business are primarily aimed at providing sub-orbital flights, others have more ambitious plans to provide longer-term access to low Earth orbit. Some are seeking to provide commercially booked flights to the International Space Station under contract with NASA and Roscosmos. Yet others, such as InterOrbital Systems (IOS) and Bigelow Aerospace, even plan to operate private space tourist facilities in low Earth orbit—and do so within a time-frame of another decade or less. Bigelow has already deployed an experimental inflatable Spacehab mission called Genesis I, with further experimental flights to follow soon.

These enterprises thus represent a wide range of technical approaches. They include single or multiple stages to orbit, vertical or horizontal take-off and landing systems, and solid, bipropellant and liquid-fueled propulsion systems. They also involve spaceport-, aircraft-, and balloon-launched spacecraft. Some have abort and escape systems and others do not. The safety features of these various launch systems, in short, are quite diverse and rocket shows and races add yet a new dimension. Among the US systems, some are currently Federal Aviation Administration Office of Commercial Space Transportation (FAA/AST) licensed and others are not. In light of this diversity of technical

Space Safety Regulations and Standards. DOI: 10.1016/B978-1-85617-752-8.10023-6

approaches being taken, the FAA/AST procedures and formal rulemaking that came into force as of February 2007 actually operate on a case-by-case basis.

Some ventures are backed by billion-dollar corporations with thousands of employees and others have only modest capital backing. In addition, there are dozens of spaceports being developed around the world in locations that include Australia, Canada, Europe, the Middle East, Russia, and the USA. Ocean-, aircraft-, and balloon-launched systems can, of course, operate from virtually anywhere. The safety standards and regulations that govern spaceports and launch and re-entry operations vary widely. Some spaceports have been governmental licensed and others not.

Efforts have been initiated within the United Nations, the UN Committee on the Peaceful Uses of Outer Space (COPUOS), and the International Civil Aviation Organization (ICAO) to regulate private space ventures and space tourism enterprises. The US government has enacted legislative and administrative controls. In particular, the USA, under Executive Order 12465 and Title 49 of the United States Code, Subtitle IX, Chapter 701, has authorized the FAA/AST to license such private space ventures and set health and safety standards for such initiatives. Currently, the FAA/AST mission is to "encourage, facilitate, and promote commercial space launches and re-entries" while also ensuring "public health and safety, the safety of property and protecting the national security and foreign policy interests of the United States". These objectives can, of course, be in conflict.

The Futron Corporation, the European Space Agency (ESA), and others who have tried to examine the anticipated market for sub-orbital and orbital space tourism, and to explore the extent to which market forces and critical "price points" will influence the development of the space tourism business, have discovered that these factors may be in conflict with safety standards and regulatory controls.

23.1 US Vs. EUROPEAN APPROACHES

In many ways, the current evolution of the space tourism business can be compared to the early days of aviation and the development of airplane technology. This was a time where a thousand flowers were allowed to bloom. A host of different approaches were undertaken to develop successful new airplane designs—some quite successful while others ended in failure. Escape systems involved everything from parachutes to flying close to the ground and at low speeds. Today, there is a great diversity of approach to developing a successful space tourism business, but escape systems must be much more sophisticated to return crew and passengers from outer space or from the extreme environment of the stratosphere. Safety and safety regulation is thus of the essence if this industry is to succeed. This is understood not only by regulatory bodies, such as the Federal Aviation Administration (FAA) in the USA, but also by entities such as the Private Spaceflight Federation. Table 23.1 provides a summary of the approaches currently being pursued by the many different entities striving to establish successful space tourism businesses.

This diversity of approach is largely a US phenomenon, which could be seen as being very much in the tradition of the wide-open "settling of the Wild West". There are literally dozens of enterprises working on space planes and other space vehicle concepts in the USA. Some of these are well-financed enterprises of established aerospace companies and others are true start-up ventures. As noted above, the US FAA is charged under federal law to achieve two potentially diverging responsibilities of "protecting public safety" while also "promoting the industry". In Europe, a more centralized approach is being taken to developing a new space tourism business.

Table 23.1 Inventory of Private Space Ventures

Company	Rocket-Launch Vehicle	Intended Markets	Capabilities	Launch Site
Advent Launch Services	Advent 1 stage (VTHL from ocean)	Sub-orbital. 300 kg to 100 km	Full-scale liquid engine tests	Ocean launch and landing
Aera Space Tours/Sprague Corp.	Altairis (VTHL)	Sub-orbital. Space tourism	2 stage, RP-1/LOX propulsion. 7 passengers to 100 km sub-orbital flights in 2007	US Air Force Cape Canaveral Launch Facility—5-year agreement
Alliant ATK	Pathfinder ALV X-1 (VTVL) http://www.astroexpo.com/news/newsdetail.asp?ID=27882&ListType=TopNews&StartDate=10/9/2006&EndDate=10/13/2006	Orbital. Launch to LEO of scientific packages. Upgradable to manned flight in time. ORS mission	Upgraded Alliant sounding rocket	Mid Atlantic Spaceport, Wallops Island
American Astronautics	Now renamed Sprague Corp. See Aera Space Tours http://www.lunar.org/docs/LUNARclips/v11/v11n1/xprize.shtml	Sub-orbital. Formed to seek X-Prize. Crew to LEO	Status unclear. Spirit of Liberty 1 stage to orbit vehicle	Unclear
Andrews	Gryphon Aerospace plane	Sub-orbital space tourism	6360 kg to 100 km and return. LOX/RP-1. Less than $1m/flight (in design)	N.A.
ARCA Space	Romanian project—now defunct http://www.lunar.org/docs/LUNARclips/v11/v11n1/xprize.shtml	Formed to seek X-Prize. Crew to LEO	Status unclear	Unclear
Armadillo Aerospace	Black Armadillo	Sub-orbital space flight	One stage. LOX/ethanol engine (limited capital investment). Vertical take-off and land (like Delta Clipper design)	White Sands, New Mexico
Benson Space Company	See Space Dev and Dreamchaser Benson Space Company will market Dreamchaser vehicle. http://www.spacedev.com/newsite/templates/subpage_article.php?pid=583	Marketing company for Dreamchaser	N.A.	N.A.
Blue Origin (backed by Jeff Bezos)	New Shepard (VTVL) http://www.blueorigin.com/index.html	Sub-orbital. Space tourism to 100 km	Reusable launch vehicle. Hydrogen peroxide and kerosene fuel. Abort system	Culberson County, Texas. HQs in Seattle, Washington
Blue Ridge Nebula	Small family enterprise	Sub-orbital. Space tourism to 100 km	Status unclear	Unclear
Bristol Space Planes Ltd.	Ascender (Subscale flight models) Space bus (concept only) 50 persons or 110 tons Space cab (concept only) 8 persons or 2 + 750 kg	Sub-orbital. Three people or 400 kg on space tourism flight	Jet. Two turbofans to 8 km. RL-10 liquid rocket engine to 100 km	UK and USA

(Continued)

Table 23.1 Inventory of Private Space Ventures *Continued*

Company	Rocket-Launch Vehicle	Intended Markets	Capabilities	Launch Site
C & Space (of Rep. of Korea) R and AirBoss Aerospace Inc. (AAI)	Proteus space plane (VTHL) http://www.hobbyspace.com/nucleus/index.php?itemid=207	Sub-orbital. Space tourism. Three crew members	LOX/methane engines. ITAR approval pending	To be decided
DaVinci Program	DaVinci (Balloon launch and vertical landing) http://www.davinciproject.com/	Sub-orbital. Space tourism. Three crew members	Balloon to 40,000 ft. Twin LOX/kerosene engines to 120,000 ft. Parachute landing	Can be launched from any balloon launch site
DTI Associates	Terrier-Orion (Terrier is surplus Navy missile motor and Orion is surplus Army missile motor)	Sub-orbital. Cargo to LEO (290 kg to 190 km)	Motors and vehicle FAA/AST licensed	Woomera, Australia
EADS Space Transportation	Phoenix space plane (VTHL) and Pre-X	Sub-orbital	Effort sponsored by CNES and ESA with development by EADS Space Transportation Corp.	Toulouse, France
Energia Rocket and Space Corporation	Clipper (VTOL) http://www.astroexpo.com/news/newsdetail.asp?ID=25688&ListType=TopNews&StartDate=5/15/2006&EndDate=5/19/2006	Orbital	In conjunction with Soyuz and Agara launch vehicles	Baikonor and Russian Northern Cosmodrome
HARC Space	Balloon launch reusable vehicle	Sub-orbital. Sounding and targeting vehicle to sub-orbital	Balloon and liquid fuel rocket engines	Can be launched by balloon at many sites
IL Aerospace (Israel)	Balloon launch and then Negev vehicle to sub-orbital space http://web1-xprize.primary.net/teams/ilat.php	Sub-orbital. 10 km by ballon and then Negev rocket launch to 120 km	Balloon and Negev solid fuel rocket with parachute to water landing	Can be launched by balloon at many sites. Israel base
Inter Orbital Systems (Mojave, California)	Sea Star (13 kg to LEO) Neptune (4500 kg to LEO) http://www.interorbital.com/	Sea Star. Microsat launch vehicle	Stage and a half. Liquid bi-propellant rocket. FAA/AST licensed	Offshore, Pacific Ocean. Los Angeles and Tonga
Japanese Aerospace Exploration Agency (JAXA)	HII transfer vehicle (unmanned but in time might be upgraded to manned and pressurized vehicle)	Unmanned cargo resupply to the ISS. Launched on the HII vehicle	Conceptual studies	To be decided
Japanese Rocket Society	Kankoh Maru (latest version of earlier Phoenix design)	Orbital. 50 passengers to 200 km LEO	Single stage to orbit. Vertical take-off	No hardware designs
JP Aerospace (Rancho Cordova, CA 95742)	Access to orbit-ascender balloon system (note this is different system than that of the Bristol Space planes). http://www.jpaerospace.com	Sub-orbital. High-altitude experiments or rocket launch	Very-high-altitude balloon. Can be used as launch platform to LEO using ion engines	California sites
Kelly Space and Technology Inc. (San Bernadino, California)	Space plane http://www.kellyspace.com/	Crew and satellite and cargo launch. Sub-orbital	Tow launch of reusable space plane	San Bernadino Airport

Lockheed Martin-EADS	Autonomous transfer vehicle (ATV) (unmanned but could be upgraded to manned and pressurized vehicle). Venturestar project canceled	Unmanned Cargo resupply to ISS. Ariane 5 launched but also on Atlas 5	Could be upgraded to become a manned vehicle. Not yet funded	Atlas 5 site at Cape Canaveral
Lorrey Aerospace (Grantham, NH 03753)	X 106 Hyper Dart Delta http://www.lorrey.biz	Orbital. Pilot + passenger and 220 kg. To LEO or Bigelow space station. Orbital data haven	(Conversion of F 106 Delta Dart to include ramjet to create a space plane)	To be decided
Masten Space	XA 1.0 (VTVL) XA 1.5 (VTVL) XA 2.0 (VTVL) Http://www.masten-space.com/products.html	Sub-orbital XA 1.0 100 kg to 100 km, XA 1.5 200 kg to 500 km, 2000 kg (5 people) to 500 km	Liquid reusuable internalized engines	To be decided
Planet Space (see also Canadian Arrow)	Silver Dart space plane and lifting body (VTHL) http://www.thestar.com/NASApp/cs/ContentServer?pagename=thestar/Layout/Article_Type1&c=Article&cid=1155678611503&call_pageid=968332188492 http://www.planetspace.org/lo/index.htm Canadian Arrow http://www.canadianarrow.com	Orbital. Crew of 8 Sub-orbital. Crew of 3	First stage liquid propellant + OX. Second stage 4 JATO rockets-abort	Nova Scotia, DaVinci Spaceport
Rocketplane-Kistler	K-1 (5700 kg to LEO, 900–1400 kg to GTO) Falcon Rocketplane XP Pathfinder	Orbital. Payloads to LEO, MEO, GTO, ISS. Cargo resupply and return missions. Sub-orbital. Cargo and microsats Sub-orbital. Four-seat fighter-sized vehicle. Up to 4 or 410 kg to 100 km. Or microgravity experiments	Various propulsion systems for K-1, Falcon, and rocket plane XP Pathfinder	Woomera, Australia and Nevada test site for K-1
Scaled Composites/ Spaceship Corporation/ Virgin Galactic (Mojave, California)	SpaceShip Two-SS (HTHL)	Sub-orbital. Space tourism Seven people to 100 km	Neoprene and NO_2 as oxidizer	Mojave Airport, South West Regional Spaceport (SRS)
Space Adventures with Myasishchev Design Bureau and Federal Russian Space Agency	Explorer space plane (C-21) and MX-55 high-altitude launcher plane (HTHL)	Sub-orbital. Space tourism	Liquid fuel motors. Horizontal take-off and horizontal landing (lifting body with parachute landing)	To operate from a number of international spaceports, including Dubai, Singapore, USA etc.
Space Dev (California)	Dreamchaser (VLHL)	Sub-orbital. Space tourism (1 stage, 6 passengers Orbital (2 stage manned access to ISS)	Single hybrid engine (neoprene and NO_2) for sub-orbit. Launch of space plane on the side of 3 large hybrid boosters to reach LEO orbit and ISS	To be decided

(Continued)

Table 23.1 Inventory of Private Space Ventures *Continued*

Company	Rocket-Launch Vehicle	Intended Markets	Capabilities	Launch Site
Space Hab Contract agreement with NASA to support COTS	Apex 1 Apex 2 Apex 3 http://www.astroexpo.com/news/newsdetail.asp?ID=27197&ListType=TopNews&StartDate= 8/21/2006&EndDate=8/25/2006	Orbital. Launch to LEO orbit (300 kg (Apex 1) to 6000 kg (Apex 3) Apex 1 and 2 unmanned. Apex 3 manned	Open architecture to support different missions and NASA's COTS Program	To be decided
Space Transport Corp. (Forks, Washington)	Rubicon 1 and 2 and N-SOLV now to be replaced by Spartan vehicle	Sub-orbital. Two passengers to 80–100 km. Spartan can launch 5 kg to LEO	Design and status of project, and financing not clear	To be decided
Space Exploration Technologies (Space X)	Falcon 9, Dragon space plane http://www.spacex.com/	Orbital. Commercial orbital transport service to ISS	Cluster of 9 Merlin engines on Falcon 9	Kwajalein Atoll launch complex
Starchaser Industries (UK and Rocket City, New Mexico)	Thunderstar-Starchaser 5	Sub-orbital. Space tourism. Launch to 60 km	Bi-liquid. LOX and kerosene rockets. Parachute recovery	To be decided
Sub-Orbital Corp. and Myasishchev Design Bureau	M-55X and Cosmopolis XXI	Sub-orbital. Two-stage to 100 km. Pilot and 2 passengers. Space tourism	First stage M-55X Geophysika. Second stage C-21, a rocket-powered lifting body with parachute landing	Flexible launch and takeoff sites
TALIS Institut and DLR of Germany	Enterprise Space Plane. Project. In cooperation with ESA and European Aviation Safety Agency	Sub-orbital. Space plane to support European space tourism bus	Under development	European spaceport to be decided
Transformation Space Corp. T/Space (Allied with Scaled Composites)	CXV (crew transfer vehicle)	Orbital. Crew of 4 to LEO or ISS and ISS resupply missions	Launches at high altitude from a large cargo carrier aircraft	To be decided
TGV Rocket	Michelle B Rocket (modular incremental compact high-energy low-cost launch experiment)	Sub-orbital. Small crew or scientific instruments	Single stage to orbit. Modular	White Sands, New Mexico
Triton Systems	Stellar-J (HTHL)	Orbital. 440 kg of cargo to LEO	Launches via a cargo jet and LOX/kerosene	To be decided
UP Aerospace	SI-1 Carrier Rocket-Space Loft XL	Launch of small scientific packages of 50 kg in 220 km LEO	Liquid fueled rocket. (Licensed by FAA/AST)	Spaceport America, New Mexico
4 Vela Technologies	Spacecruiser	Sub-orbital. Space plane. Up to 8 people	Jet plus propane/NO_2	To be decided
Wickman Spacecraft and Propulsion	WSPC small launch vehicle SHARP space plane (VTHL) http://www.space-rockets.com/sharp.html	Orbital. Cargo and in time crew to LEO. Eventually space plane to carry passengers	Phase-stablized ammonium nitrate solid fuel rocket. 900 kg to LEO	To be decided
4 XCOR Aerospace	Sphinx (Sub-orbital space) (HTHL) Xerus (Sub-orbital space) (HTHL)	Sub-orbital. Space tourism and nanosatellite launch. Xerus can also launch 10 kg microsatellite to LEO	Isopropyl alcohol/LOX. Sphinx is FAA/AST licensed	White Sands, New Mexico

Sources: http://rocketdungeon.blogspot.com/; http://home.comcast.net/~rstaff/blog_files/space_projects.htm; http://www.spacefuture.com/vehicles/designs.shtml; http://www.hobbyspace.com/Links/RLV/RLVTable.html; and numerous other sources as listed above.

The ESA is working with EADS on the Phoenix Space Plane and Pre-X project, as well as working on Project Enterprise with the TALIS Institut, in tandem with the German Space Agency (DLR) and the Swiss Propulsion Laboratory, to develop this technology in close association with the European Aviation Safety Agency (EASA). Certainly there are other independent ventures such as Bristol SpacePlane and Virgin Galactic in the UK, but these are the exception to the rule. In short, one could say that there is a "US model" of a free and open development, with the FAA applying a case-by-case review of each independent venture, and there is a "European model" of a more controlled and focused set of development activities with government funding funneled through the ESA and safety oversight being directly applied through the EASA. No one can say with certainty at this point which approach will be more successful and which course will provide the safest access to space. Nevertheless, two observations about the most likely outcomes are as follows:

1. The US approach will likely spawn more innovative ideas and offer more alternative concepts, but the diversity of approach and disparity in expertise and funding may also increase the danger of accidents and catastrophic loss of life (i.e. it is assumed that with more and more "links" in the chain, and with some of the "links" being undercapitalized and having less expertise, one of the weak links will fail).
2. The European approach, with stricter governmental controls and safety oversight, is likely to produce slower results and higher-cost systems, but perhaps is also more likely to produce vehicles with increased reliability and better escape and safety systems that will better protect against lethal accidents (i.e. it is supposed that more resources for tests and performance trials and more resources for safety and escape systems will likely produce vehicles of higher reliability).

23.2 DIVERSITY OF APPROACH TO DEVELOPMENT OF NEW SPACE VEHICLES AND ITS SAFETY IMPLICATIONS

There are always trade-offs when venturing into a new field and it is possible that both approaches might prove successful. Certainly the recent experience with NASA in the USA, that has seen very large amounts of monies being invested in "manned space vehicles" without many successes and the spectacular failures of the *Challenger* and *Columbia* space shuttles, may well have conditioned the US public and Congress to think that a more wide open and diverse approach to space tourism might indeed be the preferred course to pursue.

[Table 23.2]

23.3 SPACEPORTS

There are a remarkable number of so-called spaceports around the world today. Commercial ventures to develop such facilities are virtually exploding, since many governmental entities see this as an important new industry that represents the future in which they would like to be involved. In short, a spaceport is, to some extent, seen as a "status symbol" as well as a way to recycle and upgrade an aging or abandoned airport facility, and a way to create new jobs. Thus, many governmental agencies are offering tax credits and other incentives to those that would develop such spaceport facilities. To

Table 23.2 Various Technical Design Approaches for US and International Private Commercial Space Systems

Various Approaches	Company or Venture Using this Particular Approach
Lighter-than-air ascender vehicles and ion engines with high-altitude lift systems providing access to LEO	JP Aerospace (commercial venture with volunteer support)
Balloon-launched rockets with capsule parachute to ocean	Da Vinci Project, HARC, IL Aerospace
Vertical take-off and vertical landing	Armadillo Aerospace, Blue Origin, DTI Associates, JAXA, Lockheed Martin/EADS, Masten Space
Vertical take-off and horizontal landing (spaceport)	Aera Space Tours, Bristol Space Plane, C & Space, Air Boss, Aerospace Inc., Energia, Lorrey Aerospace, Phoenix and Pre-X by EADS Space Transportation, Planetspace, SpaceDev, Space Transportation Corp., Space Exploration Technologies (SpaceX), Sub-Orbital Corp, Myasishchev Design Bureau, t/Space, TGV Rocket, Vela Technologies, Wickman Spacecraft and Propulsion
Vertical take-off and horizontal landing (from ocean site)	Advent Launch Site, Rocketplane/Kistler
Horizontal take-off and horizontal landing	Andrews, Scaled Composites, the Spaceship Corporation, Virgin Galactic, XCOR, and Project Enterprise by the TALIS Institute, DLR and the Swiss Propulsion Laboratory
Tow launch/horizontal landing	Kelly Space and Technology Inc.
Vertical launch to LEO from spaceport	Alliant ATK, Inter Orbital Technologies, Rocketplane/Kistler, SpaceHab, UP Aerospace

date, the US FAA regulatory authorities have kept pace with this process, but, shortly, the task of re-certifying spaceports in the USA (on a 5-year cycle, which may well be too long), plus certifying new facilities, may create a problem in carrying out a thorough and effective job. Overseas, where commercial spaceports are being proposed in a growing number of countries, the problem of sufficient expertise and thoroughness of inspection processes may also become a problem. At this point, there is no single set of national or international standards for the certification of a spaceport, nor is there agreement on when a re-inspection would be required, or on what degree of regulatory oversight should be provided on an ongoing basis. Some of the planned spaceports, such as the one announced for the United Arab Emirates, involve infrastructure that resembles a theme park (see Figure 23.1).

23.4 INTERNATIONAL REGULATORY PROCESSES

Currently, there is no agreement as to what entity or entities should provide oversight of the space tourism business. At this point, there has been some analysis of the role that might be effectively played by the ICAO, the COPUOS, the EASA, and even the International Association for the

FIGURE 23.1

Artistic Concept of an UAE Spaceport.

Advancement of Space Safety (IAASS). Studies carried out by the McGill Air and Space Law and by Booz Allen Hamilton have indicated, in some detail, the role that the ICAO might play.

Others have suggested that a UN agency would provide too bureaucratic, cumbersome, and expensive oversight mechanisms for a newly emerging, and largely entrepreneurial, international business. At this stage, it is most clear that some form of international oversight will be needed, but that premature action before more experience is gained might be a mistake.

The development of a more agile and responsive regulatory process that is carried out at the national and regional level, and perhaps coordinated under the auspices of the IAASS, might be desirable at least during the first decade of the industry's development. It would seem very useful for an international conference to be organized in the near future that would seek to develop an international convention on the setting of standards and establishment of initial minimal regulatory processes for the space tourism business. This conference should include representation from the major national and regulatory space agencies, the major national and regional aviation safety regulatory agencies, the ICAO, UN COPUOS, IAASS, the Private Spaceflight Federation, the International Astronautical Federation, and the International Academy of Astronautics, together with invited participants from the relevant academic community.

The purpose of this conference would be simply to regularize the exchange of information about safety regulation of the space tourism business at the national and regional levels. This would allow all relevant governmental safety agencies to carry out a systematic search of information in the field within a standardized thesaurus of words. It would also seek to agree on the standardized posting of this information in an agreed format, to allow searches of this international database using perhaps no more than four or five languages.

23.5 POSSIBLE METHODS TO ENHANCE SAFETY WITH REGARD TO THE SPACE TOURISM INDUSTRY

The current rulemaking, as devised by the FAA, would appear to be an acceptable basis for proceeding with the development of space plane systems and for the initial operation of spaceports and the space tourism business not only for the USA, but that could likely be applied as an initial "best practices" approach for other countries as well.

Clearly, as experience is gained in the USA and abroad, these regulatory processes and licensing operations will need to be adjusted and perfected. It is our belief that a process should be started now that directly parallels the work of the White House Commission on Aviation Safety. Thus, we recommend that the FAA, in cooperation with NASA, should be directed to work with a White House Commission on Personal Space-flight Travel Safety and Security that would be a direct parallel to the White House Commission on Aviation Safety and Security.

This Commission would be directed to complete its work by 2010 and be given a specific charge to carry out the following program:

1. **Set targeted safety objectives.** Devise ways to reduce personal space-flight accidents and enhance flight safety by developing targeted and realistic objectives.
2. **Develop process for continuous safety improvement.** Develop a charge for the FAA (and NASA as appropriate) in terms of creating standards for continual safety improvement, and these goals should create targets for its regulatory resources based on performance against those standards.
3. **Set standards for certification of space planes, spaceports, operations, and training facilities.** Develop improved and more vigorous standards for certification and licensing of space planes and their operations, spaceports, and training and simulation facilities.
4. **Implement performance-based regulations by 2010.** The Federal Aviation Rules for Private Space Flight should, by 2010, be rewritten with statements in the form of performance-based regulations wherever possible.
5. **Develop improved quantitative models and analytic techniques.** The FAA should develop better quantitative models and analytic techniques to assess space plane performance and to monitor safety enhancement processes. These should be based on best industry practices as new operational vehicles and safety and emergency escape systems come online.
6. **Implement "whistleblower" protections and incentives.** The FAA Office of Commercial Space and the Department of Justice should work together to ensure that full protections are in place, including new legislation if required, so that employees of the space tourism business, including but not limited to manufacturers and/or operators of space planes, maintenance and other ground-based crew, owners and operators of spaceports, and owners and operators of personal space-flight training and simulator facilities, can report safety infractions or risk factors of concern regarding safety violations or security infractions to government officials without fear of retaliation or loss of employment—i.e. full "whistleblower" protections for such employees.

This effort should use the White House Commission on Aviation Safety and Security as a model for this activity.

23.6 POTENTIAL SHOW-STOPPERS TO BE ADDRESSED

The field of space safety is usually very narrowly and technically defined. There are at least three elements that might develop into major problems for the evolving space tourism and space plane industry that might be considered "safety concerns" if safety is conceived in a broader sense. These "broadly defined safety issues" for the field of space safety are: (i) the environmental concerns about the potentially dangerous impact on the Earth's ozone layer due to sustained and high-volume flights to the stratosphere and beyond; (ii) the increasing trends toward the weaponization of space, both in terms of possibly restricting private enterprise's ability to fly with only limited constraints into space and also increase apprehensions among prospective space tourism passengers; and (iii) orbital space debris, particularly in low Earth orbit (LEO). It is recommended that further, and indeed urgent, attention needs to be given to these subjects as the space tourism business continues to evolve quickly in coming years.

23.7 LONGER-TERM APPROACH TO PRIVATE SPACE MISSIONS

The current approach to space tourism and the development of vehicles to access space are virtually all based on extensions of current rocket launch vehicle and rocket plane development. If one takes a longer-term view of space travel and the development of safer ways to lift humans into Earth orbit and travel into space, new technologies will be needed. The options that do not involve lighting a chemically exploding bomb under crew and passengers are actually quite numerous. These advanced methods include such concepts as the use of tethers, complete space elevators to geostationary orbit (GEO), advanced ion engine thrusters, nuclear propulsion of various types, lighter-than-air craft together with ion engines, and solar electric propulsion. Once GEO is achieved, concepts such as ion engines, solar sail technology, nuclear propulsion, and solar electric propulsion, become increasingly more interesting and plausible. As private ventures continue to demonstrate the viability of the near-Earth space tourism businesses, space agencies should devote more resources to research in these follow-on systems that will ultimately be more cost-effective, more environmentally friendly, and more reliable and safer ways to access space.

CONCLUSIONS

There are rapid advances occurring in the development of a new space tourism business. This effort will be increasingly difficult to regulate as new participants evolve, new technologies are developed, more and more spaceports and training facilities begin to operate, and additional countries become involved in this industry. The sooner that standards and certification processes are established and responsible regulatory agencies assume their various functions with an effective sharing of information among participating nations, the more effective safety regulation and certification processes will become. It is not too early to think in terms of the various actions that need to be taken at the national, regional, and international levels, as outlined in this chapter.

Bibliography

The Economic Impact of Commercial Space Transportation on the US Economy, Futron Corporation, Federal Aviation Administration, Office of Space Transportation, Washington, DC, February 2006.

FAA Office of Commercial Space Transportation, 2006 Commercial Space Transportation Developments and Concepts: Vehicles, Technologies and Spaceports. January 2006. FAA, Washington, DC.

Federal Aviation Administration (2006). *2006 US Commercial Space Transportation Developments and Concepts*. FAA, Washington, DC, available at http://ast.faa.gov

Federal Aviation Administration (2006). *2006 Commercial Space Transportation Forecasts*. Office of Space Transportation, Washington, DC, May.

Futron Corporation (2002). Space Tourism Market Study, http://www.futron.com

National Security Policy Directive of 31 August 2006 on US Space Policy, http://www.fas.org/irp/offdocs/nspd/space.html

Pelton, J. with Novotny, E. (2006). *NASA's Space Safety Program: A Comparative Assessment of Its Processes, Strengths and Weaknesses, and Overall Performance*, September, Washington, DC, www.spacesafety.org

Pelton, J. N. (2007). *Space Planes and Space Tourism: The Industry and the Regulation of Its Safety*, www.spacesafety.org

White House Commission on Aviation Safety and Security: The DOT Status Report, http://www.dot.gov/affairs/whcsec1.htm

Websites

http://rocketdungeon.blogspot.com/
http://home.comcast.net/~rstaff/blog_files/space_projects.htm
http://spacefuture.com/vehicles/designs.shtml
http://www.hobbyspace.com/Links/RLV/RLVTable.html
http://www.astroexpo.com/news/newsdetail.asp?ID=27882&ListType=TopNews&StartDate=10/9/2006&EndDate=10/13

PART

Creating Technical and Regulatory Standards for the Future

5

CHAPTER

Regulations for future space traffic control and management

24

Kai-Uwe Schrogl

CHAPTER OUTLINE

INTRODUCTION

Space Traffic Management (STM) received its first full conceptualization in a study by the International Academy of Astronautics (IAA), published in 2006 [1]. Following this, numerous academic studies, as well as practitioner's activities, have been undertaken to further refine and substantiate this concept. In parallel, major policy initiatives (the EU Draft Code of Conduct, the proposal by the former UN Committee on the Peaceful Uses of Outer Space (COPUOS) Chairman Gerard Brachet on the sustainable use of outer space, and the proposal by the current COPUOS Chairman Ciro Arévalo Yepes on a UN space policy) all meet with the content of STM. This article will set out the policy and legal framework for STM and provides the links to the current policy initiatives.

24.1 THE CONCEPT OF "TRAFFIC" IN OUTER SPACE

Compared with road traffic or air traffic, use of the term "space traffic" seems bold. On first view, there are no congested roads, where a variety of traffic participants fight for their rights, and there are no take-offs and landings by the minute. In fact, since the beginning of the space age, only 30,000 man-made objects larger than 10 cm have been observed and registered. Today, there are 12,000 objects in Earth orbits, 1100 of these in the geostationary satellite orbit (GSO). This means that there are only 10^{-7} objects per cubic km. This does not sound very dramatic. But regarding "space traffic" from another angle might change this initial evaluation. The first aspect to be considered in more detail is the distribution of space objects. In fact, there are areas with a particularly high density of activities. These

Space Safety Regulations and Standards. DOI: 10.1016/B978-1-85617-752-8.10024-8

are the low orbits of up to 400 km, the polar orbits at around 800–1000 km, and the GSO at 35,800 km above sea level. With some notable exceptions (like medium Earth orbits, where navigation satellites can be found), these orbits host the largest number of main application satellites, such as Earth observation and telecommunication. This uneven distribution has already led to types of congestion, which has been known for years with regard to the GSO.

Currently, at least 600 active satellites are in orbit (300 of these in GSO). They move with a typical velocity of around 7500 m/s. Only few of them have maneuvering capabilities. They are surrounded by a growing number of space debris objects. Currently, only objects larger than 10 cm can be tracked (their number is larger than 10,000) but the millions of smaller objects can still do harm to satellites (or humans in outer space) if they are larger than 1 cm. This debris population is constantly rising, in particular through explosions of upper stages, which happen on average at the rate of five per year. The Chinese anti-satellite test of January 2007 additionally created a population of more than 2000 trackable pieces of debris, and this in a highly valuable orbit plain, where debris remains for decades or centuries. There have also been three collisions recorded between active satellites and space debris, and the need to fly debris avoidance maneuvers by the space shuttle and the International Space Station has become routine. So, this second view paints a more urgent picture of the situation around the Earth. This is why, for a couple of years, research on how to cope with the problem of maintaining space for safe use has been on the rise.

24.2 ELEMENTS OF STM: FINDINGS FROM THE STUDY BY THE INTERNATIONAL ACADEMY OF ASTRONAUTICS OF 2006

The term "traffic" was being used for space activities in the early 1980s, with the resulting need for regulation [2]. In a more comprehensive way, the American Institute of Aeronautics and Astronautics (AIAA) took up the issue at two workshops they organized at the turn of the century [3]. Emanating from these workshops, the International Academy of Astronautics (IAA) established a working group on the issue of STM in order to prepare an in-depth multidisciplinary study (a "Cosmic Study" in the IAA's nomenclature). This study was published in 2006 and is the first comprehensive work in this field.

The study defines STM as: "the set of technical and regulatory provisions for promoting safe access into outer space, operations in outer space and return from outer space to Earth free from physical and radio-frequency interference" [4]. The study acknowledges that, today, the need for STM is not yet so pressing that immediate action has to be taken. But it clearly identifies perspectives that make it seem reasonable to start now with conceptualizing a future regime. The study refers to a slow and steady decline of launches since 1980 but, on the other hand, stresses the increased number of countries with their own launching capacities and launch facilities. Due to space debris, the number of cataloged objects is also steadily rising, but the number of active satellites remains at 6–7% of the total cataloged objects. The precision of current space surveillance systems has to be improved and data sharing has to be developed further. Also, information on "space weather" is still limited and needs to improve.

The study also shows that the prospects for reusable space transportation systems are still open, that human space flight will roughly remain at 10–15% of all launches (a proportion that has been almost constant over the past 20 years), and that, following the successful flight of *SpaceShipOne*, there might be—if safety is guaranteed—a growing number of sub-orbital human flights. The picture is supplemented with a view on novel technologies like tethers, stratospheric platforms, or space elevators, which might be introduced in the future and which will have to be taken into account as well. This enumeration of perspectives shows the variety of trends and developments that will make space activities more diverse, regarding technologies as well as actors. This not only poses technological challenges, but also challenges in the regulatory field.

An STM regime will comprise four areas: the securing of the information needs, a notification system, concrete traffic rules, and mechanisms for implementation and control. The first area is the basis for any kind of traffic management in outer space. In order to manage traffic, a sound information basis regarding the space situational awareness (SSA) has to be established. Today, only the US Strategic Command possesses such a capability and shares some of its information with external users. A global STM system has to be open and accessible to all actors. The task will be to exactly define the necessary data, and to establish rules for data provision and data management, as well as rules for an information system on space weather. Only on such a basis, via shared knowledge about what is going on in Earth orbits, can traffic rules become meaningful.

The second area is a notification system. The current system of registration based on the Registration Convention of 1975 is in no way sufficient. A pre-launch notification system together with a notification system of in-orbit maneuvers has to be established. To this end, parameters for the notification of launches and operations of space objects have to be worked out. This must be complemented by rules for the notification of orbital maneuvers and for re-entries. In addition, provisions for the notification of the end-of-lifetime of space objects are necessary.

The third area comprises concrete traffic rules, which come to mind—in analogy to road traffic—when traffic management in outer space is mentioned. Here, we will find actual analogies, but also completely different rules. It starts with safety provisions for launches, then deals with space operations with right-of-way rules (comparable to "sail before motor" in maritime traffic), prioritization with regard to maneuvers, specific rules for the protection of human space flight, zoning (e.g. keep-out zones, providing special safety to military space assets [5]), specific rules for the GSO, specific rules for satellite constellations, debris mitigation rules, safety rules for re-entry (e.g. descent corridors), and environmental provisions (e.g. the prevention of the pollution of the atmosphere and the troposphere).

The fourth area will have to deal with mechanisms for implementation and control. The "modern" way of law- or rulemaking by international organizations like the International Telecommunications Union (ITU) or the International Civil Aviation Organization (ICAO) can provide an example to be followed. Basic provisions can be laid down in an international treaty (either drafted by an ad hoc assembly of States, or in the framework of an existing organization like the ICAO or an existing forum like COPUOS), and subsequent rules of the road and standards can be developed in a routine way in the format of soft law. Since space lawmaking in COPUOS is overly traditional, such an innovation would be a real culture change [6]. Another innovation would be the introduction of enforcement and arbitration mechanisms, ultimately leading to a kind of policing in outer space and sanctions such as the renouncement of access to information or the use of frequencies. This might sound utopian to those who adhere to traditional space law, but it is part of the systemic approach taken by STM, which makes it such a revolutionary concept.

Finally, STM touches upon an issue which has been on the agenda for decades now and sheds a completely new light on it. While the idea for a World Space Organization (WSO) has been around for more than 30 years, so far no convincing answer has been provided regarding what role such an organization should play. STM requires a strong operative oversight. This could be such a WSO, but the authors of the IAA study have made it clear that it might be more adequate to broaden the mandate of the already existing, and efficiently operating, ICAO than to establish a new big bureaucracy. The philosophy behind this proposal is that space traffic might ultimately (but only in some decades) evolve into air traffic in another dimension, where States and private actors with their "spacelines" operate side by side under one regulatory umbrella. But even without this optimistic vision, STM is an idea whose time has come to shape the debate on how to overcome the regulatory deadlock we are facing today.

24.3 RESEARCH ON STM FOLLOWING THE IAA STUDY

The IAA study was the first comprehensive approach to shape an STM system. Its results have been presented at conferences and published in articles, and have been brought to the attention of COPUOS. Dedicated sessions on STM are starting to be held at symposia and an institution like the International Space University (ISU) conducted a student project on this issue in 2007. In the USA, the Center for Defense Information (CDI)—a Washington-based thinktank—also is concerned with STM in its research on space security [7]. Another very notable initiative in this field was the establishment of the International Association for the Advancement of Space Safety (IAASS) some years ago, which in May 2007 published a thorough report where STM is also reflected as a cornerstone for space safety [8]. The European Space Policy Institute (ESPI) has been involved in most of these initiatives of the ISU, CDI, and IAASS, and will continue to play a leading role in the research related to the conceptual approach of STM combined with research on specific related topics [9]. The IAA itself will also continue its research in the field. Its Commission on Policy, Economics and Regulation is currently identifying topics for further related study projects. Increased visibility was also gained by private initiatives, aiming at establishing STM as a commercial service.

All these follow-ups and initiatives show that the concept of STM has its place in the framework of commemorating the anniversaries of the first space flight and the entering into force of the Outer Space Treaty. Its mission, however, is to show that we have already entered a new era of using outer space. It is an era that is characterized by new technologies and, even more important, by a growing number and type of actors. Preparing for the regulatory "big bang", leading to an effective framework for safe and equitable use of outer space, may take more than a decade. Therefore, it is encouraging to note that the debate—to a great extent initiated by the IAA—is seriously on.

24.4 STM AS A NEW CONCEPTUAL APPROACH TO REGULATING SPACE ACTIVITIES

While the space technologies rapidly change, space law is still applied on the basis set by the Outer Space Treaty 40 years ago [10]. Current space law shows strong traces of its emergence in the Cold War era. It is still characterized by a primary focus on States as actors in outer space, leading to a situation where the growing need for effective mechanisms to regulate the activities of non-governmental, private actors is not being met. Cases where this has become apparent have been the reviews of the concept of the "Launching State" and the registration practice, in particular with regard to private actors [11]. While problems have been highlighted through these agenda items and proposals for remedies have been worked out, these deliberations again proved that COPUOS is a forum that is characterized by an extremely slow decision-making process and a considerable reluctance of its Member States to accept any changes in the current regulatory framework.

This extreme fear of the Member States in COPUOS has led to results during the past 15 years that have not added any new binding provisions to international space law. They did not even produce authoritative interpretations of existing provisions, which urgently need to be re-evaluated in the light of new developments. The consequence has been that, on the one hand, other international organizations like the ITU have started to regulate areas of space activities and, on the other hand, soft law (regulations, standards etc.), instead of binding international law, has developed in technical forums like the Inter-Agency Space Debris Coordination Committee (IADC) or the Committee on Earth Observation Satellites (CEOS) or through initiatives like the Hague Code of Conduct Against Ballistic Missile Proliferation (HCOC). These developments have meant that COPUOS is slowly losing control over the regulation of outer space activities.

In this situation, STM challenges the current condition of space law and the way it will be developed further. While this present state could be regarded as "piecemeal engineering", STM would provide to the law a regulatory "big bang". STM would not tackle single issues, but would regard the regulation of space activities as a comprehensive concept. This concept is based on functionality, aiming at the provision of a complete set of rules of the road for the present and future. Space activities have to be regarded as a traffic system and not as disconnected activities of States. This would require not only new, interacting levels and forms of regulation (binding treaty provisions/technical standards, international/national provisions) but also new ways of organizing the supervision and implementation.

It is clear that specific provisions of the current space law can and will find their way into such a regime. This will certainly be the case for principles like freedom of use, non-appropriation, or peaceful uses. The comprehensive approach will also make it possible to integrate existing regimes for specific areas, such as the ITU regulations on using the orbit/frequency spectrum of the GSO, the emerging space debris mitigation regime developed in the IADC, or even the rocket pre-launch notification regime of the HCOC, one more area where COPUOS, with its post-launch registration regime, has been outpaced. But STM will be more than only the sum of these single parts. STM will develop all provisions in one coherent way from the overarching principle of guaranteeing safe operations in the space traffic system.

24.5 STM AND ITS RELATION TO CURRENT POLICY ACTIVITIES

Establishing such a comprehensive STM regime is, however, a task for the more distant future. It would require an effort comparable to the decade-long global negotiation process of the new Law of the Sea up to the 1980s. First approaches for restricted issues in the frame of STM have already been taken. Four of these are particularly remarkable and will be briefly presented in this section.

In December 2008, the EU Council Presidency presented a Draft Code of Conduct (CoC) for space activities [12]. This initiative emerged from the frustrating deliberations in the Conference on

Disarmament, which will not be discussed here. This Draft CoC was the first governmental positioning on the issue of regulation of space activities in a way that can be regarded as comprising elements of an STM regime. It comprises notification and registration and also provisions for space operations. While it is a relatively short document, it has opened the door for serious discussions on managing activities in outer space in ways other than those foreseen by the existing legal regime, and leading to STM, as outlined, as a regulatory "big bang". The EU Council Presidencies, which had been involved in the preparation of the Draft CoC since 2007, had been briefed on the 2006 IAA study on STM in order to find a basis for their initiative.

In 2007, the Chairman of COPUOS, Gerard Brachet, referring to the results from the IAA study, suggested making STM an agenda item in the Committee [13]. After further refinement of his proposal, the Committee in 2009 adopted a new agenda item, "Long-term sustainability of outer space activities", which will be dealt with under a multi-year workplan in the Scientific and Technical Subcommittee between 2010 and 2013. The goal is the preparation of a report on long-term sustainability of outer space activities and the examination of measures that could enhance it, together with the preparation of a set of best practice guidelines. It is contemplated that the Legal Subcommittee might become involved and that such best practice guidelines might develop into a UN General Assembly Resolution. This effort will certainly lead to the identification of elements for an STM regime but can only be regarded as a first step to check the interests of the States in this field and to prepare common ground in the practical technical area.

In parallel to this proposal, which will start to formally seek concrete results in 2010, the current Chairman of COPUOS, Ciro Arávalo Yepes, proposed to establish a "UN Space Policy" [14], which also contains reference to the UN's role "to regulate the orbital environment for the fair and responsible use of space". This initiative supports the workplan on the sustainable use in that it seeks to establish a policy framework.

Besides COPUOS, the then president of the ICAO Council proposed in 2005 that the organization should think about the issue and start to play a role in regulating space activities [15].

These four initiatives show that the management of space operations became an inter-governmental issue shortly after the IAA study on STM was published in 2006. The presented initiatives will now proceed rather slowly through formal, international, negotiating processes or institutional formats like COPUOS' multi-year workplan. They are, however, only covering single elements and do not yet provide for the comprehensive approach for a new regulatory concept based on STM. This might only emerge after first experience with the outcomes of the current initiatives in one or two decades.

Notes and references

[1] IAA (2006). "Cosmic Study on Space Traffic Management", Paris, edited by Corinne Contant-Jorgensen (Secretary of the Study Group), Petr Lala and Kai-Uwe Schrogl (Coordinators of the Study Group). The Study Group consisted of 16 contributors from numerous countries covering engineering, policy, and legal aspects. Online at http://iaaweb.org/iaa/Studies/spacetraffic.pdf

[2] Perek, L. (1982). Traffic Rules for Outer Space. International Colloquium on the Law of Outer Space by the International Institute of Space Law (IISL), 82-IISL-09.

[3] AIAA Workshop Proceedings. "International Cooperation: Solving Global Problems" 1999, 35–39 and "International Cooperation: Addressing Challenges for the New Millennium" 2001, 7–14.

[4] Cf. Schrogl, K.-U. (2008). Space Traffic Management: The New Comprehensive Approach for Regulating the Use of Outer Space. Results from the 2006 IAA Cosmic Study. *Acta Astronautica, 62*, 272–276.

[5] Such keep-out zones could also be a topic for the blocked negotiations in the Geneva Conference on Disarmament's Committee on the Prevention of an Arms Race in Outer Space (PAROS). Since the threat to military space assets is one of the drivers for a possible weaponization of outer space, STM could, through such specific means, also contribute to arms control.

[6] Early ideas on such an approach were presented by Jasentuliyana, N. (1999). Strengthening International Space Law. *Proceedings of the Third ECSL Colloquium on International Organisations and Space Law*, Paris (ESA SP-442), 87–96.

[7] See the activities of the CDI's Space Security Program, led by Theresa Hitchens, at http://www.cdi.org/program/index.cfm?programid=68

[8] IAASS (2007). "An ICAO For Space?" Online at http://www.iaass.org/pdf/ICAO%20for%20Space%20-%20White%20Paper%20-%20draft%2029%20May%202007.pdf

[9] See ESPI Report 10, Europe's way to Space Situational Awareness (SSA) by Wolfgang Rathgeber, which deals with data policy issues related to SSA. Online at http://www.espi.or.at/images/stories/dokumente/studies/ssa.pdf

[10] See the recently published first comprehensive commentary on the Outer Space Treaty: Hobe, S., Schmidt-Tedd, B. and Schrogl, K.-U. (eds), Goh, G. (assoc. ed.) (2009). *Cologne Commentary on Space Law*, Vol. 1: *Outer Space Treaty*. Heymanns, Cologne.

[11] Both topics have recently been dealt with in working groups under multi-year work plans of the COPUOS. See Schrogl K.-U. and Davies, C. (2002). A New Look at the Concept of the "Launching State". The Results of the UNCOPUOS Legal Subcommittee Working Group 2000–2002. *German Journal of Air and Space Law ZLW*, 51(3):359–381. Schrogl, K.-U. and Hedman, N. (2008). The UN General Assembly Resolution 62/101 of 17 December 2007 on "Recommendations on Enhancing the Practice of States and International Organizations in Registering Space Objects". *Journal of Space Law*, 34(1):141–161. The author has been the chairman of both these working groups.

[12] See Rathgeber, W. and Remuß, N.-L. (2009). Space Security—A Formative Role and Principled Identity for Europe. *ESPI Report, 16*, 58–64. February. Online at http://www.espi.or.at/images/stories/dokumente/studies/espi%20report%2016.pdf

[13] Brachet, G. (2007). UN Doc. A/AC.105/L.268 of 10 May, "Future role and activities of the COPUOS. Working paper submitted by the Chairman", para. 28.

[14] Arévalo Yepes, C. (2009). Towards a UN Space Policy. *ESPI Perspectives*, 23, June. Online at http://www.espi.or.at/images/stories/dokumente/Perspectives/espi%20perspectives%2023.pdf

[15] See van Fenema, P. (2005). Suborbital Flights and ICAO. *Air and Space Law, 30*(6), 396–411.

CHAPTER

25 Spacecraft survivability standards: enhancing traditional hazard control approaches

Michael K. Saemisch, Meghan Buchanan
Lockheed Martin Corporation

CHAPTER OUTLINE

25.1 SPACECRAFT SURVIVABILITY ENGINEERING STANDARDS FOR THE NEXT GENERATION OF SPACE VEHICLES

Lockheed Martin has been studying the application of system safety to the next generation of NASA human space-flight vehicles planned for use after retirement of the space shuttle. Through a sequence of small contracts leading up to the current Orion contract, Lockheed Martin was performing these studies and analyses to support concept development of the next generation of vehicle. One of the primary system design goals for the next generation vehicle was the advancement of space safety through a simpler and safer design. To that end, many of the studies were aimed at improving crew safety through implementation of such system features as abort and crew escape, and looking for innovative approaches to enhance safety. Concepts such as providing separate escape vehicles that could survive a major mishap and providing an abort system to safely remove the crew vehicle from a mishap were all studied and relevant trade-off analysis was performed.

It was during these studies that an observation was made that there was an overlooked aspect for improving crew safety of spacecraft designs. An opportunity was identified to potentially implement new techniques with potential major increases in safety with small or no impacts to the system design through optimization of the vehicle designs with this new consideration in mind. This chapter describes the process through which this observation was made and how it grew into Spacecraft Survivability Engineering, the process now being applied to the Orion project and being studied with NASA for application outside of Orion. Spacecraft Survivability was an element of the Lockheed

Space Safety Regulations and Standards. DOI: 10.1016/B978-1-85617-752-8.10025-X

Martin winning Orion proposal and was cited as a specific strength of the proposal by NASA. This chapter is the first public presentation of this idea.

25.2 CREW SAFETY ANALYSIS THROUGH DECOMPOSITION OF REQUIREMENTS

One of the analysis techniques applied by Lockheed Martin (in conjunction with the Massachusetts Institute of Technology) was the detailed decomposition of the functional requirements for future human transport vehicles. Through this decomposition it is possible to achieve a greater understanding of the elements that contribute to the various function requirements. This allows focused requirements to be developed that address all elements of the "decomposed" functional requirements. One of the functional requirements was deemed to be "crew survival", which was then decomposed by a multi-discipline team. The decomposition process resembles fault tree analysis to some degree, but is aimed more at the functions that comprise the higher level requirements. This decomposition process continues at each level to become more and more detailed until the lowest functional level where specific requirements can be derived. The initial decomposition resulted in Figure 25.1. Two elements were identified during the decomposition of crew survivability. The first, that depicted in "Identify and Control Hazards", was further decomposed and resulted in the application of typical System Safety engineering of identification and control of hazards through such techniques as failure tolerance, redundancy, and design for minimum risk. The right leg of the decomposition, "React to Hazards", was further decomposed into provisions for assuring crew survival should one of the hazards occur anyway. Potential environments and scenarios would be identified so that functional requirements for the crew survival systems could be derived and implemented. Providing for crew survival after the occurrence of a hazard was one of the primary drivers in the new human space-flight architecture for the next-generation NASA human space-flight carrier. Survival during both ascent and descent mishaps was a stated requirement of NASA [1].

As the team studied this decomposition and began developing each of the two legs into the traditional control approaches to prevent hazards, an innovative thought occurred. The team began to think that there might be an overlooked third aspect of providing for crew safety. This resulted in the decomposition shown in Figure 25.2. This adds "Vulnerability Reduction" as a third leg. With the leg on the left minimizing the probability of occurrence of the hazard through hazard controls that are

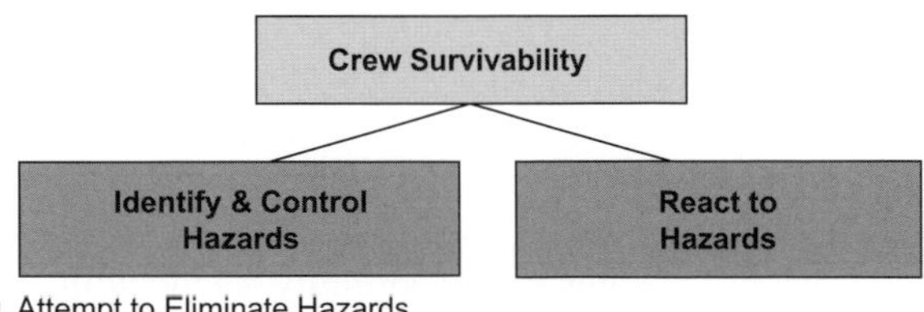

FIGURE 25.1

Crew survivability/human rating, traditionally comprised of two primary elements.

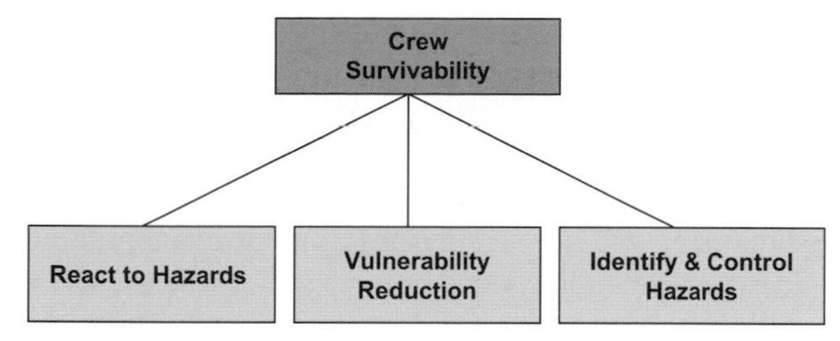

- Protect Survivability Equip. From Hazards Should They Occur (e.g. MMOD penetration)
- Protect Crew From Hazards Should They Occur
- Vulnerability Reduction

FIGURE 25.2

Reducing the effect of hazard occurrence through vulnerability reduction increases the probability of crew survivability, complementing other approaches.

intended to eliminate or reduce the probability of occurrence to acceptable levels, and the leg on the right providing for crew survival after a hazard occurs, it was observed that nothing was being applied to assure survival of the crew and the crew survival equipment during the mishap, by reducing vulnerabilities.

25.3 THE SPACECRAFT SURVIVABILITY IDEA

As just discussed, the traditional System Safety requirements are aimed at eliminating hazards (not really possible for most space travel hazards), or reducing the probability of occurrence to acceptable levels through prevention of events or assuring essential events. If one looks at typical human spaceflight safety requirement documents, the requirements provide for this hazard control through features such as designing triple inhibits to undesired events or triple redundancy for critical events (two-failure tolerance). If the system provides the required features (e.g. three inhibits, triple redundancy), the system is deemed as providing acceptable levels of risk and the system can be certified for human flight. In addition, the new architecture will have provisions for removing the crew from the mishap, further assuring crew safety. However, what if the crew does not survive the initial mishap, or if the crew survival system itself does not survive the initial mishap risk and is not available for use? Risk is comprised of two elements, likelihood and severity. By paying attention and applying requirements only to reduce the likelihood of occurrence of a hazard complemented with a crew survival system, the severity component of the initial mishap risk is not addressed through a similar structured process. Certain aspects of the potential mishap scenarios will be defined (e.g. potential blast pressures) and included in the design requirements, but with no structured process being applied to examine the design for such things as vulnerabilities in the design that could affect the severity of the mishap risk (and consequently the likelihood of crew survival during the mishap). Current definitions of hazard severity in human space flight include “catastrophic”, which includes the potential for loss of life. Nearly all identified significant hazards during space flight will carry this assessed severity, invoking the highest levels of hazard control requirements. The only way to reduce the severity assessment of "catastrophic" is to eliminate the possibility of loss of life should the hazard occur, which is generally not possible for hazards of space flight. Regardless of the amount of controls applied to hazards, they will still carry this rough assessment of potentially "catastrophic".

However, the team observed and discussed that there are differences in concepts that would be desirable from a safety standpoint, all within the coarse classification of "catastrophic". For example, the explosion of a pressure vessel with lower stored energy could be more survivable as compared to an explosion of a pressure vessel with higher stored energy. This is true even though in traditional safety analysis both hazard types would carry the same catastrophic classification. Moreover, if the placement of the pressure vessel considered the event of potential rupture (something not addressed in current requirements), the severity of the event might be reduced by consideration of where the pressure would go and use of existing spacecraft mass to act as a natural shield to the effects of the hazard, thereby increasing the likelihood of the crew and vehicle surviving the event.

From this discussion, the group wondered if there were existing approaches that could be applied to derive functional design requirements to address this newly identified component of crew survivability. Initial inquiries with the Lockheed Martin Aeronautics business sector quickly revealed that there were unique techniques and processes applied to military aircraft to do exactly that—reduce the severity of a mishap to give the flight crew an increased chance of surviving the mishap and returning or escaping to safety. The potential of cross-application of these techniques to spacecraft design became readily apparent.

25.4 AIRCRAFT VULNERABILITY/SURVIVABILITY HERITAGE

Aircraft vulnerability, according to Robert Ball in his landmark book on aviation safety, is "the inability of an aircraft to withstand (the hits by the damage-causing mechanisms created by) the man-made hostile environment". The evolution of aircraft survivability dates back to World War I, where pilots fashioned steel armor protection nailed to their seats. Through World War II, the south-east Asia conflict, the Korean conflict, Desert Storm, and many others, aircraft survivability/vulnerability has grown into an effective analysis tool and design driver [2].

Each component on an aircraft has a level of vulnerability varying from ineffective to critical, critical meaning the kill of this component is directly or indirectly (by cascading effects) responsible for the vehicle or mission kill/abort. The first vulnerability reduction techniques were implemented in World War II by identifying the most critical components and using proven techniques to reduce the vulnerability of catastrophic loss after damage:

$$\text{Survivability} = 1 - \text{Killability} = 1 - \text{Susceptibility} \cdot \text{Vulnerability} \tag{25.1}$$

Vulnerability reduction for aircraft has evolved into six techniques, all of which could be applied to spacecraft if "threats" could be identified:

1. Component redundancy with separation
2. Component location
3. Passive damage suppression
4. Active damage suppression
5. Component shielding
6. Component elimination.

These six techniques are implemented after a vulnerability analysis to mitigate or allow smooth degradation to the aircraft after a hit has occurred. This is accomplished by driving design to address these conditions before they occur. According to Robert Ball, and currently accepted methodology, a vulnerability assessment is performed following these three tasks:

- **Task 1: Identify critical components and their kill modes.** Select the type of kill, determine flight and mission essential and system/subsystem functions, complete failure modes and effects analyses (FMEA), complete an analysis called damage modes and effects analysis (DMEA), and complete fault tree analysis.
- **Task 2: Perform vulnerability assessment.** Select threat, determine critical component kill criteria, and compute vulnerability of critical component.
- **Task 3: Design for low vulnerability using vulnerability reduction (VR) techniques.** Consider the kill modes of the critical components, apply VR features to reduce vulnerability of aircraft, and conduct trade studies if necessary [3].

25.5 ORION SPACECRAFT SURVIVABILITY (SCS)

After the introduction to the aircraft survivability world, the initial concept of Spacecraft Survivability evolved for the Orion project. The concept involves the following essential elements:

1. **Threat identification.** Where aircraft vulnerability describes threats as damage-causing mechanisms created by the man made hostile environment, Spacecraft Survivability considers threats from damage-causing mechanisms created by the naturally occurring and self-induced man-made hostile environments (hazard occurrence). These threats range from system failures to penetration by micrometeoroids. The following threats are considered on a grand scale:
   a. *Induced threats*
      i. Leak
      ii. Fire
      iii. Venting
      iv. Process failure
      v. Docking
      vi. Penetration
      vii. Overpressure/under-pressure
      viii. Cascading effects
      ix. Operations error.
   b. *Natural threats*
      i. MMOD penetration
      ii. Radiation
      iii. Charged particles
      iv. Weather/lightning
      v. Temperature
      vi. Flora/fauna.

For the purposes of initial study, the threat definition list is narrowed down to a manageable number, three to five threats. These threats are can be carried from the FMEA on to the DMEA.

2. **Trade study support.** The most significant contribution of SCS to the initial Orion design has been in support of trade studies that affect configuration. Through the assessment of survivability differences in potential options, and considering these differences when scoring and making trade study selections, SCS has driven many design choices.
3. **Design assessment/metric.** A Spacecraft Survivability metric was established to produce quantitative results of the design changes made during configuration changes. Each design change was scored with a derived approach. Aircraft Survivability began by collecting data by estimates made by industry experts. Eventually, enough actual data was collected that a true quantitative analysis could be regulated. Since there is no formal method of scoring spacecraft survivability, the score is derived through discussions between SR&QA and the design team. Understandably, events of greater impact to survivability contributed greater points towards the final score. It must be stressed that this analysis was based purely on Survivability/Vulnerability differences after the design had met requirements. SCS assessed configuration changes after (or assuming) requirements had been completely satisfied. Even though a design change might have scored negatively or showed degradation in survivability, the design had met requirements (the negative scores often affect decisions during trade study selections). SCS is the study of dealing with the "bad day scenario"—failure even though two-fault tolerance or design for minimum risk requirements compliance had been achieved. Through application of SCS, the Orion design reduces spacecraft vulnerabilities, therefore increasing the probability of crew survivability should a mishap occur. An example of a survivability feature added to the design was in the propulsion system. The initial design included propellant storage in two tanks with a connecting manifold (i.e. to become essentially one tank). However, the potential isolation of one of the tanks to retain half the remaining propellant, should a leak occur, could be achieved through the addition of an isolation valve. Retaining half of the propellant could assure crew survivability in some scenarios.
4. **Damage modes and effects analysis (DMEA).** The DMEA process begins where the failure modes and effects analysis (FMEA) ends. The FMEA identifies the cause of the component or system failure. The DMEA uses this information along with the cascading effects of the failure. Derived from MIL-STD-1629A, the purpose of the DMEA is to provide early criteria for survivability and vulnerability assessment. The DMEA provides data related to damage caused by specified threat mechanisms and the effects on Flight and Mission Essential functions. The DMEA is maintained for the subsystem design during conceptual, validation, and full-scale development. This process considers all failure modes and damage modes that can occur to each item and the effect each has on the subsystems. The relationship between the subsystems' essential functions, mission capabilities, fault tolerance, and risk are analyzed to provide design criteria for survivability enhancement [4].
5. **Requirements development.** From the initial assessments, an important task is the derivation and implementation of new design requirements that increase crew survivability through application of SCS techniques, while living within the project constraints. The process is in the initial phases and will be reported in later papers.
6. **Reporting.** To capture the results of the SCS efforts (DMEA and trade study scoring), existing safety data deliverables will be utilized to document and present this data to NASA. A detailed scoring reporting system provides the results against all of the above elements.

25.6 POTENTIAL FUTURE APPLICATIONS

The Orion SCS concept is readily adaptable to other applications where a systematic approach to analyzing vehicles, to enhance mission success or safety, is desired. It is foreseen that such analyses will become even more critical as systems are designed for longer-term space travel where safety depends even more on the robustness of the system designs. To that end, NASA has expressed the desire to immediately expand the program to include all of Constellation and possibly institutionalize SCS into the NASA processes. The SCS could also be applied to commercial flight systems as well. Overall, it is believe that the SCS process could be applied as a general survivability engineering standard for human space-flight design in future years.

CONCLUSIONS

It became readily apparent that once the concept emerged, opportunities for enhancing survivability suddenly became apparent utilizing design features already planned, or by recognizing differences of proposed options during trade studies that could drive final selection. In fact, trade selection criteria are where the SCS concept has made the largest impacts on the current design. Concepts being equal from a safety standpoint (e.g. propellant stored in single or multiple linked tanks) suddenly became different from a survivability standpoint, which was weighted significantly in the trade study and often determined the outcome. Further, as the project engineers were exposed to the concept, it was embraced as a project initiative and became part of the project design lexicon.

It is thus believed that the SCS process has value in enhancing crew safety and the application of this process could potentially prevent a future catastrophic human space-flight mishap. It is also apparent that this program will be more valuable and essential for longer space exploration further from Earth and safety where designs should be optimized to utilize inherent features in the most robust system design possible. The Orion SCS concept and initiative is thus considered to be only the beginning of a potentially more defined process that should be developed with NASA and industry partners as well as overseas space agencies to benefit all future human space-flight vehicles and passengers.

References

[1] NPR 8707.2A (2005). Human-Rating Requirements for Space Systems. NASA Offices of Safety and Mission Assurance, 7 February.

[2] Ball, R. E. (2003). *The Fundamentals of Aircraft Combat Survivability Analysis and Design*, 2nd ed. AIAA Education Series.

[3] *Ibid.*

[4] MIL-STD-1629A (1980). Procedures for Performing a Failure Mode, Effects and Criticality Analysis. Department of Defense, Washington, DC, 24 November.

CHAPTER

Safety guidelines for space nuclear reactor power and propulsion systems

26

Mohamed El-Genk

Chemical and Nuclear Engineering Department, Institute for Space and Nuclear Power Studies, Mechanical Engineering Department, University of New Mexico

CHAPTER OUTLINE

Space Safety Regulations and Standards. DOI: 10.1016/B978-1-85617-752-8.10026-1

INTRODUCTION

Past and present space exploration missions have relied on solar and radioisotope power systems for data transmission and the operation of the spacecraft, both en route to, and at, destination. The solar option is either impractical or limited for missions to distant planets of Mars and beyond and in Earth orbits within the Van Allen radiation belts. The decrease in the solar intensity inversely with the square of the distance from the Sun (Figure 26.1) limits uses of solar photovoltaic power systems in deep space exploration and for generating high power for outposts on the surface of Mars, the Moon, and other celestial bodies. For example, a solar power system for generating 200 W_e at Saturn (9.51 AU), Jupiter (5.18 AU), or Mars (1.52 AU) would generate ~18.09, 5.37, or 462 W_e in Earth orbit respectively.

More than 23 NASA missions to the Sun and deep space relied on radioisotope power systems for meeting electrical power and thermal management needs (Figure 26.2). Though electrical power provided ranges from tens to a few thousands W_e for more than 5 years, these power systems had operated for many years past their design life, some for more than 20 years. Radioisotope power systems typically have a specific mass >200 kg/kW_e, operating continuously and independent of the Sun. For a designated class of space missions, solar panel power systems could be hundreds to thousands of kilograms more massive than radioisotope power systems. In addition, the degradation of the photovoltaic cells due to exposure to energetic space radiation and debris, using batteries for energy storage, and the additional propellant needed, limit uses of solar systems to low-power missions and destinations where the solar intensity is viable (Figure 26.1a,b). The solar brightness at Mars and Jupiter is ~45% and <5% of that at Earth, and practically nil beyond Jupiter (Figure 26.1a).

For distant plants in the solar system, NASA had used radioisotope power systems (RPSs) to provide electrical power (<900 W_e) to the payload and radioisotope heater units (RHUs) (1.0 W_{th} each) for thermal management of the spacecraft and payload in the bitter cold of space. During the last four to five decades, NASA used 41 radioisotope power systems on 23 missions [1–5] to the Sun and most planets in the solar system (Figure 26.2).

In addition to the Sun, NASA missions powered with RPSs had visited every planet in the solar system, except Neptune and Pluto. The *New Horizon* spacecraft launched in January 2006, and expected to arrive at Pluto in 2015, will be the first to that destination (Figure 26.2). However, the current shortage and lack of a domestic supply of the ^{238}Pu radioisotope for the general heat source of the RPSs and the recent advances in solar cell efficiency compelled NASA to send the solar-powered *Juno* spacecraft to Jupiter in 2011 [6,7]. It will be the first mission to Jupiter that uses a solar power system instead of radioisotope thermoelectric generators (RTGs). It will conclude in 2018, after 32 orbits around Jupiter. The total area of the advanced solar panels for the *Juno* spacecraft will be >60 m^2, enough to produce >18 kW_e in Earth orbit, but only ~200 W_e at Jupiter. Although solar power systems continue to be a useful option for many space missions with limited power requirements and to destinations where the solar brightness is deemed practical (Figure 26.1), the use of RPSs is critical to many missions requiring

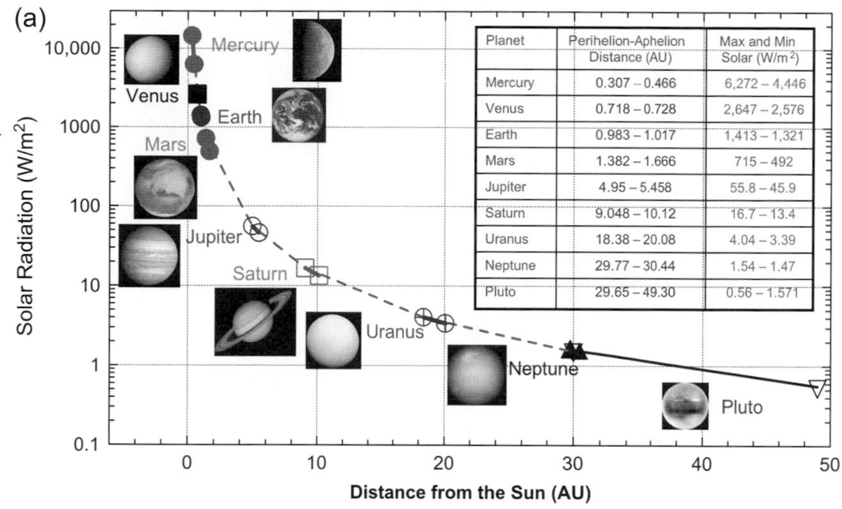

Planet	Perihelion-Aphelion Distance (AU)	Max and Min Solar (W/m^2)
Mercury	0.307 – 0.466	6,272 – 4,446
Venus	0.718 – 0.728	2,647 – 2,576
Earth	0.983 – 1.017	1,413 – 1,321
Mars	1.382 – 1.666	715 – 492
Jupiter	4.95 – 5.458	55.8 – 45.9
Saturn	9.048 – 10.12	16.7 – 13.4
Uranus	18.38 – 20.08	4.04 – 3.39
Neptune	29.77 – 30.44	1.54 – 1.47
Pluto	29.65 – 49.30	0.56 – 1.571

Solar radiation with distance from the Sun.

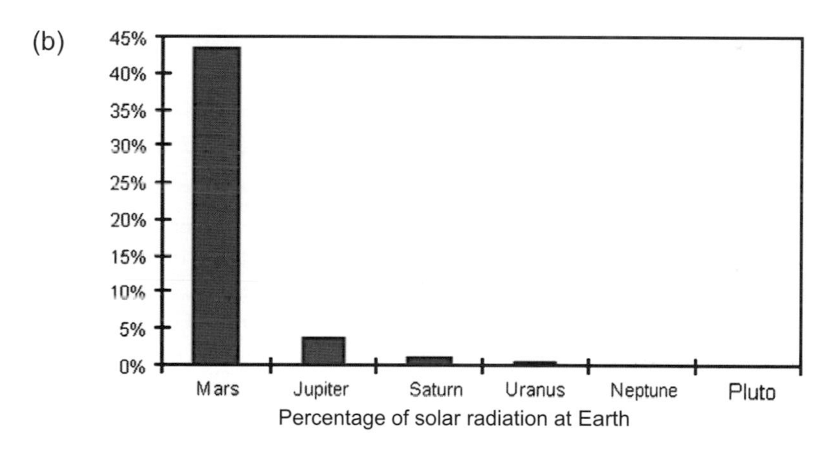

Percentage of solar radiation at Earth

FIGURE 26.1

The solar intensity at various planets and its decrease with distance from the Sun.

electricity from a few mW_e to hundreds of W_e. The US Department of Energy is in the process of developing the capabilities to resume the domestic production of ^{238}Pu, in an attempt to meet NASA's future needs for RPSs. For electrical power requirements of tens to thousands of kW_e and fast travel in space, nuclear reactor power and propulsion systems are currently the only enabling option [1–4].

26.1 RADIOISOTOPE POWER SYSTEMS

The radioisotope power systems used (Figure 26.3) typically generate $<300\ W_e$ each and operate independently of the Sun, and many had lasted far beyond their design lifetime of 5–10 years. Some

FIGURE 26.2

Twenty-three NASA space exploration missions since 1961 using radioisotope power systems.

Courtesy of US DOE and NASA

continued to operate, communicating data for up to 20 years, or even longer. They use a radioisotope source that generates heat by the natural radioactive decay of selected isotopes, emitting primarily alpha or beta particles or gamma photons. For space missions of ~6 years, ^{238}Pu in an oxide of carbide form has been the material of choice for the general-purpose heat source (GPHS). It has a half-life of ~87 years and decays by emitting mostly alpha particles (helium nuclei), requiring practically no shielding. The half-life of ^{238}Pu limits the decrease in the rate of heat generation by radioactive decay to <5% after 6 years (Figure 26.4).

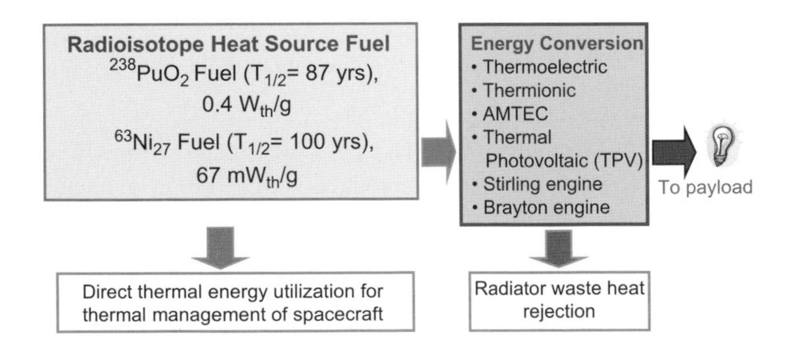

FIGURE 26.3

A schematic of a radioisotope power system.

The GPHS radioisotope thermoelectric generators (GPHS-RTGs) had used solid-state, thermoelectric elements to partially convert the heat generated by the GPHS to electricity up to ~280 W_e, at a thermal efficiency <5% [1–4]. Future radioisotope power systems with Free Piston Stirling Engines (FPSEs) for achieving a thermal efficiency of 20–30% are currently being developed at NASA Glenn Research Center. The high thermal efficiency significantly decreases the amount of ^{238}Pu needed for a given electrical power requirement, but the specific mass of these power systems remains comparable to, or slightly lower than, state-of-the-art GPHS-RTGs with SiGe thermoelectric elements for energy conversion. The latter had been used with excellent results on the *Ulysses* and *Galileo* missions to the Sun and Jupiter, and are currently being used on the *Cassini* spacecraft at Saturn and the *New Horizon* to Pluto (Figure 26.2). Advanced thermoelectric, thermionic, alkali metal thermal-to-electric energy conversion (AMTEC), and thermal photovoltaic (TPV) have also been investigated for increasing the thermal efficiency of RPSs and decreasing the amount of ^{238}Pu needed. These static energy conversion options may also decrease the specific mass of RPSs. To increase the generated electrical power by a single RPS to a few kWe, or more, NASA had investigated using organic Rankine cycle and closed Brayton cycle for energy conversion in conjunction with a GPHS. These power systems are known as dynamic isotope power systems (DIPs) [8,9].

Faster travel and higher power requirements in the tens to hundreds of kW_e could not be provided by either solar power systems or RPSs, but are possible using fission reactor power systems. These could be operated continuously for 10–15 years, or even longer, independent of the Sun. Depending on the type and the design of the reactor heat source, the power system can employ different energy conversion options. Space reactor power systems are compact with much lower specific mass (<40 kg/kW_e) than both RPSs and solar systems, and could also be used for electrical and thermal propulsion, with much higher specific impulse than chemical rockets for faster travel in space. Because of the harmful energetic space radiation, faster travel is a critical element to enabling human exploration.

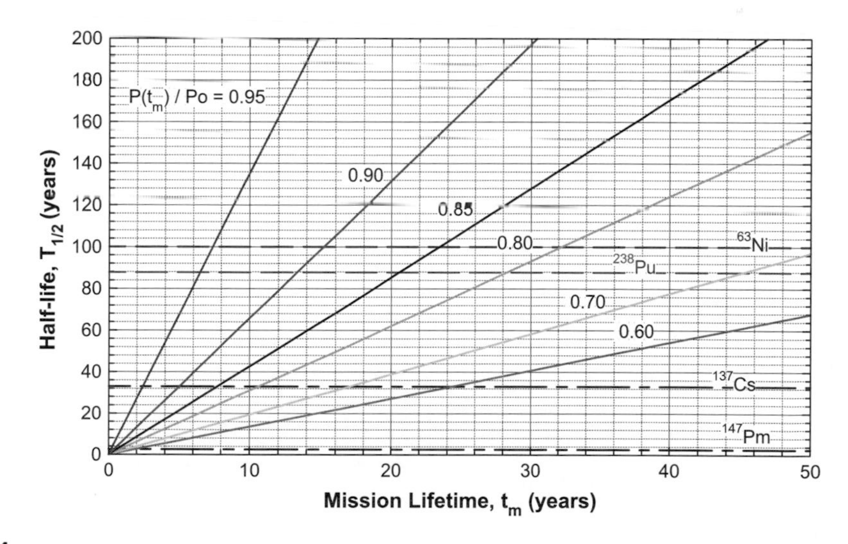

FIGURE 26.4

Mission lifetime for selected radioisotope fuel materials and different end-of-mission power ratios.

26.2 THE CASE FOR SPACE REACTOR POWER AND PROPULSION SYSTEMS

Faster travel to destination and electrical power requirements of tens to thousands of kW_e for future space exploration missions can be enabled by space reactor power and propulsion systems. These systems are much lighter and more compact, and are the only option available when the solar option is impractical or nonexistent. In addition to operating either continuously or intermittently for 10−15 years, independent of the Sun, they could be designed for multiple shutdowns and restarts and for operating at variable power levels. The electricity generated by these power systems could support a multitude of housekeeping needs and science experiments as well as operate a number of ion thrusters, propelling the spacecraft at a specific impulse of 3000−15,000 s. Such high specific impulse could halve the travel time to distant planets in the solar system (Figure 26.5). Space reactors could also provide for high thrust thermal propulsion, with more than twice the specific impulse of chemical rockets, and bimodal operation of thermal propulsion and electrical power generation (Figure 26.6).

In addition to the fast travel, compactness, and operation flexibility, the low specific mass of reactor power systems (<40 kg/kW_e) makes them very attractive for surface power on Mars and the Moon. For these systems, shielding avionics, equipment, electronic, and humans from the hostile and harmful space environment is a primary safety and operation issue. The dose from exposure to the energetic space protons and charged heavy particles and galactic cosmic rays should be kept to reasonable values, particularly on long-duration missions. Another important issue is to mitigate the impact on the spacecraft by asteroids and space debris in Earth orbits. In general, the exposure threats to the equipment and astronauts from natural space radiation could far outweigh that from operating fission reactors. For unmanned missions, a shadow shield typically made of layered tungsten and LiH is used to reduce the neutrons' fluence, and the dose from the energetic gamma photons to the payload from the operating reactor, to acceptable levels. The payload is located a distance (10−25 m) away from the shadow shield (Figure 26.5).

Future interest by spacefaring nations in space reactor power and propulsion systems to enable deep space exploration and provide surface power to outposts on the Moon and Mars may materialize within two to three decades. This necessitates the development of globally agreed-to safety standards

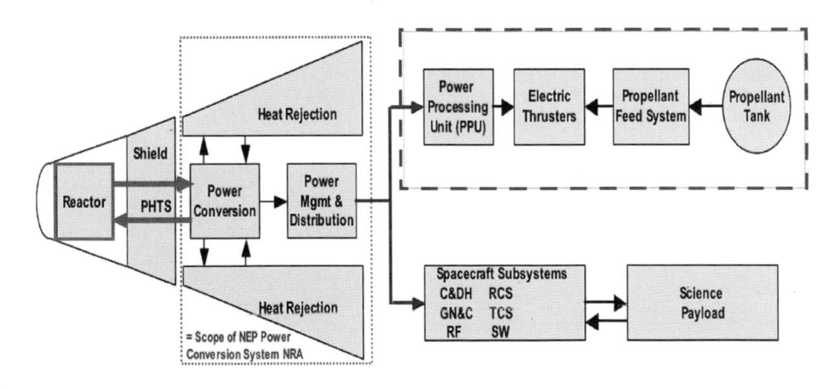

FIGURE 26.5

A generic layout of a space reactor system for electrical power and propulsion.

Adapted from NASA

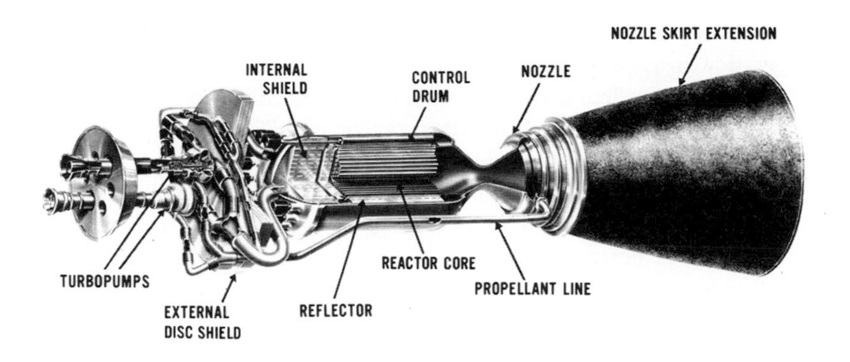

FIGURE 26.6

Nuclear thermal propulsion rocket.

Courtesy of NASA

and deployment guidelines to avoid any undue radiation exposure for individuals and astronauts beyond permissible levels. For a start, it is this author's opinion to limit future uses of space nuclear power and propulsion systems to strictly peaceful and civilian missions that involve contributions among many nations and that would benefit the citizens of the world. In addition, ensuring safe and acceptable end-of-life storage of the deactivated nuclear reactors should be addressed through acceptable, safe, and practical options that are agreeable to all nations.

The objectives of this chapter are to provide basic knowledge on the design and integration of space reactor power systems and present design choices and options for enhancing the safety and operation reliability of these systems. Examples of power systems developed by the author and members of his research team over the years, with specific features for avoidance of single-point failures in reactor cooling and energy conversion and enhanced safety and reliability, are presented. Safety issues relevant to the design, development, and future uses of space nuclear reactor power and propulsion systems are also discussed. Toward that end, lessons learned from past deployments of space reactor power systems by the former Soviet Union and USA in the 1960s to the 1980s are summarized. Using sufficiently high orbits for end-of-life storage of nuclear reactors is also addressed and the results of a global civilian mission in 1000–300 km orbit are presented and discussed.

In the absence of explicit safety standards and regulatory guidelines on future peaceful uses of reactor power and propulsion systems in outer space, it is hoped that this article stimulates future discussions of the topic. This effort is not, by any means, a comprehensive coverage of all relevant issues, but could be the beginning of an evolving effort to eventually establish agreed safety standards and guidelines.

The next section provides basic information on the design, system components, and materials choices and the integration of space reactor power systems, and discusses the inherent conflict among the various choices.

26.3 SPACE REACTOR POWER SYSTEMS DESIGN

As indicated earlier, space reactor power systems are enabling future deep space exploration and outposts on Mars and the Moon, for which the solar option is either impractical or nonexistent. It is strongly recommended that, for the peaceful uses of space, nuclear reactor power and propulsion systems be limited to deep space exploration and surface power missions. However, should they be considered for operation in Earth orbits, they should be limited only to those civilian missions for whom the solar option is impractical and uses should be prohibited in low orbits with altitudes <1000 km. For surface power on the Moon and/or Mars, practical and acceptable options for end-of-life disposal of the nuclear reactor heat source should be addressed.

The current launch cost is prohibitively expensive (~$10,000/kg, $100,000/kg and $1m/kg to Earth orbit, the Moon, and Mars respectively). Thus, reducing the size and mass of space reactor power and propulsion systems is a prime design driver. It affects not only the choices of the reactor type, system components, and structure materials, but also the operation life and performance of the system. Reducing future launch cost by at least an order of magnitude from the current levels would significantly impact the development of safe and reliable space reactor power and propulsion systems for enabling future space exploration, within and beyond the solar system.

Safety is the sole guiding principle during various phases of space reactor power system development, integration, and testing, launch, deployment into orbit or an appropriate trajectory to destination, startup and shutdown, and end-of-life storage. The guiding principle is avoiding any undue exposure of Earth inhabitants and astronauts to unacceptable levels of radiation. This can be effectively assured by design and employment of viable end-of-life-storage options. The implementation of autonomous operation schemes and adaptive control capabilities, and the avoidance of single-point failures in the design and integration of these power and propulsion systems, would effectively enhance operation reliability and safety and ensure mission success.

It is inevitable that space exploration will increasingly become more international, requiring global cooperation among spacefaring, and other, nations. A global effort is critical to alleviate undue damage to properties and avoid unacceptable levels of radiation exposure. In addition to sharing the prohibitively high cost of the missions, for a single nation to bear, global cooperation is hoped to stimulate discussions and collaborative effort to ensure the safety of space nuclear power and propulsion systems, and help develop practical and acceptable options for end-of-life storage.

The following section reviews several design considerations and different choices of reactor types, operation temperature, structural materials, and components for space reactor power systems. Potential and inherent conflicts among the different choices are also discussed.

26.4 DESIGN CHOICES AND CONSIDERATIONS

Space reactors could be designed with either a fast or thermal neutron energy spectrum and cooled with either circulating liquid metal or high-pressure gas. Space reactors could also be cooled using liquid metal heat pipes with sodium or lithium working fluids, depending on the reactor operating temperature. Fast spectrum reactors are generally more compact and lighter than thermal neutron spectrum reactors. This is because of the additional mass of the moderator used in the latter for slowing down the neutrons. The larger size reactors also increase the size and mass of the radiation shadow shield (Figure 26.5). The neutron fission cross in a fast spectrum reactor is order of magnitudes lower than in a reactor with a thermal energy spectrum. Thus, the fissile loading in a fast spectrum space reactor is as much as 20 times that required for generating thermal power over the operation life of the reactor. This is in contrast to a thermal spectrum reactor, in which the fissile loading could be only 2–2.5 times that required for generating thermal energy through its end of life.

Small size and lower mass favor a fast spectrum reactor, particularly at thermal power levels in excess of 100–200 kW_{th}. At lower power levels, fast spectrum reactors may not achieve criticality because of the high neutron leakage and the low fission cross-section. For these power levels, thermal spectrum reactors are an appropriate choice for operation lives $<1-3$ years. Examples are the former Soviet Union's "BUK" and TOPAZ reactors and the US SNAP-10A reactors, discussed later in some detail. While thermal reactors could still be used in high-power space reactor power systems, the specific mass of the system will be progressively higher than when using fast energy spectrum reactors. Other inherent limitations to these choices are discussed later.

In order to realize a low specific mass, small volume and a long operation life of 7–10 years, or even longer, space reactors use uranium fuel that is highly enriched to 90–95 wt% in fissile ^{235}U. This enriched concentration of ^{235}U is utilized despite a potential proliferation concern against the possibility of a re-entering reactor falling into the wrong hands, or landing in the territories of a hostile nation. The small concentration of depleted uranium (^{238}U) in the fuel results in a small amount of transuranic and minor actinides. As a result, within 200–300 years of storage in sufficiently high orbit (SHO), with long decay life of >1000 years, the radioactivity in the deactivated reactors would decay to almost background levels on Earth. Thus, an eventual re-entry of the reactors would not represent a radiological concern to the Earth's inhabitants.

The compact design and small volume of space reactors, however, present practical challenges to cooling the reactor and integrating it into the power system. Depending on the choice of energy conversion technology, the size of the heat rejection radiator could be by far the largest and heaviest component in the system (Figure 26.5). The size of the heat rejection radiator depends on its average surface temperature and the heat rejection load. Other important parameters are the effective emissivity and the geometrical view factor of the radiator.

The electrical power generated by a system is directly related to that rejected, the effective thermal efficiency, and the thermal power of the reactor. However, since the rate of heat rejection increases proportionally to the surface area and the average surface temperature of the radiator to the fourth power, increasing the heat rejection temperature is very effective in reducing the size and mass of the radiator. On the other hand, increasing the radiator surface temperature increases the reactor operation temperature and decreases the thermal efficiency of the power system, which in turn increases the heat rejection load of the radiator.

26.4.1 Power system integration

The layout and integration of a space reactor power system is a complex and iterative task that directly affects the system's operation and safety. Successful design and integration of a space reactor power system require employing advanced computation, iterative modeling and design, advanced simulation methods and tools, and proper selection and extensive development and testing of components. Figure 26.5 presents a layout of a generic space nuclear power system that provides electrical power to a number of ion thrusters that propel the spacecraft and a science payload for an unmanned mission to Mars or Jupiter. The height and diameter of the reactor, the thickness of the radiation shadow shield and its separation distance from the reactor, the surface area of the heat rejection radiator, and the diameter of the payload and its separation for the radiation shadow shield dictate the half-cone angle, q, typically 12–17°, minor and major diameters of the radiator and the total length of the power system (Figure 26.5).

The reactor heat source at the apex of the power system has the smallest volume and the third largest mass, after the radiator and the radiation shield. The reactor is followed by the truncated cone of the radiation shadow shield for protecting the electronic and other system components from exposure to the fast neutrons and high-energy gamma photons from the reactor (Figure 26.5). The radiation shadow shield also limits the neutrons' fluence and the dose from the gamma photons from the reactor to specified values at the payload, located 10–25 m away from the shield. The radiation shield is the second largest and most massive component after the radiator. The thickness of the radiation shadow shield, typically made of layered tungsten and both natural and depleted LiH, depends on the steady-state thermal power of the reactor and its operation life.

The primary heat transport system (PHTS) in Figure 26.5 removes the generated heat in the reactor by fission to the power conversion modules within the radiator cavity, behind the radiation shield. The power management and distribution subsystem is also placed in the radiator cavity, protected by the radiation shield. The radiator rejects the waste heat, primarily from its outer surface and partially from its inner surface, through the rear opening of the radiator cone. To enhance the radiator's performance and protect it from potential damage by the impact by space debris, its outer surface should be properly armored, for example using a strong carbon–carbon composite structure with an emissivity of 0.85–0.95. Increasing the reactor operation temperature affects the choice of the structural materials and energy conversion technology, directly impacting the power system's operation, reliability, safety, and the total size and mass. The next section addresses the effect of the reactor temperature on the selection of structure material.

26.4.2 Materials selection and consideration

Increasing the reactor operation temperature shifts the choice of the structure materials from steel alloys, to super-steel alloys, and finally to refractory alloys, with progressively higher density and less fabrication and operation experience (Figures 26.7 and 26.8). In addition to the compatibility with the reactor coolant and nuclear fuel material and the strength at the system's operation temperatures, structure materials should experience the least creep and embrittlement in the reactor's radiation environment. Potential coolants are either liquid metals such as NaK-78, sodium, or lithium in order of increasing temperature of binary mixture of noble gases such as He–Xe. The radiation embrittlement of structural materials, and the decrease in their strength with temperature, limits the practical range of their operating temperatures, particularly for the refractory alloys of niobium, molybdenum, tantalum, and tungsten (Figure 26.8).

The lower temperature limits for the radiation embrittlement of niobium (Nb), molybdenum (Mo), tantalum (Ta), and tungsten (W) alloys are relatively high, as much as ~ 800, 1000, 1100, and 1200 K respectively [10]. Such high temperatures mean that decreasing the reactor's thermal power and hence its operating temperature will compromise the strength and ductility refractory alloy structure. The induced stresses could fail the structure. For liquid-metal-cooled space reactors, a small rupture in the structure of the pipes could result in significant amounts of droplets being ejected into space. Similar events have been associated with the ejection of the fuel cores of the former Soviet Union's Radar Ocean Reconnaissance Satellite (RORSAT) reactors, causing significant pollution of Earth orbits with NaK-78 droplets. Many of these droplets are still in orbit [11,12].

In Figure 26.7, the triangular solid symbols indicate space reactor power systems that have either been launched, such as SNAP-10A, or have undergone major hardware development and testing (e.g.

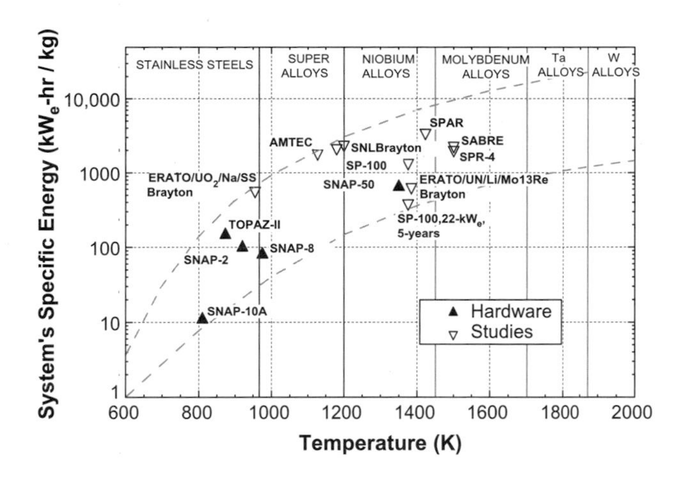

FIGURE 26.7

Effects of reactor temperature on choices of structural material and specific energy of space reactor power systems [10].

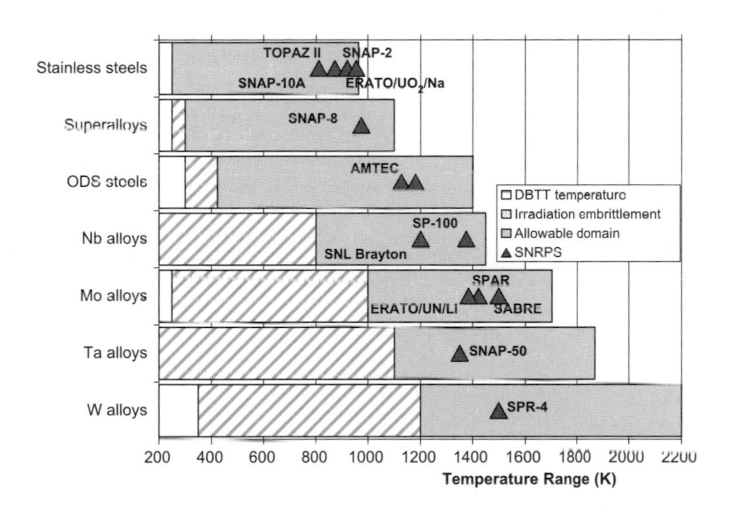

FIGURE 26.8

Operation temperature of structural materials for space reactor power systems [10].

the USA's SNAP-2, SNAP-8, and SNAP-50, and the former Soviet Union's TOPAZ II) [3,4]. The triangular open symbols indicate conceptual designs of space power systems developed in the USA (e.g. SP-100, SPAR, SABER, and SNL Brayton) [13–18], France (e.g. ERATO) [19, 20], and at the University of New Mexico (e.g. SAIRS with AMTEC energy conversion) [21].

Below 1000 K, a fast energy spectrum space reactor could be cooled using circulating liquid metal NaK-78 and employ stainless steel structure (Figures 26.7 and 26.8). Increasing the reactor's operation temperature to 1000–1200 K rapidly increases the specific power of the system and dictates using a structure made of super-steel alloys, with well-known properties and vast operational experience. Increasing the reactor's operation temperature beyond 1200 K requires using a structure made of refractory alloys that are heavier and have limited fabrication and no space operation experience. In the radiation environment of a fission reactor, only steel and super-steel alloys remain ductile at room temperature.

The oxide dispersed steels (ODSs), either iron or nickel based, experience radiation embrittlement at low temperatures, <400 K, and their upper operation temperature limit is higher than those of conventional steel and super-steel alloys. Remaining fabrication issues with ODS include rolling and welding; however, they are widely used in the fabrication of high-temperature gas-turbine blades and casing [10]. In addition to the high radiation embrittlement temperatures, some refractory alloys (such as those of Nb) lose ductility when exposed to oxygen of only a few to tens of ppm, thus they cannot be used in the CO_2 atmosphere of Mars.

The space reactor power systems deployed into low Earth orbits by the former Soviet Union in the 1970s to the 1980s to power their RORSATs, and the SNAP-10A, launched by the USA in 1965, both used stainless steel structures, circulating liquid NaK-78 coolant and thermoelectric (TE) energy conversion, and operated at or below 900 K. The next section reviews various choices and appropriate combinations of space reactor types and energy conversion technologies.

26.4.3 Reactor types and energy conversion options

There are many possible matching combinations of space reactor types and energy conversion technologies, static or dynamic, for best performance and integration of the power system and enhanced operation reliability and safety (Table 26.1):

1. **Liquid metal heat pipe (LMHP)-cooled reactors** are a good match for the energy conversion technologies of TE [9,14,22–26], alkali metal thermal-to-electric conversion (AMTEC) [21,27–32], thermionic (TI) [15,16,33–37], liquid metal Rankine cycle [3], and Free Piston Stirling Engines (FPSEs) [38–40].
2. **Circulating liquid metal (LM)-cooled reactors** are also a good match for all energy conversion technologies in (1) above [3,4,9,13,23,41–45]. Although both LMHP- and LM-cooled space reactors could also be coupled to closed Brayton cycle (CBC) energy conversion, with noble gas binary mixture working fluids, they would require employing heat pipes/gas or liquid metal/gas heat exchangers and increase the complexity of the power system integration. This, in turn, increases the mass and size of the power system.
3. **Gas-cooled reactors** are a good choice for indirect or direct coupling to CBC energy conversion using single-shaft, centrifugal flow turbo-machines [18,46–50]. These space power systems could be designed with multiple loops for redundancy and enhanced reliability, but at the expense of the increased complexity of the system's integration, mass, and volume.

Other important considerations in the selection and assembly of an energy conversion technology in space reactor power systems include modularity, operation reliability, safety, scalability, load-following characteristics, radiation hardness, thermal efficiency, the reactor exit temperature, and the average surface temperature of the radiator for heat rejection. The next section briefly reviews various static and dynamic conversion technologies for use in space nuclear reactor power systems. Some have, and could be, used in radioisotope power systems [51–55].

Table 26.1 Matching Choices of Space Reactor Types and Energy Conversion Technologies

Reactor Type: Coolant Type	Reactor Exit Temperature (K)	Energy Conversion	Thermal Efficiency/Radiator Type/Temperature (K)
1. Circulating liquid-metal-cooled reactors:			
NaK-78 alloy Sodium (Na) Depleted lithium (Li)	<1000 <1200 <1400	Thermoelectric, thermionic, thermal photovoltaic (TPV), potassium Rankine cycle, Free Piston Stirling Engines, alkali metal thermal-to-electric conversion (AMTEC)	5–15%/potassium heat pipes/<800 5–20%/potassium heat pipes/<800 8–18%/water heat pipes/<400 12–20%/rubidium heat pipes/<600 18–30%/water heat pipes/<400 20–30%/rubidium heat pipes/<600
2. Liquid metal heat-pipe-cooled reactors:			
Sodium (Na) Depleted lithium (Li)	<1200 <1400	Thermoelectric, thermionic, thermal photovoltaic (TPV), potassium Rankine cycle, Free Piston Stirling Engines, AMTEC	5–15%/potassium heat pipes/<800 5–20%/potassium heat pipes/<800 8–18%/water heat pipes/<400 12–20%/rubidium heat pipes/<600 18–30%/water heat pipes/<400 20–30%/rubidium heat pipes/<600
3. Gas-cooled reactors:			
He–Xe binary mixture (15–40 g/mol)	<1200	Closed Brayton cycle (CBC) with single-shaft centrifugal flow turbo-machines, with shaft rotation speed of 35–45 krpm	15–30%/water heat pipes and circulating liquid NaK-78 loops/<500

26.4.4 Energy conversion options and structure materials

In addition to the absence of moving parts, static conversion technologies are inherently modular and load-following [3,4,13,21,26,29,55]. The SiGe thermoelectric converters, when operating between 1273 and 790 K, have a thermal efficiency of ~6%, representing only ~16% of Carnot. Thus, for a 100 kW_e space reactor power system, the surface area of the heat rejection radiator could exceed 100 m^2 and the reactor exit temperature could be as high as 1373 K, requiring refractory alloy structures such as PWC-11 (Nb–1% Zr, 0.1% C), Mo–xRe, and TZM (<99.5% Mo, 0.5% Ti, 0.08% Zr, and traces of carbon for carbide formation) [10].

The PWC-11 alloy was used as structural material in SNAP-50 space reactor power system developed in the 1960s [3,43] and as cladding materials of the UN fuel pins in the SP-100 space reactor, developed in the USA in the 1980s and 1990s [4,13,17,41,42]. Neither reactor had been flown. The Mo–xRe alloys are much heavier than either PWC-11 and TZM, but much stronger. The addition of Re also improves ductility but increases the cold work needed for shaping the alloy. Alloys with as much as 50% Re are available commercially.

The TZM alloy offers better weldability and twice the strength of pure molybdenum (Mo) at temperatures exceeding 1300°C and its recrystallization temperature is ~250°C higher than that of Mo.

The finer grain structure of the TZM alloy, and the formation of TiC and ZrC at the grain boundaries, inhibit grain growth and related failure as a result of fractures along the grain boundaries, and improve weldability. This alloy is widely used in high-temperature applications such as turbine blades and compact heat exchangers. As indicated earlier, these alloys are heavy, and those of niobium are incompatible with oxygen, and so could not be used on Mars without a protective coating.

The SiGe thermoelectric converters had been very reliable, with flight experience on 23 NASA missions powered by RPSs and GPHS-RTGs [1–5]. These converters typically operate at a hot-side temperature ~1273–1300 K and a thermal efficiency <5% [51,52]. The skutterudite, segmented thermoelectric (STE) converters are more efficient than SiGe because the materials of the segments in the n- and p-legs operate in the temperature range in which they possess the highest figure of merit (FOM), or Z values [51–54]. For example, an SiGe unicouple operates between a hot-side temperature $T_h = 1273$ K and a heat rejection temperature $T_R = 700$ K, $Z \sim 0.7 \times 10^{-3}$ K^{-1}, while STE operates between $T_h = 973$ K and $T_R = 300$ K, and has $Z \sim 1.15 \times 10^{-3}$ K^{-1}, and a thermal efficiency ~14.8% (Figure 26.9) [51,52].

A number of STE converters have been fabricated, using p-type $CeFe_4CoSb_{12}$ and Bi_2Te_3-based alloys and n-type $CoSb_3$ and Bi_2Te_3-based alloys, and tested at cold and hot shoe temperatures of 300 and 973 K respectively [51–54]. For a cold shoe temperature of 300 K, the thermal efficiency of a space reactor power system that uses STE could be ~13%, but the radiator's specific area would as much as ~15.8 m^2/kW$_e$ (Figures 26.9 and 26.10). Increasing the radiator temperature to 373 and 573 K decreases the thermal efficiency of the reactor power system with STE converters to 12.4% and 7.8%, but also decreases the radiator's specific area to 8.0 and 2.45 m^2/kW$_e$ (Figures 26.9 and 26.10). The estimates in these figures assume that the thermal efficiency of the power system efficiency is 90% of that of the converter [23].

A cascaded TE converter with STE bottom unicouples and SiGe top unicouples would have higher thermal efficiency [51]. The top SiGe elements will operate at a hot-side temperature of 1273 K and the

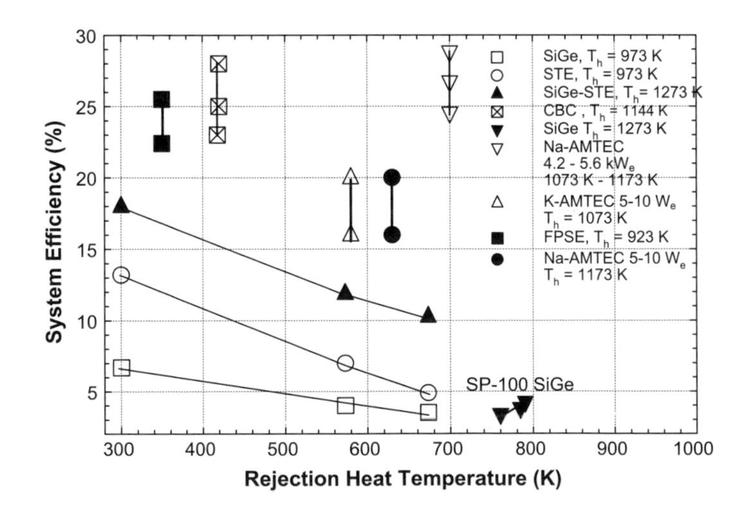

FIGURE 26.9

Thermal efficiency of space reactor power systems with different energy conversion [23].

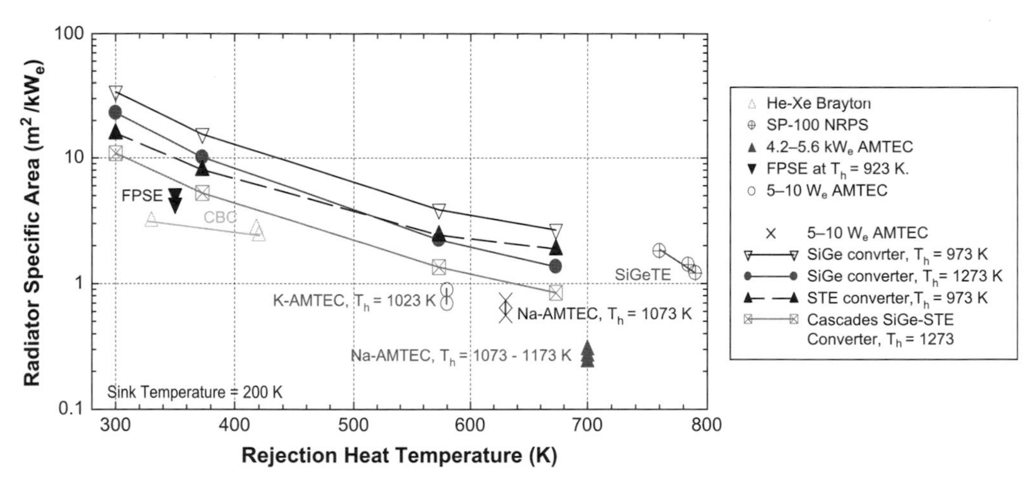

FIGURE 26.10

Radiator specific area for space reactor systems with different energy conversion [23].

STE bottom elements will operate at a hot-side temperature of 973 K. For an average radiator temperature of 680 K, the space reactor power system with cascaded SiGe/STE converters could have a thermal efficiency of more than 10% and the specific area of the radiator would be ~0.85 m^2/kW_e (Figures 26.9 and 26.10) [23].

The AMTEC energy conversion offers great operation and integration advantages. This static conversion technology has no moving parts, operating at the highest fraction of Carnot cycle efficiency, and relatively low hot-side (<1150 K) and relatively high cold-side temperature for heat rejection (~550–650 K). The relatively low hot-side temperature for AMTECs decreases the reactor exit temperature to <1150 K, making it possible to use relatively lighter super-steels, mechanically alloyed oxide-dispersed steels (MA-ODS) or titanium alloy structure materials [10], enhancing the operational reliability of the power system.

Though AMTEC technology is currently at a Technology Readiness Level-3 (TRL-3) and has never been flown into space, recently designed units could have a thermal efficiency in excess of 20% [21,29]. Thus, in order to be considered for future deployment, within a decade, AMTEC technology needs to be advanced to TRL-5. Potassium and sodium AMTEC units typically operate at moderate hot-side temperatures, T_h, of 1000 and 1123 K and radiator surface temperature of 550 and 650 K respectively. An Na-AMTEC unit (4–6 kW_e), operating between 1123 and 650 K, could have an efficiency ~6% [21,29], representing ~60% of Carnot. The radiator area for a 100 kW_e space reactor power system with these units would be <25 m^2 (Figures 26.9 and 26.10). With TE and TI energy conversion, the specific power of a space reactor power system ranges from 5 to 15 W_e/kg and may exceed 30 W_e/kg with AMTEC [4,9,33–36,41,42,46].

Dynamic energy conversion technologies of CBC and FPSE, with rotating and linear alternators respectively, typically operate at low radiator temperatures (350–450 K), but high thermal efficiencies (23–35%) [39,40,46–50]. They also are inherently radiation hard, but not load-following. With dynamic energy conversion technologies, including K-Rankine cycle, CBC, and FPSE, the specific power of a space reactor power system could be 10–30 W_e/kg [3,38].

In the 1960s to the 1980s, both in the USA and the former Soviet Union, much research was performed to develop various static and dynamic energy conversion technologies, except AMTEC. Despite the extensive development of energy conversion technologies for space reactor power systems, the RORSATs reactor systems deployed by the former Soviet Union in the 1970s and 1980s, and the SNAP-10A launched by the USA in 1965, used SiGe thermoelectric elements for energy conversion. The former Soviet Union also launched two reactor power systems with in-core thermionic conversion in the late 1980s.

The next section briefly reviews the operation history of these space reactor power systems, with emphases on lessons learned relative to the design, operation, safety, and end-of life storage [56].

26.5 DEPLOYMENT HISTORY OF SPACE REACTOR POWER SYSTEMS

During most of the later half of the last century, the former Soviet Union (currently the Russian Federation) and the USA had active programs to develop nuclear reactor power systems for use in space [1,3,4,9,57]. These programs had two basic components: (a) developing space reactor power systems for generating electrical power from a few to tens of kW_e for missions lasting from a few months up to several years, and (b) developing nuclear reactors for thermal propulsion (NTP) to enable future missions to Mars. Additional research focused on developing ion and plasma thrusters that could be coupled with a space reactor power system for interplanetary propulsion and fast travel to Mars and beyond. Limited and intermittent efforts in both countries investigated bimodal space reactor systems for the generation of electricity and NTP for missions requiring tens of kW_e, and tens to thousands of newtons of thrust at a specific impulse of 600–9000 s.

26.5.1 The BUK power system

The former Soviet Union's "BUK" reactor systems, with SiGe thermoelectric conversion, had generated ~3 kW_e each and powered a total of 31 RORSATs in the 1970s to 1980s. The BUK reactor had a fast neutron energy spectrum and the RORSATs were launched into 260 km Earth orbit. Two of the former Soviet Union's TOPAZ power systems, with in-core thermionic conversion and epithermal neutron energy spectrum reactors, for generating 5.5 kW_e each, powered two *Cosmos* missions that were launched in 1987 in ~800 km Earth orbit. The USA's SNAP-10A power system, with SiGe energy conversion and a thermal neutron energy spectrum reactor, generated ~0.5 kW_e and was launched in 1965 in 1300 km Earth orbit [56,58].

The three space reactor power systems (BUK, TOPAZ, and SNAP-10A) used circulating liquid metal NaK-78 for cooling the nuclear reactor, stainless steel structure, and highly enriched uranium fuel (90–96 wt%), and operated at reactor exit temperatures of 833–973 K [36,56–61]. The BUK reactors used U–Mo fuel rods, the TOPAZ reactor used UO_2 fuel rods and four ZrH moderator disks for moderating the neutrons, and SNAP-10A used moderated U–ZrH fuel rods. These low-power space reactor systems were designed for short missions (~1 year SNAP-10A, <5 months BUK, and up to 1 year TOPAZ).

The BUK reactor core comprised 37 2-cm-diameter rods of highly enriched (95 wt% in ^{235}U) U–Mo alloy fuel arranged in a triangular lattice [56,59,60]. The hexagonal vessel of the BUK reactor was surrounded by a circular radial beryllium reflector (Figures 26.11 and 26.12). The fuel height in the rods was ~15 cm and there were 10-cm-long Be axial reflector segments at both ends of the rods. The BUK reactor's nominal exit temperature and thermal power were of 973 K (700°C) and $\leq$100 kW_{th} [56–61]. The BUK reactor core was 0.2 m in diameter, 0.6 m long, and weighed ~53 kg, including ~30 kg of uranium. It was controlled using six sliding beryllium cylinders, 10 cm in diameter and 15 cm long. They moved axially within the radial Be reflector to adjust the neutron leakage from the reactor core throughout its operation life (Figure 26.12). During launch, at startup in orbit, and when the reactor is shut down, the sliding Be cylinders would be fully withdrawn; they would be fully inserted at the end of operation life of the BUK reactor.

The BUK reactor power system had two liquid NaK-78 loops. The primary loop cools the reactor core and transports the heat generated in it by fission to the thermoelectric (TE) energy conversion modules in the cavity of the heat rejection radiator. The secondary loop of circulating liquid NaK removes the heat rejected by the TE models at ~623 K and transports it to the heat rejection radiator. The circulation of the liquid NaK in the primary and secondary loops was accomplished using thermoelectric–electromagnetic (TEM) pumps. The TE modules in the power system were divided into two groups: one supplied electrical power to the spacecraft and the other supplied power for operating the EM pumps. The beginning-of-life thermal efficiency of the BUK reactor power systems was ~3%. At the end of operation life of 440 hours, the generated electrical power decreases by 10%, primarily due to performance degradation of TE elements.

The initial BUK power system had a safety system for boosting the deactivated reactor at the end-of-life into a circular storage orbit with an altitude greater than 850 km. In such a storage orbit with a decay life of ~1000 years, the radioactivity of the fission products in the reactor core would eventually decay to a low level. The RORSATs BUK reactors operated for periods from a few hours to ~4.5 months.

The two BUK reactor power systems for *Cosmos 954* and *Cosmos 1402* failed to boost and re-entered the earth's atmosphere. *Cosmos 954* re-entered on 24 January 1978 with the shutdown reactor

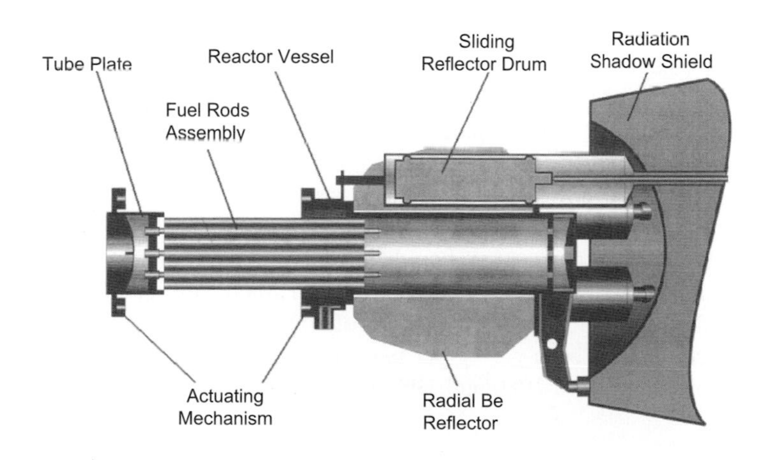

FIGURE 26.11

A schematic axial cross-section of the ejection of the BUK reactor fuel core in storage orbit.

Adapted from Ref. [57]. Original source: Kurchatov Institute, Russian Federation

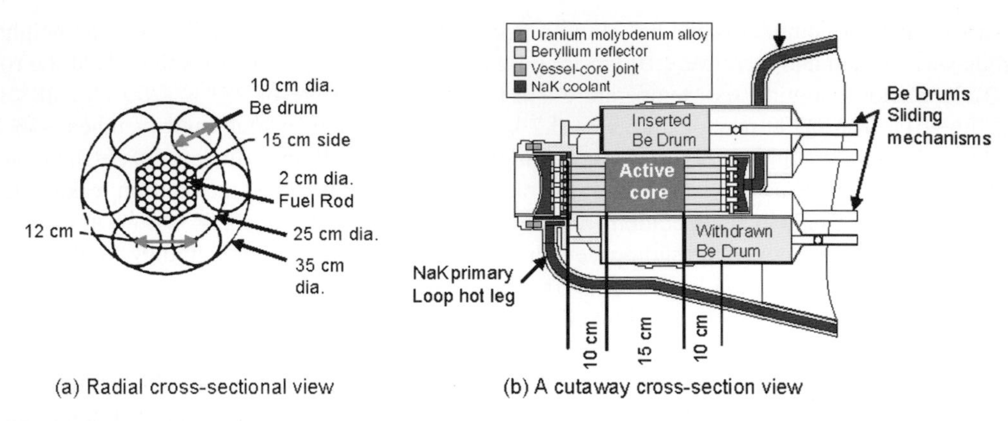

FIGURE 26.12

Radial (a) and axial (b) cross-sections of BUK reactor.

Adapted from Refs [59–61]

still attached over Queen Charlotte Island in northern Canada. The reactor disintegrated during re-entry, but an estimated ~20% of the fuel debris did not burn and came to Earth, spreading over 124,000 km^2 just east of the Great Slave Lake [59–62].

By 1978, a total of 14 RORSATs with BUK reactor power systems had been deployed, and all, except *Cosmos 954*, were boosted successfully into a storage orbit. Subsequent to the *Cosmos 954* incident, RORSATs launches were halted until 1980. When resumed, the BUK boost system was redesigned with a backup to avoid re-entry of deactivated reactors [56,59–62]. Between 1980 and 1988, the BUK reactors on 16 RORSATs with the redesigned safety system were boosted successfully into a storage orbit. These are *Cosmos 1178* through *Cosmos 1932*, excluding *Cosmos 1402*, which failed to boost and re-entered on 7 February 1983 when the reactor fuel elements re-entered over the South Atlantic [56,59–62].

Radar observations and continuous monitoring facilities of space objects in Earth orbits had firmly confirmed that the ejection of the 16 BUK fuel cores between 1980 and 1988 had populated Earth orbits with tens of thousands of NaK droplets that varied in size from 100 μm to 5 cm [11,12]. The failure of the primary coolant loop seal that accompanied the ejection of the fuel assembly may have caused most of the liquid NaK-78 in the BUK core, and some of that in the primary loop, to be discharged in the form of droplets, at an average velocity of 7–13 km/s [11,56]. The secondary loop of the BUK power system contained ~26 kg of liquid NaK-78, but was designed to stay sealed following the ejection of the reactor core fuel in storage orbit. The total amount of liquid NaK in the BUK reactor core and the primary loop was ~13 kg. The total mass of liquid NaK discharged into Earth orbits is not actually known, but could have been ≤5 kg per reactor, for a total of ≤80 kg for the 16 BUK cores ejected into storage orbits between 1980 and 1988 [56].

26.5.2 The TOPAZ power system

The former Soviet Union had also developed higher-power (>5 kW_e) reactor systems with in-core thermionic conversion [4,13,36,57,63–65]. The first reactor design, know as TOPAZ [36], used multi-cell thermionic fuel elements (TFEs) (Figure 26.13) and the second design, known as YENISEI (or TOPAZ-II), employed single-cell TFEs [13,35–37,63,64]. Two TOPAZ reactor power systems had been flown on two missions (*Cosmos 1818* and *Cosmos 1867*) of 6 months and 1 year. *Cosmos 1818* and *1867* were launched on 2 February and 10 July 1987 into an Earth orbit with an altitude between 789 and 802 km to test ion engines. The TOPAZ power system (Figure 26.13) nominally generated ~5.5 kW_e and had a thermal efficiency <5.0%.

The TOPAZ reactor core contained 79 TFEs with five thermionic cells each and four monolithic ZrH disks with penetration for the TFEs. The reactor was cooled with circulating liquid NaK-78 and operated at an exit temperature of ~970 K. A total of 60 TFEs supplied the electrical power to the load, and 19 powered the induction electromagnetic pump for circulating the liquid NaK in the reactor and the primary loop. The TFEs were loaded with ~96 wt% enriched UO_2 fuel pellets. The BeO pellets at both ends of the TFE rods served as the axial neutron reflector [13,35,36,57,63]. The TOPAZ reactor had an epithermal neutron energy spectrum.

During operation, the cesium vapor pressure in the inter-electrode gap of the thermionic cells was maintained at 0.4–1.5 torr to enhance electron emission and neutralize the space charge. The cesium to the TFEs was provided from a reservoir of a predetermined capacity, and then vented into space. Thus, the operation life of the TOPAZ system was limited by the amount of cesium on board. The fission gases generated in the TFEs during reactor operation were also readily vented into space.

The fully integrated TOPAZ reactor power system was 4.7 m long with a radiator major diameter of 1.3 m (Figure 26.13). The TOPAZ reactor had 12 beryllium rotating drums in the radial beryllium reflector. The drums had 120° thin angular segments of B_4C neutron absorber for regulating the reactivity in the reactor core. The control drums maintain the reactor subcritical during launch, start up the reactor in orbit, regulate reactor operation, and shut down the reactor at the end of the mission or in the case of an emergency. The radial Be reflector disassembles in the case of an emergency, forcing a reactor shutdown. The radiation shadow shield consists of alternating layers of LiH and depleted uranium for neutrons and gamma photons respectively [13,63], and the shield truncated cone was encased in stainless steel (Figure 26.13).

Instead of the ZrH disks in the TOPAZ reactor core, the YENISEI reactor [4,13,63] employed a single monolithic block of ZrH moderator with penetrations for 37 single-cell TFEs, 34 dedicated to the electrical load and three providing high-current, low-voltage DC power to the electromagnetic

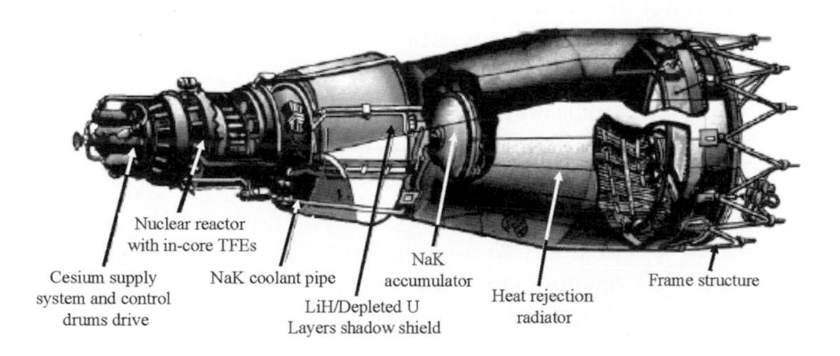

FIGURE 26.13

An isometric view of the TOPAZ space reactor power system.

Adapted from Ref. [57]. Original source: Kurchatov Institute, Russian Federation

induction pump [35]. The YENISEI reactor power system was also cooled with circulating liquid NaK-78. It generates almost the same electrical power and has similar conversion efficiency to the TOPAZ power system. As indicated earlier, while two TOPAZ systems had been deployed and operated successfully in space, no flight test was conducted for the YENISEI reactor system, designed for longer missions of 3 years. A number of YENISEI power systems had undergone extensive ground testing and performance evaluation for more than 25,000 cumulative hours [65].

26.5.3 SNAP-10A power system

The SNAP reactors systems, developed by the USA in the 1960s and 1970s, used moderated fuel elements [3,43,58]. The SNAP-10A reactor power system was the only one deployed by the USA and operated in Earth orbit. The reactor core comprised 37 fuel rods, 3.175 cm in diameter and 33 cm long. The rods were arranged in a triangular lattice with a pitch of 3.2 cm (Figure 26.14a), loaded with homogeneous U–ZrH pellets containing 10 wt% ^{235}U, and clad in 0.381-mm-thick Hastelloy with an internal hydrogen retention barrier [66] (Figure 26.14b). The SNAP-10A reactor had a thermal neutron energy spectrum and was cooled with circulating liquid metal NaK-78. The temperature of the liquid NaK exiting the reactor at full power operation was ~833 K. The spacing between the hexagonal core and the cylindrical steel vessel of the reactor was filled with beryllium (Be) wedges, for enhancing the structural integrity of the reactor core and reflecting escaping neutrons. The Be wedges also reduced the amount of water that could enter the core when submerged in wet sand and flooded with seawater, following a launch abort accident (Figure 26.14a). The reactor vessel was surrounded by Be reflector, 6.35 cm thick [3,43,58].

As a safety measure, the cold clean reactor core was significantly subcritical during all phases, including handling, transportation, integration into the launch vehicle, and accidental submersion in wet sand and flooding with seawater. The SNAP-10A reactor was controlled using four ex-core elements of half-cylindrical Be sections in the radial Be reflector. The rotation of these reactivity control elements changed the radial reflector parameters for controlling neutron leakage from the reactor core. The radial Be reflector and the control assembly were held in place using a retaining band anchored by an explosive bolt. This bolt was severed by a command from the ground or re-entry heating [3,43,58].

The nominal thermal power of the SNAP-10A reactor was 34 kW_{th}. The power system employed SiGe thermoelectric elements for partially converting the reactor thermal power to DC electrical power for the payload. The SNAP-10A power system was designed to generate 500–600 W_e at a thermal efficiency of ~1.47% for 1 year. Such low efficiency is because of the low hot-junction temperature (777 K) and the small temperature differential across the SiGe elements (167 K). All coolant ducts in SNAP-10A were made of stainless steel and the radiator's emitting surface was made of thin aluminum fins with high emissivity coating (0.9).

A thermoelectric–electromagnetic (TEM) pump with a permanent magnet circulated the liquid NaK-78 through the SNAP-10A reactor core and the ducts for transporting the reactor thermal power to the SiGe conversion elements. The electrical power for operating the TEM pump was provided by PbTe thermoelectric elements. Connected in parallel, they supplied several hundred amperes of current at very low voltage (~0.2 V) to the TEM pump [3,43,58]. The TEM pump has its own heat rejection radiator and begins operation when adequate voltage is developed across the PbTe thermoelectric elements. This pump, with no moving parts, continues to operate as long as the temperature of the

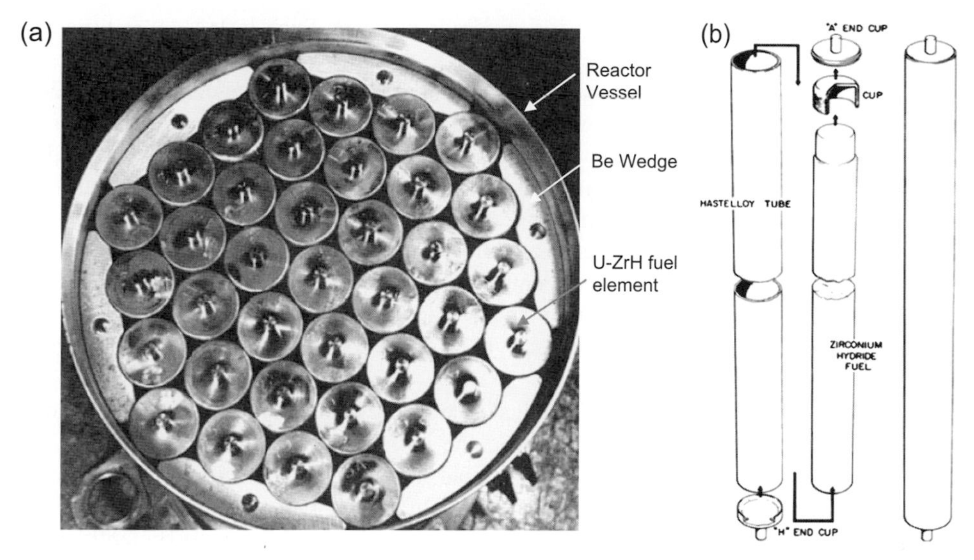

FIGURE 26.14

SNAP-10A reactor core and fuel element.

Adapted from Ref. [66]

liquid NaK in the reactor is higher than that of the cold side of the PbTe thermoelectric elements (or the pump's radiator). Thus, the operation advantage of using a TEM pump is the continuous circulation of the liquid metal coolant through the reactor after shutdown, passively removing the decay heat from the reactor core. This is a desirable design, operation, and safety feature for liquid-metal-cooled space reactor power systems.

The SNAP-10A power system, launched on 3 April 1965 on board an Atlas-Agena vehicle, was deployed into a circular orbit with an average altitude of ~1300 km. The inlet and exit temperatures of the liquid NaK in the reactor during full-power operation were ~560°C (833 K) and 487.8°C (760.8 K), and the average surface temperature of the radiator was ~315.6°C (588.6 K). After 43 days of operation, the power system was generating ~535 W_e, at which time (10 May 1965) the reactor was shut down in a normal response to an erroneous signal caused by the failure of a voltage regulator on board the spacecraft [43,58]. To this day, SNAP-10A is still stored in its deployment orbit with a decay life >1000 years.

26.6 DISCUSSION AND REMARKS

The deployment history of space reactor power systems in Earth orbits by the former Soviet Union and the USA in the 1960s to the 1980s provides the basics for future development of guidelines and recommendations to enhance the safety of these systems. *Cosmos 1818* and *Cosmos 1867* were deployed and operated in an 800 km orbit. All RORSATs with "BUK' reactor systems were deployed and operated in 250 km, circular Earth orbit, with the intention to boost and store deactivated reactors in higher Earth orbits (>800 km altitude) for hundreds of years. Such long storage time allows the radioactivity of the fission products in the reactor fuel to decay to very low levels so as not to present a radiological concern to the Earth's inhabitants, in the case of eventual re-entry and dispersion.

Space reactors use highly enriched uranium fuel (93–95 wt% in ^{235}U) in order to realize the compact size, low mass, and long operation lifetime of many years without refueling. The radioactivity resulting from the natural decay of ^{235}U is significantly small (half-life ~7.04×10^8 years). It decays into ^{234}U with a relatively shorter half-life of 2.44×10^5 years. Thus, the radioactivity in a space reactor at launch, and before being operated at full power, is mostly due to the radioactive decay of ^{234}U in the fuel. Owing to the very small concentration of ^{234}U and the long half-lives of both ^{235}U and ^{234}U, the specific activity of the fresh fuel in space reactors is of the order of 80 μCi/g or less. For a ^{235}U fuel loading of 100 kg, the total radioactivity in the reactor core fuel would be of the order of ~8 mCi [67].

Thus, nuclear reactors do not pose radiological risk during launch and until deployed and operated at a steady power level. In the operating reactor, the radioactivity in the fuel builds up rapidly to millions of curies due to the accumulation of fission products. Thus, orbiting space reactor power systems represent a radiological concern in the case of an accidental re-entry. To avoid such an occurrence, orbiting space reactor power systems should be deployed in sufficiently high orbits (SHOs), with a natural decay life >1000 years. These orbits are also suitable for end-of-life storage of deactivated reactors until an eventual re-entry, at which time the radioactivity in the fuel core drops close to background.

Past experience with deployment and use of space reactor power systems has been limited to low-power (0.5–5.5 kW_e) and short-duration missions (several hours to 1 year). Future power needs for deep space exploration missions and outposts on the Moon and Mars range from tens to thousands of kW_e, and present additional challenges to the design and integration of these systems. Initial uses of these power systems would likely be for robotic missions that would naturally by followed by human excursions, posing a new set of challenges. Chief among these is protecting the astronauts from harmful energetic space radiation. Developing lightweight effective shields and shortening the travel time to destination, which is possible using nuclear electrical or nuclear thermal propulsion systems, will effectively reduce the astronaut's exposure to radiation.

Based on past deployments of space reactor systems in Earth orbits in the 1960s to the 1980s, the following suggestions are offered for future discussions on developing global safety guidelines and regulations to ensure the safety of reactor power systems in future missions:

1. **Militarization of space, including Earth orbits, should not be allowed.** Space reactor power and propulsion systems are enabling civilian missions in deep space and selected Earth orbits, when there is no other practical or viable alternative. The use of these systems for military purposes, either in deep space or Earth orbits, should be prohibited. Practical and environmental limitations favor reactor power and propulsion systems for lunar and Martian outposts and deep space exploration for scientific purposes, and thus these should be allowed. For these missions, the solar option is either nonexistent or impractical.
2. **Uses of space reactor power systems in Earth orbits should be strongly discouraged**, unless justified for potential civilian missions of global benefit and where the solar option is impractical, such as within the Van Allen radiation belts. Orbits with altitudes of 1000–3000 km, or even higher, are also suitable for end-of-life storage of deactivated reactors for hundreds of years. In

these orbits, the radioactivity of the fission products in the deactivated reactor fuel will naturally decay to background level, thus presenting no risk to Earth's inhabitants in the case of an eventual re-entry.

3. **Passive and safe removal of the decay heat from the reactor core after shutdown should be required.** For circulating liquid-metal-cooled space reactors, this could be achieved using TEM pumps [43,58,67,68]. For liquid metal heat-pipe-cooled reactors, the heat pipes provide passive and effective means for cooling the reactor both during operation and after shutdown. Gas-cooled space reactor power systems with multiple CBC loops for energy conversion may require maintaining coolant circulation for several hours or a few days after reactor shutdown.
4. **Operate reactors at exit temperatures <1000 K**, in order to use conventional structure materials, such as stainless steel or super-steel, with extensive fabrication experience, prior uses in space and well-known properties for operation in a radiation environment. Raising the operation temperature beyond 1000 K raises compatibility and weight challenges, and may require materials development and increase the lead time for testing and validation of strength and reliability in the space radiation environment. Having to use heavier and relatively less ductile refractory alloys at higher operating temperatures (>1200 K) increases the launch mass of the power system and the radiation embitterment of the structure. Some refractory alloys may require special coatings for protection from the environment on the surface of some planets. For example, the presence of CO_2 in the atmosphere of Mars will cause oxygen embrittlement of all niobium-based alloys.
5. **Ejecting the fuel core into orbit at the end of mission should not be allowed.** The selected operation Earth orbit should be sufficient to also be used for end-of-life storage of the deactivated reactors for hundreds of years. Interplanetary missions with nuclear electric or nuclear thermal propulsion with a launch base in an SHO should also consider end-of-life storage of deactivated reactors in these orbits, without interfering with or compromising operating satellites. Lessons learned from the deployments of the RORSAT's BUK reactor power systems and their operation history are that ejecting the reactor core fuel into storage orbit significantly contaminated Earth orbits with NaK particles, compromising the safety of space assets and causing radiological contamination of the ionosphere [69]. Man-made debris is increasingly becoming a major threat to space assets in Earth orbits and thus should be minimized by all means possible [11,12,70].
6. **Designing space reactor power system with no single-point failures in reactor cooling and energy conversion should be encouraged.** As detailed in the next section, liquid metal heat-pipe-cooled, and circulating liquid- metal- and gas-cooled space reactor power systems could be designed with no-single point failures in reactor cooling and energy conversion. In addition to enhancing the safety and reliability of the system operation, this design feature ensures mission success.
7. **Development and use of high thermal efficiency energy conversion technologies should be strongly encouraged.** Most static energy conversion technologies are inherently load-following and reliable due to the absence of moving parts. However, their low thermal efficiency (except AMTEC) increases the reactor thermal power, with significant radiological and system integration consequences. For a given mission power requirement, low thermal efficiency would certainly increase the overall system size and mass, and the launch cost. In addition to increasing the inventory of fission products built up in the core of operating reactors, low

thermal efficiency depletes the inventory of ^{235}U fuel in the reactor core at a higher rate, increases the thickness and mass of the radiation shadow shield for protecting the electronics and control equipment onboard and the payload, and increases the size and mass of the heat rejection radiator. The final outcome is an increase in the power system's size and mass and, depending on the launch vehicle to be used, additional work will be needed to fit the stowed power system into the pay of the launch vehicle. Dynamic energy conversion technologies typically operate at relatively higher thermal efficiencies than static options (except AMTEC), but they involve moving parts and are inherently non-load-following. They have a limited modularity, add complexity to system integration and operation, and none has been tested or operated in space. The latter is also true for static energy conversion technologies, except thermoelectric. Therefore, a cooperative international effort to develop and qualify high thermal efficiency technology options, both static and dynamic, for future use in conjunction with space reactors *is strongly encouraged*. Such an effort would be a smart option to effectively reduce nuclear fuel inventory and the radioactivity source term in space reactors, both during operation and after shutdown at the end of operation life.

8. **The radial reflector of the space reactor should be designed to disassemble upon impact on a solid or water surface and the bare reactor should be designed to remain subcritical when submerged in wet sand and flooded with seawater or liquid propellant, following a launch abort accident.** Ensuring subcriticality in such an event favors: (a) solid core reactors with the smallest amount of voids, (b) low fuel mass, and (c) mixing a few at% of thermal neutron poisons with the highly enriched nuclear fuel in the reactor core [71].
9. **Space reactors should be designed with sufficient effective negative temperature reactivity feedback.** Thus, an inadvertent rise in temperature decreases the reactivity in the reactor and the fission power, forcing the reactor temperature to decrease. It also could eventually shut down the reactor. The SNAP-10A was designed with a negative temperature reactivity feedback, while the former Soviet Union's TOPAZ had a positive temperature reactivity feedback. No information has been found on the BUK reactors.
10. **Storage of deactivated space reactors in operation orbits of 1000–3000 km, with natural decay life >1000 years, should be required.** This will allow all accumulated radioactivity in the reactor core during operation to decay to a very low level, thus posing no radiological risk in an eventual re-entry. This end-of-life storage option may apply to earth missions in the same orbits and to propulsion missions to and from the Moon and Mars, which could use these orbits for take-off and parking for regular maintenance.
11. **Uses of space reactor power systems on the lunar and Martian surfaces should be strongly tied to the development of credible and acceptable options for end-of-life storage and eventual disposal and/or recovery of the nuclear reactors or fuel.** A proposed approach may involve designing efficient space power systems with reactors having full-power operation life of 20–30 years, or even longer. In addition, developing unique designs would enable remote and robotic refueling of the surface reactors. Reactors on the surface of the Moon and Mars may be assembled below surface to take advantage of the shielding and neutron reflection properties of the indigenous regolith. Thus, at the end of operation life, the reactor could be left in place, provided that there are adequate means for removing the decay heat from the reactor, for 20–30 years, before removing the nuclear fuel for storage in an

internationally designated site. The spent fuel could also be processed at a later date to recover the expensive ^{235}U fuel for future use. Because of the high fuel enrichment (93–95%) in space reactors, the amount of transuranic fission products (such as Pu, Am, and Cm isotopes) generated by neutron capture in ^{238}U (~5–7%) is very small as regards long-term storage and reprocessing concerns, unlike terrestrial reactor fuel with enrichments $<5\%$.

Although there are no specific design and end-of-life storage guidelines for space reactor power and propulsions systems, there are a number of international agreements that provide a framework and a process that could be exploited for future development of specific safety standards and deployment regulations. The next section presents and briefly discusses the existing United Nations agreements on the peaceful uses of outer space, including space nuclear power systems.

26.7 INTERNATIONAL AGREEMENTS ON PEACEFUL USES OF OUTER SPACE

International agreements on the peaceful uses of outer space, including space nuclear power systems, are [72,73]:

1. **The Outer Space Treaty of 1967**, on Principles Governing the Activities of States in the Exploration and Use of Outer Space, Including the Moon and Other Celestial Bodies.
2. **The Rescue Agreement of 1968**, on the Rescue and the Return of Astronauts and the Return of Objects Launched into Outer Space.
3. **The Convention of 1972**, on the International Liability for Damage Caused by Space objects.
4. **The Convention of 1976**, on the Registration of Objects Launched into Outer Space.
5. **The Moon Treaty of 1979** (entered into force 11 July 1984). This agreement governs the Activities of States on the Moon and other Celestial Bodies, declaring that they should be used for the benefit of all States of the international community and preventing the Moon from becoming a source of international conflict.
6. **The United Nations (UN) Principles of 1992**, on the Peaceful Uses of Nuclear Power Systems in Outer Space.

The first four agreements addressing the liability of a country for objects launched into space and any damage caused by either operation or return to Earth are legally binding. On the other hand, for all practical purposes, the Moon Treaty has been unsuccessful because of many objectionable restrictions, some of which are discussed below. Thus, this treaty is not directly relevant to current space activities.

The Moon Treaty has neither been signed nor ratified by the USA and many other countries. Only 13 states (Australia, Austria, Belgium, Chile, Kazakhstan, Lebanon, Mexico, Morocco, Netherlands, Pakistan, Peru, Philippines, and Uruguay) have signed and ratified the Moon Treaty; none is a major spacefaring nation. France, Guatemala, India, and Romania have signed, but not ratified, the Moon Treaty [73]. The key objections to the Moon Treaty [74,75] may be summarized as follows:

1. The Common Heritage of Mankind Statement in Article 11 stating that "the Moon and its natural resources are the common heritage of mankind" implies that all State parties have rights to the mined resources, irrespective of assuming any risk or contributing capital. It also states that "any effort to develop the lunar resources requires the consent of all States". The absence of a relationship between risk and capital investment versus potential reward and what would be

time-consuming and complex procedures to gain consent are viewed as detrimental to the process and would discourage investment in future development of lunar resources.

2. The Ban on Property Rights, in article 11 paragraph 3, also discourages investment and innovation.
3. The International Regime, in article 11 paragraph 5, gives no specifics on either the composition or the guidelines for the formation of the international regime that will govern the exploration of the natural resources of the Moon and the sharing of these resources by various countries.

The UN Principles were recommended in 1972 by the 69 state member Committee on the Peaceful Uses of Outer Space (COPUOS) and adopted by the General Assembly of the UN [72]. COPUOS and its subcommittees and working groups operate by consensus, meaning that there can be no disagreement with the text of the reports, which are not legally binding documents of a treaty or convention.

The UN Principles on the peaceful uses of nuclear reactor power systems in outer space are currently the only guidelines agreed to by member nations. These principles, however, do not offer any technical details on specific issues. Examples are the end-of-life storage and eventual disposal of nuclear power systems used on or below the surface of the Moon or other celestial bodies or in Earth orbits. Instead, they provide broad recommendations for ensuring the safe use and deployment of these systems, particularly in Principle 3. This principle states that space nuclear reactors should: (a) *only* use highly enriched uranium fuel; (b) *not* be made critical before reaching operating orbit or interplanetary trajectory; and (c) *not* become critical before reaching operating orbit and during all hypothetical events, which may occur during a launch. These may include a rocket explosion, inadvertent re-entry, impact on the ground or water, and submersion in and flooding with water or liquid propellant.

The passage (b) above has been challenged by the USA as not to exclude "zero power testing on the ground". The bases for the challenge are that a brief zero power testing of the reactor, to ensure proper functioning, will result in a negligible buildup of fission products. The proposed revision of passage (b) by the USA reads: "Nuclear reactor should not be made critical *during or after launch* until they have reached operating orbit or interplanetary trajectory." This is a technically valid revision that would not compromise the safety of space nuclear reactor power systems.

Principle 3 of the UN resolutions also recommends that:

a. States launching space objects with nuclear power sources on board shall endeavor to protect individuals, populations, and the biosphere against radiological hazards.
b. The hazards, in foreseeable operational or accidental circumstances, should be kept below the acceptable levels recommended by the International Commission on Radiological Protections.
c. The radiation exposure to a limited geographical region and to individuals should be restricted to the principal limit of 1 mSv in a year.
d. The reliability of the subsystems important for safety should be ensured by redundancy, physical separation, functional isolation, and adequate independence of components.

A sufficiently high orbit (SHO) for the operation and the storage of reactor systems has been defined [72] as that: (a) *with long enough lifetime* to allow sufficient decay of the fission products to approximately the activity of the actinides; (b) *which does not pose risks* to existing and future outer space missions; and (c) *where collision* of the reactor and power system with other space objects *is kept to a minimum.*

On the communication and transparency for enhancing the operation reliability and safety of nuclear reactor in space, Principle 4 of the UN resolutions recommends that:

a. *Thorough and comprehensive safety assessment* should be conducted, which covers all relevant phases of the mission and deals with all systems involved, including the means of launching, the space platform, the nuclear power source and its equipment, and the means of control and communication between ground and space.

b. *Results of the safety assessment*, together with, to the extent feasible, an indication of the approximate intended time-frame of the launch, *shall be made publicly available prior to each launch*, and the Secretary General of the United Nations shall be informed on how States may obtain such results of the safety assessment as soon as possible prior to each launch.

26.8 EARTH ATMOSPHERE AND SPACE REACTOR POWER SYSTEMS

The Earth's atmosphere is divided into five layers or regions, in which the changes in temperature with altitude are different. Most of the weather and clouds are found in the first layer (the troposphere; Figure 26.15), extending from sea level up to about 18 km at the equator and about 8 km at the poles. In both the troposphere and the mesosphere, extending from ~50 to ~87 km, the temperature steadily decreases with increased altitude. Conversely, in the stratosphere, which extends from ~10 to ~50 km, the temperature increases with increased altitude (Figure 26.15).

In the thermosphere (or ionosphere), which extends from ~87 to 500 km, the temperature also increases with increased altitude; however, the rate of increase depends on the altitude within this layer. In the exosphere, which extends from ~500 to 5000 km, the temperature increases very little with increased altitude and most of the particles in this layer are ionized by the high-energy, charged-particle wind from the Sun, as well as by cosmic rays. Thus, the structure of the thermosphere is strongly influenced by the level of solar activity.

The magnetosphere is enormous, extending from ~5000 km to >>60,000 km. It is the outermost shell, which is strongly influenced by the Earth's magnetic field and the solar wind and contains both the inner and outer Van Allen radiation belts, and it is where geosynchronous satellites are deployed [69].

During re-entry of a deactivated space nuclear reactor, the dispersion of the debris and particles released within the stratosphere is relatively limited. It would be strongly influenced by weather conditions (temperature, pressure, wind speed and direction, rain fall, snow, etc.) and surface topography. In the stratosphere, the released debris and particles could travel thousands of kilometers from their point of origin, before reaching the Earth's surface.

The debris and particles released in either the mesosphere or the thermosphere would engulf the entire globe and take years to finally reach the Earth's surface, depending on their size and the density of the materials. Large and dense debris typically re-enters much earlier than finer and or less dense debris. Depending on the re-entry speed and the materials' composition, the dispersed particles and debris could melt, evaporate, or burn. Thus, the re-entry conditions of deactivated space reactors, after storage for hundreds of year in an SHO (>1000 km), strongly affect the dispersion of the resulting core fuel fragments and debris.

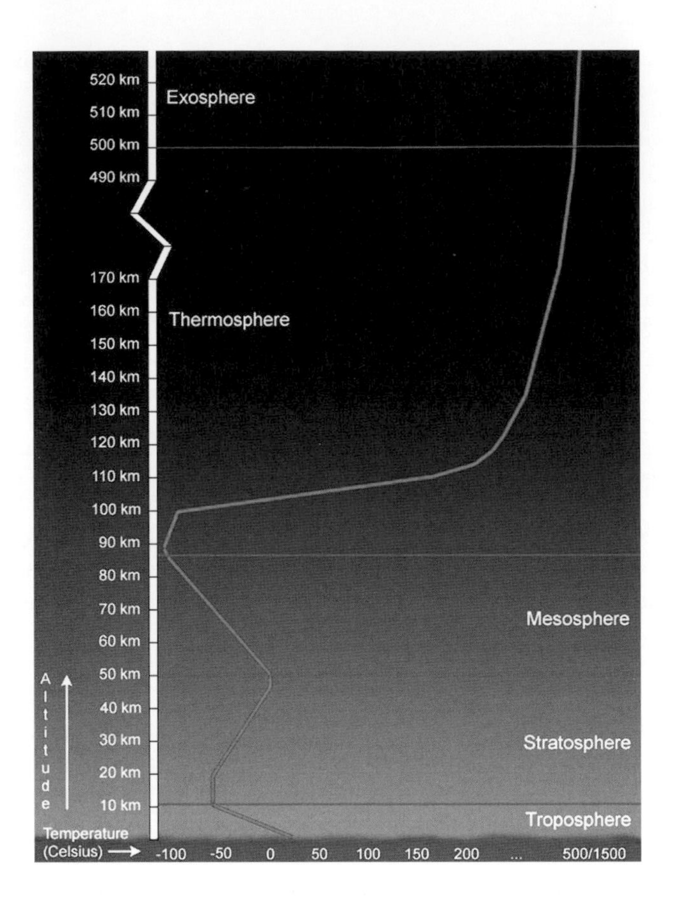

FIGURE 26.15

The Earth's atmospheric regions and temperatures [69].

Source: Windows to the Universe at: http://www.windows.ucar.edu/)

Depending on the size and density of the material, the mean residence time of the debris in the mesosphere could be 4–20 years, compared to 1 month to 2 years in the stratosphere and 0.5 days to 2 weeks in the troposphere [76]. The radiological risk from the re-entry of a space reactor after being stored in SHO orbit for hundreds of years following shutdown is likely to be negligible. This is due to the natural radioactive decay of the accumulated fission product in the reactor fuel. Nonetheless, the solid debris still represents real risk to space assets in Earth orbits.

The debris, ranging from 1 μm to 5 mm in size and traveling at a very high speed (>20 km/s), could cause serious structural damage upon impact [70]. On the other hand, the impact of a deactivated space reactor while in a storage orbit, by space debris, could generate additional debris by fragmentation and the energy of impact, potentially releasing liquid metal coolant from the reactor and piping. While this release is a real possibility for a liquid-metal-cooled reactor, as was the case with the ejected fuel cores of the BUK reactors booted into storage at the end of life, it is eliminated for gas-cooled space reactor power systems. Thus, space power systems with gas-cooled reactors offer some safety advantages over those cooled with circulating liquid metals. This advantage is also shared by liquid metal heat-pipe-cooled reactors.

The safety and reliable operation of space reactor power systems can be assured in the design of the reactor and its integration into the power system, and by making proper choices of the operation temperatures and the various system components.

26.9 ENHANCING SAFETY AND RELIABILITY OF SPACE REACTOR POWER SYSTEMS BY DESIGN

The operation reliability and safety of space reactor power systems could be effectively enhanced by design. Some desirable design features and choices are:

1. Redundancy in energy conversion, reactor cooling, and heat rejection.
2. Avoidance of single-point failures in reactor cooling and energy conversion.
3. Use passive components, to the greatest extent possible, such as thermoelectric−electromagnetic (TEM) pumps and heat pipes, for the removal of decay heat for the reactor after shutdown.
4. Operation at reactor temperatures below 1000 K, which are compatible with using steel and super-steel alloys with known properties and prior flight experience.
5. Armor the heat rejection radiator and the critical system components against impact from space debris and exposure to space ionizing radiation and electromagnetic fields.
6. Fission reactors that have a sufficient negative temperature reactivity feedback.
7. Design reactors for long operation life of 20−40 years, or even longer, which would be possible using energy conversion technology with high thermal efficiency (>20%).
8. Design reactors for potential refueling and multiple shutdowns and restart in orbit and on the surface of the Moon and Mars.
9. Make appropriate design and system integration choices—some of these are discussed next.

Though proper design choices and features would enhance the safety and reliability of space reactor power systems, some are inherently in conflict with one another, requiring rigorous optimization of the design and choices of the system components. Examples of these design choices relevant to space reactor power systems are: (a) energy conversion technology (static or dynamic); (b) nuclear fuel type (U−ZrH, UN, UC, or UO_2); (c) reactor type (fast or thermal energy spectrum); (d) reactor cooling method (circulating liquid metal (LM) or inert gas, or LM heat pipes); (e) structural materials (steel, super-steel, or refractory alloys); and (f) type and capabilities of the launch vehicle (allowable launch weight and volume for the power system and the payload).

26.9.1 Reactor operation temperature and structural materials

While a reactor operation temperature in excess of 1200 K effectively reduces the size and mass of the power system, it requires a structure made of refractory alloys, with high density and no, or limited, flight experience. These alloys also become embrittled by exposure to radiation from the reactor up to relatively high temperatures (800−1000 K). This is in addition to the difficulty in the fabrication and less favorable workability of these alloys. Thus, these alloys may decrease reliability and safety, restricting operation of the power system at multiple levels and for multiple shutdowns and restarts.

Thus, high reactor temperature may eventually increase the development cost and the time to deployment. It also increases the swelling of the nuclear fuel and the release of fission gases, further challenging the integrity of the fuel cladding and reducing the operation lifetime of the reactor.

Operating at a relatively low reactor temperature ($\leq$1000 K) decreases the thermal efficiency and the radiator surface temperature, which increases the size and mass of the radiator and of the power system. On the other hand, it is possible to use stainless steel and super-steel materials, with well-known properties, extensive databases, and very low radiation embrittlement temperatures (Figure 26.8) [10], thus offering the potential to operate for tens of years, at multiple power levels, and with multiple restart and shutdown. The latter is particularly important to reactor power systems for surface power on the Moon or Mars and nuclear electrical propulsion. For these applications, low-temperature reactor power systems can greatly benefit from employing an energy conversion technology with a high thermal efficiency (15–30%). This may be practically difficult at present, but possible in the future through a focused and sustained research effort [23].

26.9.2 Energy conversion and load-following operation

Gas-cooled space reactors are most suitable for integration to closed Brayton cycle (CBC) loops with centrifugal flow turbo-machines for energy conversion [47,48,50]. The thermal efficiency of the resulting power system could be as high as 21% and 34% when operating at reactor exit temperature of 900 and 1149 K respectively; however, the low heat rejection temperature (<500 K) increases the mass and size of the radiator [48]. Also, since CBC turbo-machines are non-load-following, the entire reactor power system would be non- or partially load-following. Nuclear reactors designed with a negative temperature feedback are inherently load-following.

Load-following is a desirable operation and safety feature for space reactor power and propulsion systems, allowing them to respond in orderly fashion to changes in the load demand, within a certain range (typically $\pm$5–15%) determined by the design, without active control of the reactor heat source. Space power systems in which the reactor is designed with a negative temperature reactivity feedback and that employ static energy conversion, such as TE, TI, or AMTEC, are inherently load-following. Recent and future advances in static energy conversion technologies could demonstrate thermal efficiencies in the range of 15–30%. This would enhance the power system performance, reliability, operation and safety, increase the operation life of the reactor, and decrease both the size and total mass of the power system. Static conversion options are suitable choices for integration to LM or LM heat-pipe-cooled reactors.

26.9.3 Reactivity control and safety

It is preferable to keep all reactivity control, as is practically possible, external to the reactor core. This will decrease the size of the reactor and minimize the number of penetrations into the reactor vessel. The smaller reactor size, in turn, decreases the size and mass of the radiation shadow shield and of the power system.

As a safety measure during launch, the neutron reflector (usually Be or BeO) of the reactor should be designed to fully disassemble upon impact on a solid surface, on soil, or water. Thus, following a launch abort accident, the bare reactor core, submerged in wet sand and flooded with water or liquid propellant, remains sufficiently subcritical. A key measure to assure subcriticality of the reactor in such an event is adding an appropriate amount (<3–20 at%, depending on the reactor design) of spectral shift absorbers (or neutron poisons) to the nuclear fuel in the core [77].

Also, depending on the design of the reactor, applying a thin layer (<0.5 mm thick) of the spectral shift absorber on the outer surface of the reactor vessel further ensures the subcriticality of the submerged and flooded space reactor, following a launch abort accident. The application of thin films of spectral shift absorbers on the outer surface of the vessel becomes less effective when the reactor is designed with a flow annulus or voids on the inside of the vessel. A conservative measure, following a launch abort accident, is assuming that the liquid propellant or water will displace the reactor coolant and fill all voids within the reactor core.

26.9.4 Launch vehicle, heat pipe radiator, and system integration

A key consideration in the development of high-power space reactor power systems is the type of launch vehicle to be used. This determines the allowable weight and volume in the launch bay for the power system and the payload. Depending on the thermal efficiency of the power system and nominal power level, the radiator could be very large (>tens to hundreds of m^2). A large radiator should be redundant, void of single-point failures, and constructed of fixed and deployable segments that fold during launch, in order to fit into the bay of the launch vehicle.

Heat pipe radiators are lightweight and because they operate at high average surface temperature their specific mass is small (<8 kg/m^2), including the carbon–carbon armor [78,79]. They enhance the operation safety and reliability of the power system by eliminating single-point failures in heat rejection. A leak or failure in one or several heat pipes will not affect the operation of others in the radiator. The heat pipes are typically designed to operate at a fraction (~60%) of their nominal power throughput. Thus, the thermal load of the failed heat pipes will simply be shared among the adjacent heat pipes, without compromising the operation or safety of the radiator or the power system.

In addition to being lightweight and self-contained, the heat pipes are high thermal conductance and passive thermal energy transport devices. They are also used to reliably cool space reactors, both during operation and after shutdown [22,78,79]. To protect radiator heat pipes from impact by meteorites and space debris, the radiator's outer surface may be made of, or covered with, a light-weight, high toughness, and thermal conductance carbon–carbon reinforced structure [78,79].

26.9.5 Reactor types

Space reactors could be designed with either a thermal, or a fast neutron, energy spectrum. Reactors with fast energy spectra are compact because they are void of a low-density moderator used to slow down the neutrons in thermal reactors [3,4,13,36,58,62,65]. Thus, fast spectrum reactors are suitable for high-power and high-temperature operation, and long operation life (>10 years), but require high fissile loading.

Though space reactors with either epithermal or thermal neutron energy spectra are larger in size, they require an order of magnitude lower fissile loading than fast spectrum space reactors and are preferable for low-power generation and shorter missions (<~2 years). Because of the high neutron leakage, a fast spectrum reactor could not achieve criticality below a certain size [44], since the surface-to-volume ratio for the reactor core increases with the decreasing length scale. The operation temperatures of thermal energy spectrum reactors are lower than in fast spectrum reactors because of the limitation imposed by the moderator materials (e.g. ZrH, Be, graphite, or water). For example, a ZrH moderator typically limits the reactor operation to <900 K to minimize hydrogen dissociation and loss of neutron moderation effectiveness [3,4,13,43,58].

The next section presents a number of reactor designs and power system integrations for enhancing operation reliability and safety, including the avoidance of single-point failures in reactor cooling and energy conversion. The next section also presents a space reactor power system for a civilian mission in 1000–3000 km earth orbit to support satellites for global sea and air traffic control [67,80]. These orbits are within the inner Van Allen radiation belts, in which solar photovoltaic would not survive the exposure to the energetic charged particles trapped in these orbits without a significant degradation in performance.

26.10 SPACE REACTOR POWER SYSTEMS FOR AVOIDANCE OF SINGLE-POINT FAILURES

This section presents examples of space reactor designs and power system integrations for enhancing operation safety and reliability through the avoidance of single-point failures in reactor cooling and energy conversion, selection of matching energy conversion technology, and use of heat pipe heat rejection radiators.

26.10.1 SCoRe–NaK–TE power system for global civilian air and sea traffic control satellites

This subsection describes a circulating, liquid-metal-cooled, fast spectrum space reactor and power system with no single-point failures in reactor cooling and energy conversion, and with passive means for decay heat removal. Also presented are the performance results of this system for powering satellites in a 1000–3000 km Earth orbit for global civilian air and ocean traffic control [24,67,81].

The space reactor power system described in this section employs a Sectored Compact fission Reactor (SCoRe) heat source, cooled by circulating liquid NaK-78 and designed with a negative temperature reactivity feedback. The SCoRe–NaK–TE power system for this civilian mission possesses a simplified design and the operation parameters were selected for enhancing safety and operation reliability. The reactor core is divided into six sectors that are hydraulically independent, but thermally and neutronically coupled (Figure 26.16a). Each sector has a separate pair of primary and secondary loops with SiGe thermoelectric energy conversion modules and TEM pumps in the secondary and primary loops [24,67,81].

The hexagonal core of the SCoRe is surrounded by a 10-cm-thick BeO radial reflector and 4-cm-thick BeO axial reflectors. The forward axial reflector is within the uranium nitride fuel pins in the reactor core, while the rear axial reflector is placed outside the reactor vessel, in front of the radiation shadow shield (Figure 26.16a, b). The six enriched B_4C/BeO rotating drums in the radial BeO reflector control the reactivity in the reactor core during startup and nominal operation and are used to shut down the reactor at the end of life. The 0.5-cm-thick, enriched B_4C segments (120° arc) in the drums face the reactor core in the shutdown mode and during launch, and face 180° away at end of life. The boron-10 isotope has high neutron absorption cross-section; thus the reactor is in the least

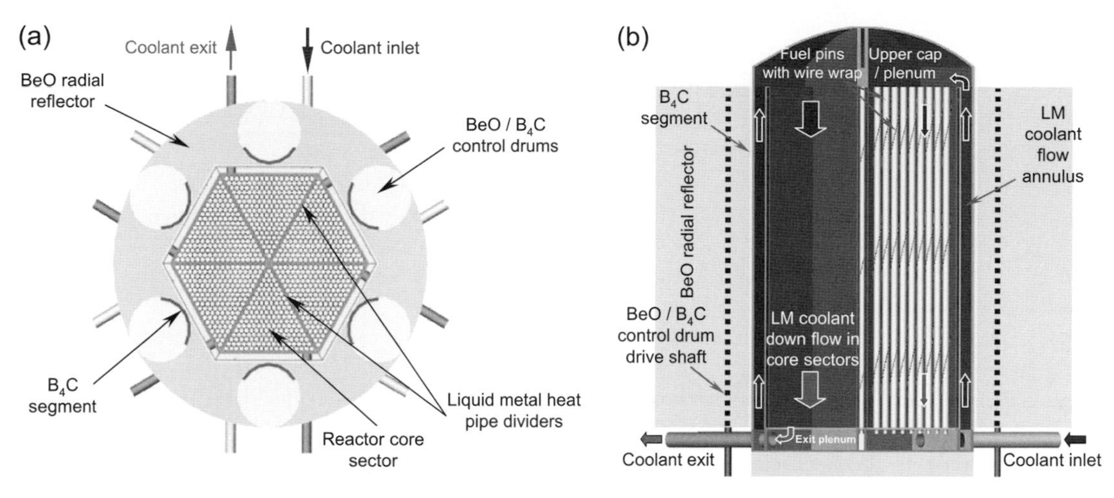

FIGURE 26.16

Cross-section views of the SCoRe core and reflector [81].

reactive condition (subcritical) when the B_4C segments in the control drums are facing inward (Figure 26.16a).

The six sectors in the SCoRe core are thermally coupled using flat potassium heat pipe dividers (Figure 26.16a). In the case of a loss of coolant (LOC) in one sector, due to a break in the inlet or exit pipe, the fission power generated in all sectors is reduced to avoid overheating the fuel pins. The heat generated in the reactor sector experiencing an LOC is transported passively by the heat pipe dividers to the adjacent sectors and circulating liquid NaK-78 in the inlet annulus on the inside of the reactor vessel (Figure 26.16a, b).

Each sector in the SCore core (Figure 26.16a) is loaded with 171 uranium nitride fuel pins clad in Mo−14Re, for a total of 1026 pins in the reactor core [81]. Each fuel pin nominally generates <2.5 kW_{th} in order to maintain the maximum fuel temperature reasonably low (<1500 K). The Mo−14Re wires wrapped on the outside surface of the cladding of the uranium nitride fuel pins maintains uniform flow area and provide excellent structural support. The 95 wt% enriched uranium nitride fuel pellets in the pins include 17.2 at% ^{157}GdN spectral shift absorber [77]. A thin (0.1 cm) coating of $^{157}Gd_2O_3$ spectral shift absorber is also applied on the outer surface of the reactor vessel [82]. The spectral shift absorber additives to the fuel and the coating ensure that the bare reactor remains sufficiently subcritical when submerged in wet sand and flooded with seawater, following a launch abort accident [44,82].

Each sector in the SCoRe core has separate inlet and exit plenums and pipes (Figure 26.16a, b), separate primary loop with a thermoelectric Power Conversion Assembly (PCA), and a secondary loop with three rubidium heat pipe radiator panel segments (Figure 26.17). Each primary and secondary loop has a separate TEM pump for circulating the liquid NaK-78 and a bellows-type accumulator for accommodating the changes in the liquid NaK volume during operation, startup, and shutdown of the power system [83]. The electrical power needed to operate the TEM pumps in the six secondary and primary loops is supplied by six separate SiGe thermoelectric conversion assemblies (TCAs). Each

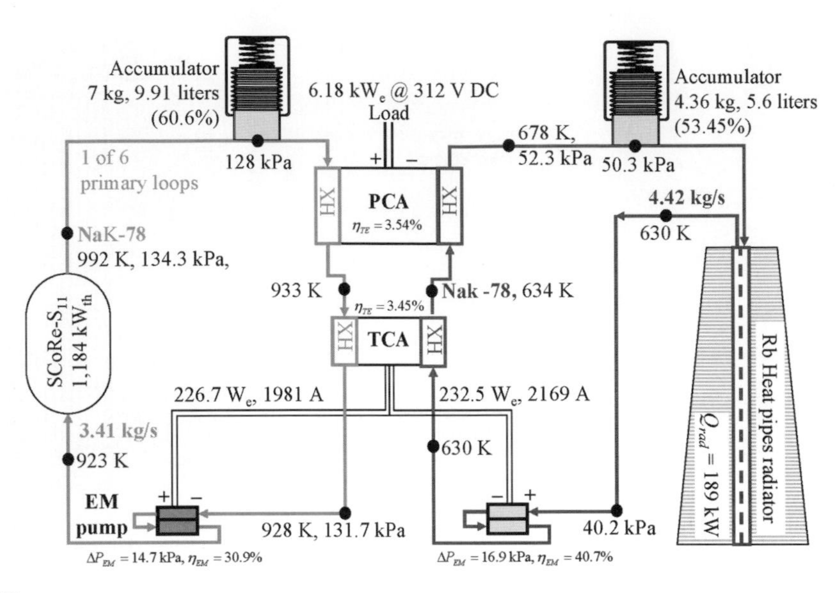

FIGURE 26.17

Line diagram of a pair of primary and secondary loops in SCoRe—NaK—TE power system with nominal operation parameters [67].

pair of primary and secondary loops is served by one fixed forward and two rear deployable rubidium heat pipe radiator panels [24,81]. The returning liquid NaK from the primary loop flows upward in the annulus on the inside of the reactor vessel, and then reverses direction at the opposite end, flowing through the reactor sectors and removing the fission heat generated in the uranium nitride fuel pins in the core sectors before exiting to the primary loops (Figure 26.16b). The reactor operates at steady-state, full power of 1183.6 kW_{th}, and exit temperature of ~992 K.

The six PCAs in the SCoRe—NaK—TE power system provide DC power to the electrical load at >300 VDC, while the six TCAs provide high-current, low-voltage power to the 12 TEM pumps in the primary and secondary loops. The TEM pumps also circulate the liquid NaK-78 in the primary and secondary loops after reactor shutdown, passively removing the decay heat from the reactor core. These pumps continue to operate so long as there is a sufficient temperature difference (several degrees) between the circulating liquid NaK in the primary and in the secondary loops (Figure 26.17). Thus, the operation of the SCoRe—NaK—TE power system is fully passive, except for the rotating control drums in the radial Be reflector (Figure 26.16a). This space reactor power system is designed to nominally generate 37.1 kW_e for up to 6 years.

26.10.1.1 Startup in orbit

After it is safely deployed into a 1000—3000 km earth orbit, the SCoRe—NaK—TE power system is started up. The deployment orbit is also used to store the deactivated reactor power system at the end of its operation life, for up to 1000 years. To start up the power system in orbit, first the control drums are rotated slowly outward to bring the reactor to a critical state ($k_{eff} = 1$) at a low thermal power of

~10 W_{th}, while the radiator panels are still folded and covered by a thermal blanket. The temperature of the liquid NaK in the power system is initially assumed to be uniform and equal to 500 K; it is much higher than the NaK freezing temperature of 261 K (or −12°C). At 500 K, the liquid NaK in the six secondary and primary loops is subcooled at 34 and 51 kPa respectively. The bellows-type accumulators in these loops (Figure 26.17) maintain the liquid NaK subcooled, not only at startup, but also during full-power operation and after shutdown [24,67,83].

The startup procedures from low-power critical condition to steady-state full power are completed in ~5000 s [67]. During the first 978 s, the auxiliary batteries on board supply 600 A to the TEM pumps (50 A to each of the 12 pumps) in the six primary and six secondary loops of the power system (Figure 26.17). After 978 s into the startup procedures, the open-circuit voltage of the pumps' TCAs becomes high enough to provide more than 50 A to each of the TEM pumps. Thus, the TCAs are connected to the pumps, while simultaneously disconnecting the auxiliary batteries.

After 3900 s and before connecting the electrical load, the reactor thermal power reaches a steady-state value of 1030 kW_{th} and the PCAs' open-circuit voltage reaches 620 VDC. The six PCAs in the SCoRe–NaK–TE power system are connected electrically in parallel for redundancy. When the electrical load is connected, the load voltage drops to 312 VDC and the reactor's thermal power increases to a nominal steady-state value of 1183.6 kW_{th} [67]. This occurs at ~5000 s, marking the end of the startup procedures. At such time, the nominal electrical power provided to the load is 37.1 kW_e at a thermal efficiency of 3.137%. When the PCAs are connected to the electric load, the current supplied by the TCAs to the TEM pumps drops momentarily, approaching steady-state values of 2169 and 1981 A respectively, to each pump in the primary and secondary loops (Figure 26.17). This figure presents the operation parameters of the SCoRe–NaK–TE power system during steady-state full-power operation for up to 6 full years [67].

26.10.1.2 Reactivity control and reactor shutdown

During the 6 years of steady-state, full-power operation, the SCoRe's hot-clean excess reactivity (+2.05) is used up gradually to compensate for the partial depletion of ^{235}U and the accumulation of fission products in the uranium nitride fuel pins in the core. The reactor's thermal power and temperatures are maintained within specified design margins, by periodically rotating the control drums incrementally outward. The period and the amount of reactivity inserted depend on the design-allowed drop in reactor temperature and the load voltage between subsequent reactivity insertions [67].

At the end of life, the reactor shutdown is initiated by rotating the control drums inward over a 24-hour period. At the end of this period, the B_4C segments in the control drums face the reactor core (Figure 26.16a). The reactor thermal power and temperatures drop precipitously during and following the reactor shutdown. At the end of the shutdown period, the reactor exit and inlet temperatures decrease to ~389.9 and 400 K, and the reactor thermal power drops to only 4.75 kW_{th}. This thermal power is mostly due to the radioactive decay of fission products accumulated in uranium nitride fuel pins in the reactor core [67].

The flow rates of the circulating liquid NaK in the primary and secondary loops also drop rapidly from 3.41 and 4.42 kg/s before reactor shutdown to only ~77 and 66 g/s at the end of the shutdown period. At such time, the electric current supplied by the TCAs to each of the TEM pumps in the primary and secondary loops drops to 5.6 and 7.6 A respectively, and the electrical power supplied by the PCAs to the load becomes negligibly small, ~146 mW_e. The exponential drop in the decay heat generation in the SCoRe core after shutdown, following 6 years of steady-state, full-power operation,

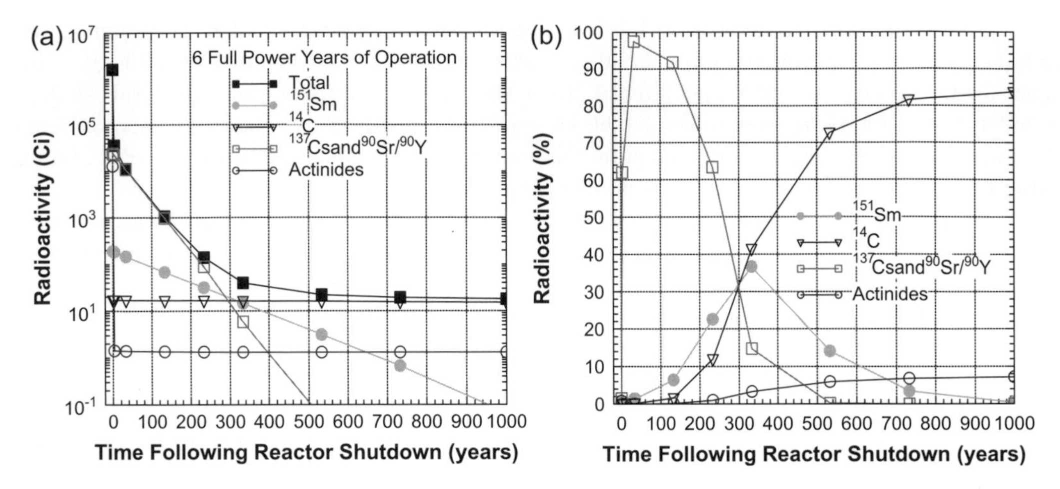

FIGURE 26.18

Estimates of radioactivity in deactivated SCoRe [67].

is indicative of the decrease in the radioactivity of the accumulated fission products in the uranium nitride fuel pins in the core (Figure 26.18) [67], discussed next.

26.10.1.3 Radioactivity in SCoRe core after shutdown

Before starting up the SCoRe—NaK—TE power system in orbit, the radioactivity in the reactor core, solely due to the natural radioactive decay of U in the fuel pins, is very small (~0.258 Ci). It increases very quickly, to as much as ~1.05×10^6 and 1.505×10^6 Ci after 10 days and 6 months of steady-state, full-power operation of the reactor respectively. It then increases very slowly with additional operation time of the reactor to reach 1.61×10^6 Ci at the end of 6 years of operation at steady thermal power of 1183.6 kW_{th} [67]. The contributions of actinides to the total radioactivity in the reactor core are very small. After 10 days, 6 months and 6 years of nominal full-power operation, the calculated radioactivity buildup of the actinides in the reactor core is only 1.108×10^4, 1.115×10^4, and 1.273×10^4 Ci respectively. Other fission products contribute the balance.

After shutdown, following 6 years of full-power operation, the radioactivity in the SCoRe core decays rapidly, from ~1.611×10^6 Ci immediately before reactor shutdown to ~4.0×10^4 Ci after reactor shutdown and storage in orbit for 3 years (Figure 26.18a). The total radioactivity in the deactivated reactor core continues to decrease exponentially, reaching ~1000 Ci and only ~31.5 Ci after 137 and 400 years of storage. The total radioactivity in the deactivated SCoRe-S core becomes very small, ~23.7 and 19.7 Ci, after 500 and 700 years of storage, and changes very little with time thereafter (Figure 26.18a).

The operation time before shutdown slightly affects the level of radioactivity in the deactivated reactor after years of storage in orbit. After 300 years of storage in orbit, the total radioactivity in the deactivated reactor that operated for 6 years at full power before shutdown is ~63 Ci, compared to ~42.6 and 21.7 Ci for operating for 4 and 2 years at full power before shutdown. After 700 years of storage in orbit, these radioactivities decrease to 19.68, 13.21, and 6.74 Ci respectively. The results in Figure 26.18a and b clearly show that after 300 years of storage in orbit, a re-entry of the deactivated

SCoRe-S in which the calculated total radioactivity is ~63 Ci should not represent a radiological hazard to the Earth's inhabitants [67].

26.10.1.4 Decay heat removal after shutdown

It is important to design space reactors with capabilities for removing the decay heat from the core after shutdown, to avoid overheating the fuel and potentially breaching the coolant loops and contaminating storage orbit with debris. As indicated earlier, TEM pumps in the SCoRe—NaK—TE space power system (Figure 26.17) continue to circulate NaK-78 coolant in the primary and secondary loops after reactor shutdown. The decay heat is removed from the reactor core and transported to the radiator panels to be rejected into space. The flow rates of the circulating liquid NaK-78 in the loops decrease with time after reactor shutdown, commensurate with the decreases in reactor temperature and decay heat generation rate.

Immediately at shutdown, the decay power generated in the reactor core, mostly due to the radioactive decay of the short-lived fission products, is 9.905 kW_{th}. It decreases to 8.153 kW_{th} an hour following shutdown, and to 4.602, 2.616 and 1.362 kW_{th} at the end of 1 day, 1 week and 4 weeks after reactor shutdown. At the end of the first year after shutdown, the short and moderate-lived nuclides decay away, reducing the decay power in the SCoRe core to 109 W_{th} [67].

The next subsection presents another space reactor power system (S^4—CBC), designed with no single-point failures in reactor cooling and energy conversion. It generates >100 kW_e continuously for more than 10 years. This power system employs a gas-cooled reactor heat source and multiple CBC loops for energy conversion.

26.10.2 High-Power S^4—CBC Space Reactor Power System

The S^4—CBC space reactor power system employs a gas-cooled, Submersion, Subcritical Safe Space (S^4) reactor heat source [71], cooled with circulating He—Xe binary gas mixture (40 g/mol), which is also the working fluid of the power system's three CBC loops for energy conversion. The hexagonal, Mo—14Re (molybdenum with 14 wt% rhenium) solid core of the S^4 reactor is divided into three sectors that are hydraulically independent, but thermally and neutronically coupled (Figure 26.19a). The S^4 reactor generates 471 kW_{th} of steady-state thermal power for 12 years; thus, each reactor sector generates 157 kW_{th} [71,77,84]. The cylindrical cavities loaded with stacks of uranium nitride fuel pellets in the reactor core are surrounded by multiple coolant channels (Figure 26.19b). Iridium foils (0.1 mm thick) are wrapped on the outside of the uranium nitride fuel stacks in the cavities to enhance heat conduction and suppress the diffusion of solid fission products from the fuel pellets to the solid core block. There are a total of 217 fuel cavities and 1977 coolant channels in the S^4 reactor core [71,84]. The 1.25-cm-diameter uranium nitride fuel stacks are arranged in a triangular lattice with a pitch of 1.779 cm and the coolant channels are 3 mm in diameter. Each UN fuel cavity is surrounded by 12 coolant channels, and effectively cooled with nine channels (Figure 26.19a, b).

26.10.2.1 Integration of S^4—CBC space power system

Each sector in the S^4 reactor core (Figure 26.19a) is thermal-hydraulically coupled to a separate CBC loop (Figure 26.20) with a Brayton Rotating Unit (BRU) designated UNM-BRU-1 [47,48,85]. The UNM-BRU-1 (Figure 26.21) has a permanent magnet alternator (PMA) and its

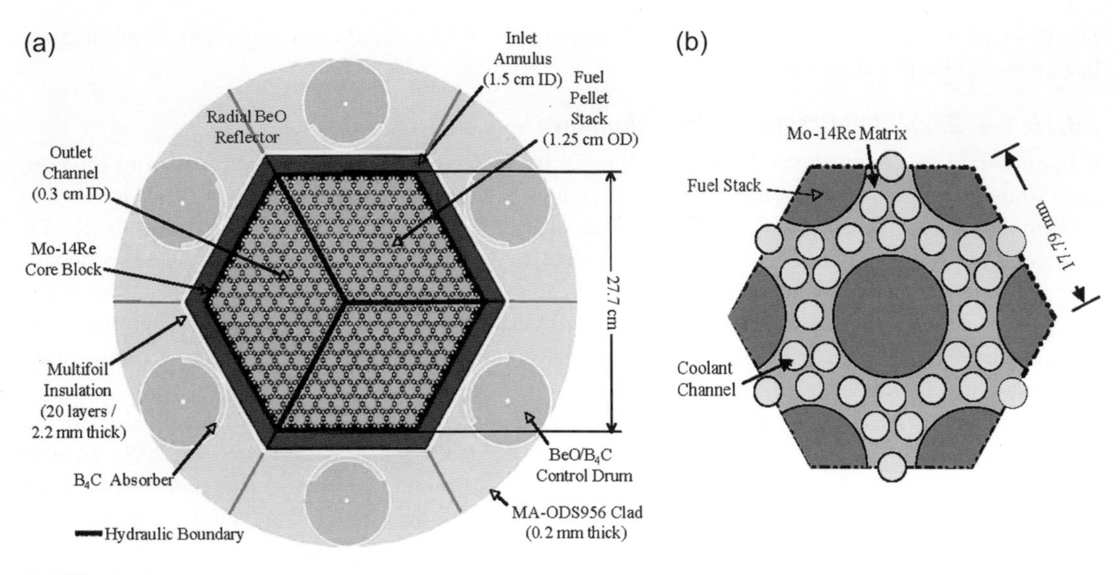

FIGURE 26.19

Radial cross-sections of (a) S^4 reactor core and (b) a uranium nitride fuel stack [84].

design has been optimized for a peak thermal efficiency and electrical power of 28.5% and generating 44.7 kW, when operating at turbine and compressor inlet temperatures of 1149 K and 400 K, shaft rotation speed of 45 krpm, and input thermal power to the turbine of 157 kW_{th}. This thermal power corresponds to the reactor's full thermal power of 471 kW_{th} [83]. Figure 26.20

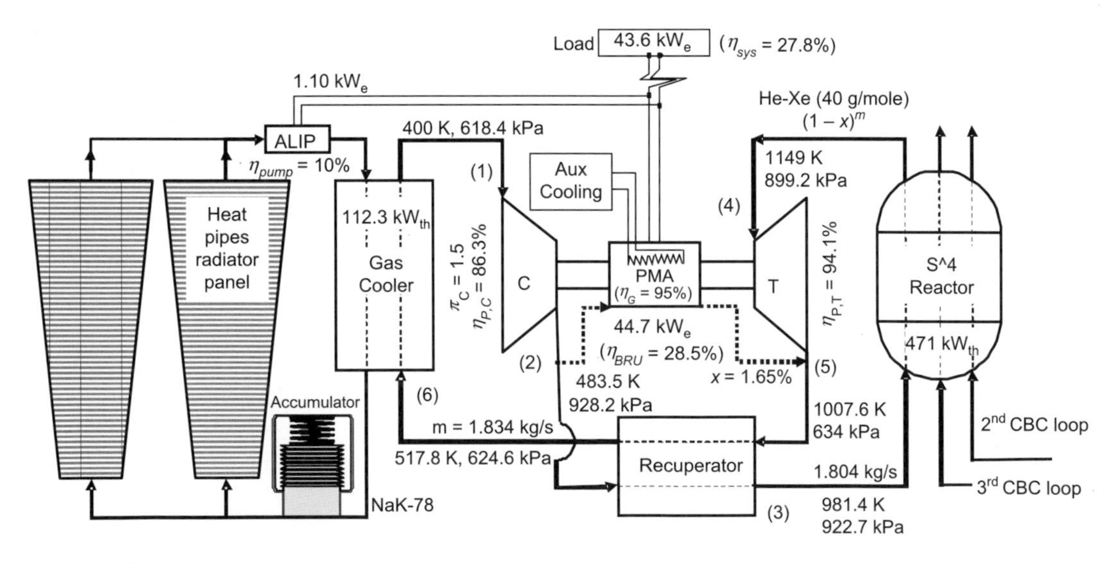

FIGURE 26.20

Full-power, steady-state operation parameters of a CBC loop in S^4—CBC space power system [47,48,85].

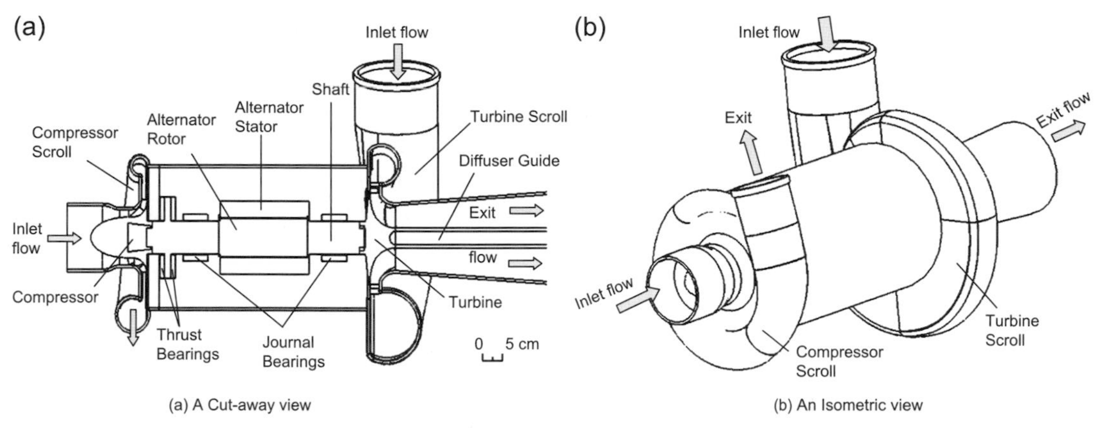

FIGURE 26.21

Cutaway (a) and isometric (b) views of the UNM-BRU-1 [48,85].

shows a line diagram of one of the three CBC loops of the S^4–CBC power system during steady-state, full-power operation.

For waste heat rejection, each of the three CBC loops in the S^4–CBC power system is thermally coupled to a circulating liquid NaK-78 secondary loop with two water heat pipe radiator panels, each comprised of one forward fixed and two rear deployable segments (Figure 26.20). The liquid NaK in each of the heat rejection loops is circulated using an alternating linear induction pump (ALIP). The six radiator panels for each CBC loop, two forward and four rear deployable panels, are hydraulically coupled in parallel in order to enhance performance and reduce pressure losses [78].

The ALIPs for circulating the liquid NaK-78 in the heat rejection loops each consume ~1.1 kW_e, decreasing the net electrical power supplied to the electrical load by the UNM-BRU-1 in each CBC loop by that amount to 43.6 kW_e (Figure 26.20). Thus, the electrical power supplied to the load by the S^4–CBC space power system at steady-state, full-power operation is ~130.8 kW_e at a thermal efficiency of ~27.8% [48,85].

26.10.2.2 Safety and reliability features

As indicated earlier, the S^4–CBC power system with a reactor core comprised of three hydraulically independent sectors, each with a separate CBC loop, avoids single-point failures in reactor cooling and energy conversion. The safety and reliability features of the system are:

1. Following a loss of cooling (LOC), or a pipe break in one of the CBC loops, the reactor fission power will be reduced, and that generated in the sector with an LOC will be transported by conduction and/or radiation to the metal dividers, then to the two adjacent sectors. It is then removed by the circulating He–Xe gas in these sectors. The high thermal conductivity of the S^4 reactor Mo–14% Re solid core (>65 W/m·K) facilitates the transfer of fission heat from the sector experiencing an LOC to the adjacent sectors. Thus, the power system may continue to operate with only two functioning CBC loops, but at a lower power.

2. The bare core of the S^4 reactor is sufficiently subcritical when submerged in wet sand and flooded with seawater, following a launch abort accident. This is because of the added 1.95 wt% ^{151}EuN, spectral shift neutron absorber to the 95 wt% enriched uranium nitride fuel pellets in the reactor core. The ^{151}EuN additives minimally affect the reactivity of the fast neutron energy spectrum S^4 reactor during nominal operation. However, it effectively absorbs thermal neutrons produced in a submerged and flooded core, following a launch abort accident, keeping the reactor subcritical [84]. In this accident, for conservative considerations, all voids and coolant channels in the reactor core are assumed to be fully filled with water.
3. The S^4 reactor is designed with a negative temperature reactivity feedback to ensure that in case of an inadvertent increase in temperature, the reactor thermal power decreases, forcing the temperature to eventually decrease. The Mo–Re solid core of the S^4 reactor provides a thermal storage mass that moderates the rise rate in the reactor temperature during an inadvertent increase in the reactor thermal power and also after the reactor shutdown.
4. After reactor shutdown, the high thermal conductivity of the Mo–Re solid core readily conducts the decay heat generated in the core to the reactor vessel, where it is eventually dissipated into outer space by thermal radiation. Thus, the passive removal of the decay heat from the S^4 core after shutdown is emphasized.
5. The number of BeO/B_4C control drums in the radial BeO reflector of the S^4 reactor (Figure 26.19a) could be increased to three to four per sector, thus enhancing the redundancy in reactor control. With three to four control drums per sector, or a total of nine to 12 control drums for the reactor, it would be possible to start up, maintain full-power steady-state operation, and shut down the reactor at the end of its 12 years of operation life, with one or more control drums per sector stuck in the shutdown position.
6. At the end of reactor operation life, if many of the control drums are stuck in the shutdown position, the negative temperature reactivity feedback will sustain the reactor and the power system operation for many months at continually decreasing power level, without any safety concern.
7. The S^4–CBC space reactor power system eliminates concerns associated with liquid-metal-cooled space reactors and power systems in so far as the potential exists for polluting Earth orbits with liquid droplets, if a leak develops in one of the CBC loops.
8. The S^4 Mo–Re solid core block will fully contain uranium nitride fuel pellets during an eventual re-entry of the deactivated reactor after >300 years in storage orbit, limiting the generation and dissipation of solid debris in Earth orbits. At such time, the radioactivity of the uranium nitride fuel is likely to drop below background on Earth [67].
9. The radiator's water heat pipe panels eliminate single-point failures in heat rejection [78]. Also, employing three hydraulically independent CBC loops, each with a separate circulating NaK loop for heat rejection and two water heat pipe radiator panels, ensures excellent redundancy and safe operation of the power system (Figures 26.20 and 26.22).

Figure 26.22 shows the S^4 space reactor power system fully deployed. The water heat pipe radiator, with six forward segments and 12 rear segments grouped into six integrated panels, is the largest and heaviest subsystem. The radiator water heat pipes have carbon–carbon armor and fins to protect against impact by space debris and extend the heat rejection surface of the panels respectively [78]. Most of the waste heat is rejected from the outer surface of the radiator panels, and only a small fraction is rejected from the inner surface to space through the rear opening of the radiator cone.

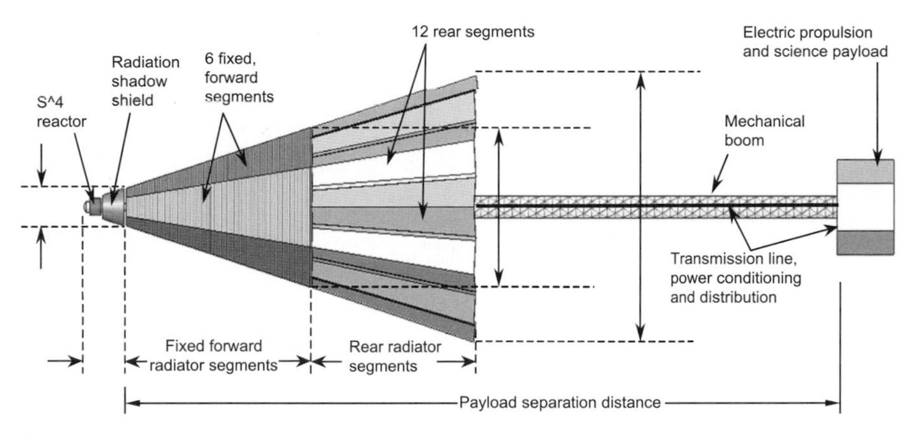

FIGURE 26.22

S^4–CBC space reactor power system, fully deployed [22,23].

During launch, the 12 rear radiator segments are folded on to the six forward segments for the stowed S^4–CBC power system (Figure 26.23) to fit into the launch bay of the *Delta-IV* heavy launch vehicle.

The next section presents another space reactor power system with no single-point failures in reactor cooling and energy conversion. The fast neutron energy spectrum reactor is cooled using liquid metal heat pipes. These heat pipes also transport the reactor thermal power to a multitude of static, alkali metal thermal-to-electric conversion (AMTEC) units.

26.10.3 Scalable AMTEC Integrated Reactor Space (SAIRS) power system

Figure 26.24a and b shows cross-sectional views of the Scalable AMTEC Integrated Reactor Space (SAIRS) power system, also designed with no single-point failures in reactor cooling and energy conversion. It employs a fast neutron energy spectrum reactor cooled with sodium (Na) heat pipes. For

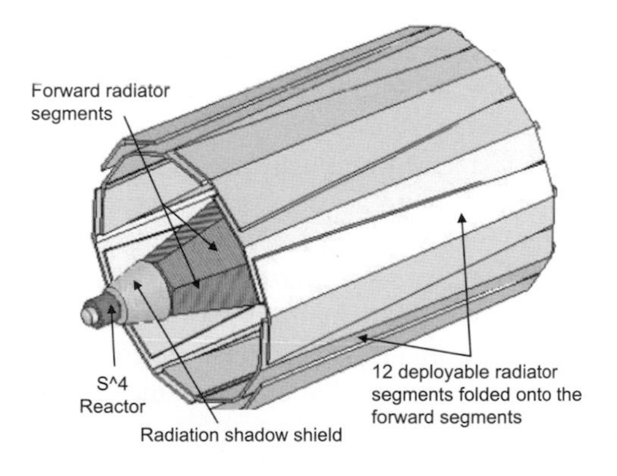

FIGURE 26.23

Stowed S^4–CBC space reactor power system during launch [79].

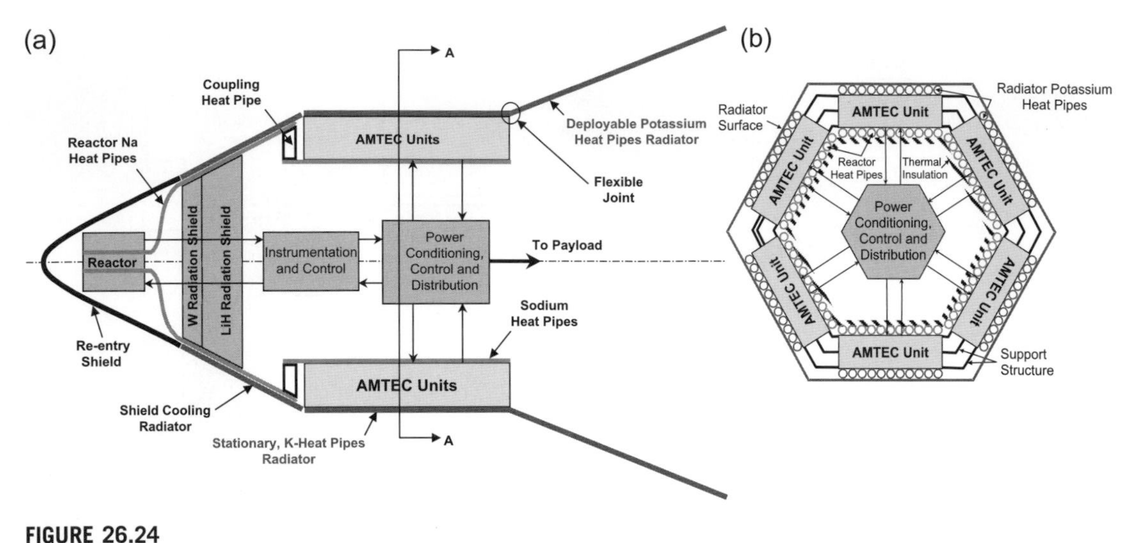

FIGURE 26.24

SAIRS, 100 kW_e power system with Na-heat-pipe-cooled reactor and AMTEC energy conversion [21,29,55].

energy conversion, the Na-AMTEC units are designed and optimized for the reactor's operation temperature and thermal power. The system may employ either 18 5.6-kW_e or 24 4.2-kW_e Na-AMTEC units, for generating a steady-state, full power of 100.8 kW_e [29,55].

The AMTEC units, placed behind the radiation shadow shield, are heated by a multitude of sodium heat pipes, which are thermally coupled to the reactor's Na heat pipes (Figure 26.24a, b). The AMTEC units in the SAIRS power system are grouped into six blocks of three or four units each. Each AMTEC block is cooled by a multitude of potassium (K) heat pipes, assembled into a separate radiator panel (Figure 26.24a, b). The heat rejection radiator panel for each AMTEC block is made of two sections, a stationary forward section attached to the AMTEC units and a rear conical deployable section, which is folded on to the stationary section in the stowed launch configuration (Figure 26.24a). The K-heat pipes in the forward fixed and rear deployable sections of the radiator panels are hydraulically coupled using flexible joints. The surface of the radiator is covered with C−C composite armor to protect against impact by meteoroids. The specific mass of the armored, potassium heat pipe radiator panels is 7.67 kg/m^2 [21,79].

The SAIRS's hexagonal core, fast spectrum nuclear reactor is cooled using a total of 60 1.5-cm-OD, Mo−14% Re sodium heat pipes (Figure 26.25a, b). Each heat pipe cools a module comprised of three uranium nitride, Re-clad fuel pins arranged in a triangular lattice and brazed lengthwise to the centrally positioned heat pipe. Six Re tri-cusps braze the heat pipe wall (0.4 mm thick) to the Re cladding of the fuel pins along their active length [29,55]. To minimize the physical penetrations through the radiation shadow shield behind the reactor (Figure 26.25a), the Na-heat pipes exiting the reactor vessel and the axial BeO reflector are bent around the shadow shield before entering the radiator cavity (Figures 26.25 and 26.26).

The reactor's Na-heat pipes are structurally supported using a 2-mm-thick graphite disk placed in front of, but thermally insulated from, the radiation shadow shield (Figure 26.25a). The heat deposited in the shield (<6 kW_{th}) by the attenuation of the escaping fast neutrons and the primary gammas from

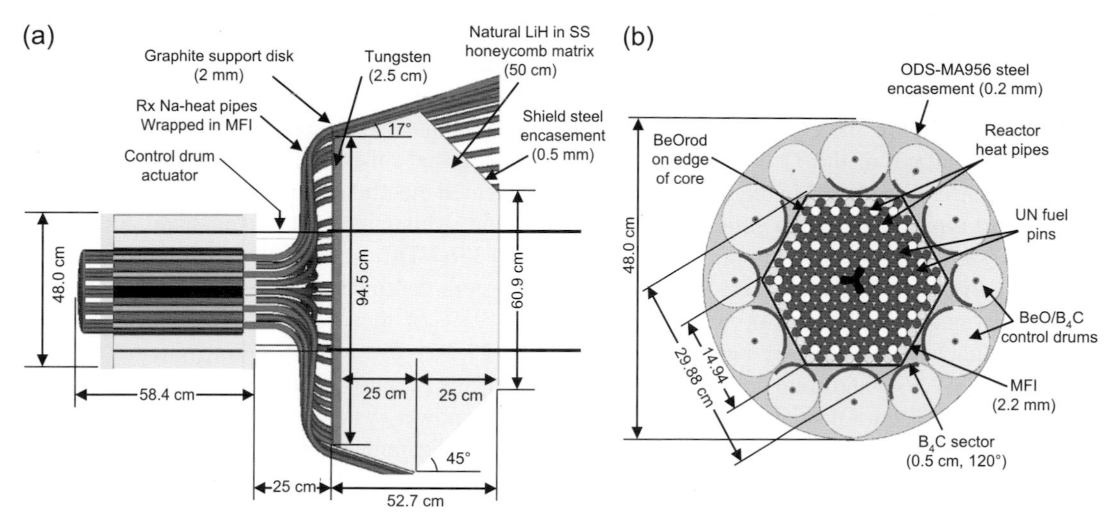

FIGURE 26.25

Cross-sectional views of the SAIRS reactor and radiation shadow shield [21,29,55].

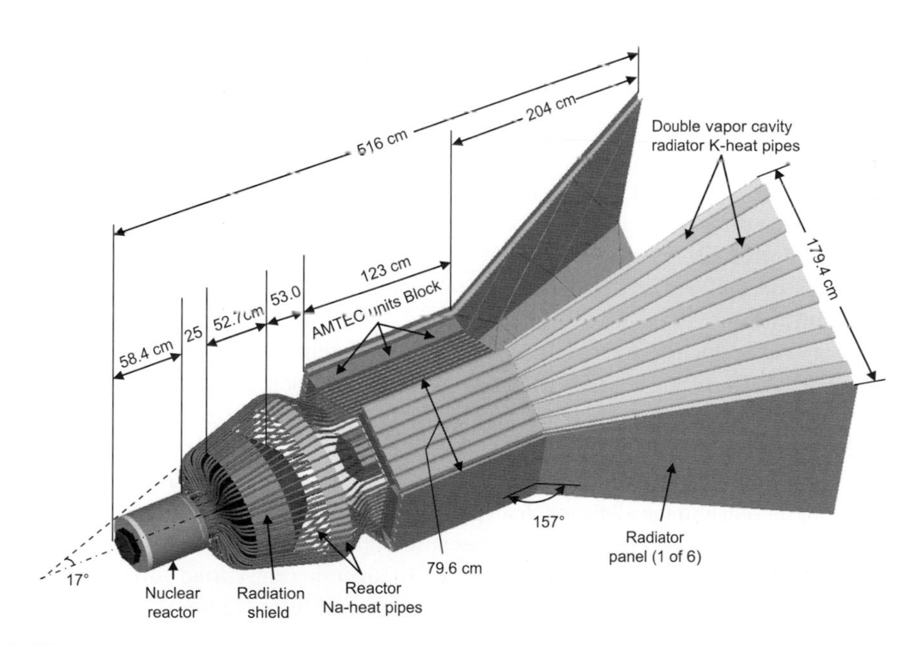

FIGURE 26.26

Isometric view of SAIRS space power system with approximate dimensions [21,55].

the reactor is dissipated by thermal radiation into space from the side surface of the truncated cone of the shield. The shield design ensures that the temperature of the LiH shield remains within the desirable range of 600–680 K [86]. The reactor heat pipes are kept 3.0 cm away from the side surface of the shield, to allow the emitted thermal radiation to stream into space, and do not obstruct the drive shafts (or actuators) of the 12 O/B_4C control drums in the radial reflector (Figure 26.25a).

The six AMTEC blocks in the power system are connected electrically in parallel, and the units in each block are connected in series, providing excellent redundancy. At steady-state, full-power operation, the electrical power to the load is provided at >176 VDC. Thus, with one or more AMTEC units failed in one or more blocks, the SAIRS power system continues to operate, but at a reduced reactor thermal power. When operating nominally at ≤95% of the peak electrical power, the net thermal efficiency of the power system (Figure 26.25) is 27.3%. The corresponding system's specific power is 34.8 W_e/kg and total geometrical surface area of the radiator is 26.9 m^2 (Figure 26.26).

The shadow radiation shield in the SAIRS space reactor power system (Figures 26.25a and 26.26) consists of a thin tungsten layer for attenuating the high-energy gamma photons, followed by relatively thick LiH, for attenuating the fast neutrons from the nuclear reactor. To minimize the diameter and mass of the radiation shadow shield as well as the minor diameter of the heat rejection radiator, the shadow shield is placed as close as possible to the reactor, while maneuvering the reactor heat pipes around the shield before entering the radiator cavity (Figures 26.25a and 26.26).

The vapor-fed, liquid anode Na-AMTEC units in the SAIRS power system have a monolithic sodium beta-alumina solid electrolyte (Na-BASE) [28,29,30–32]. Each AMTEC unit measures 410 mm × 594 mm × 115 mm and weighs 44.3 kg. In order to obtain a terminal voltage of 50 VDC or higher per AMTEC unit, the 128 elongated, dome-shaped BASE elements in the units are connected electrically in series [29]. The BASE elements are tightly arranged in four rows for compactness and to minimize parasitic heat losses from the outer surface (or cathode side) of the BASE elements to the sidewalls and the condenser [29].

Figure 26.27 compares the calculated electrical power for the Na-AMTEC and K-AMTEC versus the thermal efficiency at different BASE temperatures (or anode vapor pressure in kPa), but the same condenser temperature, T_{cd} = 700 K. The electrical power generated at the maximum thermal efficiency is ~49% lower than the maximum electrical power. Similarly, the thermal efficiency of the AMTEC units at the peak electrical power is ~6–8 percentage points lower than its maximum value. The solid portions of the performance curves in Figure 26.26 are where the AMTEC units are inherently load-following.

At the same anode vapor pressure of 76.8 kPa, the BASE temperature in the Na-AMTEC unit is 1123 K, but only 1002 K in the K-AMTEC unit. The crossed open circle symbols in Figure 26.27 indicate the maximum conversion efficiency and the thick-dashed curve connecting the open square symbols represents the locus of the steady-state operation points at different BASE temperatures and corresponding to 86% of the maximum electrical power. The remaining 14% is the design margin for accommodating an increase in the electrical load demand up to that value, without active control of the reactor. This load-following operation feature is because both the reactor and the static AMTEC units in the SAIRS space reactor power system (Figure 26.26) are inherently load-following.

At a BASE temperature of 1123 K (or a nuclear reactor exit temperature of ~1165 K), and when operating at 86% of the maximum electrical power, the Na-AMTEC units' thermal efficiency is 26.7%. Each generates 5.6 kW_e at a terminal voltage of 56 VDC. At the same anode vapor pressure of 76.8 kPa, the K-AMTEC unit has a thermal efficiency of only 20.3% and generates 2.0 kW_e at a terminal

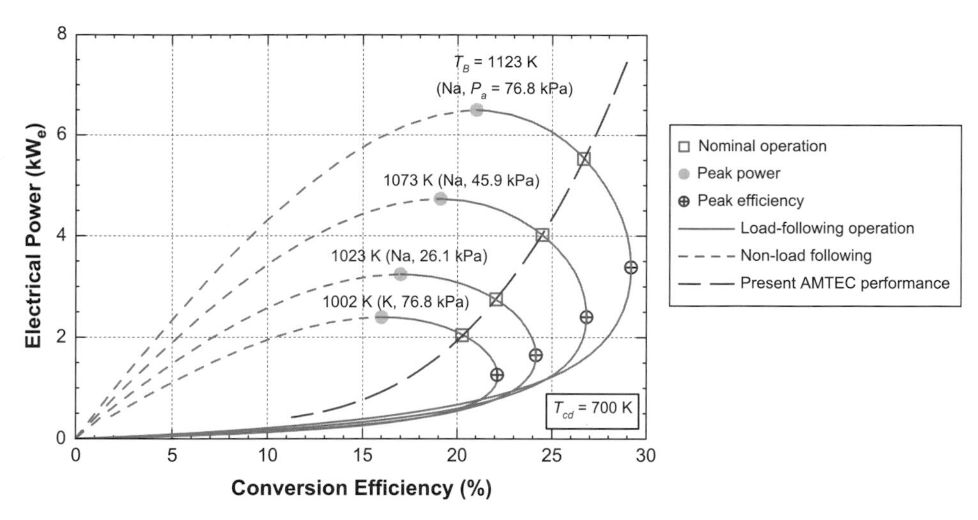

FIGURE 26.27

Performance of Na- and K-AMTEC units at a condenser temperature of 700 K [29].

voltage of only 32 VDC. Therefore, for a 100 kW_e SRPS, a total of 50 2.0-kW_e K-AMTEC units would be needed, compared to only 18 5.6-kW_e Na-AMTEC units, significantly increasing the mass of the power system, since all AMTEC units weigh the same (44.3 kg) [21,29,55].

The SAIRS heat pipe nuclear reactor is designed with enough excess reactivity for operating at steady-state full power for 10 years and the bare reactor remains sufficiently subcritical when submerged in wet sand and flooded with seawater, following a launch abort accident. To facilitate that, the Be reflectors are designed to contain many parts that are assembled together and secured using metal bands. These bands are severed upon impact on water or soil surfaces. The SAIRS space reactor power system is primarily designed for deep space exploration including electrical propulsion using a number of ion thrusters. The same nuclear reactor could easily be used for surface power on the Moon and Mars. Some other unique design features for enhancing safety and operation reliability of SAIRS are:

1. The inventory of the liquid sodium working fluid in the reactor and the AMTEC hot-side Na-heat pipes as well as in the radiator's K-heat pipes is very small, just sufficient to saturate the thin porous metal wick ($<$1.0 mm) on the inside of the heat pipe wall. The liquid metal working fluid is held firmly in the porous wick by the capillary action of the wick. Also, the thin titanium walls of the heat pipes are protected and structurally supported using carbon–carbon armor. Therefore, puncturing the wall of one or more of the heat pipes, by meteorites or space debris, would result in no, or very little, ejection of liquid sodium.
2. Na-heat pipes transport the heat generated in the reactor by fission during steady-state, full-power operation and remove the decay heat after shutdown passively and efficiently.
3. The Na- and K-heat pipes in the SAIRS space reactor power system are designed to nominally operate at 50–60% of their design power throughput. This is ~20 kW/cm^2 for the reactor heat pipes and ~5 kW/cm^2 for the radiator's K-heat pipes.

4. The heat pipes provide excellent redundancy and eliminate single-point failures. A failure of one heat pipe, or more, in the reactor core or the radiator panels would not affect system operation. The thermal load of the failed heat pipes will be carried out by adjacent ones, without exceeding their design power throughput limits.
5. The circulation of the liquid metal working fluid in the heat pipes is fully passive using the capillary pressure head generated in the porous wick on the inside of the heat pipe wall.
6. The voids on the inside of the hexagonal vessel of the SAIRS nuclear reactor are filled with BeO rods. This serves two purposes: it reduces the voids in the core that could be filled with water, following a launch abort accident, and improves the effectiveness of neutron reflection into the reactor core during nominal operation.
7. To ensure sufficient subcriticality of the submerged and flooded bare reactor, following a launch abort accident, neutron absorption additives are added to the uranium nitride fuel pellet in the reactor. For conservative considerations, it is assumed that in such an accident, not only are all voids in the reactor core flooded with water, but also all reactor Na-heat pipes are severed and also filled with water.

SUMMARY AND CONCLUSIONS

This chapter has reviewed technology and design options for space reactor power systems and presented design and system integration choices to enhance the safety and reliability of these systems. The use of these systems should not result in undue radiation exposure to people and astronauts, or damage to property, equipment or space assets. This has, and always will be, the premier safety concern. The information provided is offered to stimulate future discussion of the challenging subject of developing safety standards and guidelines for the launch and future peaceful uses of space reactor power and propulsion systems. These include those deployed in Earth orbits and used for deep space exploration and surface power on the Moon, Mars, and other celestial bodies. The history of the reactor power systems launched by the USA and the former Soviet Union in the 1960s to the 1980s provides valuable lessons. In addition, a number of space reactor designs and power system combinations, for enhancing safety and operation reliability, are presented and discussed.

In the absence of explicit safety standards and regulatory control on deployment and peaceful uses of space reactor power and propulsion systems in outer space, it is hoped that this article will stimulate and contribute to future discussions on the topic. This effort is by no means a comprehensive coverage of all relevant issues, but rather the beginning of an evolving effort to eventually establish globally agreed safety standards and guidelines. Although not a safety issue, future discussions may address non-proliferation concerns associated with using highly enriched nuclear fuel (>90 wt% ^{235}U) in space reactors.

In general, the exposure threat to equipment and astronauts from natural space radiation may far outweigh that from operating fission reactors. Passive operation and decay heat removal and the absence of single-point failures in the reactor design and the integration of space reactor power systems are desirable features for enhancing safety and operation reliability. In addition, the use and implementation of autonomous operation schemes and adaptive control capabilities would greatly enhance safety and ensure mission success, particularly in deep space exploration.

Space reactors use highly enriched uranium fuel in order to realize compact reactors with long operation lifetimes of many years without refueling. The radioactivity resulting from the natural decay of ^{235}U is insignificantly small. Thus, nuclear reactors do not pose a radiological risk during launch and until deployed and operated on the surface of the Moon or other celestial bodies, inserted safely into orbit, or placed on an interplanetary trajectory and operated at a steady power level. Earth-orbiting space reactor power systems represent a radiological concern only in the case of an accidental re-entry. To help avoid such an occurrence, Earth-orbiting systems should be deployed in SHOs, with a natural decay life of 1000 years or more. These orbits are also suitable for end-of-life storage of space reactors until eventual re-entry, at which time the radioactivity in the fuel core would have dropped close to background.

The deployment history of space reactor power systems placed in Earth orbits by the former Soviet Union and the USA in the 1960s to the 1980s provides valuable lessons for the development of guidelines and making recommendations to enhance safety in the future. Based in part on the lessons learned and in order to ensure safe operation and deployment, the following suggestions are offered for consideration in future discussions of the subject of establishing safety standards and guidelines for peaceful uses of space reactor power and propulsion systems:

1. Militarization of space, including earth orbits, *should not be allowed.*
2. Uses in Earth orbits *should be strongly discouraged*, unless justified for potential civilian missions of global benefit and where the solar option is impractical, such as within the Van Allen radiation belts.
3. Passive and safe removal of the decay heat from the reactor core after shutdown *should be required.*
4. Redundancy in energy conversion, reactor cooling, and heat rejection *should be strongly encouraged.*
5. Future uses of liquid metal heat-pipe-cooled and gas-cooled space reactor power systems with no-single point failures in reactor cooling and energy conversion *should be encouraged.*
6. Operating at nominal reactor temperatures below 1000 K *is recommended.* These temperatures are compatible with using steel and super-steel alloys, with known properties and prior flight experience.
7. Armoring the heat rejection radiator and the critical system components *should be required* to protect against impact by meteorites and space debris and exposure to space ionizing radiation and electromagnetic fields.
8. Ejection of reactor fuel core at the end of mission into orbits *should not be allowed.*
9. The development and use of high thermal efficiency energy conversion technology *should be strongly encouraged.*
10. The radial reflector of the space reactor *should be designed to easily disassemble* upon impact on a solid or water surface and the bare reactor *should be designed to remain subcritical* when submerged in wet sand and flooded with seawater or liquid propellant, following a launch abort accident. To ensure subcriticality of the submerged and flooded bare reactor, spectral shift absorber additives to the uranium fuel may be used [77].
11. Storage of deactivated space reactors in operation SHOs with natural decay life $>$1000 years *should be required.*

12. Future uses of space reactor power systems on lunar and Martian surfaces should trigger simultaneous development of credible and acceptable options for end-of-life storage and eventual disposal and/or recovery of nuclear fuel.
13. It is strongly recommended that space reactors be designed:
    a. With sufficient negative temperature reactivity feedback.
    b. For long operation life of 20–40 years, or even longer. This would be possible when using technologies with high thermal efficiencies (>20%).
    c. With the potential for refueling and multiple shutdowns and restart, on the surface of the Moon, Mars, and other celestial bodies.

Acknowledgements

This effort is sponsored by the University of New Mexico's Institute for Space and Nuclear Power Studies (http://www.unm.edu/~ISNPS/).

References

[1] Bennett, G. L. (2002). *Space Nuclear Power.* In: *Encyclopedia of Physical Science and Technology, Vol. 15.* New York: Academic Press. pp. 537.

[2] Bennett, G. L. (2006). Space Nuclear Power: Opening the Final Frontier. San Diego, CA: *Proceedings of the International Energy Conversion Engineering Conference.* Paper AIAA-2006-4191.

[3] Angelo, J., Jr., & Buden, D. (1985). *Space Nuclear Power.* Malabar, FL: Orbit Book Co.

[4] El-Genk, M. S. (Ed.) (1995). *A Critical Review of Space Nuclear Power and Propulsion 1984–1993.* New York: AIP Press, American Institute of Physics.

[5] Ottman, G. K., & Hersman, C. B. (2006). The Pluto–New Horizons RTG and Power System Early Mission Performance. San Diego, CA: *Proceedings 4th International Energy Conversion Engineering Conference.* Paper AIAA-2006-4029.

[6] NASA. (2005). NASA Selects New Frontiers Concept Study: Juno Mission to Jupiter. *NASA HQ press release.* http://www.jupitertoday.com/news/viewpr.rss.html?pid=16990 1 June.

[7] NASA (2009). New Frontiers Program, http://newfrontiers.nasa.gov/index.html

[8] Bennett, G. L. (2009). Dynamic Power for Defense and Exploration a Look at the Department of Energy's 1987 Nuclear Power Sources Assessment Team Report. *Proceedings 7th International Energy Conversion Engineering Conference, American Institute of Aeronautics and Astronautics.* Paper # AIAA-2009-4636.

[9] Hunt, M. E., & Rovang, R. D. (1992). 2.5 kWe Dynamic Isotope Power System for the Space Exploration Initiative Including An Antarctica Demonstration. *Proceedings 9th Symposium on Space Nuclear Power Systems.* 246:222–227.

[10] El-Genk, M. S., & Tournier, J.-M. (2005). Review of Refractory Metal Alloys and Mechanically Alloyed-Oxide Dispersion Strengthened Steels for Space Nuclear Power Systems. *J. Nuclear Materials, 340,* 93–112.

[11] Wiedemann, C., Bendisch, J., Klinkrad, Krag, H., Wegener, P., & Rex, D. (1998). Modeling of Liquid Metal Droplets Released by RORSATs. *Proceedings 29th IAF Congress.* Paper No. IAA 98–IAA6.3.03.

[12] Wiedemann, C., Oswald, M., Stabroth, S., KlinKrad, H., & Vorsmann, P. (2005). Size Distribution of NaK Droplets Released During RORSAT Reactor Core Ejection. *Advances Space Research, 35*(7), 1290–1295.

[13] Buden, D. (1995). Summary of Space Nuclear Reactor Power Systems (1983–1992). In *A Critical Review of Space Nuclear Power and Propulsion 1984–1993* (El-Genk, M., ed.), pp. 21–86.

[14] El-Genk, M. S., Woodall, D. W., Dean, V. F., & Louie, D. L. (1985). In M. S. El-Genk, & M. D. Hoover (Eds), *Space Nuclear Power Systems 1984. Review of the Design Status of the SP-100 Space Nuclear Reactor System, Vol. 1* (pp. 177–190). Orbit Book Co, CONF-840113.

[15] Paramonov, D. V., & El-Genk, M. S. (1994). Development and Comparison of a TOPAZ-II System Model with Experimental Data. *J. Nuclear Technology, 108*(2), 157–170.

[16] El-Genk, M. S., & Xue, H. (1994). Two-Dimensional Steady-State and Transient Analysis of Single Cell Thermionic Fuel Elements. *J. Nuclear Technology, 108*(1), 112–125.

[17] Marriot, A., & Fujita, T. A. (1994). Evolution of SP-100 System Designs. In M. S. El-Genk (Ed.), *Proceedings Symposium on Space Nuclear Power and Propulsion* (pp. 157–169). New York: American Institute of Physics, AIP CP-301.

[18] Lipinski, R. J., Wright, S. A., Sherman, M. P., Lenard, R. X., Marshall, A. C., Talandis, R. A., Poston, D. I., Kapernick, R., Guffee, R., Reid, R., Elson, J., & Lee, J. (2002). In M. S. El-Genk (Ed.), *Proceedings Space Technology and Applications International Forum (STAIF-2002). Small Fission Power for NEP, Vol. 608* (pp. 1054–1062). Melville, NY: American Institute of Physics, AIP Conference Proceedings.

[19] Carre, F. O., Proust, E., Chaudourne, S., & Keirle, P. (1990). Update of the ERATO Program and Conceptual Studies on LMFBR Derivative Space Power Systems. *Proceedings 7th Symposium on Space Nuclear Power Systems*. CONF-900109, Vol. 1, pp. 381–386.

[20] Tilliette, Z. P., Delaplace, J., & Proust, E. (1991). Brayton Cycle Conversion and Additional French Investigations on Space Nuclear Power Systems. *Proceedings 8th Symposium on Space Nuclear Power Systems, AIP Conference Proceedings*. Vol. 217, pp. 26–31.

[21] El-Genk, M. S., & Tournier, J.-M. (2004). "SAIRS"—Scalable AMTEC Integrated Reactor Space Power System. *J. Progress Nuclear Energy, 45*(1), 25–69.

[22] El-Genk, M. S. (2008). Space Reactor Power Systems with No Single Point Failures. *J. Nuclear Engineering Design, 238*(9), 2245–2255.

[23] El-Genk, M. S. (2008). Space Nuclear Reactor Power System Concepts with Static and Dynamic Energy Conversion. *J. Energy Conversion Management, 49*(3), 402–411.

[24] El-Genk, M. S., & Tournier, J.-M. (2006). DynMo-TE: Dynamic Simulation Model for Space Reactor Power Systems with Thermoelectric Converters. *J. Nuclear Engineering Design, 236*(23), 2501–2529.

[25] El-Genk, M., & Saber, H. H. (2006). Thermal and Performance Analyses of Efficient Radioisotope Power Systems. *J. Energy Conversion Management, 47*, 2290–2307.

[26] El-Genk, M. S., & Tournier, J.-M. (2004). Conceptual Design of HP-STMCs Space Reactor Power System for 110 kWe. *Proceedings of Space Technology and Applications International Forum (STAIF-04), AIP Conference Proceedings, Vol. 669*. Melville, NY: American Institute of Physics. pp. 658–672.

[27] Hendricks, T., Huang, C., & Huang, L. (1999). AMTEC Cell Optimization for Advanced Radioisotope Power System (ARPS) Design. *Proceedings of Intersociety Energy Conversion Engineering Conference, Paper No. 1999-1-2655*. Society of Automotive Engineers.

[28] Cole, T. (1983). Thermoelectric Energy Conversion with Solid Electrolytes. *Science, 221*, 915–920.

[29] Tournier, J.-M., & El-Genk, M. S. (2003). Design Optimization of High-Power, Liquid Anode AMTEC. *Proceedings of Space Technology and Applications International Forum (STAIF-03), AIP Conference Proceedings, Vol. 654*. Melville, NY: American Institute of Physics. pp. 740–750.

[30] Williams, R. M., Ryan, M. A., Homer, M. L., Lara, L., Manatt, K., Shields, V., Cortez, R. H., & Kulleck, J. (1999). The Thermal Stability of Sodium Beta-Alumina Solid Electrolyte Ceramic in AMTEC Converters. *Proceedings of Space Technology and Applications International Forum, Vol. 2*. Woodbury, NY: American Institute of Physics. pp. 1306–1311.

[31] Virkar, A. V, Jue, J.-F., & Fung, K.-Z. (2000). Alkali-Metal-P- and P″-Alumina and Gallate Polycrystalline Ceramics and Fabrication by a Vapor Phase Method. US Patent No. 6,117,807.

[32] Ryan, M. A., Williams, R. M., Lara, L., Fiebig, B. G., Cortez, R. H., Kisor, A. K., Shields, V. B., & Homler, M. L. (2001). Advances in Electrode Materials for AMTEC. *Proceedings of Space Technology and Applications International Forum, AIP-CP-552, Vol. 2*. Melville, NY: American Institute of Physics. pp. 1088–1093.

[33] Mills, J., & Van Hagan, T. (1994). S-PRIME-SNP Conceptual Design Summary. *Proceedings Symposium on Space Nuclear Power and Propulsion, AIP CP-301, Vol. 1*. New York: American Institute of Physics. pp. 695–700.

[34] El-Genk, M. S., & Luke, J. R. (1999). Performance Comparison of Thermionic Converters with Smooth and Micro-grooved Electrodes. *J. Energy Conversion Management, 40*, 319–334.

[35] El-Genk, M. S., Xue, H., & Paramonov, D. (1994). Transient Analysis and Start-up Simulation of a Thermionic Space Nuclear Reactor System. *J. Nuclear Technology, 105*(1), 70–86.

[36] Nikitin, V. P., Ogloblin, B. G., Sokolov, E. N., Klimov, A. V., Barabanshchikov, A. A., Ponomarev-Stepnio, N. N., Kukharkin, N. F., Usov, V. A., & Nikolaev, Yu., V. (2000). "YENISEI" Space Nuclear Power System. *Atomic Energy, 88*(2), 98–110.

[37] Kornilov, V. A. (2001). Low-Power Space Thermionic Nuclear Power System. *J. Atomic Energy, 91*(1), 586–589.

[38] Schreiber, J. (2001). Power Characteristics of a Stirling Radioisotope Power System over the Life of the Mission. *Proceedings of Space Technology and Applications International Forum, AIP CP-552, Vol. 1*. Melville, NY: American Institute of Physics. pp. 1011–1016.

[39] Thieme, L., Schreiber, J., & Mason, L. (2002). Stirling Technology Development at NASA GRC. *Proceedings of Space Technology and Applications International Forum, AIP CP-608, Vol. 1*. Melville, NY: American Institute of Physics. pp. 872–879.

[40] Dochat, G. (1992). Free Piston Stirling Component Power Converter Test Results and Potential Applications. San Diego, CA: *Proceedings 21st Intersociety Energy Conversion Engineering Conference*.

[41] Katucki, R., Josloff, A., Kirpich, A., & Florio., F. (1985). In M. S. El-Genk, & M. D. Hoover (Eds), *Space Nuclear Power Systems 1984. Evolution of System Concepts for a 100 kWe Class Space Nuclear Power System, Vol. 2* (pp. 149–164). Orbit Book Co, CONF-840113.

[42] Moriarty, M., & Determan, W. (1989). SP-100 Advanced Reactors Designs for Thermoelectric and Stirling Applications. *Proceedings Intersociety Energy Conversion Engineering Conference, IEEE. Vol. 2,* pp. 1245–1250.

[43] Staub, D.W. (1967). SNAP 10A Summary Report. Atomic International Report NAA-SR-12073, 25 March.

[44] Hatton, S. A., & El-Genk, M. S. (2006). How Small Can Fast-Spectrum Space Reactor Get? *Proceedings of Space Technology and Applications International Forum (STAIF-06), AIP Conference Proceedings, Vol. 813*. Melville, NY: American Institute of Physics. pp. 426–436.

[45] El-Genk, M. S., Seo, J. T., & Buksa, J. J. (1988). Load Following and Reliability Studies of an Integrated SP-100 System. *J. Propulsion Power, 4*(2), 152–156.

[46] Harty, R., & Mason, L. (1993). 100 kWe Lunar/Mars Surface Power Utilizing the SP-100 Reactor with Dynamic Conversion. *Proceedings Symposium on Space Nuclear Power and Propulsion, AIP CP-271, Vol. 2*. New York: American Institute of Physics. pp. 1065–1071.

[47] Gallo, B. M., & El-Genk, M. S. (2008). Performance Analysis of 38 kWe Turbo-Machine Unit for Space Reactor Power Systems. *Proceedings of Space Technology and Applications International Forum (STAIF-08) AIP Conference Proceedings, Vol. 969*. Melville, NY: American Institute of Physics. pp. 625–636.

[48] Gallo, B. M., & El-Genk, M. S. (2009). Brayton Rotating Units for Space Reactor Power Systems. *J. Energy Conversion Management, 50*, 2210–2232.

[49] El-Genk, M. S. (2001). A High-Energy Utilization, Dual-Mode System Concept for Mars Missions. *J. Propulsion Power, 17*(2), 340–346.

[50] Mason, L. (2004). A Power Conversion Concept for the Jupiter Icy Moons Orbiter. *J. Propulsion Power, 20* (5), 902–910.

[51] Saber, H. H., El-Genk, M. S., & Caillat, T. (2007). Test Results of Skutterudite-Based Thermoelectric Unicouples. *J. Energy Conversion Management, 48*, 555–567.

[52] El-Genk, M. S., & Saber, H. H. (2005). Performance Analysis of Cascaded Thermoelectric Converters for Advanced Radioisotope Power Systems. *J. Energy Conversion Management, 46*, 1083–1105.

[53] Fleurial, J.-P., Borshchevsky, A., Caillat, T., & Ewell, R. (1997). New Materials and Devices for Thermoelectric Applications. *Proceedings 32nd Intersociety Energy Conversion Engineering Conference*. New York: American Institute of Chemical Engineers. pp. 1080–1085.

[54] Caillat, T., Borshchevsky, A., Snyder, J., & Fleurial, J.-P. (2000). High Efficiency Segmented Thermoelectric Unicouples. *Proceedings Space Technology and Applications International Forum (STAIF-2000), AIP Conference Proceedings, Vol. 504*. New York: American Institute of Physics. pp. 1508–1512.

[55] El-Genk, M. S., & Tournier, J.-M. (2003). High Power AMTEC Converters for Deep-Space Nuclear Reactor Power Systems. *Proceedings Space Technology and Applications International Forum (STAIF-03), AIP Conference Proceedings, Vol. 654*. Melville, NY: American Institute of Physics. pp. 730–739.

[56] El-Genk, M. S. (2009). Deployment History and Design Considerations for Space Reactor Power Systems. *Acta Astronautica, 64*(9–10), 833–849.

[57] IAEA. (2005). *The Role of Nuclear Power and Nuclear Propulsion in the Peaceful Exploration of Space*. Vienna, Austria: International Atomic Energy Agency.

[58] Dieckamp, H. M. (1967). *Nuclear Space Power Systems. Atomics International Report, September.* Canoga Park, CA: Atomic International.

[59] Siddiqi, A. (1999). Starting at the Sea: The Soviet RPRSAT and EORSAT Programmes. *J. Br. Interplanetary Soc., 52*(11/12), 397–416.

[60] Perry, G. (1978). Russian Ocean Surveillance Satellites. *Royal Air Forces Quarterly, 18*, 60–67.

[61] Bennett, G. L. (1989). *Soviet Space Nuclear Reactor Incidents: Perception versus Reality. Space Nuclear Power Systems 1989, Vol. 25*. Malabar, FL: Orbit Book Co. pp. 273–278.

[62] Grasty, R. L. (1978). Estimating the Fallout on Great Slave Lake from Cosmos 954. *Trans. Am. Nuclear Soc., 29*, 116–118.

[63] Andreev, P. V., Gryaznov, G. M., Zhabotinsky, E. E., Zaritsky, G. A., Nikonov, A. M., & Serbin, V. I. (1991). The Conceptual Design and Main Characteristics of Long Lifetime Thermionic SNPS with Thermal Reactor. *Proceedings of 8th Symposium of Space Nuclear Power Systems, AIP Proceedings*. 217, Vol. 1, pp. 367–372.

[64] Paramonov, D. V., & El-Genk, M. S. (1997). Test Results of Ya-21u Thermionic Space Power System. *J. Nuclear Technology, 117*(1), 1–14.

[65] Stavissky, Y. Y. (2007). Nuclear Energy for Space Missions. *Physics-Uspekhi, 50*(11), 1179–1187.

[66] Krass, A. W., & Goluoglu, K. L. (2005). Experimental Criticality Benchmarks for SNAP 10A/2 Reactor Cores. *Oak Ridge National Laboratory Technical Report ORNL/TM-2005/54, Vol. 1.*

[67] El-Genk, M. S., & Schriener, T. M. (2009). Performance and Radiological Analyses of Space Reactor Power System Deployed into a 1000–3000 km Earth Orbit. *J. Progress Nuclear Energy.*

[68] El-Genk, M. S., & Rider, W. J. (1990). Reliability and Vulnerability Studies of the SP-100 Dual-Loop Thermoelectric–Electromagnetic Pumps. *J. Propulsion Power, 6*(3), 305–314.

[69] Russel, R. (2005). *Layers of Earth's Atmosphere*, at http://www.windows.ucar.edu/tour/link=/earth/Atmosphere/layers_activity_print.html

[70] Krisko, P. H. (2007). The Predicted Growth of the Low Earth Orbit Space Debris Environment—An Assessment of Future Risk for Spacecraft. *Proceedings Institution of Mechanical Engineers, Part G: J. Aerospace Engineering, 221*(6):795–985.

[71] King, J. C., & El-Genk, M. S. (2006). Submersion Subcritical Safe Space (S^4) Reactor. *J. Nuclear Engineering Design, 236*, 1759–1777.

[72] United Nations General Assembly (1992). *Principles Relevant to the Use of Nuclear Power Sources in Outer Space*, G.A. res. 47/68, 47 U.N.GAOR Supp. (No. 49) at 88, UN Doc. A/47/49.

[73] United Nations. (2008). *Status of International Agreements Relating to Activities in Outer Space*. Office for Outer Space Affairs. http://www.unoosa.org/pdf/publications/ST_SPACE_11_Rev2_Add1E.pdf

[74] Bashor, H. (2004). *The Moon Treaty Paradox*. Philadelphia: Xlibris.

[75] Bouvet, I. (2004). Use of Nuclear Power Sources in Outer Space: Key Technology Legal Challenges. *J. Space Law, 30*(2), 203–226.

[76] Haskin, F.E., & Marshall, A.C. (2008). Accident Consequence Modeling. In *Space Nuclear Safety* (Marshall, A.C., ed.; Haskin, F. E and Usov, V.A., co-eds), Chapter 11, pp. 385–434. Krieger, Malabar, FL.

[77] King, J. C., & El-Genk, M. S. (2006). Submersion Criticality Safety of Fast Spectrum Space Reactors: Potential Spectral Shift Absorbers. *J. Nuclear Engineering Design, 236*(3), 238–254.

[78] El-Genk, M. S., & Tournier, J.-M. (2006). High Temperature Water Heat Pipes Radiator for a Brayton Space Reactor Power System. *Proceedings of Space Technology and Applications International Forum (STAIF-06), AIP Conference Proceedings, Vol. 813*. Melville, NY: American Institute of Physics. pp. 716–729.

[79] Tournier, J.-M., & El-Genk, M. S. (2006). Liquid Metal Loop and Heat Pipes Radiator for Space Reactor Power Systems. *J. Propulsion Power, 22*(5), 1117–1134.

[80] Rosen, R., & Schnyer, A. D. (1989). Civilian Uses of Nuclear Reactors in Space. *Science Global Security, 1*, 147–164.

[81] El-Genk, M. S., Hatton, S., Fox, C., & Tournier, J.-M. (2005). SCoRe—Concepts of Liquid Metal Cooled Space Reactors for Avoidance of Single-Point Failure. *Proceedings Space Technology and Applications International Forum (STAIF-05), AIP Conference Proceedings, Vol. 746*. Melville, NY: American Institute of Physics. pp. 473–484.

[82] Hatton, A. S., & El-Genk, M. S. (2007). Low Mass SCoRe-S Designs for Affordable Planetary Exploration. *Proceedings of Space Technology and Applications International Forum (STAIF-07), AIP Conference Proceedings, Vol. 880*. Melville, NY: American Institute of Physics. pp. 242–253.

[83] Tournier, J.-M., & El-Genk, M. S. (2006). Bellows-Type Accumulator for Liquid Metal Loops of Space Reactor Power Systems. *Proceedings of Space Technology and Applications International Forum (STAIF-06), AIP Conference Proceedings, Vol. 813*. Melville, NY: American Institute of Physics. pp. 730–742.

[84] King, J. C., & El-Genk, M. S. (2007). Thermal-Hydraulic Analyses of the Submersion-Subcritical Safe Space (S^4) Reactor. *Proceedings of Space Technology and Applications International Forum (STAIF-07), AIP Conference Proceedings, Vol. 880*. Melville, NY: American Institute of Physics. pp. 261–270.

[85] El-Genk, M. S., & Tournier, J.-M. (2009). DynMo-CBC: Dynamic Simulation Model for a Space Reactor Power System with Multiple CBC Loops. *Proceedings 7th International Energy Conversion Engineering Conference (IECEC-09), Paper No. AIAA-2009-4577*. American Institute of Aeronautics and Astronautics.

[86] Barattino, W., El-Genk, M. S., & Voss, S. (1985). Review of Previous Shield Analysis for Space Reactors. In *Space Nuclear Power Systems 1984, CONF-840113, Vol. 2* (pp. 329–339). Malabar, FL: Orbit Book Co.

CHAPTER

Removal of hazardous space debris

27

Martha Mejía-Kaiser

Universidad Nacional Autónoma de México, IISL Member

CHAPTER OUTLINE

INTRODUCTION

Since 1957, humans have introduced man-made objects into outer space. Most of these objects are no longer functional, move in orbits around our planet, and are increasingly populating the Earth's

Space Safety Regulations and Standards. DOI: 10.1016/B978-1-85617-752-8.10027-3

neighborhood. These objects have been labeled as space debris [1]. Some of these space objects descend sufficiently to be captured by the Earth's atmosphere. Other objects that reach high altitudes are condemned to stay there for several years or even centuries. It is difficult to estimate the number of such objects, but Perek has considered that approximately 5000 tons of materials have been launched into outer space [2]. A significant portion of these materials is still in orbit and is non-functional. In outer space, all space debris objects, small and large, are in permanent movement at speeds that make them a hazard for operating satellites, manned spacecrafts, and astronauts' health and life. With the increasing presence of space debris, the number of collisions will grow, which will also have an impact on commercial space activities.

27.1 HAZARDOUS SPACE DEBRIS

The Inter-Agency Space Debris Coordination Committee (IADC), an international non-governmental organization which encompasses space agencies of spacefaring countries [3], has been formulating recommendations for the mitigation of space debris since 1997. In 2002 [4], the IADC adopted the Space Debris Mitigation Guidelines. In 2007, the United Nations General Assembly adopted Resolution 62/217 on Space Debris Mitigation Guidelines [5], which is based on the IADC Mitigation Guidelines.

States, international organizations, and private companies began to follow the IADC recommendations since their inception, removing satellites approaching their end-of-life using satellites' own capabilities. Surveillance systems began to be aware of a reduction of the space debris population. Nevertheless, space debris specialists have noticed that, even with the application of the IADC Mitigation Guidelines, the increase of the small space debris population (such as bolts, droplets, paint flakes, and aluminum bits) is in permanent growth [6]. Small space debris objects are generated by their release from large space objects or by fragmentation of large space objects. It is obvious that in order to reduce the increase of small space debris, measures to avoid large space debris objects are the logical step.

Specialists consider the removal of space debris objects by external mechanisms. Although the removal of small and large space debris is difficult to achieve, the development of technology points in the direction of removal of large space debris objects first, before removing small space debris.

Several solutions for the mitigation of non-functional large space debris have been proposed. One deals with active external removal. However, the costs of these operations may be too high and some questions arise: Who will pay for the removal of non-valuable space objects? Can a State remove a non-functional space object without the owner's prior consent? Possible answers to these questions can be found in another area, where similar concerns had already moved States to create an international self-enforcing mechanism.

27.2 MARITIME WRECK REMOVAL

27.2.1 Historic development

Since ancient times, ships have suffered maritime casualties and have sunk . Other ships have engaged in helping to save life and cargo. In some cases, stressed ships have been pulled out and brought to ports, where they can be refurbished. Over the course of time, the rescue of valuable objects evolved into the "maritime salvage industry" [7], which has existed for several centuries. At present, this industry is mature, has many salvage methods and technologies, and its expertise is under permanent development [8].

Companies started to offer contracts for salvage operations, using the principle "no cure, no pay"; the salvor would be rewarded only if the salvage operation was successful. The recovery costs of valuable sunken ships and cargoes were covered by owners. As such operations are expensive, ship owners decided to obtain insurance or other financial backing to fund the salvage. In other cases, when owners considered that the ship and its cargo were no longer valuable, they abandoned them at the place where they sank. No salvage operation was performed and such ships and cargo were labeled "wrecks" [9]. With time, shipwrecks started to populate traffic lanes.

In order to provide safe passage for national and foreign vessels, Coastal States started to take measures to clear traffic lanes. At first, the removal costs were covered by Coastal States. Later, shipwreck owners received demands to contribute to fund removal operations. And again, as such operations are expensive, ship owners decided to obtain insurance or other financial backing to fund the removal. Since then, maritime wreck removal has been performed by the experienced salvage industry.

27.2.2 Wreck removal in territorial waters

Many Coastal States have adopted national legislation by which all vessels entering the Coastal State's ports need insurance that, in the case of accident, covers the removal of wrecks that are a hazard to navigation or environment [10].

27.2.3 Wreck removal in the Exclusive Economic Zone

As the International Maritime Organization realized that shipwreck removal in territorial waters was developing into State practice, it devoted its attention to draft a treaty to apply such mechanisms beyond territorial waters, into the Exclusive Economic Zone (EEZ) [11]. In May 2007, the text of the "Nairobi International Convention on the Removal of Wrecks" was adopted (hereafter "Nairobi Wreck Convention") [12]. This Convention addresses the procedures to remove a wreck that hinders maritime traffic or a wreck with hazardous materials that poses a risk of damage to the marine environment. One of the main pillars of this treaty is wreck removal insurance that allows for a self-enforcing mechanism.

27.2.4 Wreck removal in international waters

There is no international legal instrument to regulate wreck removal in international waters. Ships on the high seas are likely to sink very deeply, without hindering maritime traffic. However, in the shallow waters of international straits with busy traffic lines or international rivers, wreck removal is necessary. In such cases, the same mechanisms as used in territorial waters are applied, so insurance and salvage companies play an important role [13].

Today, when a State orders the removal of a wreck from its territorial waters, from the EEZ, or from an international strait or river, salvage companies bid to get the salvage contract. The negotiations for wreck removal include representatives of ship owners, insurance underwriters, Coastal States' authorities, and interested salvage companies. Such negotiations may be complex and may take from several hours to months until an agreement is reached and the salvage is begun. But, in all cases, wrecks are removed.

27.2.5 Reporting wrecks

National legislation, the Nairobi Wreck Convention, and the practice in international waters have more or less standardized the mechanisms for wreck removal. Reporting wrecks is one element that is already an obligation in important international maritime traffic lanes. After a ship suffers a mishap in a maritime traffic lane and becomes a wreck, the ship's master and operator report to the authorities of the Affected State [14] on the event and provide information on the wreck. Such a report provides data on the location of the wreck, and its size and construction, the nature of the damage, the condition, nature and quantity of the cargo, and any hazardous and noxious substances on board. This information is also made public in important nautical publications. The wreck is marked in order to avoid other ships colliding with it.

27.2.6 Wreck removal without prior consent

Owners of hazardous wrecks contract a salvor to perform wreck removal on their behalf. Through the Nairobi Wreck Convention, ratifying States will be obliged to take measures to ensure that their registered owners arrange wreck removal from the EEZ [15].

In cases where a shipwreck is not reported, another procedure takes place. After authorities of a Coastal State spot a wreck in an important maritime lane, an investigation is undertaken. It is determined if this wreck constitutes a hazard to navigation and/or the environment. The identity of the wreck's ownership and its nationality are determined. Once a wreck has been labeled as a hazard, the Coastal State informs the State of the ship's registry and owner. A reasonable deadline is set for the removal.

If in territorial waters, if the registered owner does not remove the wreck within a deadline, Coastal States may remove the wreck at the registered owner's expense, without the wreck owner's prior consent. The costs of the wreck removal operation are covered by insurance. The same mechanism is envisioned by the Nairobi Wreck Convention, where State Parties give their implicit consent to the Affected State to remove the wreck, provided the owner was notified [16]. The Nairobi Convention recognizes that registered owners have rights in their wreck that shall be observed by the State wishing to remove it [17]. But it is clear that removal operations can be accomplished without the owner's prior consent, if the registered owner fails to comply with the procedural steps that were accepted by its State through its ratification of the Convention.

27.2.7 Maritime wreck removal insurance

With the growing practice of wreck removal, many States have already implemented national regulations that mandate owners of ships flying their flags to obtain wreck removal insurance. Coastal States demand foreign vessels entering territorial waters show insurance certificates or other financial securities for wreck removal [18]. The same States also take for granted that ship owners have wreck removal insurance when their ships navigate international waters.

Insurance underwriters commit to give financial backing based on a fair profits scenario. In such a scenario, ship owners are expected to take measures to reduce risks to navigation and the marine environment. The premium is proportional to these risks. After financing a removal operation, the insurer may recover the cost from the shipwreck's owner. The insurer may invoke the defense that the maritime casualty was caused by the owner's willful misconduct [19].

This mandatory insurance plays a crucial role as a self-enforcing mechanism; if the registered owner does not remove the wreck, third parties may do so without prior consent of the owner and with the incentive of payment by the insurer.

27.3 SPACE DEBRIS REMOVAL SYSTEMS

Space salvage activities have already started with manned space missions [20]. So far, several astronauts have recovered experimental objects and some stranded satellites. Some of these satellites have been brought to Earth for refurbishment. Other satellites, such as the Hubble Space Telescope, are captured, refurbished *in situ*, and released again in outer space. But, up to now, no removal of non-valuable space assets by astronauts has taken place.

The Inter-Agency Space Debris Coordination Committee (IADC) has designated two protected regions around our planet. One protected region is from the lowest orbit up to 2000 km altitude. The second is the geostationary orbit. The IADC has proposed that these regions should be cleared of space debris through active removal into "disposal orbits".

Space objects approaching their end-of-life in low orbits must be transferred (re-orbited) [21] into graveyard orbits above 2000 km Earth altitude or must be transferred into the upper layers of the atmosphere in order to reduce their orbital life (de-orbiting) [22]. Space objects approaching their end-of-life in the geostationary orbit should be removed to graveyard orbits, which have been designated at an altitude of approximately 300 km over the geostationary orbit [23]. At present, active removal can only be undertaken by using the existing guidance capabilities of space objects and the remaining fuel on board. In the case of space objects without maneuverable ability, future removal may be performed using external means: by astronauts or with Earth remote-controlled systems [24].

27.3.1 Manned removal

Astronauts may perform removal of space debris by hand or using remote-controlled devices from their spacecraft. Nevertheless, astronauts already face risks while retrieving valuable space assets [25]. The risks are still too high to consider manned missions as the primary method for the removal of space debris. Also, astronauts themselves may not be willing to risk their lives to remove space debris. In addition, such manned missions only take place in low orbits (around 300 km altitude) within the Earth's magnetosphere where astronauts are protected against dangerous radiation from the Sun. The geostationary orbit at 36,000 km is definitely out of reach of manned missions considering present safety concerns.

27.3.2 Earth automated removal

A second method, in the future, would be Earth remote-controlled removal systems. In 2007, the first salvage operation using an Earth remote-controlled system was successfully tested. The "Orbital Express" servicing system was developed by the US Defense Advanced Research Projects Agency (DARPA) and consists of a "chaser" satellite and a heavier target satellite. In 2007, the Orbital Express system completed "... satellite rendezvous, capture, refueling and components exchange" in low orbit (500 km altitude) [26]. Orbital Express is designed to work with compatible docking systems, so only specially constructed spacecrafts can be captured. Nevertheless, other proposals are under way to capture objects with nets, tugs, and other techniques [27]. Following the parallel of maritime salvage, it is very likely that the salvage industry in space can evolve to be used for space debris removal.

27.3.3 Existing procedures and practices applicable to future space debris removal

Various existing procedures and practices performed by States, private companies, and international organizations for wreck removal can support future commercial space debris removal.

27.3.3.1 Decommission declaration and space debris production information

Some satellite owners or operators declare their dead satellites as "decommissioned" when all the fuel is used up and they are not maneuverable, when their capabilities have been reduced and it is no longer viable for them to remain operative, or when they malfunction. An important aspect of this decommissioning notification is to warn other satellite operators that a space object will no longer be controlled and may endanger operational satellites in proximity. In other cases, States do not officially declare a space object as decommissioned, but inform the international community about the detachment of stages during launching and positioning phases. Both the decommissioning notice and information on detachment of stages are implicit declarations about new space debris. These public declarations are becoming international practice.

27.3.3.2 Space surveillance

From the beginning of the space age, it became important for the military sector of the USA and the then Soviet Union to detect and track the movements of any space object launched into outer space [28]. These countries operate space surveillance systems that not only permanently monitor functional space systems, but also the whole space debris population [29]. For continuity in their observations, the USA (USSPACECOM) and the Russian Federation (Space Surveillance System) enter all observed space objects into a catalog of their own. In addition to dead satellites and launching stages, large fragments resulting from space object collisions, explosions, and tests are also entered into these catalogs. It is important to add that both monitoring systems are able to identify the nationality of many space debris objects, which are entered into these catalogs [30].

Unclassified information on space debris is transmitted on a routine basis to satellite operators and agencies managing manned space missions (e.g. for the International Space Station). Space surveillance information has, until now, helped most space controllers to perform evasive maneuvers. If no space debris mitigation is undertaken, the number of space debris objects will increase and evasive maneuvers will become more complex. Such evasive maneuvers may even become impossible, when the space debris population reaches a level where space surveillance information is rendered worthless. Sooner or later, removal of space debris will become necessary.

27.3.3.3 Space insurance

At the moment, many States require compulsory insurance for national space activities. In most States, insurance is mandatory for the granting of licenses and authorizations for space activities. In addition to mandatory insurance coverage, most States, international organizations, and private companies subscribe to supplementary insurance in order to secure their space assets and revenues.

27.4 PROPOSAL FOR A SPACE DEBRIS REMOVAL SYSTEM

The procedures developed for maritime wreck removal and some existing practices of space actors can serve as a basis for a model for the removal of space debris, when space removal becomes a reality and where such systems become available on a commercial basis. The following is a proposal.

27.4.1 International Convention on the Removal of Hazardous Space Debris and an international technical institution

A system may be adopted to establish a self-enforcing mechanism for the removal of hazardous space debris objects, following the model of the Nairobi Wreck Convention. A draft proposal on an International Convention on the Removal of Hazardous Space Debris is included in Appendix 4 of this book. It tracks, as closely as possible, the procedures of the Nairobi Wreck Convention and, thus, elaborates on the consensus of States in the maritime field. One important requirement of such a system would be registration of space objects, before they are launched into outer space.

As the concept of "Coastal States" (found in maritime removal) cannot be used for space activities, it is proposed that States party to the space debris removal system confer powers either on an existing international body or establish an international technical institution for that purpose.

27.4.2 National regulations on space debris removal

1. At the national level, States need to introduce procedures for the removal of space objects under their nationality. These procedures may be based on an updated version of the IADC Space Debris Mitigation Guidelines. States need to mandate that the first step is removal using a space object's own capabilities before it reaches its end-of-life. The USA and Germany have already introduced such regulations [31].
2. In cases where the transfer to a disposal orbit is not possible due to a spacecraft's malfunction, States must require a mandatory external removal arrangement.
3. The operator must be required to furnish a compulsory decommission declaration and/or information on detachment of stages and other relevant information. This information may include the orbital parameters of the space debris object, its size and construction, and any danger to space traffic and hazardous elements onboard.
4. States need to establish national regulation requiring mandatory space debris removal insurance or other financial backing. Space debris removal insurance could be included in the existing space insurance packages.

27.4.3 An international technical institution

An international technical institution should perform the following tasks:

1. Engage with surveillance systems operators, in order to be informed of suspected hazardous space debris (declared or not declared as decommissioned or detached stages).

2. Determine if a particular space debris object is hazardous. Among the criteria for the determination of hazard are the size and shape, the affected traffic, the vulnerability of manned space missions (e.g. orbits of the ISS). Important aspects of the space debris hazard determination are:
   a. Kinetic energy of debris that may endanger operating spacecrafts and astronauts. Entire dead spacecraft that were not "passivated" [32] are in danger of exploding in outer space and can be particularly considered a hazard.
   b. Hazardous materials of certain spacecraft, such as nuclear power sources, do not pose a danger to the environment of outer space but to the Earth's environment, if they enter the atmosphere and survive air friction [33].
3. In cases where a space debris object is labeled as hazardous, the international technical institution needs:
   a. To identify the nationality and ownership of the space debris object.
   b. To inform the State of Registry and the owner and/or operator about the hazardous nature of its object.
   c. To request the owner and/or operator for the removal of a space debris object, and to inform the State of Registry on that request.
   d. To set a deadline for the removal.

27.4.4 Removal without prior consent

In cases where the space debris owner does not remove the space debris within the deadline:

1. States of Registry that have introduced national regulatory mechanisms and agree to grant authority to an international institution will be implicitly accepting the possibility that, in case of default by the space debris owner and/or operator, third parties can perform the removal of the hazardous space debris without the prior consent of the owner and/or operator and State of Registry, provided that the international technical institution notified the owner and/or operator, and its State of Registry. The institution shall coordinate the removal.
2. The costs of the removal operations have to be borne directly by the insurer or other person providing financial security. In order to benefit from the insurance, space debris owners and/or operators have to demonstrate that they have taken all measures to reduce risks to space navigation and/or Earth environment. The international institution may set the minimum measures to be taken by the owner and/or operator [34].

As not all large space debris objects may be transformed into a hazard, with the imperative to be removed, insurance underwriters or other financial sponsors would commit to give financial backing to space removal operations, based on a fair profits scenario. The insurance, or financial backing premium, will be proportional to such risks.

Insurance underwriters or other financial sponsors may make use of space surveillance information in order to measure and verify if owners and/or operators of space objects have complied with the space debris mitigation guidelines and the Convention. Also, contracts with satellite constructors and space controllers can provide proof that a satellite owner or operator took the required measures relating to the design of the satellite and launch vehicle, its construction, operation, and removal before the space object's end-of-life. Failure to follow minimal mitigation guidelines may be considered by the insurance or other financial sponsor as contract breach. In cases where a space debris object becomes an imminent hazard to space navigation and it is urgent to remove it from valuable orbits, insurance underwriters or other financial sponsors would have the obligation to cover space removal operations. Once the removal operation has been finalized, insurance and other financial sponsors may recover the costs from the owner and/or operator of the space debris object, invoking owners and/or operators' willful misconduct. This is another motive that may move owners and/or operators of space objects to comply with space debris mitigation guidelines.

27.4.5 Complementary systems

Old, large space debris objects under the nationality of States that are not insured are outside of this scheme. For the removal financing of such space debris, another mechanism must be introduced. An international fund may be established for this purpose [35]. The introduction of a self-enforcing mechanism, where third parties perform space debris removal, with the incentive of the payment by insurance, will not only help to clear valuable orbits from space debris, but it will also give a push to a new commercial activity: the space debris removal industry.

Notes and references

[1] According to the Inter-Agency Space Debris Committee (IADC), "space debris" are "... all man-made objects, including fragments and elements thereof, in Earth orbit or re-entering the atmosphere, that are non-functional". Inter-Agency Space Debris Coordination Committee space debris mitigation guidelines, UN Doc. A/AC.105/C.1/L.260, 29 November 2002, 3.1. Space Debris.

[2] Perek, L. (2004). Rational Space Management. ZLW No. 53, p. 574.

[3] The IADC is a non-governmental organization. Members of the IADC are the space agencies of China, India, Italy, France, Germany, Japan, Russian Federation, Ukraine, the UK, the USA, and the European Space Agency (ESA).

[4] UN Doc. A/AC.105/C.1/L.260, 29 November 2002, *supra* note 1.

[5] United Nations General Assembly Resolution 62/217 "International Cooperation in the Peaceful Uses of Outer Space", 21 December 2007 (para. 26), Official Records of the General Assembly, 62nd Session, Supplement No. 20 (A/A/62/20), 2007, paras. 117 and 118 and annex (IADC Guidelines).

[6] Portelli, C. (2009). Presentation at the International Interdisciplinary Congress on Space Debris, Montreal, 7–9 May.

[7] Salvage is defined in *Black's Dictionary* as "... a compensation allowed to persons by whose assistance a ship or its cargo has been saved, in whole or in part, from impending danger, or recovered from actual loss in cases of shipwreck, derelict or recapture ...". *Black's Law Dictionary* (1991). 932, 6th ed. Key word: salvage.

[8] E.g. pulling tugs, divers, robotic submarines, floating cranes, systems to remove oil or to cut hull in pieces on the sea bottom, use of cargo helicopters, etc.

[9] The Nairobi Wreck Convention defines "wreck" as "a sunken or stranded ship, including any parts of such ship and any object that is or has been on board ... any object that is lost at sea from a ship and that is stranded, sunken or adrift at sea". The definition includes also ships still afloat that are "... reasonable [to] be expected to sink or strand". Nairobi International Convention on the Removal of Wrecks (2007). Article 1(4). Adoption of the Final Act in Nairobi, on 18 May 2007. International Maritime Organization, IMO LEG/CONF.16/19, 23 May 2007.

[10] "Hazard" is defined as "impediment to navigation; or ... may be ... expected to result in major harmful consequence to the marine environment or damage to coastline or related interests of one or more States". Article 1(5), *Ibid.*

[11] The EEZ is the area adjacent to the territorial waters of a Coastal State but no larger than 200 miles from the baseline agreed by the United Nations Convention on the Law of the Sea (UNCLOS) (1982). States party to UNCLOS, and many States that have not yet ratified it, have accepted 12 miles as territorial water, measured from its baseline in accordance with UNCLOS. According to UNCLOS Art. 58, "… all States, whether coastal or land-locked, enjoy … freedom of navigation" in the EEZ. The first area of application proposed in the early drafts of the Nairobi Wreck Convention referred to "international waters". Griggs, P. (Immediate Past President, Committee Maritime International), Draft Wreck Removal Convention, at 3. www.comitemaritime.org/capetown/pdf/13.pdf, at 1.

[12] The text of the Nairobi Wreck Convention was open for signature in November 2007. Nairobi Wreck Convention, *supra* note 9, Art. 17.

[13] In 2002 a ship sank in the English Channel as a result of a collision. The *Tricolor*, flying the Norwegian flag, touched the sea bottom at 30 m depth. As the location of the ship was in a busy maritime traffic line in an international strait, it was properly marked. Nevertheless, some hours later, a third ship was damaged after colliding with the sunken *Tricolor* and two others made near misses. A team dove and inspected *Tricolor*'s damages. As a result of such inspections the *Tricolor* was declared as a total loss 9 days after the sinking and became a major navigational hazard for many months. The event took place in the English Channel, an international strait, but also inside the French Economic Exclusive Zone. The French authorities issued a wreck removal order 10 days after the collision. The removal operation included the cutting of the vessels into pieces, which were later transported to a Belgian port. This removal operation became the biggest ever and the wreck was finally removed 2 years after it sank. The location of the event was 55 nautical miles from Dover, UK and 25 miles from Zeebrugge, Belgium (51′ 21.9″ N 02′ 65″ E), www.tricolorsalvage.com

[14] "Affected State" means "… the State in whose Convention area the wreck is located". "Convention area" means "the exclusive economic zone of a State Party …". Nairobi Wreck Convention, art. 1(10) and 1(1), *supra* note 9.

[15] Nairobi Wreck Convention, *supra* note 9, Art. 9.

[16] "The Affected State shall … inform the registered owner in writing of the deadline … if the registered owner does not remove the wreck within that deadline, it may remove the wreck at the registered owner's expense …". Nairobi Wreck Convention, *supra* note 9, Art. 9(6b).

[17] Nairobi Wreck Convention, *supra* note 9, Art. 2(3).

[18] In the Nairobi Wreck Convention the insurance mechanism is extensively addressed. The obligation to produce such a certificate is in Art. 9(3). Nairobi Wreck Convention, *supra* note 9.

[19] Wreck removal cost recovery by the insurance has been introduced in the Nairobi Wreck Convention. Nairobi Wreck Convention, *supra* note 9, Art. 12(10).

[20] In outer space, "salvage" is the operation to rescue space objects that are valuable. Such operations may be undertaken for refueling or refurbishing purposes, for gathering scientific or military data, for studying the causes of a space object malfunction, etc.

[21] 3.4.3. Re-orbit: "Re-orbiting is the intentional changing of a space system's orbit". UN Doc. A/AC.105/C.1/L.260 (2002). *Supra* note 1.

[22] 3.4.2. De-orbit: "… intentional changing of orbit for re-entry of a space system into the Earth's atmosphere to eliminate the hazard it poses to other space systems, by applying a retarding force, usually via a propulsion system". Inter-Agency Space Debris Coordination Committee space debris mitigation guidelines, UN Doc. A/AC.105/C.1/L.260 (2002). The IADC considered a maximum of 25 years as a "reasonable and appropriate lifetime limit" to de-orbit a space object. International Academy of Astronautics (IAA), (2006). Position Paper on Space Debris Mitigation, p. 8.

[23] According to the Robotic Geostationary Orbit Restorer, Final Report, graveyard orbits are between 245 and 435 km above the geostationary ring. Robotic Geostationary Orbit Restorer, Final Report—Executive Summary, EADS Space Transportation, 10 June 2003, at 1.

[24] Most planned systems are intended for refurbishing large space objects or for retrieving experiments. Although small space debris, like bolts, droplets, paint flakes, and aluminum bits are as dangerous as large space debris, present plans for experimental removal systems are mostly for large space objects.

[25] In 1992, three astronauts of the US space shuttle *Endeavor* made a dangerous grabbing by hand of the Intelsat 603, after NASA-developed devices for the grasping failed to work. The liquid propellant continued to slosh inside the satellite in weightlessness and kept the satellite rotating. In the rescue report it was said "... the operation had not been cost-effective, and was not worth the risk of the astronaut's lives". Harland, D. and Lorenz, R. (2005). *Space System Failures*, p. 66. Springer-Praxis, UK.

[26] The mission started in March 2007 and lasted 3 months. In July 2007 both spacecraft were declared decommissioned and are in a de-orbiting modus. AW & ST, 28 May 2007, at 22; AW & ST, 9 July 2007, at 18; and Dornheim M., Express Service, AW & ST, 5 June 2006, at 48. The success of this test moved Arabsat to sign a contract in 2007 with the German/Greek private company Kosmas GEO-Ring, which works on the design of in-orbit refueling. Arabsat has committed to order geostationary orbit satellites with architecture that allows refueling trials. Early in 2007 Arabsat ordered two satellites from EADS Astrium and Thales Alenia Space. The Arabsat 5 may be the first to be constructed to allow in-orbit servicing. AW & ST, 17 September 2007, at 22.

[27] One such proposal is the "Robotic Geostationary Orbit Restorer" (ROGER). Among some of the institutions taking part in ROGER are the European Space Agency, Astrium (Germany), EADS Space Transportation (France), Technical University of Braunschweig (Germany), German Aerospace Center, MacDonald Dettweiler Space Robotics (Canada) Space Applications Services (Belgium), and Tohoku University (Japan). ROGER Final Report: *supra* note 23.

[28] USSPACECOM not only detects objects in outer space but "... characterizes them, correlates them with a launch or release event, determines their orbits and tasks its sensors for subsequent follow-up observations". Klinkrad, H. (2003). *Monitoring Space: Efforts Made by European Countries*, IAF, p. 2, www.fas.org/spp/military/program/track/klinkrad.pdf

[29] Surveillance radar systems can detect objects of 10 cm diameter and larger in low orbits. ESA (2005). *Space Debris—Assessing the Risk*, 16 March, www.esa.int. In the geostationary orbit, objects with 1 m diameter and larger are detected by radar, while smaller ones, between 10 cm and 1 m, are monitored using optical telescopes.

[30] Among the items entered in these catalogs are the non-functional *Ariane* booster, that after 10 years orbiting collided with the operational French military satellite *Cerise* (July 1996). This catalog also lists a part of a US rocket, which after 31 years collided with a Chinese space object, which had separated from its launching vehicle after an explosion. The collision between the US and Chinese objects created at least three large pieces that separated from the US rocket body. Press Announcement (15 August 1996), www.cc.surrey.ac.uk; Leonard, D. (2005). US–China Space Debris Collide in Orbit, 16 April, www.space.com

[31] The US regulation on satellite disposal is in the Code of Federal Regulations, US Government Printing Office, cite 47CFR25.283, Title 47, Vol. 2, Chapter I, §25.114 (14), Applications for space station authorizations, §25.283 End-of-life disposal (2004). The German legislation on satellite disposal is in *Verfahren zur Anmeldung von Satellitensystemen bei der Internationalen Fernmeldunion und Übertragung deutscher Orbit-und Frequenznutzungsrechte, Amts-blatt* Reg. TP No. 6/2005, 6 April 2005, S. 239 ff. Mejía-Kaiser, M. (2006). Taking Garbage Outside the Geostationary Orbit and Graveyard Orbits. *Proc. 49th Colloquium on the Law of Outer Space*, Valencia.

[32] The IADC has recommended "passivating" spacecrafts through the elimination of internal energy: "... residual propellants shall be dumped, pressurants shall be depleted, batteries safed, etc." IAA Position Paper on Space Debris Mitigation, *supra* note 22, at 4.

[33] In 1978, nuclear elements of the Soviet *Cosmos 954* survived the atmospheric entrance in north Canada.

[34] These measures may be: to limit debris released during normal operations; to minimize the potential for break-ups during operational phases; to avoid intentional destruction and other harmful activities; to minimize potential for post-mission break-ups resulting from stored energy; to limit long-term presence of spacecraft and launch vehicle orbital stages in LEO and GEO after the end of their mission. Inter-Agency Space Debris Coordination Committee space debris mitigation guidelines, UN Doc. A/AC.105/C.1/L.260 (2002).

[35] Prasad and Lochan have proposed a trust fund system, with contributions established according to past space debris generation. Prasad, M. and Lochan, R. (2007). Common but Differentiated Responsibility—A Principle to Maintain Space Environment with Respect to Space Debris. *Proc. 50th Colloquium on the Law of Outer Space*, Hyderabad.

CHAPTER

International standards enhance interoperability: a safer lunar colony

28

Sandra Coleman, Joseph Pellegrino
ATK, Arlington, Virginia

CHAPTER OUTLINE

BACKGROUND AND ACKNOWLEDGEMENTS

International standardization of spacecraft components, of astronauts and cosmonaut space suits, of electrical systems, heating and cooling systems, telecommunications systems, various tools and parts, and more could benefit space safety in almost innumerable ways. If and when there is a permanent lunar outpost standardization of as many elements and systems as possible could benefit ongoing operations in terms of backup and redundancy and be of particular value in emergency conditions. The possible set of elements, functions and operational scenarios being considered for the first lunar outpost creates a wide spectrum of possibilities for international standardization. If all nations involved in the planning, construction and use of such a lunar facility could agree to a single set of standards for this outpost this would enhance interoperability, increase reliability and redundancy and lead to a safer outpost. The Space Enterprise Council (SEC) consisting of all the major U.S. aerospace companies, collaborated with NASA Exploration Systems Missions Directorate to define the criteria for standards to enhance success of developing, deploying and utilizing such a future international lunar outpost. This paper will describe the process and results of these effort in support of NASA's planing efforts. A team of industry experts created and executed a disciplined process to determine which aspects of the lunar outpost would most benefit from standards, which standards are most urgently needed, and where these international standards may already exist. This paper will share thr disciplined methodology that has been used in this analysis of common standards for an international lunar outpost and also explore how these results could be applied not only to safety critical aspects of a lunar outpost but also other collaborative space exploration efforts.

Space Safety Regulations and Standards. DOI: 10.1016/B978-1-85617-752-8.10028-5

INTRODUCTION

The first step undertaken was to develop a standards evaluation methodology that could be applied to prioritize standards for a 2030 lunar outpost. NASA requested that this standards evaluation system be able to:

1. Evaluate, prioritize, and identify lunar standard interfaces with a focus on standards that have long-term applicability to the lunar architecture, and
2. Identify candidate sources for those standards.

The standards evaluation method first evaluated the quality of standards, and then applied a temporal assessment. The result of the evaluation process was a prioritized list of standards and information on standard sources. The standard assessment process utilized the following metrics:

1. Return on Investment
2. Standard Assessment
3. Temporal Assessment
4. Probability versus Applicability Assessment

This process utilized this standard evaluation method twice. One evaluation was based on standards assessment; the second evaluation was based on theoretical scenarios that might be foreseen to occur on the lunar surface. Two sets of results were generated. It was anticipated that the standards assessment results would be of prime interest to the NASA engineering community, while the scenario-based evaluation results were seen as likely to be of prime interest to the NASA mission operations community.

28.1 STANDARDS EVALUATION METHODOLOGY

The first step in performing this assessment was to define the scope of the assignment. Since the information generated by this exercise was to be applied to the design of a future international lunar outpost, the team had to define the "trade space" and thus to seek ways to maximize future flexibility of design. It was thus important that the team focus on standards that were at the architecture level. This effort to extend the limits of the trade space (i.e. approaches that minimally restrict the development of future solutions based on technology that currently does not exist) was a key element in the methodology. For example, a power standard should only address voltage and amperage and should not define specific generation or delivery systems.

The evaluation team defined which elements would be required for a future lunar outpost (shown in Table 28.1). A definition for each surface element was generated and agreed to by the entire team. NASA had requested that the scope of this initial exercise be limited to activities on the lunar surface. Elements in lunar or Earth orbit were not addressed in this exercise, but the methodology could very logically be applied to systems design for orbital operations and maneuvers.

The team then developed a matrix that mapped the Lunar Surface Elements to the Interface Functional Needs (shown in Table 28.2). The team evaluated where each functional need applied to the various lunar elements. Certain functional needs, such as power, were very influential and affect every lunar element. By counting the number of intersections for a given functional need, a metric called the Return on Investment (ROI) was established. The ROI for a particular functional need provides some

Table 28.1 Lunar Outpost Surface Elements	
Surface Suit	EVA suits and portable life-support units
Living Habitat	Crew living quarters. Accommodations for sleeping, washing and hygiene, food preparation and storage, personal space, etc.
Work Habitat	Allocations for mission experiments, data storage and retrieval, monitoring and maintenance
Health Habitat	Medical center and supplies, exercise equipment
Power Systems	Facilities and equipment for power generation, storage, distribution
Surface Transportation and Handling Systems	Vehicles for transport of manned and unmanned loads and associated storage and maintenance equipment and facilities
Communication and Navigation	Systems for habitat communication with Earth and outpost assets. Navigation provisions for surface vehicles, outpost sites, and EVA crew
Logistics Resupply	Outpost interface, allocation and storage systems for supply shipments
ISRU Production	Systems and equipment for processing of lunar regolith
Emergency Egress Systems	Outpost and manned vehicle contingency capabilities
Surface Construction and Maintenance	Hardware and facilities for outpost construction, site preparation, habitat maintenance
Scientific Instruments and Equipment	Value items, payloads, or equipment for accomplishing mission objectives

insight into the level of influence and priority. For example, power has a high ROI, while waste management and navigation have lower ROI values.

The team then needed to establish a scale that could be utilized to evaluate which standards had desirable qualities. Three assessment scales were defined:

1. Depth of Decomposition
2. Degree of Maturity
3. Degree of Commonality

A numeral rating was assigned to each assessment scale. As can be seen in Figure 28.1, those standards that are at the architecture level, are long standing and globally accepted, and provide identical solutions were rated the highest (identified with a green box in Figure 28.1). The scale for the Depth of Decomposition was biased high (with a maximum score of 4) because the team felt that it was imperative that the standards be at the architecture level. This is the level that would be most useful to NASA and to other space agencies that would seek to apply this methodology and still not limit future innovative solutions.

This scale was utilized to evaluate the intersection of each functional need and lunar element shown in Table 28.3. During this assessment, the team was careful to document all assumptions. This was important as the ratings changed based on the assumptions made by the team. For example, for the Health Habitat, it was assumed that telemedicine would be utilized. Tele-medicine would utilize medical staff on Earth to review x-rays and perhaps even remotely perform surgery on the lunar surface, utilizing robotics. This would require greater bandwidth capability and an increased data transfer rate. This assumption was documented and the rating for communications protocol was increased in importance in order to provide this capability. The score for each function need evaluation is obtained by multiplying the values for each of the evaluation criteria. Using the Health Habitat example shown in Table 28.3, the score for

Table 28.2 Lunar Surface Elements Versus Interface Functional Needs

Interface Functional Needs	Lunar Surface Element											
	Surface Suit	**Living Habitat**	**Work Habitat**	**Health Habitat**	**Power Systems**	**Surface Transport**	**Comm. and Nav.**	**Logistics Resupply**	**ISRU Production**	**Emergency Egress**	**Surface Construction**	**Scientific Equip.**
Pressurized mechanical interfaces	X	X	X	X		X		X	X	X	X	X
Unpressurized mechanical interfaces	X	X	X	X	X	X	X	X	X	X	X	X
Atmosphere/ environmental	X	X	X	X		X	X	X		X	X	X
Water	X	X	X	X		X		X	X	X		
Power	X	X	X	X	X	X	X	X	X	X	X	X
Communications protocol	X	X	X	X		X	X	X		X	X	X
Diet	X	X	X	X		X		X		X		
Reactants and working media	X	X	X	X	X	X		X	X	X	X	X
Materials	X	X	X	X	X	X	X	X	X	X	X	X
Human factors	X	X	X	X		X	X	X	X	X	X	X
Waste management and recycling	X	X	X	X	X	X		X		X	X	
Navigation	X				X	X	X	X		X	X	X

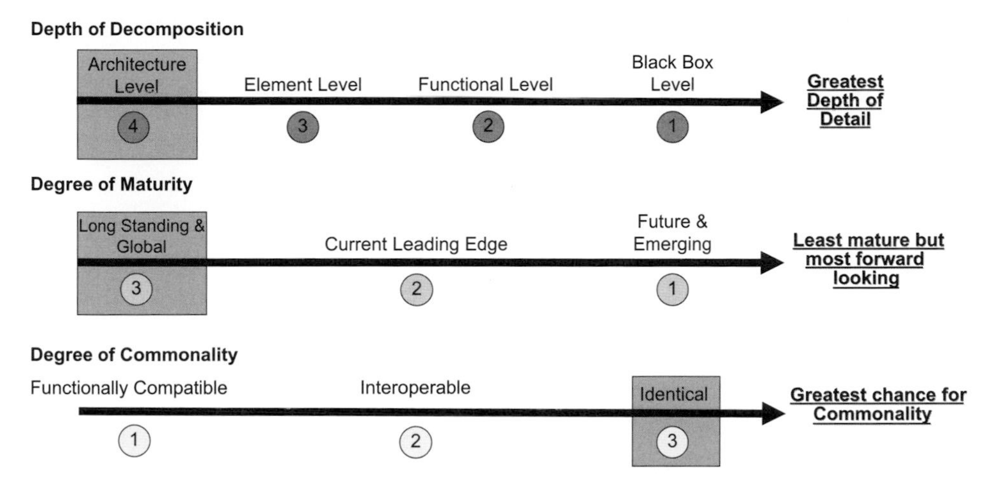

FIGURE 28.1

Standards Assessment Scale.

Table 28.3 Standard Assessment Example/Health Habitat

Interface Functional Needs	Decomp.	Maturity	Common.	Score	Assumptions
Pressurized mechanical interfaces	4	2	3	24 (High)	
Unpressurized mechanical interfaces	4	2	3	24 (High)	
Atmosphere/ environmental	3	3	2	18 (High)	
Water	4	3	2	24 (High)	
Power	4	2	3	24 (High)	Driven by special provision for intensive care
Communications protocol	3	2	2	12 (Medium)	Special needs for tele-medicine
Diet	2	2	1	4 (Low)	Special diet needs, medical storage, handling
Reactants and working media	3	2	2	12 (Medium)	Special needs for plasma, gas storage Handling of toxic gas/fluids
Materials	4	2	2	16 (Medium)	Special needs for plasma, gas storage
Human factors	3	3	2	18 (High)	Identical materials for use on humans
Waste management and recycling	3	2	2	12 (Medium)	Driven by operating room and isolation wards

Table 28.4 Temporal Assessment Example/Health Habitat

High	Power Water	Unpressurized and pressurized mechanical I/F Atmosphere/ environment Materials	Diet Reactants/working media Waste management Human factors	Communication protocol
Medium	Unpressurized and pressurized mechanical I/F Atmosphere/ environment Materials	Diet Reactants/working media Waste management Human factors	Communication protocol	
Low	Diet Reactants/working media Waste management Human factors	Communication protocol		
	In 2008	2009–2011	2011–2015	After 2015

communication protocol is obtained by multiplying the Level of Decomposition value (3) times the Level of Maturity value (2) times the Level of Commonality (2), to obtain a score of 12.

The team felt that it was important to add a temporal assessment to this exercise so that NASA could gain some insight into which standards need to be addressed in the near term, and which standards could be defined in the future without negatively impacting the architecture development schedule. The temporal assessment for the Health Habitat is shown in Table 28.4. The team placed each functional need onto the matrix according to importance and need date. As can be seen in Table 28.4, functional needs that have a low priority in the near term (for example, Human Factors) increase in priority over time. The placement of the functional needs into the various locations on the temporal assessment matrix was made utilizing the collective knowledge of the team. Often team members had insight into specific NASA milestones and priorities.

NASA had indicated that it would like the standards identified through this process to be compatible with future international partner elements and systems, as well. To address this request, the team developed the Probability versus Applicability Assessment shown in Table 28.5. This standard evaluation tool places the functional needs in a matrix according to likelihood of existence versus applicability to international partner long duration needs. Referring to Table 28.5, it can be seen that items such as power and communication protocol have a very high likelihood of standards existing when needed and are very applicable to the international partner long-duration needs. The functional needs shown as "High" in Table 28.5 should be easy to locate and are very influential. Those functional needs shown as "Medium" or "Low" may be more difficult to find and may have a relatively high level of influence (for example, pressurized and unpressurized mechanical interfaces).

The SEC team compiled all of the standard assessment data and generated a final assessment for each lunar surface element. This assessment took into account all of the standards evaluation data and the engineering judgement of the SEC team members. Table 28.6 shows the final assessment for the Health Habitat. When developing standards for the Health Habitat, functional needs of water, power,

Table 28.5 Probability versus Applicability Assessment Example/Health Habitat

Lowest			
Likely to exist when needed	Pressurized and unpressurized mechanical I/F (Medium)	Waste management (Low)	
	Atmosphere/environment Materials (High)	Reactants/working media (Medium)	
	Power Communication protocol (High)	Water Human factors (High)	Diet (Medium)
Highest	Broadest applicability		Lowest applicability
Applicable to ESMD international partner long-duration needs			

Table 28.6 Final Assessment Example/Health Habitat

Interface Functional Needs	High/Medium/Low
Pressurized mechanical interfaces	High
Unpressurized mechanical interfaces	High
Atmosphere/environmental	Medium
Water	High
Power	High
Communications protocol	Medium
Diet	Low
Reactants and working media	Medium
Materials	Medium
Human factors	High
Waste management and recycling	Medium

human factors, and pressurized/unpressurized mechanical interfaces scored highly and should be addressed in the near term. Standards for atmosphere/environmental, communication protocol, reactants/working media, materials, and waste management/recycling are of medium importance and should be addressed in the next 3–5 years. Diet is a functional need that scored low in the standards assessment, and is a standard that could be addressed in the 2015 time-frame.

In April 2008, the study team sponsored a workshop to undertake an independent standards evaluation by carrying out completely independent scenario exercise utilizing the process previously described. The team was divided into three groups;

1. Exploration and Science (ES)
2. Emergency Scenarios (E)
3. Living and Resupply (L)

Each group developed lunar surface scenarios and performed a standards-evaluation. NASA had requested that the evaluation take into account the distance from the outpost (at outpost, 50 k from outpost, and 500 k from outpost). A scenario example from each of the groups can be found in Table 28.7.

Table 28.7 Sample Scenarios

Scenario Identifier	Scenario	Assumption and Ground Rules	Location of Scenario		
			At Colony	50 k From Colony	500 k From Colony
E-1	Electrical fire in habitat module	i. In a group area ii. In a private area where person is gone iii. Habitat occupied	X		
ES-2	Four crew members are exploring the lunar surface on a 7-day excursion	i. The crew is using a rover that serves as living quarters ii. Geological sample gathering iii. Core drilling		X	
L-3	Crew members inside the habitat need to communicate with members out on EVA	i. Capability needs to exist regardless of EVA distance ii. Baseline point-to-point comm., habitat to EVA iii. Communication can occur directly or through lunar or Earth relay iv. Visual communication for a loss of comm. contingency v. Personal communication should allow privacy	X	X	X

The results from the scenario based evaluation closely resembled those produced by the standards assessment exercise. However, communication protocol was the highest rated functional need in the scenario-based evaluation, while power rated the highest for the standards evaluation. The data that the SEC team provided to NASA can be utilized to perform trade studies and provides a roadmap for standards development.

The SEC team used the results of all the standards evaluations to perform a search of standards organizations that could be contacted to start the process of developing standards for a lunar outpost. A standardized worksheet was developed and provided to NASA. A Standard Source Information Sheet is shown in Figure 28.2.

The team developed a scale to evaluate the standard sources shown in Figure 28.3.

Using the scale described in Figure 28.3, the team performed an assessment of each standard source. In many cases, broad categories were divided into specific functions. For example, within the area of Human Factors, sub-categories such as anthropometry, engineering, environment, and physiology were established. An example of a standard source evaluation can be found in Table 28.8.

28.2 CONCLUSIONS AND RESULTS

All of the data generated for both the Standards Evaluation Exercise and the Scenario Based evaluation were compiled and ranked. For the Standards Evaluation Exercise, Power and Water ranked as the most important standards as shown in Figure 28.4 in page 394.

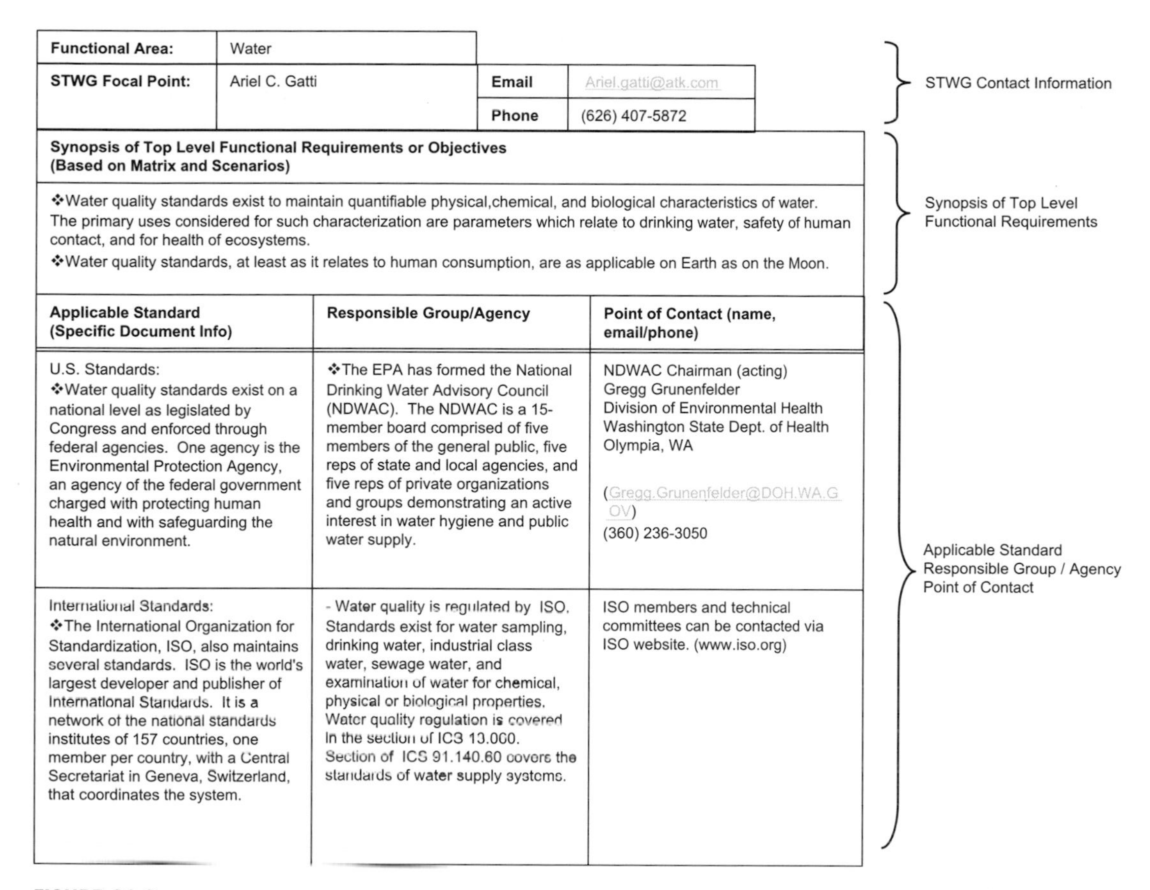

Functional Area:	Water		
STWG Focal Point:	Ariel C. Gatti	Email	Ariel.gatti@atk.com
		Phone	(626) 407-5872

Synopsis of Top Level Functional Requirements or Objectives (Based on Matrix and Scenarios)
❖Water quality standards exist to maintain quantifiable physical,chemical, and biological characteristics of water. The primary uses considered for such characterization are parameters which relate to drinking water, safety of human contact, and for health of ecosystems. ❖Water quality standards, at least as it relates to human consumption, are as applicable on Earth as on the Moon.

Applicable Standard (Specific Document Info)	Responsible Group/Agency	Point of Contact (name, email/phone)
U.S. Standards: ❖Water quality standards exist on a national level as legislated by Congress and enforced through federal agencies. One agency is the Environmental Protection Agency, an agency of the federal government charged with protecting human health and with safeguarding the natural environment.	❖The EPA has formed the National Drinking Water Advisory Council (NDWAC). The NDWAC is a 15-member board comprised of five members of the general public, five reps of state and local agencies, and five reps of private organizations and groups demonstrating an active interest in water hygiene and public water supply.	NDWAC Chairman (acting) Gregg Grunenfelder Division of Environmental Health Washington State Dept. of Health Olympia, WA (Gregg.Grunenfelder@DOH.WA.GOV) (360) 236-3050
International Standards: ❖The International Organization for Standardization, ISO, also maintains several standards. ISO is the world's largest developer and publisher of International Standards. It is a network of the national standards institutes of 157 countries, one member per country, with a Central Secretariat in Geneva, Switzerland, that coordinates the system.	- Water quality is regulated by ISO. Standards exist for water sampling, drinking water, industrial class water, sewage water, and examination of water for chemical, physical or biological properties. Water quality regulation is covered in the section of ICS 13.060. Section of ICS 91.140.60 covers the standards of water supply systems.	ISO members and technical committees can be contacted via ISO website. (www.iso.org)

FIGURE 28.2

Standard Source Information Sheet Example/Water

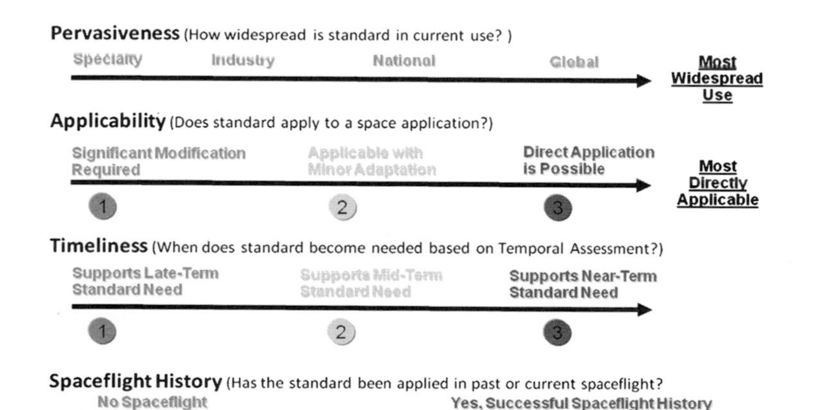

FIGURE 28.3

Standard Source Evaluation Metrics

Table 28.8 Standard Source Evaluation Example/Human Factors

Human Factor Standards			STWG POC: Dr Adam F. Dissel (Lockheed Martin)					
Subcategory	**Standard ID**	**Standard Title/ Description**	**Country**	**Organization**	**Pervasiveness**	**Applicability**	**Timeliness**	**Space-Flight History**
Anthropometry	DOD-HDBK-743A	Anthropometry of US Military Personnel	USA	US Department of Defense	National	High	Medium	Low
Anthropometry	ISO 7250-1:2008	Description of anthropometric measurements	International	ISO	Global	High	Medium	High
Anthropometry	ISO 15536-2:2007	Requirements for the verification of the functions and validation of dimensions of computer manikins	International	ISO	Global	High	Medium	Low
Human Factors Engineering	MIL-STD-1801	Human Engineering Requirements for Used Computer Interface	USA	US Department of Defense	National	Medium	Low	Low
Human Factors Engineering	MIL-HDBK-759	Human Engineering Design Guidelines	USA	US Department of Defense	National	Medium	Medium	High
Human Factors Engineering	MIL-STD-1472F	Design Criteria Standard: Human Engineering	USA	US Department of Defense	National	Low	Medium	Low
Human Factors Engineering	Human Factors Design Standard (HFDS)		USA	FAA	National	Low	Medium	Low
Human Factors Engineering	FAA-HF-002	Provides a source of data to evaluate the extent to which equipment having an interface with operators meets human performance requirements and human engineering criteria	USA	FAA	National	Low	Low	Low
Human Factors Engineering	FAA-HF-004	Results of analyses of critical tasks performed to provide a basis for evaluation of the design of the system	USA	FAA	National	Medium	Medium	Low

Human Factors Engineering	ISO 17399:2003	Defines all generic requirements for manned space-flight vehicle/habitat structures and flight crew training or simulation facilities and the related equipment that directly interfaces with manned space system flight crew	International	ISO	Global	High	High	High
Human Factors Engineering	ISO/TR 16982:2002	Provides information on human-centered usability methods that can be used for design and evaluation	International	ISO	Global	Low	Medium	Low
Human Factors Engineering	ISO 14619:2003	Specifies the procedure for preparing and carrying out space experiments and processing the resulting data	International	ISO	Global	High	High	High
Human Environment	MIL-STD-1474D	Establishes military noise limits	USA	US Department of Defense	National	Medium	High	Low
Human Environment	ISO 14505-3:2006	Gives guidelines and specifies a standard best method for the assessment, using human subjects, of thermal comfort in vehicles	International	ISO	Global	Medium	Medium	Low
Human Environment	C148 and R156	Working Environment (air pollution, noise, and vibration)	International	International Labour Organization	Global	Medium	Medium	High
Human Environment	C115	Radiation Protection Convention, 1960	International	International Labour Organization	Global	High	High	High
Physiology	ISO/TS 16976-1:2007	Information related to respiratory and metabolic responses to rest and work	International	ISO	Global	Low	Medium	Low
Physiology	IEC 80000-14:2008	Names, symbols, and definitions for quantities and units of telebiometrics related to human physiology	International	ISO	Global	High	Low	High

FAA, Federal Aviation Administration.

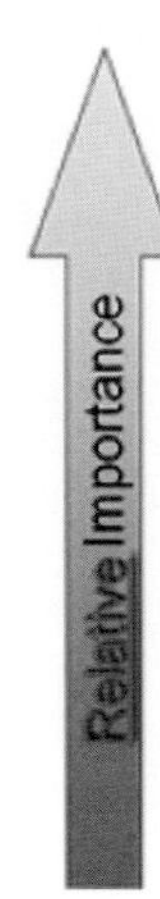

Relative Importance
Power
Water
Human Factors
Unpressurized Mechanical Interfaces
Communication Protocol
Atmosphere / Environmental
Materials
Reactants / Working Media

FIGURE 28.4

Standard Evaluation Exercise Prioritized Results

Relative Importance
Communication Protocol
Power
Atmosphere / Environmental
Pressurized Mechanical Interfaces
Unpressurized Mechanical Interfaces
Human Factors
Reactants / Working Media
Materials
Waste Management / Recycling

FIGURE 28.5

Scenario Based Evaluation Prioritized Results

For the Scenario Based Evaluation, Communication Protocol and Power were the most important. The results for the Scenario Based Evaluation can be found in Figure 28.5.

As a final exercise, the team consolidated both the standards assessment data and the scenario-based standards assessment information into a single matrix. This matrix is designed to provide a quick overview of which standards should be prioritized for the development of a lunar outpost. These consolidated results can be found in Table 28.9 in page 395.

The team produced a video that demonstrated how international standards could improve mission safety in quite important ways. In this video, an astronaut was depicted as roving the lunar surface on an exploration sortie in the year 2020. The rover in this scenario had an accident within a lunar meteor crater and the vehicle became inoperable. The astronaut was injured during the accident and pinned

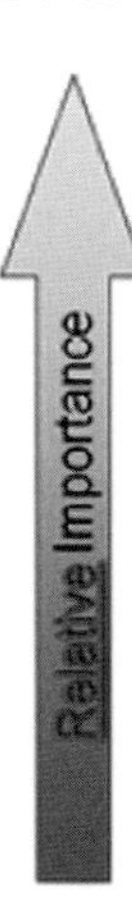

Table 28.9 Consolidated Results

Interface Functional Needs	Lunar Surface Element											
	Surface Suit	Living Habitat	Work Habitat	Health Habitat	Power Systems	Surface Transport	Comm.	Logistics Resupply	ISRU Produc-tion	Emerg-ency Egress	Surface Constru-ction	Scientific Equip.
Pressurized mechanical interfaces	Medium	Medium	Medium	High		High		High	Medium	Medium	Medium	Medium
Unpressurized mechanical interfaces	Medium	Medium	Medium	Medium	High	High	Low	High	Medium	Medium	Medium	Medium
Atmosphere/ environmental	High	Medium	Medium	Medium	Medium	Medium	Medium	Medium		Low	Low	High
Water	High	High	Medium	High		Medium		Medium	Low	Medium		Medium
Power	Low	Medium	High	Medium	High	High	High	High	Medium	High	High	High
Communications protocol	Medium	Medium	Medium	High		High	Medium	Medium		Medium	Medium	High
Diet		Low	Medium	Low		Low		Medium		Low		
Reactants and working media	Medium	Low	Low	Medium	Medium	High		High	Low	Low	Low	Medium
Materials	Medium	Medium	Medium	Medium	Low	Medium	Low	Low	Medium	Medium	High	High
Human factors	High	Medium	Medium	Medium		High	Medium	High		Medium	Medium	Medium
Waste management and recycling	Medium	Medium	Medium	Medium	Medium	Medium		Medium		Low	Low	Medium
Navigation	High	Low	Medium			High	Medium	High	Medium	High	Medium	High

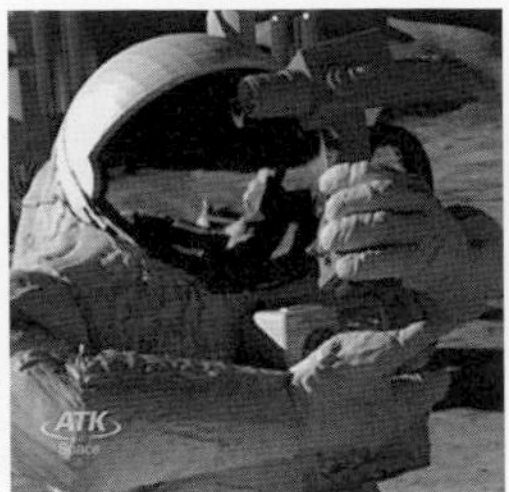

FIGURE 28.6

Screenshots from Lunar Exploration Rescue Mission Video

under one of the rover wheels. The injured astronaut then sent out a distress signal by depressing a button on the exterior of the Extravehicular Activity (EVA) suit. The astronaut's power level for the EVA suit's environmental, control, and life-support systems was shown to be critically low. The astronaut removed a battery from a power drill and inserted it into the EVA suit. The power level is raised to a safe level. After a couple of hours, the injured astronaut received an alarm stating that the oxygen supply in the suit was dangerously low. An international astronaut driving a rover soon, thereafter arrived at the scene of the accident. The international partner astronuat plugged a cable onto the chest plate of the injured astronaut's EVA suit. The oxygen level on the injured astronaut's suit was increased by sharing the compatible oxygen supply of the international astronaut. A digital display on the arm of the international astronauts's EVA suit then provided key medical data about the condition of the injured astronaut. Based on this information, the international astronaut inserted a canister containing liquid food, water, and medicine into a port of the injured astronaut's EVA suit. The injured astronaut was then able to drink these items through the straw positioned near his/her mouth in the helmet of the EVA suit. The international astronaut connected the two rovers together by engaging a mechanical connection and a hose to share hydraulic fluid. The injured astronaut was finally loaded onto the international rover which towed the disabled rover behind it. During this rescue mission, many of the functional needs (mechanical interfaces, communication protocol, diet, water, power, working fluids) addressed in this study were able to effectively interact with the various lunar elements (EVA suits, tools, surface transportation, logistics resupply) because appropriate standards were established. Some screenshots of the video are provided in Figure 28.6 above.

SUMMARY

The need for and utility of international standards to establish compatibility of various support systems for lunar operations was clearly established by the Space Enterprise Council study described above. A ranking system to establish the highest priorities for compatible standards indicates that power and communications are perhaps the most urgent. The various scenarios, however indicate that universal compatibility of virtually all systems and subsystems would likely add utility and increase the safety of astronauts associated with an international lunar outpost. The approach to planning for future compatible standards focused on: architectural and generic concepts; and avoidance of tying standards to specific technologies or types of equipment to allow for the evolution of new capabilities.

CHAPTER

29 Future space safety technology, standards, and regulations

Joseph N. Pelton

Former Dean, International Space University

Over the long run the safety of all human beings in the global commons of outer space is a responsibility that must be shared by all spacefaring nations.

G. Rodney, Associate Administrator for Safety and Mission Assurance, 40th IAF Conference, Beijing, China, 1989 [1]

CHAPTER OUTLINE

INTRODUCTION

Changes for the future that can be anticipated now include such capabilities as:

- **New forms of propulsion and launching systems.** Tether systems, space elevator systems, advanced electrical ion systems, solar sails, nuclear propulsion, electromagnetic propulsion derived from the solar wind, and even mass drivers and rail guns for non-human launching systems (or possible use for off-Earth surface transport) will be the subject of increasing research and development [2].

Space Safety Regulations and Standards. DOI: 10.1016/B978-1-85617-752-8.10029-7

- **Advanced life-support and radiation protection systems.** New types of ultraviolet (UV) and radiation protection capabilities, solar storm protection systems, combined life-support and information and communications technology (ICT) systems for off-world conditions, etc.
- **New materials and systems for space habitats.** Inflatable habitat systems, various types of improved materials to provide improved thermal, radiation protection, tensile strength, and lightweight systems will be developed. There will also be new types of systems designed to power and protect space habitats and to cope with orbital debris, etc.
- **New approaches to space safety standards, risk management and education.** The future will not only be defined by new technology, new propulsion systems, and improved materials but also by new approaches to risk management, safety systems, education, risk management, and safety standards and regulation.

Any attempt to develop specific safety standards or regulatory controls for such new space technologies would be premature at this time. Progress continues apace in developing new sub-orbital and orbital space plane flights. New commercial space ventures continue to evolve. Designs for new types of space habitats are being deployed and tested. Plans are made to create permanent space colonies on the Moon and elsewhere, and robotic devices are being tested to substitute for human presence in space in many ways. All of these activities and more give rise to the need for new types of safety standards and regulation. New types of escape and rescue systems will become possible. Safety monitoring systems similar to those used on aircraft can be adapted to space missions. The correct balancing between safety standards, safety regulation, and safety margins will undoubtedly change as we move to different types of space propulsion systems and longer-term missions in space. Critical success factors change with longer endurance space missions [3].

This pattern of ongoing change in space technology and propulsion, space habitat systems and man–machine interfaces, and redefinition of safety margins and techniques implies the need for a constantly evolving international mechanism and process to agree to new safety standards and material specifications, new forms of space safety educational and training programs, and more. This chapter explores possible "processes" to develop new space safety standards and regulations that respond to evolving new forms of space systems and technology.

This "evolving regulatory and standards process" will require a dynamic range of capabilities and likely the delegation of national to international authority. This will include monitoring, assessing, and creating measurement techniques for safety and performance—both relating to public space agencies and private space industry. This activity could include the systematic assessment and ongoing evaluation of new technologies. This could be done by perhaps establishing a range of "meta-standards" related to new capabilities such as space tethers, inflatable habitats, and UV screens. It could also include performance testing of new types of materials and capabilities.

In many cases, the key will be to develop appropriate safety standards and measurement and testing systems for subsystems in addition to seeking overall systems standards for a complete spacecraft, vehicle, or habitat. For decades, the emphasis in space safety has been primarily with regard to the launch and re-entry of chemically powered vehicles. For the future, this may well become an increasingly less important aspect of space safety standards. Also, independent verification and validation (IV&V) in acceptance testing has been perhaps the prime safety process in many public space programs. But in the future, design engineering, appropriate performance margins, and escape systems may assume greater importance.

29.1 INTERNATIONAL REGULATORY ROLES RELATED TO SPACE SAFETY

Regulatory processes ultimately will follow in terms of radiation safety standards, due diligence related to reduction and elimination of orbital debris, oversight and reduction of atmospheric emissions, etc. In many cases such regulatory actions and standards will be critical to commercial space ventures being able to obtain insurance coverage, meeting national and regional environmental requirements, and to reassure investors that risk mitigation and appropriate due diligence procedures have been accomplished to a reasonable degree. This may mean, among other things, that insurance and risk management enterprises may assume a greater degree of participation and standard setting in the space safety field. It also means that there may be regulatory and legal “conflicts”.

A possible role in these areas may also evolve for the International Association for the Advancement of Space Safety (IAASS), the International Space Safety Foundation (ISSF), or other groups such as the Secure World Foundation (SWF). The IAASS might serve to establish and support the operation of an Independent Space Safety Board (ISSB) for Commercial Human Space-flight Vehicle Certification [4]. This ISSB could, for instance, serve the role of providing an independent testing capability to examine commercial space systems or subsystems for safety performance and reliability.

The IAASS, ISSF, and SWF are among a number of private organizations in addition to public international organizations and the various space agencies that might play a role going forward. The major unanswered question is whether one or more specialized agencies of the United Nations, and in particular the International Civil Aviation Organization (ICAO), as well as possibly the International Telecommunications Union (ITU), the UN Environmental Program (UNEP), and/or the World Meteorological Organization (WMO) might play a role with regard to the emerging commercial space-flight industry. At this stage, the commercial space-flight industry and the Commercial Spaceflight Federation (formerly the Personal Spaceflight Federation (PSF)) have resisted, during these early years, governmental intervention or any international oversight to the maximum degree possible.

Finally there is the educational perspective to consider. The exploration of the future of space systems and longer-term safety technologies and systems will depend, in part, on the training of younger people today. Advanced curricula and innovative safety modeling and simulation that embrace a wide diversity of future concepts and approaches could prove to be critical to designing effective, reliable, and safe new space technology and systems many decades into the future. This seems to be particularly the case for new types of launch and space transport systems and new environmental systems to protect humans in space over sustained periods of time.

29.2 MANAGEMENT OF AIR AND SPACE TRAFFIC CONTROL SAFETY AT THE NATIONAL, REGIONAL, AND INTERNATIONAL LEVELS

Today, air traffic control relies on a combination of national, regional, and international agencies and institutions to regulate the safe operation of private and civilian public aviation. In Europe, there is a transition in process from national air traffic control systems to an integrated European Air Safety Agency (EASA) that appears poised to assume responsibility for at least some types of commercial space flight as well. In short, the EASA appears likely to follow the model of the Federal Aviation Administration (FAA) in assuming control for commercial aviation with an expanding role with regard to commercial space flight with passengers aboard.

The aviation air traffic control system has operated with a remarkable record of safety—over a billion passenger miles of operation per human fatality. This aviation safety record has continued to improve over time. Some would argue that improved aviation technology is the key. Others would claim that it is regulatory oversight and well-enforced safety practices, and yet others would claim it is a combination of the two. The hope is that the regulatory processes that have made commercial aviation and private aviation increasingly safe can be effectively applied to commercial space flight. What is not unclear is the role that international organizations should play. The ICAO has international responsibility under the so-called Chicago Convention to set international standards for aviation safety [5].

The ICAO does not have acknowledged parallel responsibility with regard to commercial space flight. Further, national governments, at least in Europe and the USA, have been reluctant to move to grant such an extension of regulatory power to the ICAO. Instead, there has been the thought to allow national entities, specifically in the USA and Australia, to provide such oversight and regulation. In Europe, the regulatory control is still to be sorted out between the European Space Agency (ESA), the European Union (EU), and the EASA, but here a regional approach seems likely. The reluctance to move to an international treaty or convention with regard to commercial space flight seems to stem from the view that the ICAO's international oversight and regulations might be more extensive and structurally rigid than needed at this stage—as well as more expensive. This viewpoint, of keeping regulatory control at the national level (or in Europe at the regional level), is significantly based on the idea that all, or virtually all, flights would originate and terminate at the same national spaceport for some time to come. In light of this fact, and also the truly start-up and entrepreneurial nature of the emerging "space tourism" business at this stage, many have expressed the view that national regulation might be sufficient until much more experience is gained and until the viability of commercial space flight is more clearly established. Certainly, if capabilities should evolve to the point where hypersonic space planes are providing transportation services as a common carrier from one country to another, ICAO treaty responsibilities and regulatory obligations would then likely come into play. This, however, could be a decade or more away.

A paper entitled "An ICAO for Space?" has been prepared by knowledgeable researchers from within the IAASS, but the practical answer for the time being seems to be no—at least until some years to come [6].

Conditions that might change the current consensus would seem to include: future plans to offer international flights between countries via space planes; an accident that has international implications; or specific liability claims, made at some future date, that invoke or seek the jurisdiction of the ICAO, the Chicago Convention, or some otherwise parallel legal claim.

There are several other international organizations that might find some form of regulatory role with regard to future commercial space-flight operations. In this respect one might cite:

- The International Telecommunication Union (ITU), which has responsibilities related to radio frequencies used for relaying information from Earth to space and space to Earth.
- The UN Environmental Program (UNEP) and the World Meteorological Organization (WMO) that have international responsibilities related to the Earth's atmosphere, environmental pollution, and curtailing global warming.

Although other international organizations could be listed, these are the UN specialized agencies that have some pertinent international authorities. Currently the ITU World Radio Conference (WRC) has allocated frequencies for national and international use including for space communications [7]. At present, one set of radio frequencies is allocated for air traffic control and another set of frequencies is allocated for space communications and tracking, telemetry, and command of space vehicles and satellites. If there comes a time when a vehicle is used for international transport that operates as both an airplane and a space vehicle, then some alteration of the frequency allocation or accommodation would be appropriate in order to be able to use a common set of frequencies to promote easy interconnection and hybridization of the equipment used by air—and space—traffic controllers.

The UNEP and the WMO, in cooperation with national aviation authorities, have carried out studies of emissions into the atmosphere of pollutants by aircraft, and further studies of the stratosphere and high-altitude contrails are contemplated. Again, if a regular and high volume of space plane flights through the stratosphere and beyond were to be contemplated in the future, the environmental impact of such flights on the ozone layer, on UV radiation levels for pilots and crew, etc. might be considered necessary. At such a time it would need to be determined where national regulation would start and end and where action by relevant UN international organizations might be appropriate. In an age of climate change and global warming, the divide of the authority between national regulatory bodies for environmental controls and international agencies might be particularly sensitive. The intersecting concerns between safe launch systems on one hand and environmental and global warming concerns on the other could become an increasingly important issue in coming decades. One of the safest solid rocket motors that allows immediate "throttle-able" shutdown such as will be used on *SpaceShipTwo* also creates one of the more environmentally undesirable residues. The choice of "space plane passenger safety" versus longer-term environmental safety for people on the ground is a potentially awkward future dilemma.

The proper coordination at national, regional, and international levels could actually become complex. The interest here would be a longer-term strategic approach so that development of new hydrogen-fueled craft (or other environmentally less harmful systems) could be developed in a timely way so as to promote advanced transportation systems that are safe and environmental sound.

29.3 SAFETY TESTING AND CERTIFICATION OF PUBLIC AND PRIVATE SPACECRAFT

The early stage of development of any new industry attracts a number of people and companies that seek to pioneer a new technology or a new system that becomes the successful and dominant player. In the early days of the automobile there were literally over 100 companies that sought to develop vehicles that were safe, attractive, durable, and affordable. After a number of decades, the number of successful automobile manufacturers dwindled to about a dozen around the world. The same was true of the aircraft industry as well. A recent study performed for the FAA and the US Congress, as well as an earlier study at the George Washington University, identified of the order of some 40 organizations currently active in seeking to develop a space plane or space plane service, and noted that over two dozen organizations in this area are now defunct [8]. In light of this "churn" in the number of entities seeking to develop commercial space-flight capabilities, it would seem that an independent laboratory that might test the reliability and safety of commercial space systems (and subsystems) would be highly desirable. Such an independent laboratory, independent of governments, international organizations, commercial developers, etc. that could carry out separate tests of components, subsystems, and complete systems would appear something that would be beneficial to risk management and insurance companies, consumers, and the commercial space-flight industry. This entity that would have a high level of expertise in space systems, safety, and independent verification and validation (IV&V) processes that could undertake performance testing on a cost-effective and independent basis would seem to have advantages as follows:

- Independence
- International impartiality
- Cost-efficiency
- Ability to adapt to new and emerging technology and systems
- Ability to address not only launch systems and space planes but also habitats, life-support systems, radiation shielding, space power systems, and man–machine interfaces
- New types of spacecraft—i.e. tethered habitable pods.

The explicit model that such a unit could very usefully follow is that of the Underwriters' Laboratories, Inc. (UL), whose reliability and certification testing labs have proven valuable to industry for many decades [9].

29.4 LAUNCHER CERTIFICATION

It is anticipated that various official space agencies around the world and private space-flight companies will continue to undertake test firings of rocket motors, and carry out IV&V of their various launch vehicles and launch systems for some time to come. Thus, it is foreseen that public agencies will play a role with regard to certification of spaceports and space launch operations for at least a number of decades into the future. Nevertheless, there are a number of roles that an independent laboratory could play in calibration or certification of test processes, test equipment, or launch range safety procedures and launch controls. Over time, as new launch systems are developed using new technologies, it might be possible for such an independent lab to play an even greater role in certification of launch systems as safe. It is currently somewhat of an anomaly that space launch systems are developed, tested, and operated by public space agencies and now private commercial space-flight developers rather than having one entity develop the launch system and other separate and independent agencies provide the oversight and safety regulation of the launchers. In the nuclear power industry, for instance, the industry now implements the system and government regulators oversee the installations' safety. In the aviation field, companies build and certify aircraft under the regulation of air safety agencies. It could be that new technology that uses tethers, nuclear energy, ion engines, etc. might well evolve into this type of regulatory model. The concepts set forth above could allow this type of model to evolve.

29.5 SUBSYSTEM CERTIFICATION

The first important step to evolve toward a new type of regulation where the builders of launch systems, space habitat operators, etc. are separate from independent regulators could come in the area of subsystems inspections and certification. During such environment safety inspections, IV&V and, ultimately, certifications could be overseen and regulated in a variety of ways. This might be either by governmental entities, independent inspectors/laboratories, or other designated entities that would

BOX 29.1 CRITICAL SUBSYSTEMS

- Environmental control and life support system
- Main propulsion system and fuels
- Guidance
- Navigation and control system
- Avionics and software
- Main structural system
- Thermal protection system
- Thermal control system
- Health monitoring system
- Electrical power system
- Mechanical systems
- Flight safety system and crew system

offer separate inspections and certification of safety to the builder of launch systems, potential "space tourism customers", and the risk management and launch insurance industries [10].

The objective at the systems and subsystems level would be to separate more completely the "players" and the "referee" in the arena of space safety. This could apply to both commercial space-flight organizations and even public space agencies [11]. It would be the move beyond chemical rockets where each launch was based on a short, explosive, high-risk event, to a new environment where reusable systems that did not involve dangerous and highly explosive systems would make such a transition much more viable. Although it makes sense to evolve a system that separates the regulators or independent testing authorities from the industry participants, a process for collaboration and information exchange also contains a good deal of logic so that critical data (or metadata) to help safety engineering is captured [12].

It seems regrettable that longer-term planning for vehicles to go to the Moon, Mars, and beyond is still focused entirely on not only chemically explosive rocket systems, but fuels that are environmentally harmful and highly dangerous systems as well.

This planning process seems to ensure that we are locked into launch systems with high danger indexes for 15–20 years to come. Meanwhile, progress might be made at the subsystem-level safety testing and certification process, starting with the subsystems identified in Box 29.1.

29.6 ADDITIONAL SAFETY MEASURES AND DIFFERENCE IN APPROACHES BETWEEN PUBLIC AND PRIVATE SPACE-FLIGHT SYSTEMS

The future of space safety is highly dependent on innovative thought. The advent of a commercial space-flight industry is dedicated to finding new ways to find access to space and finding new systems that are better, safer, and more cost-effective. For instance, one of the prime areas of focus is to allow re-entry without having astronauts subjected to large thermal gradients. The key to the future may be finding ways around "assumed safety constraints" that have been embedded in public space agency programs for decades. New ways to access space and to make space systems safe could lead to new paradigms. Diversification of space efforts has already shown positive results. In order to achieve orders of magnitude improvements in space safety that take us from today's one in 100 fatal accidents to performance at the level of one in 10,000 or one in 100,000, breakthrough concepts will be needed.

29.7 NEW APPROACHES TO RISK MANAGEMENT IN THE FIELD OF SPACE FLIGHT, EXPLORATION, AND SPACE TOURISM

Moving from today's regulatory systems in space safety to a new environment where fatalities are greatly reduced and access to space is greatly reduced in cost is perhaps most dependent on commercialization of space-flight programs. This means not only the development of new types of launch systems by commercial companies, but also the entry of risk management and insurance companies into the planning, engineering, and implementation process. Space insurance companies will not be willing to underwrite commercial space-flight programs if 1–2% of the flights end in fatalities [13].

Again, the business models require the use of new technologies and orders of magnitude increases in safety. International collaboration by all of the entities and organizations cited above will be essential, but the key seems to hinge on breakthrough innovation with the use of certifiable space systems that can be used safely time and time again. At this stage, no one knows whether the way forward will involve tethers, space elevators, nuclear propulsion, ion propulsion, lighter-than-air craft, or a new technology yet to be conceived. The contribution of the risk management industry is quite simple. They provide business models that indicate that the conventional chemical rocket systems approach to space launches as used by public space agencies is not a viable approach to the longer-term future. Their bottom line indicates that new technologies, systems, and standards are essential in making human space access a business.

29.8 LONGER-TERM LAUNCH SYSTEMS

It seems clear that one cannot specify safety standards, reliability testing requirements, or international regulations for new space transport systems that are today unknown or at very early stages of research and development. Nevertheless, Table 29.1 gives some indication of what some of the safety requirements of the future might be. In short, there is merit in monitoring and trying to anticipate the future rather than simply ignoring it. Further, safety practices such as independent verification and validations can often be transferred from one transportation or propulsion system to another. In the longer-term future, it seems likely that safety concerns will be transferred from explosive transport systems to environmental concerns such as cosmic radiation and space weather.

29.9 EDUCATION AND TRAINING OF SPACE SAFETY PROFESSIONALS

In light of the above, education and training for the next generation of space safety professionals is a huge challenge. This is because all of the past highly technical skills and engineering expertise is relevant to the space industry of the future. But the future also seems likely to hinge on new technical breakthroughs and new systems to provide space access. Today's training programs and university programs are very heavily dependent on the technologies and safety regulatory systems of the past. They give only passing attention to the new technologies of the future. In part, this is because there are

Table 29.1 Longer-Term Planning for New Initiatives in Space Safety

Technological Approach	Application	New Standards Required?	Reliability Testing	Regulatory Issues
Nuclear propulsion	Transport from GEO outward to Moon and beyond	Yes. Rapid reactor shutdown, fail-safe measures, reactor shielding	Extended reliability testing	Non-use in near space (i.e. below GEO)
Enhanced ion thrusters	Spiral deployment from LEO to GEO, cislunar transport	No	Extended reliability testing	Rules/guidelines for longer-term spiral deployments to GEO
Space elevators or tether lift systems	Lift from Earth to LEO or GEO	Yes. Material strength, resilience to Van Allen belt radiation	Extended reliability testing for tether strength, radiation, etc.	International deployment regulations. Coordination with air traffic control
Mass driver	For non-human transport or for human transport off Moon surface	Yes. Standards for specified level of near vacuum for operation	Different requirements for non-human and human transport	Creation of international mass driver "corridors" for space launch systems
Rail gun systems	For non-human transport or possibly for human transport off Moon surface	Yes. Standards for specified level of near vacuum for operation	Different requirements for non-human and human transport	Creation of international mass driver "corridors" for space launch systems
Solar sail systems	For interplanetary transport	Yes. For astronaut shielding from longer-term cosmic radiation	Depending on materials and resilience to longer-term cosmic radiation	No requirements obvious at this time

GEO, geostationary Earth orbit; LEO, low Earth orbit.

so many disparate technologies that may prove the critical path to the future. A creative blend of the past, present, and future is clearly not easy. Any educational program that provides students with a useful base of knowledge will need to address the technologies, standards, and safety regulations of the past while also seeking to address key ways forward as well.

29.10 THE WAY FORWARD: BOTH BUILDING ON EXISTING INSTITUTIONS AND REGULATION, AND CREATING NEW PROCESSES AND PROCEDURES

In light of the above discussion, specific recommendations are:

- Developing new initiatives to achieve improved international cooperative arrangements, treaty agreements, and international organizational efforts to promote enhanced space safety processes.

- Exploring the creation of new independent laboratories to test the safety and designs of components of launch systems—particularly for commercial space-flight systems.
- Evolving testing procedures and certification processes for components and subsystems for launch systems that could be ancillary to testing and certification for complete space launch systems.
- Recognizing that new and significant longer-term gains in space safety may well depend on new technological breakthroughs; thus, space safety standards and regulation must be able to adapt flexibly to new technologies.
- Pursuing fully the implications that arise from the introduction of commercial launch companies, risk management, and the insurance industry into the "manned" space launch programs—this will give focus to new ways to launch people into space and new space safety concepts.
- Examining how the dynamics of all of the above impact the design, content, and even the purpose of training and education programs in the area of space safety.

References

[1] Paper presented by Rodney, G. (1989). Associate Administrator for Safety and Mission Assurance, US National Aeronautical and Space Administration (NASA) at the 40th IAF Conference held in Beijing, China.

[2] See Pelton J. N. and Marshall, P. (2006). *Space Exploration and Astronaut Safety.* American Institute of Aeronautics and Astronautics, Reston, VA.

[3] Pelton, J. N. and Marshall, P. (2009). *License to Orbit: The Future of Commercial Space Travel.* Apogee Books, Canada.

[4] See Appendix 2 to this book. Also see International Association for the Advancement of Space Safety (2006). International Space Safety Board: Space Safety Standards—Commercial Manned Spacecraft, IAASS-ISSB-1700.

[5] Chicago Convention of 1944, Document 7300, www.icao.int/icaonet/dcs/7300.html

[6] Bahr, N., Sgobba, T., Jahku, R., Schrogl, K.-U., Haber, J., Wilde, P., Kirkpatrick, P., Trinchero, J. P. and Baccini, H. (2007). "An ICAO for Space?", an IAASS White Paper.

[7] The International Telecommunication Union and the World Radio Conference, www.itu.int

[8] Seibold, R. W., Vedda, J. A., Penn, J. P., Barr, S. E., Kephart, J. F., Law, G. W., Richardson, G. G., Pelton, J. N., Hertzfeld, H. R., Logsdon, J. M., Hoffman, J. A. and Leybovich, M. (2008). *Analysis of Human Spaceflight Safety: Report to Congress. Independent Study Mandated by the Commercial Space Launch Amendments Act of 2004.* Prepared for the FAA Office of Commercial Space Transportation, September. Also see Pelton, J. N. with Marshall, P. (2007). *Space Planes and Space Tourism: The Industry and the Regulation of Its Safety,* George Washington University Research Study, April (see online at www.issfoundation.org)

[9] Underwriters' Labs, Inc., www.ul.com/

[10] Patel, N. R., Martin, J. C., Francis, R. J. and Seibold, R. W. (2000). *Human Flight Safety Guidelines for Reusable Launch Vehicles.* Volpe National Transportation Systems Center, Cambridge, MA, 31 July.

[11] GAO Report to Congress (2006). "Commercial Space Launches: FAA Needs Planning and Monitoring to Oversee the Safety of the Emerging Space Tourism Industry", GAO-07-16, October.

[12] Wong, K. (2007). "Working With COMSTAC To Develop An Appropriate Space Flight Safety Performance Target", COMSTAC RLV Working Group, 10 October.

[13] Sgobba, T. (2008). International Space Safety Standards. Trismac Conference, April.

PART

Conclusions and Next Steps

6

CHAPTER

Conclusions and next steps

30

Joseph N. Pelton*, Ram S. Jakhu†

* *Former Dean, International Space University,* † *Associate Professor, Institute of Air and Space Law, McGill University*

CHAPTER OUTLINE

30.1 NEXT STEPS FORWARD IN SPACE SAFETY REGULATIONS AND STANDARDS

What is common to all of the parts of this book is the fact that space safety remains the prime responsibility of national and regional regulatory bodies and the space agencies themselves. As outer space activities continue to mature in the twenty-first century, some clear transitions seem likely. One transition that is foreseen is that outer space-related activities, whether at the governmental level or at the commercial level, will continue to become safer and more international and that informal standards and regulatory processes will be strengthened in form, and become more structured and backed by either "incentives" or "disincentives" to increase compliance, especially at the international level. Commercial activities will likely serve as a particular stimulus to this transition within the next two to three decades. As these regulatory processes and standards-making activities move forward, technology will also continue to move forward. The most important transition in this regard may well be the shift from chemical rocket propulsion to more efficient, environmentally "greener", "more economic" and "safer" space transportation systems, as outlined in Part 6 of the book.

30.2 THE DRIVERS OF CHANGE

There is no way of predicting, at this point, the extent to which various factors will change the global space safety regime. We can, however, anticipate that driving forces will be space systems' technical progress, space commercialization, improved metadata systems to collect space safety systems data, better independent verification and validation procedures, space safety certification procedures, and

Space Safety Regulations and Standards. DOI: 10.1016/B978-1-85617-752-8.10030-3

more broadly agreed upon global regulations and standards. To get to where one wishes to be in the future, one must have clear and specific goals of where it is that one wishes to go.

30.3 DEFINING THE GOALS FOR NEXT STEPS FORWARD

In this respect, the International Association for the Advancement of Space Safety (IAASS) has devoted a good deal of its strategic planning efforts to defining clear and broadly agreed objectives. These goals are specified in the "Manifesto" that is provided below [1]. We would note that none of these objectives can effectively be achieved unless there are clear-cut international agreements to space safety regulations and standards in each and every one of these areas. Progress will thus be measured by the degree to which specific international space safety policy agreements are reached in the six areas outlined in the Manifesto. This will be particularly so when these agreements on space safety regulations and standards are not only adopted, but also backed up by globally agreed enforcement mechanisms that either reward desired behavior or punish, through disincentives, actions that run counter to space safety goals—as broadly defined. This is to say that these regulations and standards must give regard to both manned and unmanned space activities, to commercial and governmental space programs, and to people and vehicles in space, as well as to people and the property on the ground, including protecting all people from hazardous environmental effects from space travel, exploration, and applications. This can be succinctly stated as that the global goals for space safety regulations and standards must be to make space systems and space activities sustainable for the longer-term future.

30.4 MANIFESTO FOR SAFE AND SUSTAINABLE OUTER SPACE AS ADOPTED BY THE INTERNATIONAL ASSOCIATION FOR THE ADVANCEMENT OF SPACE SAFETY (VERSION 1—OCTOBER 2009)

The International Association for the Advancement of Space Safety (IAASS) expresses serious concern about the safety and sustainability of civil and commercial space activities and calls upon all nations to actively cooperate with determination and goodwill to enhance access to, and promote the safe use of, outer space for the benefit of present and future human generations by committing to:

1. Equally protect the citizens of all nations from the risks posed by launching, overflying, and re-entering of space systems.
2. Develop, build, and operate space systems in accordance with common ground and flight safety rules, procedures and standards based on the status of knowledge, and the accumulated experience of all spacefaring nations.
3. Establish international traffic control rules for launch, on-orbit, and re-entry operations to prevent collisions or interference with other space systems and with air traffic.
4. Protect the ground, air, and space environments from chemical, radioactive, and debris contamination related to space operations.
5. Ban intentional destruction of any on-orbit space system or other harmful activities that pose safety and environmental risks.
6. Establish mutual aid provisions for space mission emergencies.

References

[1] "Manifesto for Safe and Sustainable Outer Space" as adopted by the International Association for the Advancement of Space Safety (IAASS) (Version 1—October 2009). This document was developed through the efforts of the Technical Committees of the IAASS and particularly their Chairs. This document was then reviewed and approved by the Executive Board of the IAASS and its Full Board. It may be amended and revised over time. It serves as a guiding document to establish the goals and objectives of this unique international body devoted to creating global standards for space safety.

PART

Appendices

7

Memorandum of Understanding Concerning Cooperation on Civil and Commercial Space Safety Standards and the Establishment of an International Space Safety Standardization Organization

BACKGROUND

By comparing the current status of civil and commercial space developments and operations with the early times of civil aviation, striking similarities can be seen in terms of poor safety record and lack of international traffic management. Today civil aviation is considered to be the safest mode of transportation. This has primarily been achieved through intergovernmental cooperation under the aegis of the International Civil Aviation Organization (ICAO), the specialized agency of the United Nations that was established in 1944 in Chicago, Illinois, in the United States.

The creation of a body similar to the ICAO for regulating civil and commercial space would require an intensive diplomatic initiative and depend upon the political willingness of major spacefaring countries. This is apparently still in the distant future. The foundation of such a regulatory framework, which could be achieved earlier, would be a set of international space safety standards developed through technical consensus and coordination between institutional national stakeholders.

Such international standards should primarily cover those space safety risk matters that are international in nature and scope and that need to be effectively coordinated (agreed upon) only at the international level. The Cooperating Parties subscribing to this Memorandum of Understanding (MOU) would commit to use such standards as the main/preferred reference for their own national regulations. In addition, optional space safety standards would be developed whenever international harmonization could help to remove unwanted barriers to space commerce or to allow assistance and aid in the case of a space emergency.

This international space safety standardization initiative is promoted by two non-profit organizsations, the International Association for the Advancement of Space Safety (IAASS) of the Netherlands, and the International Space Safety Foundation (ISSF) of the USA.

ARTICLE 1: PURPOSE AND OBJECTIVES

1.1 The purpose of this Memorandum of Understanding (MOU) is to establish arrangements between ... TBD, hereafter referred to as the Subscribing Parties (SP) for a genuinely open, and as wide as possible,

Space Safety Regulations and Standards. DOI: 10.1016/B978-1-85617-752-8.10037-6

international partnership in developing civil and commercial space standards, in accordance with international law, in order to: (a) ensure safe access to, use of, and transit through "near-space" by all countries, and (b) safeguard the functional and physical integrity of any space object operating therein.

Note: For the purpose of this MOU, near-space is defined as the region of outer space extending up to and including geostationary orbit and highly elliptical orbits.

1.2 The objectives of this MOU are specifically to:

a. provide the basis for cooperation between Subscribing Parties and establish their roles and responsibilities;

b. establish the management structure as interfaces necessary to ensure effective planning, funding and coordination; and

c. provide a general description of the civil and commercial standards within the scope of this MOU and the main groupings comprising it.

ARTICLE 2: MEMBERSHIP, INTERNATIONAL SPACE SAFETY STANDARDIZATION ORGANIZATION AND STANDARDS GROUPINGS

1.1 The Subscribing Parties hereby establish the International Space Safety Standardization Organization, herein referred to as "the ISSSO".

1.2 A recognized national institution (organization) with at least 5 years of activity in space safety standardization is entitled to be a Subscribing Party and to participate in the activities as well as the management of the ISSSO.

1.3 The ISSSO shall have the following four coordinated groupings of standards:

(I) *Public Safety Risk of Space Missions*. Standards dealing with public safety risk management, including launch and re-entry operations, safe use of NPS (Nuclear Power Sources), health hazards in proximity of launch sites, as well as interfaces between airspace and outer space bound traffic.

(II) *Ground Processing of Commercial Space Vehicles and Payloads*. Standards establishing common design and operations safety requirements for ground processing of Commercial Space Vehicles and Payloads at international spaceports, including certification of ground personnel. [In perspective, uniform requirements would allow a regime of mutual recognition of safety certificates for payloads and vehicles which may be granted by a national safety authority and accepted for international operations.]

(III) *On-Orbit Space Traffic Control*. Standards establishing exchange of space situational awareness data and operational traffic management rules to prevent on-orbit physical and functional interferences between functional spacecraft and to prevent collision with orbital debris.

(IV) *On-Orbit Safe and Rescue and Servicing Operations*. Standards establishing international rendezvous and docking requirements and minimum systems interoperability requirements, for on-orbit safe and rescue and servicing operations. It also includes requirements for interoperability of EVA (Extravehicular Activity) suits.

ARTICLE 3: ORGANIZATION

2.1 The supreme body for guiding, coordinating, and managing all aspects of the standardization activities is the **International Space Safety Standardization Steering Board** (ISSSB). Each Subscribing Party shall have the right to nominate one representative as a member of the Steering Board. The Steering Board can invite qualified observers to attend its meetings.

The Steering Board is supported by four Sub-Boards dealing with the following specific areas of standardization:

- The **Public Space Safety Board** (PSSB), for standards dealing with public safety risk management of space missions.
- The **Ground Space Safety Board** (GSSB), for standards dealing with ground processing safety risk management of space systems.
- The **Space Traffic and Situational Awareness Board** (STSB), for standards dealing with space traffic management and space situational awareness.
- The **Safe and Rescue Board** (SRB), for standards dealing with interoperability of on-orbit safe and rescue systems and servicing systems.

In addition, there shall be the **International Space Safety Standardization Secretariat**, which under the supervision and control of an Executive Secretary shall provide the secretariat and overall management function. A Secretary, acting in accordance with the policies and directives of the Steering Board, shall be the chief executive and the legal representative of the ISSSO and shall be directly responsible to the Steering Board.

The Steering Board shall appoint a Secretary (for the period of ... years), members, and chairs of the Sub-Boards on the basis of proven knowledge and experience in the specific field. Each Sub-Board nominates experts to be the members and chairs of each Working Group to which the development of one or more standards is assigned.

The Steering Board decisions are taken on the basis of consensus. The decisions of the Sub-Boards and Working Groups can be made either by consensus or by a qualified two-thirds majority, if no consensus is reached. In the latter case the Sub-Board and Working Group decisions will need to be ratified by a decision of the Steering Board.

2.2 *International Space Safety Steering Board—Terms of reference*

The Steering Board is the international body responsible for the overall coordination of the space safety standardization efforts. It is responsible for:

- establishing a 4-year strategic implementation plan, including funding profile and sources, and submitting it to the approval of the head representatives of the Subscribing Parties at dedicated meetings;
- approving the annual budget prepared by the Secretary;
- deciding, on the basis of Sub-Board assessment and recommendation, when a standard has been approved by a Working Group with only a two-thirds majority decision;
- nominating members and chairs of the Sub-Boards.

The Steering Board Chair and the Sub-Board Chairs will be elected by consensus by the Steering Board members for a period of 4 years, which can be renewed two times.

2.3 *International Space Safety Standardization Secretariat—Terms of reference*

The International Space Safety Standardization Secretariat under the lead of the Secretary provides the secretariat and overall management support functions to the Steering Board and its subordinate Boards and Working Groups. It is responsible for:

- detailed annual planning of the standardization activities;
- issuing of operating procedures;
- monitoring the progress of working group activities;
- publishing the standards;
- maintaining the website of the organization;

- issuing a detailed annual report to the Sub-Boards concerning the status of Working Group activities including updating of the annual planning and recommendations for future work;
- ensuring performance of all administrative duties.

2.4 *Sub-Boards—Terms of reference*

Each Sub-Board is responsible for the overall coordination of the standardization efforts of the Working Group for its assigned grouping. It is responsible for:

- providing to the Steering Board input for the 4-year strategic implementation plan;
- approving the detailed annual plan prepared by the Secretariat;
- providing assessment and recommendation to the Steering Board when a standard has been approved by a working group with only a two-thirds majority decision;
- nominating members and chairs of the Working Group (after confirmation of sponsorship availability by the relevant Subscribing Party; see Article 4);
- issuing an annual summary report to the Steering Board concerning the status of activities and future direction.

ARTICLE 4: FUNDING

Funding to the International Space Safety Standardization Organization is provided directly and indirectly by the Subscribing Parties.

The direct funds are those provided to the International Space Safety Standardization Organization to cover all costs of operating the Secretariat, including staff, office rentals, etc. Such costs are equally shared among all the Subscribing Parties.

The indirect funds are those that the Subscribing Parties will internally allocate to sponsor the participation of their nationals (staff, contractors, consultants, etc.) in the standardization activities, including travel costs.

ARTICLE 5: TRANSITIONAL RULES

4.1 *Initial Standards Baseline*

Each Subscribing Party will propose an initial list of candidate international space safety standards among those already formally issued in the past by the Subscribing Party as national standards.

The Sub-Boards will determine if overlaps exist between Subscribing Parties standards lists, and initiate working groups with the participation only of members and chairs sponsored by the Subscribing Parties of the overlapping standards. Observers from other Subscribing Parties can attend the meetings, but not vote. Each resulting standard will be baselined as international standard by the relevant Sub-Board.

If no overlap exists and only a single national standard exists, the national standard will be automatically adopted as the international standard, save for text adaptations or reformulation necessary for international use.

4.2 *Initial Funding*

The annual initial direct funds for operating the Secretariat are established to be US $800,000.

4.3 *Seat of the Organization*

The International Space Safety Standardization Organization will be registered as a non-profit organization with its seat in Montreal, Canada.

Commercial Human Space-flight Vehicle "Certification" by an International Space Safety Board

Note: This Appendix should be read in the context of Chapter 8: Certification of New Experimental Commercial Human Space-Flight Vehicles by Tommaso Sgobba and Joseph N. Pelton

1. GENERAL

100 Purpose

This document addresses the creation of an International Space Safety Board for Commercial Human Space-flight Vehicle Certification. The following text sets the process that would be followed to establish safety requirements applicable to "initial certification" for a Commercial Human Space-flight Vehicle (CHSV). This initial standard would serve to cover any spacecraft commercially developed and operated to perform sub-orbital or orbital flights with humans aboard, including transport vehicles such as capsules, winged bodies or lighter-than-air craft, commercial orbital stations, and unmanned cargo transport vehicles intended to dock with a manned station. To the extent these vehicles involve the use of balloons, parafoils, parachutes, carrier vehicles or other ancillary equipment, relevant safety standards or certification will also be addressed. The International Space Safety Board would be entirely independent of any commercial enterprise engaged in space flight or of any government. Its services would be available to any entity wishing to achieve some form of independent "initial certification" of a CHSV designed and built to carry commercial passengers or to support governmental programs (such as to support something like a Commercial Orbital Transport Service (COTS-D)) vehicle.

101 Scope

These requirements, as set forth in this document, are intended to protect the flight personnel (crew and flight participants), ground personnel, the CHSV and relevant launcher or carrier, and any other interfacing spacecraft, from CHSV-related hazards. This document contains technical and system safety requirements applicable to CHS during ground and flight operations. Safety requirements for expendable launchers or winged carriers used for CHS transport during part of a mission are outside the scope of this standard, but only on the understanding that these vehicles' safety certification will be separately addressed (see Note below). Furthermore, all issues related to public safety are outside the scope of this standard, in particular during launch, air-carried, and re-entry phases.

Note: In the case of a spacecraft that operates as an aircraft (powered and not) during part of its flight, the relevant civil aviation requirements and certification authority should in principle apply. In other words, a spacecraft also operating as an aircraft should require two complementary certifications, as spacecraft and aircraft respectively (only the former being addressed by this standard).

Space Safety Regulations and Standards. DOI: 10.1016/B978-1-85617-752-8.10038-8

101.1 Ground Operations and Ground Support Equipment (GSE) Design. For additional safety requirements that are unique to ground operations and for requirements on GSE design, the CHSV Operator (CO) shall refer to applicable national safety and health regulations, as well as to spaceport ground safety regulations.

101.2 Flight Rules. Flight rules will be prepared for each CHSV flight. These flight rules will indicate preplanned decisions designed to minimize the amount of real-time rationalization or emergency decision-making that may be required when anomalous situations occur. These flight rules are not additional safety requirements, but do define actions for the execution of the flight consistent with flight personnel safety.

102 Responsibility

102.1 CHSV Operator. It is the responsibility of the CHSV Operator (CO) to assure the safety of its spacecraft and to implement the requirements of this document.

102.2 Launcher or Carrier Operator. It is the responsibility of the Launcher or Carrier to interface with the national regulatory body(ies) to obtain the necessary licenses. It is also the responsibility of the Launcher or Carrier operator to assure that interaction between CHS and the Launcher or Carrier, and the integrated system does not create a hazard for the general public and ground personnel.

103 Implementation

This document identifies the safety policy and requirements that are to be implemented by the CO. The implementation of safety requirements by the CO will be assessed by the ISSB during the safety review process and must be consistent with hazard potential. The ISSB assessment of safety compliance will include a complete review of the safety assessment reports (paragraph 301) and may include audits and safety inspections of flight hardware. The detailed interpretations of these safety requirements will be by the ISSB, and will be determined on a case-by-case basis consistent with the CHSV actual architecture and hazard potential. The following supplementary documents are meant to assist the CO in complying with the requirements of this document.

103.1 Implementation Procedure. A detailed "Implementation Procedures" document will be published to assist the CHSV Operator in implementing the system safety requirements and to define further the safety analyses, data submittals, and safety assessment review meetings.

103.2 Interpretations of Requirements. A detailed "Interpretation of Requirements" document will be issued as a collection of interpretations of requirements relative to specific CHSV detailed designs. Additional interpretations will be generated as necessary.

104 Glossary of Terms

For definitions applicable to this document, see Addendum A.

105 Applicable Documents and Tables

A list of documents that are referenced in this document is in Addendum B and for tables see Addendum C.

2. TECHNICAL REQUIREMENTS

200 Scope

The following requirements are applicable to a CHSV as determined by the safety analysis performed by the CO. When a requirement that is identified as applicable by the safety analysis cannot be met, a non-compliance report must be submitted to the ISSB in accordance with the "Implementation Procedures" document for resolution (see para. 103.1).

200.1 Design to Tolerate Failures. Failure tolerance is the basic safety requirement that shall be used to control most CHS hazards. The CHSV must tolerate a minimum number of credible failures and/or crew errors determined by the hazard level. This criterion applies when the loss of a function or the inadvertent occurrence of a function results in a hazardous event.

200.1a Critical Hazards. Critical hazards shall be controlled such that no single failure or crew error can result in damage to Launcher or Carrier, a non-disabling flight personnel injury, or the use of unscheduled safety recovery procedures that affect operations of the integrated system. Failure of de-orbiting an unmanned cargo spacecraft (used for servicing an on-orbit manned vehicle) would also be considered a critical hazard.

200.1b Catastrophic Hazards. Catastrophic hazards shall be controlled such that no combination of two failures or crew errors can result in the potential for a disabling or fatal personnel injury or loss of the Launcher or Carrier, CHSV or other interfacing spacecraft.

200.2 Design for Minimum Risk. CHSV hazards that are controlled by compliance with specific requirements of this document other than failure tolerance are called "Design for Minimum Risk" areas of design. Examples are structures, pressure vessels, pressurized line and fittings, functional pyrotechnic devices, mechanisms in critical applications, material compatibility, and flammability. Hazard controls related to these areas are extremely critical and warrant careful attention to the details of verification of compliance on the part of the CO. Minimum supporting data requirements for these areas of design will be identified in a detailed "Implementation Procedures" document (see para. 103.1).

200.3 Environmental Compatibility. A CHSV shall be certified safe in the applicable worst case natural and induced environments, including those defined by the Launcher or Carrier manufacturer interface control document (ICD).

200.4 Launcher or Carrier Services

200.4a Safe Without Services. The CHSV should be designed to maintain fault tolerance or safety margins consistent with the hazard potential without Launcher or Carrier flight services.

200.4b Critical Launcher or Carrier Services. When Launcher or Carrier services are to be utilized to control CHSV hazards, the integrated system must meet the failure tolerance requirements of paragraph 200.1 and adequate redundancy of the Launcher or Carrier services must be negotiated. The CO must provide a summary of the hazards being controlled by Launcher or Carrier services in the safety assessment report (see para. 301), and document in the individual hazard reports those Launcher or Carrier interfaces used to control and/or monitor the hazards. CHSV hazards that are controlled by Launcher or Carrier provided services shall require post-mate interface test verification for both controls and monitors. In addition, the CO shall identify in the CHSV/Launcher or Carrier ICD those Launcher or Carrier interfaces used to control and/or monitor the hazards.

201 Control of Hazardous Functions

201.1 General. Hazardous functions are operational events (e.g. motor firings, appendage deployments, active thermal control) whose inadvertent operations or loss may result in a hazard.

201.1a "Inhibits". An "inhibit" is a design feature that provides a physical interruption between an energy source and a function (a relay or transistor between a battery and a pyrotechnic initiator, a latch valve in the plumbing line between a propellant tank and a thruster, etc.). Two or more inhibits are independent if no single credible failure, event, or environment can eliminate more than one inhibit.

201.1b Controls. A device or function that operates an inhibit is referred to as a control for an inhibit. Controls do not satisfy the "inhibit or failure tolerance" requirements for hazardous functions. The "electrical inhibits" in a liquid propellant propulsion system (para. 202.2a(3)) are exceptions in that these devices operate the flow control devices (i.e. mechanical inhibits to propellant flow), but are referred to as inhibits and not as controls.

201.1c Monitors. Monitors are used to ascertain the safe status of CHSV functions, devices, inhibits, and parameters. Monitoring circuits should be designed such that the information obtained is as directly related to the status of the monitored device as possible. Monitor circuits shall be current-limited or otherwise designed to prevent operation of the hazardous functions with credible failures. In addition, loss of input or failure of the monitor should cause a change in state of the indicator. Monitoring shall be available to the launch site when necessary to assure safe ground operations. Notification of changes in the status of safety monitoring shall be given to the flight crew in either near real time or real time.

201.1c(1) Near-Real-Time Monitoring. Near-real-time monitoring (NRTM) is defined as notification of changes in inhibit or safety status on a periodic basis.

201.1c(2) Real-Time Monitoring. Real-time monitoring (RTM) is defined as immediate notification to the crew. RTM shall be accomplished via the use of the CHSV failure detection and annunciation system. Real-time monitoring of inhibits to a catastrophic hazardous function is required when changing the configuration of the applicable CHSV system or when the provisions of paragraph 203 are implemented for flight crew control of the hazard.

201.1c(3) Unpowered Bus Exception. Monitoring and safety assurance of inhibits for a catastrophic hazardous function will not be required if the function power is de-energized (i.e. an additional fourth inhibit is in place between the power source and the three required inhibits) and the control circuits for the three required inhibits are disabled (i.e. no single failure in the control circuitry will result in the removal of an inhibit) until the hazard potential no longer exists.

201.1d Use of Timers. When timers are used to control inhibits to hazardous functions, a reliable physical feedback system must be in place for the initiation of the timer. If credible failure modes exist that could allow the timer to start prior to the relevant physical event a safety assurance capability must be provided to the flight crew.

201.1e Computer-Based Control Systems

201.1e(1) Active Processing to Prevent a Catastrophic Hazard. While a computer system is being used to actively process data to operate a CHSV system with catastrophic potential, the catastrophic hazard must be prevented in a two-failure tolerant manner. One of the methods to control the hazard must be independent of the computer system. A computer system shall be considered zero-fault tolerant in controlling a hazardous system (i.e. a single failure will cause loss of control), unless the system utilizes independent computers, each executing uniquely developed instruction sequences to provide the remaining two hazard controls.

201.1e(2) Control of Inhibits. The inhibits to a hazardous function may be controlled by a computer-based system used as a timer, provided the system meets all the requirements for independent inhibits.

201.2 Functions Resulting in Critical Hazards. A function whose inadvertent operation could result in a critical hazard must be controlled by two independent inhibits, whenever the hazard potential exists. Requirements for monitoring (para. 201.1c) of these inhibits and for the capability to restore inhibits to a safe condition are normally not imposed, but may be imposed on a case-by-case basis. Where loss of a function could result in a critical hazard, no single credible failure shall cause loss of that function.

201.3 Functions Resulting in Catastrophic Hazards. A function whose inadvertent operation could result in a catastrophic hazard must be controlled by a minimum of three independent inhibits, whenever the hazard potential exists. One of these inhibits must preclude operation by an RF command or the RF link must be encrypted. In addition, the ground return for the function circuit must be interrupted by one of the independent inhibits. At least two of the three required inhibits shall be monitored (para. 201.1c). If loss of a function could cause a catastrophic hazard, no two credible failures shall cause loss of that function.

202 Specific Catastrophic Hazardous Functions

In the following subparagraphs, specific requirements related to inhibits, monitoring, and operations are defined for several identified potentially catastrophic hazardous functions.

202.1 Solid Propellant Rocket Motors. Premature firing of a CHSV solid propellant rocket motor is a catastrophic hazard. Unless the solid propellant rocket motor can be "throttled" in near real time as in the case of a "neoprene-fueled" system with an oxidizer that can be turned on and off in near real time, the use of such propellants is contraindicated.

202.1a Safe Distance. The safe distance for firing a solid rocket motor shall be defined by the Launcher or Carrier ICD.

202.1b Safe and Arm (S&A) Device. All solid propellant rocket motors shall be equipped with an S&A device that provides a mechanical interrupt in the pyrotechnic train immediately downstream of the initiator. The S&A device shall be designed and tested in accordance with provisions of MIL-STD-1576. If the S&A device is to be rotated to the arm position prior to the CHSV achieving a safe distance from the Launcher or Carrier, rotation must be a flight crew function and must be done as part of the final deployment activities of the CHSV, and the initiator must meet the requirements of paragraph 209. The S&A must be in the safe position during the launch or carry phase. There must be a capability to "re-safe" the S&A device: (i) if the S&A device is to be rotated to the arm position while the CHSV is attached to the Launcher or Carrier; or (ii) if the solid rocket motor propulsion subsystem does not qualify for the unpowered bus exception of paragraph 201.1c(3). In determining compliance with paragraph 201.1c(3), the S&A device in the "safe" position shall be counted as one of the required inhibits.

202.1c Electrical "Inhibits". In addition to the S&A, there shall be at least two independent electrical inhibits, to prevent firing of the motor if the S&A device will be in the "safe" position until the safe distance is reached. There shall be at least three independent electrical inhibits, in addition to the S&A, if the S&A device will be rotated to the arm position prior to the CHSV reaching a safe distance from the Launcher or Carrier Orbiter.

202.1d Monitoring. Monitoring requirements are a function of the design and associated operations as follows:

202.1d(1) No Rotation of the S&A Prior to a Safe Distance. The capability to monitor the status of the S&A device and one electrical inhibit in near real time is required until final separation of the CHSV from the Launcher or Carrier. No monitoring is required if the payload qualifies for the unpowered bus exception of paragraph 201.1c(3).

202.1d(2) S&A Will be Rotated to Arm Prior to a Safe Distance. Prior to rotation of the S&A and separation of the CHSV from the Launcher or Carrier, the flight or ground crew must have continuous real-time monitoring to determine the status of the S&A and to assure that two of the three electrical inhibits are in place (paragraph 201.1c(2)).

202.2 Liquid Propellant Propulsion Systems

202.2a Premature Firing. The premature firing of a liquid propellant propulsion system is a catastrophic hazard. Each propellant delivery system must contain a minimum of three mechanically independent flow control devices in series to prevent engine firing. A bipropellant system shall contain a minimum of three mechanically independent flow control devices in series both in the oxidizer and in the fuel sides of the delivery system. These devices must prevent contact between the fuel and oxidizer as well as prevent expulsion through the thrust chamber(s). Except during ground servicing, and as defined in paragraph 202.2a(2)(a), these devices will remain closed during all ground and flight phases until the time firing is foreseen. A minimum of one of the three devices will be fail-safe, i.e. return to the closed condition in the absence of an opening signal.

202.2a(1) Safe Distance Criteria. The hazard of engine firing close enough to inflict damage to the Launcher or Carrier due to heat flux, contamination, and/or perturbation of the Launcher or Carrier, is in proportion to the total thrust imparted by the CHSV in any axis and shall be controlled by establishing a safe distance for the event. The safe distance shall be determined using appropriate references in IAASS-ISSB-S-1700 at www.iaass.org. For large thruster systems with greater than 10 pounds total thrust, the collision hazard with the Launcher or Carrier must be controlled by considering the safe distance criteria cited above, together with the correct attitude at time of firing. For small reaction control system (RCS) thrusters with less than 10 pounds (4.5 kg) total thrust, the collision hazard must be controlled by the safe distance criteria with consideration of many variables such as deployment method, appendage orientation, and control authority.

202.2a(2) Isolation Valve. One of the flow control devices shall isolate the propellant tank(s) from the remainder of the distribution system.

202.2a(2)(a) Opening the Isolation Valve. If a CHSV with a large liquid propellant thruster system also uses a small reaction control thruster system for attitude control, the isolation valve in a common distribution system may be opened after the CHSV has reached a safe distance for firing the reaction control thrusters provided the applicable requirements of paragraphs 202.2a(3) and 202.2a(4) have been met and two mechanical flow control devices remain to prevent thrusting of the larger system.

202.2a(2)(b) Pyrotechnic Isolation Valves. If a normally closed, pyrotechnically initiated, parent metal valve is used, fluid flow or leakage past the barrier will be considered mechanically non-credible if: (i) the valve has an internal flow barrier fabricated from a continuous unit of non-welded parent metal; or (ii) the valve integrity is established by rigorous qualification and acceptance testing. When the valve is used as a flow control device, the number of inhibits to valve activation determines the failure tolerance against fluid flow.

202.2a(3) Specific Number of Electrical "Inhibits". While the CHSV is closer to the Launcher or Carrier than the minimum safe distance for engine firing, there shall be at least three independent electrical inhibits that control the opening of the flow control devices. The electrical inhibits shall be arranged such that the failure of one of the electrical inhibits will not open more than one flow control device. If the isolation valve will be opened under the conditions of paragraph 202.2a(2)(a) prior to the payload achieving a safe distance for firing a large thruster, three independent electrical inhibits must control the opening of the remaining flow control devices for the large thruster system.

202.2a(4) Monitoring. At least two of the three required independent electrical inhibits shall be monitored by the flight crew until final separation of the CHSV from the Launcher or Carrier. The position of a mechanical flow control device may be monitored in lieu of its electrical inhibit, provided the two monitors used to meet the above requirement are independent. Either near-real-time or real-time monitoring will be required as defined in paragraphs 201.1c(1) and 201.1c(2). One of the monitors must be the electrical inhibit or mechanical position of the isolation valve. Monitoring will not be required if the CHSV qualifies for the unpowered bus exception of paragraph 201.1c(3). If the isolation valve will be opened prior to the payload achieving a safe distance from the Launcher or Carrier, all three of the electrical inhibits that remain after the opening of the isolation valve must be verified safe during final pre-deployment activities by the flight crew.

202.2b Adiabatic/Rapid Compression Detonation. While the CHSV is attached to the Launcher or Carrier, the inadvertent opening of isolation valves in a hydrazine (N_2H_4) propellant system shall be controlled as a catastrophic hazard unless the outlet lines are completely filled with hydrazine or the system is shown to be insensitive to adiabatic or rapid compression detonation. Hydrazine systems will be considered sensitive to compression detonation unless insensitivity is verified by testing on flight hardware or on a high-fidelity flight type system that is constructed and cleaned to flight specifications. Test plans must be submitted to the ISSB as part of the appropriate hazard report. If the design solution is to fly wet downstream of the isolation valve, the hazard analysis must consider other issues such as hydrazine freezing or overheating, leakage, single barrier failures, and back-pressure relief.

202.2c Propellant Overheating. Raising the temperature of a propellant above the fluid compatibility limit for the materials of the system is a catastrophic hazard. Components in propellant systems that are capable of heating the system (heaters, valve coils, etc.) shall be two-failure tolerant to heating the propellant above the material/fluid compatibility limits of the system. These limits shall be based on test data derived from NASA-STD-6001 test methods or on data furnished by the CHSV manufacturer and accepted by the ISSB. Propellant temperatures less than the material/fluid compatibility limit, but greater than 200°F, must be approved by the ISSB. Inhibits, cut-off devices, and/or crew safety assurance actions may be used to make the system two failure tolerant to over-heating. Monitoring of inhibits (paras 201.1c and 201.3) or of propellant temperature will be required.

202.2d Propellant Leakage. A CHSV shall be two-failure tolerant to prevent leakage of propellant if the leak has a flow path to the storage vessel. If the leak is in an isolated segment of the distribution system, failure tolerance to prevent the leak will depend on the type and quantity of propellant that could be released. As a minimum, such a leak will be one-failure tolerant. The CHSV shall provide data related to pressure, temperature, and quantity gauging of the CHSV propulsion system tanks, components, and lines to the crew to monitor system health and safety.

202.2e Hazardous Impingement and Venting. The CHSV attitude control shall be designed to prevent hazardous thrusters impingement on another spacecraft (e.g. interfacing vehicle, visiting vehicle). The CHSV propulsion system vents (relief valves, turbo pump assemblies, etc.) shall perform the venting function without causing an additional hazard to another interfacing spacecraft.

202.3 Inadvertent Deployment, Separation, and Jettison Functions. Inadvertent deployment, separation, or jettison of a CHSV, CHSV element, or appendage is a catastrophic hazard unless it is shown otherwise. The general inhibit and monitoring requirements of paragraph 201 shall apply.

202.4 Planned Deployment/Extension Functions

202.4a Cannot Withstand Subsequent Loads. If during planned operations an element of a CHSV is deployed, extended, or otherwise unstowed to a condition where it cannot withstand subsequent induced loads, there shall be design provisions to safe the CHS with redundancy appropriate to the hazard level. Safety assurance procedures may include deployment, jettison, or provisions to change the configuration of the CHS to eliminate the hazard.

202.5 RF Transmitters. Allowable levels of radiation from CHSV vehicle transmitting antenna systems shall be defined in the CHSV to Launcher or Carrier or other interfacing vehicles' ICDs.

202.5a Fluid Release from a Pressurized System Inside of a Closed Volume. Release of any fluid from pressurized systems shall not compromise the structural integrity of any closed volume in which the hardware is contained, such as habitable volumes. Pressurized systems that are two-fault tolerant to release of fluid through controlled release devices do not require analysis. Also, pressurized systems that are two-fault tolerant or designed for minimum risk, as applicable, to prevent leakage, do not require analysis. Systems that do not meet the above shall be reviewed and assessed for safety on a case-by-case basis.

202.7 On-Orbit Rendezvous and Docking

202.7.1 Safe Trajectories. The trajectory of an active CHSV during rendezvous and proximity operations shall be such that the natural drift including three-sigma dispersed trajectories ensures that: (i) prior to the Approach Initiation (AI) burn, the CHSV stays outside the Approach Ellipsoid (AE) for a minimum of 24 hours; (ii) after the AI burn and prior to the CHSV stopping at the arrival point on V-bar inside the AE, the CHSV stays outside the keep-out sphere (KOS) for a minimum of four orbits; (iii) during any retreat out of the Approach Ellipsoid, the CHSV maintains a positive relative range rate until it is outside the Approach Ellipsoid and thereafter stays outside the Approach Ellipsoid for a minimum of 24 hours.

202.7.2 Use of Dedicated Rendezvous Sensors. Relative navigation during rendezvous shall be based on the use of rendezvous sensors for docking operations on the active CHSV (where relative GPS data may be corrupted by multi-path effects and/or will not provide sufficient accuracy) and corresponding target pattern on the passive interfacing CHSV.

202.7.3 Collision Avoidance Maneuver. The active CHSV shall implement collision avoidance maneuver strategies in addition to safe free drift trajectories, as a means to avoid collision with a passive CHSV, in case of contingencies up to docking.

203 Hazard Detection and Corrective Safety

The need for hazard detection, annunciation, and corrective safety actions (CSA) by the flight crew to control time-critical hazards will be minimized and implemented only when an alternate means of reduction or control of hazardous conditions are not available. When implemented, these functions will be capable of being tested for proper operations during both ground and flight phases. Likewise, CHSV designs should be such that real-time monitoring is not required to maintain control of hazardous functions. In exceptional circumstances and after review by certification personnel, real-time monitoring and hazard detection and CSA may be utilized to support control of hazardous functions provided that adequate crew response time is available and acceptable, and CSA procedures are developed.

204 Abort, Escape, and Safe Haven

The system design and operations shall allow for safe abort, including as necessary flight personnel escape and rescue capabilities, for all flight phases starting with on-pad operations or lift-off operations as the case may be. Flight personnel escape capability is achievable by various means (e.g. ejection seats, capsule tractor rocket, winged spacecraft). The escape system may be either manual or automatic. The escape system, including any sensor, equipment, and circuitry that is part of the Launcher or Carrier vehicle, shall comply with the requirements in 200.1 and 200.2. Safe aborts and contingency return shall include design provisions for rapid CSA. Hazard controls may include deployment, jettison, or design provisions to change the configuration of the CHSV. Safe-haven capabilities shall be included in the design to cope with uncontrollable emergency conditions (e.g. fire, depressurization). The safe haven is meant to sustain flight personnel life until escape or rescue can be accomplished. Safe haven is normally used on-orbit and achieved by isolation of orbit habitable volumes.

205 Failure Propagation

The design shall preclude propagation of failures from the CHSV to the environment outside the CHSV.

206 Redundancy Separation

Safety-critical redundant subsystems shall be separated by the maximum practical distance, or otherwise protected, to ensure that an unexpected event that damages one is not likely to prevent the others from performing the function. All redundant functions that are required to prevent a catastrophic hazard must not be routed through a single connector.

207 Structures

207.1 Structural Design. The structural design shall provide ultimate factors of safety equal to or greater than 1.5 for all CHSV mission phases except emergency landing. This includes loads incurred during CHSV and Launcher or Carrier operations for all CHSV configurations or while changing configuration. A Structural Verification Plan will need to be submitted to the certifying review and approval team. When failure of structure can result in a catastrophic event, the design needs to be based on fracture control procedures to prevent structural failure because of the initiation or propagation of flaws or crack-like defects during fabrication, testing, and service life. Requirements for fracture control are specified in NASA-STD-5003 or ECSS-E-30. Safety-critical fasteners should be procured in accordance with international aerospace standards. To meet certification requirements, safety-critical fasteners need to be designed to include redundant features (e.g. torque and self-locking helicoids) to prevent inadvertent back-out.

207.2 Emergency Landing Loads. The structural design to meet certification requirements needs to comply with the ultimate design load factors for emergency landing loads that are specified in the ICDs between the Carrier and the CHSV. Structural verification for these loads may be certified by analysis only.

207.3 Design "Allowables". Material design "allowables" and other physical properties to be used for the design/analysis of flight hardware should be taken from MIL-HDBK-5G to meet certification requirements. For all applications of metals, material "A" as specified in MIL-HDBK-5G should be used. Likewise for non-metallic materials, material equivalent "A allowables" as defined in MIL-HDBK-5G should be used.

207.4 Stress Corrosion. Materials used in the design of CHSV structures should be rated for resistance to stress corrosion cracking (SCC) in accordance with the tables in MSFC-HDBK-527/JSC 09604 and MSFC-STD-3029. Alloys with high resistance to SCC should be used whenever possible. When failure of a part made from a moderate- or low-resistance alloy could result in a critical or catastrophic hazard, a Stress Corrosion Evaluation Form from MSFC-HDBK-527/JSC 09604 is to be attached to the applicable stress corrosion hazard report contained in the safety assessment report (see para. 301). When failure of a part made from a moderate- or low-resistance alloy would not result in a hazard, rationale to support the non-hazard assessment is to be included in the stress corrosion hazard report. Controls that are required to prevent SCC of components after manufacturing must be identified in the hazard report and closure on this action is required to be documented in the verification log (see para. 306.2).

207.5 Pressure Systems. The maximum design pressure (MDP) for a pressurized system is defined by maximum relief pressure, by maximum regulator pressure, and/or maximum temperature. Transient pressures will also be considered. Design factors of safety will apply to MDP. Where pressure regulators, relief devices, and/or a thermal control system (e.g. heaters) are used to control pressure, collectively they need to be two-fault tolerant to ensure the pressure exceeds the MDP of the system. Pressure integrity will be verified at the system level.

207.5a Pressure Vessels. Safety requirements for CHS pressure vessels are listed in the paragraphs below. Particular attention will be given to ensure compatibility of vessel materials with fluids used in cleaning, test, and operation. MDP as defined in paragraph 207.5 shall be substituted for all references to maximum expected operating pressure (MEOP) in the pressure vessel standards. Data requirements for pressure vessels are listed in the "Implementation Procedures" document (see para. 103.1).

207.5a(1) Metallic Pressure Vessels. Metallic pressure vessels should comply with the pressure vessel requirements of MIL-STD-1522A as modified by subparagraphs (a), (b), and (c) below. (a) Approach "B", as specified in MIL-STD-1522A, is not considered acceptable. (b) Non-destructive evaluation (NDE) of safe-life pressure vessels will require inspection of welds after proof testing. (c) A proof test of each flight pressure vessel to a minimum of 1.5 $\times$ MDP and a fatigue analysis showing a minimum of 10 design lifetimes can be used in lieu of testing a certification vessel to qualify a vessel design that in all other respects meets the requirements of this document and MIL-STD-1522A, Approach A. ANSI/AIAA S-080 can be used in lieu of MIL-STD-1522A and the above subparagraphs for metallic pressure vessels.

207.5a(2) Composite Over-Wrapped Pressure Vessels (COPVs). COPVs will be required to meet the intent of the pressure vessel requirements in ANSI/AIAA S-081. A damage control plan and stress rupture life assessment will be needed for each COPV.

207.5b Dewars. Dewar/cryostat systems are a special category of pressurized vessels because of unique structural design and performance requirements. Pressure containers in such systems will be subject to the requirements for pressure vessels specified in paragraphs 207.5 and 207.5a as supplemented by the requirements of this section.

1. Pressure containers need to be leak-before-burst (LBB) designs where possible, as determined by a fracture mechanics analysis. Containers of hazardous fluids and all non-LBB designs need to employ a fracture mechanics safe-life approach to assure safety of operation.
2. MDP of the pressure container will be as determined in paragraph 207.5 or the pressure achieved under maximum venting conditions, whichever is higher. Relief devices must be sized for full flow at MDP.
3. Outer shells (i.e. vacuum jackets) need pressure relief capability to preclude rupture in the event of pressure container leakage. If pressure containers do not vent external to the "dewar" but instead vent into the volume contained by the outer shell, the outer shell relief devices must be capable of venting at a rate to release full flow without outer shell rupture. Relief devices must be redundant and individually capable of full flow.
4. Pressure relief devices that limit maximum design pressure must be certified to operate at the required conditions of use. Certification shall include testing of the same part number from the flight lot under the expected use conditions.
5. Non-hazardous fluids may be vented into closed volumes if analysis shows that a worst-case credible volume release will not affect the structural integrity or thermal capability of the system.
6. The proof test factor for each flight pressure container needs to be a minimum of 1.1 times MDP. Qualification burst and pressure cycle testing is not required if all the requirements of paragraphs 207.5, 207.5a, and 207.5b are met. The structural integrity for external load environments will also need to be demonstrated.

207.5c Pressurized Lines, Fittings, and Components

1. Pressurized lines and fittings with less than a 1.5-inch (4 cm) outside diameter and all flex-hoses shall have an ultimate factor of safety equal to or greater than 4.0. Lines and fittings with a 1.5-inch (4 cm) or greater outside diameter shall have an ultimate factor of safety equal to or greater than 1.5.
2. All line-installed bellows and all heat pipes shall have an ultimate safety factor equal to or greater than 2.5.
3. Other components (valves, filters, regulators, sensors, etc.) and their internal parts (bellows, diaphragms, etc.) that are exposed to system pressure shall have an ultimate factor of safety equal to or greater than 2.5.
4. Secondary compartments or volumes that are integral to or attached by design to the above parts and which can become pressurized as a result of a credible single barrier failure must be designed for safety consistent with structural requirements. These compartments shall have a minimum safety factor of 1.5 based on MDP. If external leakage would not present a catastrophic hazard to the CHS, the secondary volume must either be vented or equipped with a relief provision in lieu of designing for system pressure.

207.5d Flow-Induced Vibration. Flexible hoses and bellows need to be designed to exclude flow-induced vibrations that could result in a catastrophic hazard.

207.6 Sealed Compartments. Sealed compartments within a habitable volume, including containers that present a safety hazard if rupture occurs, shall be capable of withstanding the maximum pressure differential associated with emergency depressurization of the habitable volume. Sealed compartments and containers located in any other region of the CHSV need to be designed to withstand the decompression and re-pressurization environments associated with ascent or descent.

208 Materials

MSFC-HDBK-527/JSC 09604 contains a listing of materials (both metals and non-metals) with a "rating" indicating acceptability for each material's characteristic. For materials which create potential hazardous situations as described in the paragraphs below and for which no prior test data or rating exists, the CHSV Operator (CO) will need to present other test results for ISSB review. The CHS material requirements for hazardous materials, flammability, and for off-gassing and out-gassing are as follows:

208.1 Hazardous Materials. Hazardous materials shall not be released or ejected near manned systems (interfacing or in close proximity). The CHSV Operator will need to submit independent toxicological assessments for all CHS hazardous materials.

208.1a Fluid Systems. Particular attention shall be given to materials used in systems containing hazardous fluids. These hazardous fluids include gaseous oxygen, liquid oxygen, fuels, oxidizers, and other fluids that could chemically or physically degrade the system or cause an exothermic reaction. Those materials within the system exposed to oxygen (liquid and gaseous), both directly and by a credible single barrier failure, will need to meet the requirements of NASA-STD-6001 at MDP and temperature. Materials within the system exposed to other hazardous fluids, both directly and by a credible single barrier failure, must pass the fluid compatibility requirements of NASA-STD-6001 at MDP and temperature. CHSV manufacturers' compatibility data on hazardous fluids may be used to accept materials in this category if accepted by the ISSB.

208.1b Chemical Releases. The use of chemicals that would create a toxicity problem (including irritation to skin or eyes) or cause a hazard to the CHSV and other manned systems (interfacing or in close proximity) if released should be avoided. If use of such chemicals cannot be avoided, adequate containment shall be provided by the use of an approved pressure vessel as defined in paragraph 207.5 or the use of two or three redundantly sealed containers, depending on the toxicological hazard for a chemical with a vapor pressure below 15 psia (or the equivalent in the metric system). The CO needs to assure that each level of containment will not leak under the maximum use conditions (vibration, temperature, pressure, etc.). Mercury is an example of such a chemical, since it produces toxic vapors and can amalgamate with metals or metal alloys used in spacecraft hardware. Documentation of chemical usage, along with the containment methods, will be supplied for review and approval.

208.2 Flammable Materials. CHS materials must not constitute an uncontrolled fire hazard. The minimum use of flammable materials shall be the preferred means of hazard reduction. The determination of flammability shall be in accordance with NASA-STD-6001. Guidelines for the conduct of flammability assessments are provided in NSTS 22648. A flammability assessment shall be documented in accordance with the "Implementation Procedure" document (see para. 103.1).

208.2a Habitable Areas. Materials used in habitable areas need to be tested in accordance with NASA-STD-6001 in the worst-case atmosphere (i.e. oxygen concentration). Fire propagation path considerations also apply.

208.2b Outside Habitable Areas. Materials used outside the CHSV shall be evaluated for flammability in an air environment at 14.7 psi. Propagation path considerations of NSTS 22648 apply for material usages of greater than 1 pound (0.45 kg) and/or dimensions exceeding 12 inches (30 cm).

208.3 Material Off-Gassing in Habitable Areas. Usage of materials that produce toxic levels of out-gassing products shall be avoided in habitable areas. CHSV elements in such areas need to be subjected to out-gassing tests (black-box levels) for safety validation. Rigorous material control to insure that all selected materials have acceptable "off-gassing" characteristics is a negotiable alternative to black-box level testing. The off-gassing test specified in NASA-STD-6001 or an approved equivalent is to be used for the black-box level out-gassing test.

208.4 Material Out-Gassing. Materials used in the design and construction of CHSV hardware exposed to the vacuum environment need to have low out-gassing properties, whenever out-gassing products may be detrimental to safety-critical devices and functions (e.g. fogging of optics).

209 Pyrotechnics

If premature firing or failure to fire will cause a hazard, the pyrotechnic subsystem and devices need to meet the design and test requirements of MIL-STD-1576.

209.1 Initiators. NASA Standard Initiators (NSIs) are the preferred initiators for all safety-critical pyrotechnic functions. MIL-STD-1576 qualification and acceptance test requirements, or equivalent, apply if other initiators are used.

209.2 Pyrotechnic Operated Devices

209.2a Debris Protection. Pyrotechnic devices that are to be operated in the proximity of the Launcher, Carrier, or another spacecraft that do not meet the criteria of this document to prevent inadvertent operation shall be designed to preclude hazards due to effects of shock, debris, and hot gases resulting from operation. Such devices need to be subjected to a "locked-shut" safety demonstration test (i.e. a test to demonstrate the capability of the devices to safely withstand internal pressures generated in operation with the movable part restrained in its initial position).

209.2b Must Function Safety-Critical Devices. Where failure to operate will cause a catastrophic hazard, pyrotechnic operated devices need to be designed, controlled, inspected, and certified to criteria equivalent to those specified in NSTS 08060. The data needed for ISSB review are identified in the "Implementation Procedure" document (see para. 103.1). If the device is used in a redundant application where the hazard is being controlled by the use of multiple independent methods, then, in lieu of demonstrating compliance with criteria equivalent to NSTS 08060, sufficient margin to assure operation needs to be demonstrated. When required, pyrotechnic operated devices need to demonstrate performance margin using a single charge or cartridge loaded with 85% (by weight) of the minimum allowable charge or other equivalent margin demonstrations.

209.2c Electrical Connection. Pyrotechnic devices which, if prematurely fired, may cause a hazard shall be designed such that these devices can be electrically connected to the Launcher or Carrier after all electrical interface verification tests have been completed. Ordnance circuitry needs to be verified safe prior to connection of pyrotechnic devices.

209.3 Traceability. The CO shall maintain a list of all safety-critical pyrotechnic initiators installed or to be installed on the CHSV, giving the function to be performed, the part number, the lot number, and the serial number.

210 Mechanisms

210.1 Design Factors. Safety-critical mechanisms need to be sized to provide actuation forces that exceed the predicted worst-case resistance torques/forces by a factor of at least 2. The following minimum factors are applicable for the components of resistance:

a. Friction: 3.
b. Hysteresis: 3.
c. Spring: 1.2.
d. Inertia: 1.1.

When the contributing sources of the components of resistance are multiple and independent, these factors need only to be applied to the two worst sources in each category.

210.2 Lifetime Testing. The lifetime of safety-critical mechanisms needs to be demonstrated by test in an operationally representative environment, using the sum of the predicted nominal ground test cycles and the flight and on-orbit operation cycles. For the test demonstration, the number of the predicted cycles shall be multiplied by the following factors:

a. Ground Testing cycles ×4 (with 10 as minimum number of cycles)
b. Flight and on-orbit cycles
   - 1–10 actuations ×10
   - 11–1000 actuations ×4
   - 1001–10,000 actuations ×2
   - over 10,000 actuations ×1.25.

A full output cycle or full revolution of the mechanism is defined as one actuation. In order to determine the lifetime to be demonstrated by test, an accumulation of actuations multiplied by their individual factors shall be used. Any element in the chain of actuation (motor, bearing, gear, etc.) has to be compliant with the maximum number of cycles applicable to any of the remaining elements in the chain.

211 Radiation

211.1 Ionizing Radiation. CHSVs containing or using radioactive materials, or that generate ionizing radiation, shall be identified and approval obtained for their use by the relevant national regulatory body(ies).

211.2 Emissions and Susceptibility. CHSV emissions shall be limited to those levels identified in the CHSV to Launcher or Carrier or other interfacing spacecraft ICDs. CHSVs with unintentional radiation (EMI) above the levels identified in ICDs will be assessed for hazardous impact. The CHSV must demonstrate that its safety-critical equipment is not susceptible to the electromagnetic environment defined in the ICDs.

211.3 Lasers. Lasers used need to be designed and operated in accordance with American National Standard for Safe Use of Lasers, ANSI-Z-136.1.

211.4 Optical Requirements. Optical instruments shall prevent harmful light intensities and wavelengths from being viewed by operating and flight personnel. Quartz windows, apertures or beam stops and enclosures shall be used for hazardous wavelengths and intensities. Light intensities and spectral wavelengths at the eyepiece of direct viewing optical systems need to be below the Threshold Limit Values (TLVs) for physical agents as defined by the American Conference of Governmental Industrial Hygienists (ACGIH) in Threshold Limit Values and Biological Exposure Indices.

212 Electrical Systems

212.1 General. Electrical power distribution circuitry needs to be designed to include circuit protection devices to guard against circuit overloads: that could result in distribution circuit damage and generation of excessive hazardous products in habitable volumes; to prevent damage to other safety-critical circuits, to Launcher or Carrier or other interfacing spacecraft; or that present a hazard to the flight personnel by direct or propagated effects. Electrical faults shall not cause ignition of adjacent materials. Bent pins or conductive contamination in an electrical connector will not be considered a credible failure mode if a "post-mate" functional verification is performed to assure that shorts between adjacent connector pins or from pins to connector shell do not exist. If this test cannot be performed, then the electrical design needs to insure that any pin if bent prior to or during connector mating cannot invalidate more than one "inhibit" and that conductive contamination is precluded by proper inspection procedures. Circuit protective devices shall be sized such that steady-state currents in excess of the de-rated values for wires and cables are precluded. Electrical equipment shall be designed to provide protection from accidental contact with high voltage and generation of molten metal during mating and/or de-mating of power connectors in accordance with the "Interpretations of Requirements" document (see para. 103.2). Wire/cable insulation constructions shall not be susceptible to arc tracking. All selected wire/cable will need to be tested for arc tracking unless they are polytetrafluoroethylene (PTFE), PTFE-laminate, or silicone-insulated wires.

212.2 Batteries. Batteries shall be designed to control applicable hazards caused by buildup or venting of flammable, corrosive or toxic gases, and reaction products; the expulsion of electrolyte; and failure modes of over-temperature, shorts, reverse current, cell reversal, leakage, cell grounds, and overpressure. Safety guidelines for batteries are contained in JSC 20793.

212.3 Lightning. Electrical circuits may be subjected to electromagnetic fields due to a lightning strike. If circuit upset could result in a catastrophic hazard, the circuit design shall be hardened against the environment or insensitive devices (relays) shall be added to control the hazard.

213 Safe Return and Landing

213.1 Winged system. The civil aviation airworthiness regulations and certification requirements shall apply for such use, as determined by the relevant national civil aviation authority.

213.2 Capsule and Hybrid. These requirements are currently under development and will be provided in a later version of this document.

214 Hazardous Operations

214.1 Hazard Identification. The CHSV Operator shall assess all flight and ground operations and determine their hazard potential. The hazardous operations identified shall be assessed in the applicable flight or ground safety assessment report.

214.2 Exposure to Risk. Those ground operations (armplug installation in a CHSV pyrotechnic system, final ordnance connection, etc.) that may place the CHSV in a configuration of increased hazard potential shall be accomplished as late as practicable during the CHSV processing flow at the spaceport.

214.3 Access to Moving Parts. Moving parts such as fans, belt drives, and similar components that could cause personnel injury or equipment damage due to inadvertent contact or entrapment of floating objects shall be provided with guards or other protective devices.

215 Reserved. This section is to be provided in an updated version of this document.

216 Hazardous Commands

All hazardous commands shall be identified. Hazardous commands are those that can remove an "inhibit" to a hazardous function or activate an unpowered hazardous system. Failure modes associated with CHSV flight and ground operations including hardware, software, and procedures used in commanding must be considered in the safety assessment to determine compliance with the requirements of paragraphs 200.1, 201, and 202.

216.1 Rejection of Commands. Software shall reject hazardous commands with catastrophic or critical consequences, when these commands do not meet predefined checks for execution.

216.2 Removal of Software Controlled Inhibits. Command messages to change the state of "inhibits" shall be unique for each inhibit.

217 Habitable Volume

This paragraph establishes specific additional safety requirements applicable to flight personnel habitable volume. A flight personnel habitable volume is defined as the volume in a spacecraft (cabin, capsule or module) that is capable of supporting intravehicular activity (IVA).

217.1 Atmosphere

217.1a(1) Off-gassing. The CHSV design should assure the off-gassing load to the internal manned compartment will not exceed the spacecraft maximum allowable concentrations (SMACs) of atmospheric contaminants at the time of ingress. These levels are specified in Table 1 of NASA-STD-6001. All flight personnel habitable volumes will be tested for off-gassing characteristics according to NASA-STD-6001 as required by paragraph 208.3 of this document and will include measurement of the internal atmosphere of a full-scale, flight configured CHSV as a final verification of acceptability. Time periods prior to flight personnel ingress during which the system does not have active atmospheric contamination control must be considered.

217.1a(2) Payload/Cargo Leakage. Payload/Cargo flown in the CHSV habitable volume must meet the containment requirements of paragraph 208.1b. Payload/Cargo configurations during unmanned operations are not restricted; however, the manned compartment must be environmentally safe for flight personnel ingress during any revisit.

Safe conditions for entry may be established by review of the containment design features, proof of adequate atmospheric scrubbing for the chemical involved, vacuum evacuation, use of equipment capable of detecting toxic chemicals prior to crew exposure, or other techniques suitable for the particular experiment involved.

217.1b Internal Environment. A safe and habitable internal environment shall be provided within the CHSV throughout all manned operational phases.

217.1c Cross Contamination. The CHSV habitable volume shall be designed so as not to create a contamination hazard in the atmosphere being shared with an interfacing spacecraft (e.g. visiting vehicle). SMACs of atmospheric contaminants are specified in Table 1 and Table IV of NASA-STD-3000.

217.2 Habitability. The habitability of the CHS habitable volume directly affects the flight personnel's ability to perform efficiently and safely. CHSV design features related to habitability need to comply with NASA-STD-3000 guidelines for the design of manned systems.

217.2a Acoustic Noise. The acoustic noise environment shall be within the limits acceptable in a ground industrial environment, in accordance with national health regulations.

217.2b Ionizing Radiation. The CHS shall include the radiation protection features/mass shielding required to insure that the crew member dose rates from naturally occurring space radiation are kept as low as reasonably achievable (ALARA). Exposure level limits are defined in Figure 5.7.2.2.1-2 of NASA-STD-3000.

217.2c Mechanical Hazards. CHSV and equipment design should be undertaken to protect flight personnel from sharp edges, protrusions, etc. during all flight operations. Translation paths and adjacent equipment need to be designed to minimize the possibility of entanglement or injury to flight personnel crew members.

217.2d Thermal Hazards. During normal operations, flight personnel should not be exposed to high or low surface temperature extremes. Protection needs to be provided against continuous skin contact with surfaces above 49°C (113°F) or below 4°C (39°F). Safeguards such as warning labels, protective devices or special design features to protect the flight personnel from surface temperatures outside these safe limits need to be provided for both nominal and contingency operations.

217.2e Electrical Hazards. Grounding, bonding, and insulation need to be provided for all electrical equipment to protect the flight personnel from electric shock during nominal and contingency operational phases while the crew is in the CHSV.

217.2f Lighting. The lighting illumination level provided throughout the CHSV should be designed to permit planned crew activities without injury. A backup/secondary lighting system needs to be provided consistent with emergency egress requirements, or in case of failure of the primary lighting system.

217.3 Fire Protection. A fire protection system comprised of fire detection, warning, and suppression devices need to be provided in the CHSV. The fire protection system needs to encompass both hardware and flight personnel procedures for adequate control of the fire hazard within the habitable volume. The fire protection system needs to incorporate test and checkout capabilities such that the operational readiness of the entire system can be verified by the crew members. The fire protection system shall have redundant electrical power sources and shall incorporate redundant detection and warning capability and redundant activation of suppressant devices. Fire detection annunciation and control of the CHS fire protection system need to be provided to the crew. Fire suppressant shall be compatible with CHSV life-support hardware. The fire suppressant need not exceed 1 hour SMAC levels in any isolated elements and needs to be non-corrosive. Fire suppressant by-products need to be compatible with the CHSV contamination control capability.

217.4 Emergency Safety Response

217.4a Flight Personnel Egress. The CHSV design shall be compatible with emergency safety response procedures and rapid flight personnel escape. Flight personnel shall be provided with clearly defined escape routes for emergency egress in the event of a hazardous condition. Where practical, dual escape routes from all activity areas shall be provided. Equipment location shall provide for protection of compartment entry/exit paths in the event of an accident. Routing of hard-lines, cables, or hoses through a tunnel or hatch that could hinder crew escape or interfere with hatch operation for emergency egress is not considered an acceptable design. Hatches that could impede flight personnel escape must remain open during all manned operations.

217.4b Electrical System. Separate safety response systems need to be used for nominal CHSV functions and for essential/emergency functions (fire protection, caution and warning, and emergency lighting, etc.). Essential/emergency functions shall be powered from a dedicated electrical power bus with redundant power sources.

217.5 Hatches. CHSV hatch design shall be compatible with emergency flight personnel egress. CHSV hatches between different habitable modules need to provide a capability to allow a visual inspection of the interior of the CHSV prior to hatch opening and flight personnel ingress. All operable hatches that could close and latch inadvertently, thereby blocking an escape route, need to have a redundant (backup) opening mechanism and shall be capable of being operated from both sides. External pressure hatches (i.e. interfacing directly to space vacuum) need to be self-sealing (inward opening). Hatches shall have a pressure difference indicator, clearly visible to the crew member operating the hatch, and a pressure equalization device. All hatches shall nominally be operable without detachable tools or operating devices and need to be designed to prevent inadvertent opening prior to complete pressure equalization. Hatches at docking locations need to provide the capability to verify that the environment is within the oxygen, nitrogen, and carbon dioxide levels as well as within the SMAC levels (of selected compounds), and provide visual inspection of the interior of the pressurized volume prior to crew ingress into an unmanned cargo transportation spacecraft.

217.6 Caution and Warning. The CHSV needs to incorporate a caution and warning system. All safety caution and warning parameters need to be redundantly monitored and need to support an annunciation capability. As a minimum, CHSV total pressure, fan differential pressure, fire detection, oxygen partial pressure, and carbon dioxide partial pressure need to be monitored. The status of all monitored parameters need to be available to the crew prior to in-flight entry into a CHSV habitable module. The caution and warning system needs to include test provisions to allow the crew members to verify proper operation of the system.

217.7 Windows

217.7a Structural Design. The number of windows shall be minimized and all assemblies shall provide a redundant pressure pane. The pressure panes shall be protected from damage by external impact. The structural design of window panes in the pressure hull shall provide a minimum initial ultimate factor of safety of 3.0 and an end-of-life minimum factor of safety of 1.4. Window design shall be based on fracture mechanics considering flaw growth over the design life of the CHS.

217.7b Transmissivity. The transmissivity of CHS windows needs to be based on protection of the flight personnel from exposure to excess levels of naturally occurring non-ionizing radiation. Exposure of the skin and eyes of flight personnel to non-ionizing radiation shall not exceed the threshold limit values (TLVs) for physical agents as defined by the American Conference of Governmental Industrial Hygienists (ACGIH) in Threshold Limit Values and Biological Exposure Indices. Window design needs to be coordinated with other shielding protection design to comply with the ionizing radiation limits specified in paragraph 217.2b.

217.8 Communications. Protected voice communications, with backup capability, are essential.

217.9 Pressure Hull. The design of the habitable volume shall comply with the structural design requirements of paragraphs 207.1 and 207.2. The hull maximum design pressure (MDP) shall be determined as defined in paragraph 207.5. The ultimate factor of safety of hull design needs to be equal to or greater than 2.0 for both the MDP and the maximum negative pressure differential the hull may be subjected to during normal and contingency operations or as the result of two credible failures. The pressure hull needs to be designed to leak-before-burst criteria.

217.10 Life Support System. The CHSV life-support system needs to be able to provide the following functions in any configuration (e.g. open/closed hatches to different habitable volumes or interfacing spacecraft) in response to metabolic consumption and loss of cabin atmosphere to space:

a. Monitor total pressure in the range of 0–16.0 psia with an accuracy of ±0.01 psia and report cabin atmospheric pressure once per minute. The system shall alert the crew within 1 minute when the cabin atmosphere pressure drops below 13.9 psia for longer than 3 minutes.
b. Controlled release of gaseous nitrogen and gaseous oxygen into the habitable volume for maintenance and restoration of habitable volume pressure, and to maintain the habitable volume pressure in response to loss of atmosphere to space.
c. Remote and manual on/off control of introduction of gaseous nitrogen and gaseous oxygen into the internal atmosphere at a flow rate for each of 0.1–0.2 lbm per min (or equivalent in the metric system).
d. Capability to maintain cabin total pressure at greater than 14.1 psia. This maintenance of cabin pressure shall not cause nitrogen partial pressure to exceed 11.6 psia, or cabin total pressure to exceed 14.9 psia (or equivalent in the metric system).
e. Capability to maintain oxygen partial pressure above 2.83 psia. This maintenance of oxygen partial pressure shall not cause the oxygen partial pressure to exceed 3.35 psia or 24.1% by volume (or equivalent in the metric system).
f. Control the maximum internal-to-external differential pressure of the CHSV to less than 15.2 psia. Venting of atmosphere to space should not occur at less than 15.0 psia (or equivalent in the metric system).
g. Monitor atmosphere temperature over the range of 60–90°F (15–38°C) with an accuracy of ±1°F (±0.5°C).
h. Detect combustion products over the entire range.
i. Monitor the atmosphere of carbon dioxide partial pressure over a range of 0–15 mmHg with an accuracy of ±1% of full scale.
j. Remove gaseous contaminants to maintain contaminant concentrations in the atmosphere below acceptable limits, which are defined as less than or equal to the Spacecraft Maximum Allowable Concentration (SMAC) levels.

217.11 Depressurization and Repressurization

217.11a Pressure differential tolerance. Equipment located in pressurized volumes needs to be capable of withstanding the differential pressure of depressurization, repressurization, and the depressurized condition without resulting in a hazard.

217.11b Operation during pressure changes. Equipment expected to function during depressurization or repressurization needs to be designed to operate without producing hazards.

218 On-Orbit Resistance to Micro-meteoroids/Orbital debris

The probability that the exposure to meteoroid and debris environment will not lead to penetration of, or spall detachment from, M/OD critical items needs to be higher than 0.9946 over the mission.

3. DESIGN AND CERTIFICATION

300 General

The following requirements are applicable to any CHSV development.

301 Safety Analysis

A safety analysis needs to be performed in a systematic manner on the CHSV, and on related software, and ground and flight operations to identify hazardous subsystems and functions. The safety analysis needs to be initiated early in the design phase and should be kept current throughout the development phase. A safety assessment report that documents the results of this analysis, including hazard identification, classification, and resolution, and a record of all safety-related failures, should be prepared, maintained, and submitted in support of the safety assessment reviews conducted by the IAASS-ISSB in accordance with paragraph 304. Detailed instructions for the safety analysis and safety assessment reports will be provided in the IAASS-ISSB Implementation Procedures document (see para. 103.1).

302 Hazard Levels

Hazards are classified according to potential as follows:

302.1 Critical Hazard. A critical hazard can result in the form of damage to CHSV equipment, or to Launcher or Carrier equipment. It can also apply to interfacing spacecraft equipment, a non-disabling personnel injury, or the use of unscheduled safety response procedures that affect operations.

302.2 Catastrophic Hazard. Catastrophic hazards—or their potential—can result in the form of a disabling or fatal personnel injury, loss of the CHS, of the Launcher or Carrier, or of another interfacing spacecraft or ground facilities.

303 Hazard Reduction

Action for reducing hazards should be conducted in the following order of precedence:

303.1 Design for Minimum Hazard. The major goal throughout the design phase needs to be to insure inherent safety through the selection of appropriate design features. Damage control, containment, and isolation of potential hazards need to be included in design considerations.

303.2 Safety Devices. Hazards that cannot be eliminated through design selection next need to be reduced and made controllable through the use of automatic safety devices as part of the system, subsystem, or equipment.

303.3 Warning Devices. When it is not practical to preclude the existence or occurrence of known hazards or to use automatic safety devices, then devices need to be employed for the timely detection of the condition and the generation of an adequate warning signal, coupled with emergency controls of corrective action for operating personnel to make safe or shut down the affected subsystem. Warning signals and their application need to be designed to minimize the probability of wrong signals or of improper reaction to the signal.

303.4 Special Procedures. Where it is not possible to reduce the magnitude of an existing or potential hazard through design or the use of safety and warning devices, special procedures need to be developed to counter hazardous conditions for enhancement of personnel safety.

304 Safety Assessment Reviews and Safety Certification. Safety assessment reviews can be conducted by the IAASS-ISSB to determine compliance with the requirements of this document, excluding any aspect of public safety and ground personnel safety which are regulated by national bodies and by the spaceport safety authority. An initial contact meeting can be held at the earliest appropriate time and can then be followed by formal review meetings spaced throughout the development phase. The depth, number, and scheduling of reviews will be negotiated with the CHSV designer or CHSV Operator and this will be dependent on the complexity, technical maturity, and hazard potential.

305 Safety Compliance Data

Safety compliance data packages will be prepared by the CHSV Operator.

305.1 Data. The data listed below need to be submitted as part of the data package for the phase III flight safety review.

a. A safety assessment report for CHSV design and flight and ground operations (see para. 301).
b. A CHSV safety verification tracking log.
c. Approved waivers and deviations.
d. A summary and safety assessment of all safety-related failures and accidents applicable to CHSV processing, test, and checkout.
e. A list of all pyrotechnic initiators installed or to be installed on the CHSV, giving the function to be performed, the part number, the lot number, and the serial number. Submittal of this list may be delayed to be concurrent with the submittal of the flight safety certification statement.
f. A log book template for each limited life item needs to be kept current over the CHSV lifetime.

305.2 Post-Phase III Compliance. When the flight certification statement of paragraph 304 is submitted, it needs to be included with an updated CHSV safety verification tracking log that documents the closeout of all required safety verification. The verification tracking log and the certification statements need to reflect the final configuration of the CHSV that includes all post-phase III safety activity.

306 Verification

Test, analysis, and inspection are common techniques for verification of design features used to control potential hazards. The successful completion of the safety process will require positive feedback of completion results for all verification items associated with a given hazard. Reporting of results by procedure/report number and date will be required.

306.1 Mandatory Inspection Points (MIPs). When procedures and/or processes are critical steps in controlling a hazard and the procedure and/or process results will not be independently verified by subsequent test or inspection, it will be necessary to insure the procedure/process is independently verified in real time. Critical procedure/process steps must be identified in the appropriate hazard report as MIPs requiring independent QA observation.

306.2 Verification Tracking Log. A safety verification tracking log (see IAASS-ISSB Implementation Procedure (para. 103.1)) is required to properly indicate the status of the completion steps associated with hazard report verification items.

307 Reusable Systems

"Reusable systems" are those CHSV elements that are made up of hardware items that are foreseen for reuse.

307.1 Recertification of Safety. Reusable systems must be recertified safe and must meet all the safety requirements of this document. Caution should be exercised in the use of previous safety verification data for the new flight.

307.2 Previous Flight Safety Deficiencies. All anomalies during the previous flight must be assessed for safety impact. Those anomalies affecting safety-critical systems need to be reported and corrected. Rationale supporting continued use of the affected design, operations, or hardware must be provided for ISSB review and approval.

307.3 Limited Life Items. All safety-critical age-sensitive equipment must be refurbished or replaced to meet the requirements of the new flight.

307.4 Refurbishment. Safety impact of any changes, maintenance, or refurbishment made to the hardware or operating procedures must be assessed and reported in the safety assessment reviews (para. 304). Hardware changes include changes in the design, changes of the materials of construction, etc.

308 Mishap/Incident/Flight and Mission Failures Investigation and Reporting. Mishap/incident/flight and mission failures investigation and reporting will be handled under the provisions of the applicable national regulations.

ADDENDUM A: GLOSSARY OF TERMS AND ACRONYMS

Abort. A specific action or sequence of actions initiated by an onboard automated function, by crew, or by ground control that terminates a flight process.

Adiabatic compression detonation. An observed phenomenon whereby the heat obtained by compressing the vapors from fluids (e.g. hydrazine) is sufficient to initiate a self-sustaining explosive decomposition. This compression may arise from advancing liquid columns in sealed spacecraft systems.

AIAA. American Institute of Aeronautics and Astronautics.

ANSI. American National Standards Institute.

Certificate of Safety Compliance. A formal written statement by the CHSV Operator attesting that the CHSV is safe and that all safety requirements for this document have been met and, if not, what waivers and deviations are applicable.

CHSV. A Commercial Human Space-flight Vehicle—a spacecraft commercially operated to perform sub-orbital or orbital flights with humans aboard. It includes spacecraft such as capsules or winged bodies, performing transportation to and from orbit operations. It also includes orbital stations. A CHSV may be, for a part of its flight, a payload for a launcher (expendable launch vehicle) or for a winged carrier.

CHS elements. Subsystems, equipment, and any other item which are subsets of a CHSV.

CHSV Operator. The company that provides commercial human space-flight transportation services and on-orbit operations services. Also, a company that provides commercial unmanned cargo transportation services to on-orbit "manned" vehicles.

Control. A device or function that operates an inhibit is referred to as a control for an inhibit and does not satisfy inhibit requirements. The electrical devices that operate the flow control devices in a liquid propellant propulsion system are exceptions in that they are referred to as electrical inhibits.

Corrective action. Action taken to preclude occurrence of an identified hazard or to prevent recurrence of a problem.

Credible. A condition that can occur and is reasonably likely to occur. For the purposes of this document, failures of structure, pressure vessels, and pressurized lines and fittings are not considered credible failure modes if those elements comply with the applicable requirements of this document.

Credible single barrier failure (material/fluid compatibility). Potential leaks within a component that permit fluid to directly contact the materials behind the barrier or expose secondary compartments to system pressure conditions shall be considered in single barrier failure analysis (e.g. leaks from a fluid enclosure to an adjacent enclosure such as through mechanical joints, O-rings, gaskets, bladders, bellows, and diaphragms). Redundant seals in series that have been acceptance pressure tested individually prior to flight shall not be considered credible single barrier failures. Failures of structural parts such as pressure lines and tanks, and properly designed and tested welded or brazed joints, are not considered single barrier failures. Metallic bellows and diaphragms designed for and tested to demonstrate sufficiently high margins can be considered for exclusion from the category of credible single barrier failure. In order to be classified as a non-credible failure, the item needs to be designed for a safety factor of 2.5 on the maximum design pressure, pass appropriate manufacturing inspections (such as dye penetrant, radiographic, and visual inspections) and leak checks, and be certified for all the operating environments, including fatigue conditions.

Deviation. Granted use or acceptance for more than one flight of a CHSV aspect that does not meet the specified requirements. The intent of the requirement should be satisfied and a comparable or higher degree of safety should be achieved.

Electromagnetic interference (EMI). Any conducted or radiated electromagnetic energy that interrupts, obstructs, or otherwise degrades or limits the effective performance of electronic or electrical equipment.

Emergency (Flight Personnel). Any condition that can result in flight personnel injury or threat to life and requires immediate corrective action, including predetermined flight personnel response.

Factor of safety. The factor by which the limit load is multiplied to obtain the ultimate load. The limit load is the maximum anticipated load or combination of loads which a structure may be expected to experience. Ultimate load is the load that a payload must be able to withstand without failure.

Failure. The inability of a system, subsystem component, or part to perform its required function under specified conditions for a specified duration.

Failure tolerance. The number of failures that can occur in a system or subsystem without the occurrence of a hazard. Single failure tolerance would require a minimum of two failures for the hazard to occur. Two-failure tolerance would require a minimum of three failures for a hazard to occur.

Final separation. Achieved when the last physical connection between the CHSV and its Launcher or Carrier is severed and the CHSV becomes autonomous.

Fire event. Localized or propagating combustion, pyrolysis, smoldering or other thermal degradation processes, characterized by the potentially hazardous release of energy, particulates, or gases.

Flight abort. An abort of a flight wherein the CHS returns to a landing site.

Flight crew. Any flight personnel on board the CHS engaged in flying the CHSV and/or managing resources onboard, e.g. commander, pilot.

GSE. Ground Support Equipment.

Ground control personnel. The people to handle in-flight monitoring. This term includes any personnel supporting the flight from a console in a flight control center or other support area.

Hazard. The presence of a potential risk situation caused by an unsafe act or condition. A condition or changing set of circumstances that presents a potential for adverse or harmful consequences, or the inherent characteristics of an activity, condition, or circumstance that can produce adverse or harmful consequences.

Hazard controls. Design or operational features used to reduce the likelihood of occurrence of a hazardous effect.

Hazard detection. An alarm system used to alert the crew to an actual or impending hazardous situation for which the crew is required to take corrective or protective action.

Hazardous command. A command that can create an unsafe or hazardous condition that potentially endangers the crew or station safety. It is a command whose execution can lead to an identified hazard or a command whose execution can lead to a reduction in the control of a hazard such as the removal of a required safety inhibit to a hazardous function.

IAASS. International Association for the Advancement of Space Safety.

ICD. Interface Control Document.

Independent inhibit. Two or more "inhibits" are independent if no single credible failure, event, or environment can eliminate more than one inhibit.

Inhibit. An "inhibit" is a design feature that provides a physical interruption between an energy source and a function (a relay or transistor between a battery and a pyrotechnic initiator, a latch valve between a propellant tank and a thruster, etc.).

Interlock. A design feature that ensures that any conditions perquisite for a given function or event are met before the function or event can proceed.

ISSB. Independent Space Safety Board

Manned pressurized volume. Any module in which a person can enter and perform activities in a shirt-sleeve environment.

MDP. Maximum Design Pressure.

Metereoroid/orbital debris (M/OD) critical item. An item is deemed to be M/OD critical when effects resulting from a meteoroid or orbital debris penetration will endanger the flight personnel or CHS survivability.

Mishap/incident. This is an unplanned event that results in personnel fatality or injury, damage to or loss of the system, environment, public property, or private property, or could result in an unsafe situation or operational mode. A mishap refers to a major event, whereas an incident is a minor event or episode that could lead to a mishap.

Monitor. To ascertain the safety status of CHSV functions, devices, inhibits, or parameters.

Non-compliance report. A report documenting a condition in which a requirement cannot be met. It is the report used to request a waiver or deviation. See IAASS-ISSB Implementation Procedure document (see para. 103.1).

NSI. NASA Standard Initiator (pyrotechnic). The NSI is provided to the payload customer by NASA.

Off-gassing. The emanation of volatile matter of any kind from materials into habitable areas.

Operator error. Any inadvertent CHS action by either flight or ground personnel that could eliminate, disable, or defeat an inhibit, redundant system, or other design features that are provided to control a hazard. The intent is not to include all possible actions by a crew person that could result in an inappropriate action but rather to limit the scope of error to those actions that were inadvertent errors, such as an out-of-sequence step in a procedure, or a wrong keystroke, or an inadvertent switch throw.

Out-gassing. This is the emanation of volatile matter of any kind from materials to outside the CHSV habitable areas.

Personnel injury. With respect to catastrophic hazard levels personnel injury will be limited to loss of life or major injury that can lead to either temporary or permanent incapacitation of personnel (flight or ground) (e.g. bone fractures, second- or third-degree burns, severe lacerations, internal injury, severe (>1 Gy) radiation exposure, and unconsciousness). Other personnel injuries are related to a critical hazard level provided the injury does not impact the flight crew's capability to accomplish safety-critical tasks.

Pressure vessel. A container designed primarily for pressurized storage of gases or liquids and: (1) contains stored energy of 14,240 footpounds (0.01 pounds trinitrotoluene (TNT) equivalent) or greater based on adiabatic expansion of a perfect gas; or (2) will experience a design limit pressure >100 pounds per square inch absolute (psia); or (3) contains a fluid in excess of 15 psia that will create a hazard if released.

RLV. Reusable Launch Vehicle.

RF. Radio frequency.

Risk. Exposure to the chance of injury or loss. Risk is a function of the possible frequency of occurrence of an undesirable event, of the potential severity of the resulting consequences, and of the uncertainties associated with the frequency and severity.

Safe. A general term denoting an acceptable level of risk, relative freedom from, and low probability of: personal injury; fatality; damage to property; or loss of the function of critical equipment.

Safety analysis. The technique used to systematically identify, evaluate, and resolve hazards.

Safety critical. Safety risk requiring a "safing action that is necessary to prevent a hazard or minimize a risk".

Safety response action (sometimes referred to as "safing"). An action or sequence of actions necessary to place systems, subsystems, or component parts into predetermined safe conditions.

Sealed container. A housing or enclosure designed to retain its internal atmosphere and that does not meet the pressure vessel definition (e.g. an electronics housing).

Structure. Any assemblage of materials which is intended to sustain mechanical loads.

Waiver. Granted use or acceptance of a payload aspect that does not meet the specified requirements; a waiver is given or authorized for one mission only. Safety waivers could include acceptance of increased risk. (Note: Most accidents that have resulted in the loss of life in space can be traced back to one or more waivers being granted.)

ADDENDUM B: APPLICABLE DOCUMENTS

The latest revision and changes of the following documents form a part of this document to the extent specified herein. In the event of conflict between the reference documents and the contents of this document, the contents of this document will be considered superseding requirements.

Document Numbers and Titles Referenced in Paragraph

IAASS-ISSB Implementation Procedure Document, Safety Review and 103.1, 200, 200.2, Data Submittal Requirements. 207.5a, 208.2, 209.2b, 211.1, 306.2, 301, 304.

IAASS-ISSB 18798, Interpretation of Requirements Document para. 103.2, 212.1

CHSV Safety Requirements.

MIL-STD-1576, Electroexplosive 202.1b, 209, 209.1.

Subsystem Safety Requirements and Test Methods for Space Systems.

NASA-STD-6001, Flammability, Odor, Off-gassing, and Compatibility Requirements and Test Procedures for Materials in Environments that Support Combustion (formerly NHB 8060.1C).

202.2c, 208.1a, 208.2, 208.2a, 208.2b, 208.3, 217.1a(1)

NASA-STD-5003, Fracture Control Requirements for 207.1 Payloads Using the Space Shuttle.

ECSS-E-30, European Cooperation for Space Standards—Fracture Control Requirements 207.1

MSFC-STD-3029, Guidelines for the Selection of Metallic Materials for Stress Corrosion Cracking Resistance in Sodium Chloride Environments (superseded MSFC-SPEC-522B) 207.3

Document Numbers and Titles Referenced in Paragraph

MSFC-HDBK-527/JSC 09604, Materials Selection 207.3, 208, 208.3 List for Space Hardware Systems.

MIL-STD-1522, Revision A, Standard General 207.5a(1) Requirements for Safe Design and Operation of Pressurized Missile and Space Systems.

ANSI/AIAA S-080, Space Systems—Metallic Pressure Vessels, Pressurized Structures, and 207.5a(1).

ANSI/AIAA S-081, Space Systems—Composite Overwrapped Pressure Vessels (COPVs). 207.5a(2).

NSTS 22648, Flammability Configuration Analysis.

208.2, 208.2c for Spacecraft Applications.

NSTS 08060, Space Shuttle System Pyrotechnic 209.2b Specification.

ANSI-Z-136.1, American National Standard for 211.3 Safe Use of Lasers.

American Conference of Governmental Industrial Hygienists, 211.4, 217.7b, Threshold Limit Values and Biological Exposure Indices.

JSC 20793, Manned Space Vehicle Battery 212.2 Safety Handbook.

JSC 20584, Spacecraft Maximum Allowable 217.1a(1), 217.1c, Concentrations for Airborne Contaminants.

NASA-STD-3000, Volume 1, Man-Systems 217.2, 217.2b, Integration Standards.

MIL-HDBK-5G 207.3.

ADDENDUM C: TABLES

See: IAASS-ISSB-S-1700 at www.iaass.org

Compound Range (ppm) at www.iaass.org

Carbon Monoxide (CO) 5–400

Hydrogen Chloride (HCl) 1–100

Hydrogen Cyanide (HCN) 1–100

Hydrogen Fluoride (HF)/Carbonyl Fluoride (COF_2) 1–100

Model code of conduct for responsible spacefaring nations

Released by the Stimson Center, 24 October 2007

CENTRAL OBJECTIVE OF THIS CODE OF CONDUCT

To preserve and advance the peaceful exploration and use of outer space.

PREAMBLE

We the undersigned:

Recognizing the common interest of all humankind in achieving progress in the exploration and use of outer space for peaceful purposes;
Reaffirming the crucial importance of outer space for global economic progress, commercial advancement, scientific research, sustainable development, as well as national, regional and international security;
Desiring to prevent conflict in outer space;
Reaffirming our commitment to the United Nations Charter;
Taking into consideration the salience of Article 2(4) of the Charter, which obliges all members to refrain in their international relations from the threat or use of force against the territorial integrity or political independence of any state, or in any other manner inconsistent with the purposes of the United Nations;
Taking special account of Article 42 of the Charter, under which the United Nations Security Council may mandate action by air, sea, or land forces as may be necessary to maintain or restore international peace and security;
Recognizing the inherent right of self-defense of all states under Article 51 of the Charter;
Reinforcing the principles of the Outer Space Treaty of 1967, including:

- the exploration and use of outer space, including the Moon and other celestial bodies, shall be carried out for the benefit and in the interests of all countries,
- outer space, including the Moon and other celestial bodies, shall be free for exploration and use by all States without discrimination of any kind, on a basis of equality and in accordance with international law,
- outer space, including the Moon and other celestial bodies, is not subject to national appropriation by claim of sovereignty, by means of use or occupation, or by any other means, in the exploration

Space Safety Regulations and Standards. DOI: 10.1016/B978-1-85617-752-8.10039-X

and use of outer space, States party to the Treaty shall be guided by the principle of cooperation and mutual assistance and shall conduct all their activities in outer space with due regard to the corresponding interests of all other States party to the Treaty,
- States party to the Treaty undertake not to place in orbit around the Earth any objects carrying weapons of mass destruction,
- the Moon and other celestial bodies shall be used by all States party to the Treaty exclusively for peaceful purposes;

Recalling the importance of space assets for non-proliferation, disarmament and arms control treaties, conventions and regimes;
Recognizing that harmful actions against space objects would have injurious consequences for international peace, security and stability;
Encouraging signature, ratification, accession, and adherence to all legal instruments governing outer space, including:

- 1967 Outer Space Treaty
- 1968 Rescue Agreement
- 1972 Liability Convention
- 1976 Registration Convention
- 1984 Moon Agreement;

Recognizing the value of mechanisms currently in place related to outer space, including the 1994 Constitution of International Telecommunications Union, the 1963 Partial Test Ban Treaty, the 1988 Intermediate-Range Nuclear Forces Treaty, the 1994 Strategic Arms Reduction Treaty, and the 2003 Treaty on Strategic Offensive Reductions;
Recognizing the dangers posed by space debris for safe space operations and recognizing the importance of the 2007 Space Debris Mitigation Guidelines of the Scientific and Technical Subcommittee of the Committee on the Peaceful Uses of Outer Space;
Recognizing the importance of a space traffic management system to assist in the safe and orderly operation of outer space activities;
Believing that universal adherence to this Code of Conduct does not in any way diminish the need for additional international legal instruments that preserve, advance, and guarantee the exploration and use of outer space for peaceful purposes;
Declare the following rights and responsibilities:

RIGHTS OF SPACEFARING STATES

1. The right of access to space for exploration or other peaceful purposes.
2. The right of safe and interference-free space operations, including military support functions.
3. The right of self-defense as enumerated in the Charter of the United Nations.
4. The right to be informed on matters pertaining to the objectives and purposes of this Code of Conduct.
5. The right of consultation on matters of concern and the proper implementation of this Code of Conduct.

RESPONSIBILITIES OF SPACEFARING STATES

1. The responsibility to respect the rights of other spacefaring states and legitimate stakeholders.
2. The responsibility to regulate stakeholders that operate within their territory or that use their space launch services in conformity with the objectives and purposes of this Code of Conduct.
3. Each state has the responsibility to regulate the behavior of its nationals in conformity with the objectives and purposes of this Code of Conduct, wherever those actions occur.
4. The responsibility to develop and abide by rules of safe space operation and traffic management.
5. The responsibility to share information related to safe space operations and traffic management and to enhance cooperation on space situational awareness.
6. The responsibility to mitigate and minimize space debris in accordance with the best practices established by the international community in such agreements as the Inter-Agency Debris Coordination Committee guidelines and guidelines of the Scientific and Technical Subcommittee of the United Nations Committee on the Peaceful Uses of Outer Space.
7. The responsibility to refrain from harmful interference against space objects.
8. The responsibility to consult with other spacefaring states regarding activities of concern in space and to enhance cooperation to advance the objectives and purposes of this Code of Conduct.
9. The responsibility to establish consultative procedures to address and resolve questions relating to compliance with this Code of Conduct, and to agree upon such additional measures as may be necessary to improve the viability and effectiveness of this Code of Conduct.

The Model Code of Conduct was completed by experts from NGOs in Canada, France, Japan, Russia and the United States in October 2007. The group included Setsuko Aoki of Keio University, Alexei Arbatov of the Carnegie Moscow Center, Vladimir Dvorkin of the Center for Policy Studies in Russia, Trevor Findlay of the Canadian Centre for Treaty Compliance, Katsuhisa Furukawa of the Japan Science and Technology Agency, Scott Lofquist-Morgan of the Canadian Centre for Treaty Compliance, Laurence Nardon of the French Institute of International Relations, and Sergei Oznobistchev of the Institute of Strategic Studies and Analysis. NGO participants worked on this project in a personal capacity. Their support for the model Code of Conduct therefore does not reflect endorsements by their institutions or governments.

Draft International Convention on the Removal of Hazardous Space Debris

INTRODUCTION

Attached is a draft proposal of an International Convention on the Removal of Hazardous Space Debris (and should be read in conjunction with Chapter 27). It is based on the wording of the Nairobi International Convention on Wrecks Removal of 2007.

The present Convention Draft proposal has been created considering a future scenario where the removal of "large" space debris is possible. It is proposed to determine in a separate annex the minimum size of space debris to be considered under the present convention, which will be updated with the development of removal technologies and best practices (not included here).

Like the Nairobi Convention, this Draft Proposal considers only the removal of space debris of private organizations (companies). Although the wording addresses State parties, it does not ban international organizations to be part of this system.

Article 1 of the Draft presents some basic definitions, which are to be completed by an annex (not included here).

This Convention Draft is intended to involve an International Technical Institution. It is proposed that under the Convention, State parties convey powers to the International Technical Institution as to require owners and/or operators to remove their hazardous space debris objects and to take measures in case the owner and/or operator fails to comply with the removal request.

Insurance, or financial security, and a corresponding certificate are an important part of the self-enforcing mechanism for the removal of space debris. An annex to this convention may contain a sample of such a certificate.

This Draft Convention proposal presents only core provisions. General provisions as settlement of disputes, time limits, entry into force, etc. are omitted and may be added at a later stage.

DRAFT INTERNATIONAL CONVENTION ON THE REMOVAL OF HAZARDOUS SPACE DEBRIS

Preamble

The States party to the present Convention,

Noting the importance of the United Nations General Assembly Resolution 62/217 on Space Debris Mitigation Guidelines, adopted on 21 December 21 2007,

Space Safety Regulations and Standards. DOI: 10.1016/B978-1-85617-752-8.10040-6

Conscious of the fact that large space debris, if not removed, may pose a hazard to space navigation or the Earth's orbital environment,

Taking account that even with the application of the UN COPUOS Space Debris Mitigation Guidelines the population of small space debris objects, generated from large space debris objects, is continuously growing and will constitute an obstacle to space activities around the Earth,

Considering that the orbital life of space debris at 600 km and above is 25 years and longer,

Noting that space debris can enfold large amounts of kinetic energy and is permanently moving in outer space in areas of interest for the international community,

Convinced of the need to adopt uniform international rules and procedures to ensure the prompt and effective removal of hazardous space debris and payment of compensation for the costs therein involved,

Recognizing the benefits of a uniform legal regime governing responsibility and liability for removal of hazardous space debris,

Bearing in mind the importance of:

The Treaty on the Principles Governing the Activities of States in the Exploration and Use of Outer Space, including the Moon and Other Celestial Bodies, done on 27 January 1967,

The Convention on International Liability for Damage Caused by Space Objects, done on 29 March 1972,

The Convention on the Registration of Space Objects Launched into Outer Space, done on 14 January 1975,

The Convention on Early Notification of a Nuclear Accident, done on 26 September 1986,

The UN Resolution 33/16 No. 9 Relating to the Information to be Furnished by States about the Malfunctioning of NPS in Outer Space, adopted on 10 November 1978,

The UN Resolution 47/68 on the Principles Relevant to the Use of Nuclear Power Sources in Outer Space, adopted on 14 December 1992,

and the consequent need to implement the present Convention in accordance with such provisions,

Have agreed as follows:

ARTICLE 1: DEFINITIONS [1]

For the purposes of this Convention:

1. "Convention area" includes the protected regions around the Earth, as defined by the Inter-Agency Space Debris Coordination Committee [2].
2. "Disposal orbit" is the orbit designated in which to place space objects that have reached their end of life [3].
3. "Space object" means all man-made orbiting objects designed to perform a specific function or mission (e.g. communications, navigation, or Earth observation). A space object that can no longer fulfill its intended mission is considered non-functional. Space objects in reserve or standby modes, as well as space objects waiting for refueling *in situ*, refurbishing *in situ*, or removal outside the Convention area for refurbishing purposes, are considered functional [4].

4. "Space debris" or "space debris object" means all man-made objects, including fragments and elements thereof in Earth orbit or which re-enter the atmosphere, that are non-functional and for which no efforts are undertaken to refuel or to salvage them [5].
5. "Large space debris objects" are the space debris objects that may be safely removed into disposal orbits using space removal technologies.
6. "Hazard" means any condition or threat in the Convention area that:
   a. constitutes a danger or impediment to space navigation; or
   b. may reasonably be expected to result in major harmful consequences to the Earth's orbital environment.
7. "Removal" means any form of prevention, mitigation or elimination of the hazard created by space debris.
8. "State of Registry" means the State that has registered the space object under its nationality with the Secretary-General of the International Technical Institution for the removal of space debris.
9. "Owners and/or operators of the space object" means any private organization who own a space object as well as entities that are contracted to operate a space object.
10. "Institution" means the International Technical Institution for the Removal of Space Debris.
11. "Secretary-General" means the Secretary-General of the Institution.
12. Other definitions are included in Annex 1 to this Convention.

ARTICLE 2: OBJECTIVES AND GENERAL PRINCIPLES

1. A State party, or any legal entity under the jurisdiction or control of a State party, may take measures in accordance with this Convention in relation to the removal of space debris that poses a hazard in the Convention area. Such measures shall be in accordance to the specific requirements addressed by the Institute.
2. Measures taken by the State party, or any legal entity under the jurisdiction or control of a State party in accordance with paragraph 1, shall be proportionate to the hazard.
3. Such measures shall not go beyond what is reasonably necessary to remove space debris that poses a hazard and shall cease as soon as the space debris has been removed; they shall not unnecessarily interfere with the rights and interests of other States including the State of Registry, and of any international organization and private company concerned.
4. The application of this Convention within the Convention area shall not entitle a State party to claim or exercise sovereignty or sovereign rights over any part of outer space.

ARTICLE 3: SCOPE OF APPLICATION

1. Except as otherwise provided in this Convention, this Convention shall apply to space debris in the Convention area.
2. The International Technical Institution will determine which space debris objects are hazardous. Characteristics of the space debris to be removed are established in Annex 2 to this Convention.

ARTICLE 4: EXCLUSIONS

1. This Convention shall not apply to any military space object or other space object owned or operated by a State party, unless the State decides otherwise.
2. If a State party decides to apply this Convention to its military space objects or other space objects owned or operated by it, it shall notify the Secretary-General of the Institution thereof, specifying the terms and conditions of such application.
3. A State party that has made a notification under paragraph 2 may withdraw it at any time by means of a notification of withdrawal to the Secretary-General.

ARTICLE 5: REGISTRY, DECOMMISSION DECLARATION, AND REMOVAL USING THE SPACE OBJECT'S OWN CAPABILITIES

1. State parties should mandate owners and/or operators to register all their space objects at the Institution and to provide it with information, including size, shape, expected useful life and the orbital parameters, at least 6 months before the space object's launch. The State party will be deemed to be the State of Registry. When there is more than one owner and/or operator of one space object of the same State party or of different State parties, they may agree that only one registers the object with the Institution or they may jointly register the space object (States of Registry) indicating the amount of their participation by percentage.
2. State parties should require owners and/or operators to furnish a compulsory space object decommission declaration, information on detaching of stages and other relevant information to the Institution. This information should include the orbital parameters of the decommissioned space debris object, its size and construction, any danger to space traffic, and hazardous elements on board.
3. State parties shall mandate owners and/or operators to remove all their space objects from valuable orbits, using the space object's own capabilities, before they reach their end of life. Intentions of removal shall be reported to the Institution and they shall take place in coordination with the Institution. Once space objects have reached a disposal orbit, they shall be passivated before the communication is terminated.

ARTICLE 6: REPORTING LARGE SPACE DEBRIS

1. A State party shall require the owner and/or operator of a space object under its nationality to report to the Institution without delay when a space object:
   a. is approaching its end of life and will be removed into a disposal orbit using the space object's own propulsion systems;
   b. has reached its end of life and will be refueled *in situ*;
   c. has reached its end of life and/or has malfunctioned and will be refurbished *in situ*;

d. has reached its end of life or has malfunctioned and will be removed outside the Convention areas for refurbishing purposes by an external removal system provider;
e. is approaching its end of life and will not be removed by the space object's own propulsion system;
f. has reached its end of life and was not removed;
g. has malfunctioned and is no longer maneuverable;
h. has been damaged due to a collision or other accident with another space object and is no longer maneuverable; or
i. has been involved in a space collision or other accident and has been severed into several fragments.

2. Such report shall provide the name of the State of Registry and the principal place of business of the registered owner and/or operator and all the relevant information necessary for the Institution to determine whether the space debris object constitutes a hazard in accordance to Art. 7 (Determination of Hazard), including:
   a. the orbital parameters of the space object;
   b. the type, size, and construction of the space object;
   c. the condition of the space object;
   d. in the case of collision or accident, the nature of the damage; and
   e. the nature of energy still on board, in particular any hazardous substances; and
   f. any other relevant information.

ARTICLE 7: DETERMINATION OF HAZARD

1. States party to this Convention agree to recognize the International Technical Institution for the Removal of Space Debris as authority to apply the criteria for determining that a space debris object poses a hazard.
2. The following criteria should be taken into account by the Institution:
   a. the type, size, and construction of the space object to be labeled as a space debris object;
   b. the orbital parameters of the space debris object and its kinetic energy;
   c. the traffic density and the type of traffic in the affected orbits of the Convention:
      i. particularly sensitive areas as the orbits of manned space objects;
      ii. if applicable, traffic density in north—south crossings of Earth's poles, where conjunctions are most likely for Sun-synchronic space objects;
      iii. prevailing transit and density of space debris in the particular orbit likely to be followed by the space debris object under consideration;
   d. in the case where the space debris object was not passivated timely, the nature and quantity of onboard energy elements that are likely to be released and/or are in danger of resulting in the space object's break-up;
   e. presence of hazardous substances and nuclear power sources on board, especially the ones that may survive atmosphere re-entry;
   f. space weather conditions;
   g. any other circumstances that might necessitate the removal of the space debris object.

ARTICLE 8: LOCATING HAZARDOUS SPACE DEBRIS OBJECTS

1. The Institution should collect information to determine if a space debris object constitutes a hazard to the Convention area [6].
2. If the Institution has reason to believe that a space debris object constitutes a hazard, it shall ensure that all practicable steps are taken to establish the characteristics of the space debris object, its orbital parameters, and the identification of the space debris object.
3. In a case where the Institution has determined a space debris object as hazardous and such object has not been declared as space debris by its owner and/or operator or by its State of Registry, the Institution has to undertake all efforts to identify the space debris object. State parties that have information on the identity of such space debris object shall inform the Institution without delay.
4. If a State party has reason to believe that a space debris object of another State party poses a hazard, it shall communicate this to the Institution without delay and provide it with relevant information that may help to establish the characteristics and orbital parameters of the space debris object.

ARTICLE 9: PROMULGATION OF INFORMATION ABOUT HAZARDOUS SPACE DEBRIS OBJECTS

Upon determining that a space debris object constitutes a hazard, the institution shall immediately promulgate the particulars of the space debris object by use of all appropriate means, including the good offices of States and organizations, to warn owners and/or operators of functional space objects of the characteristics and trajectory of the hazardous space debris objects as a matter of urgency.

ARTICLE 10: MEASURES TO FACILITATE THE REMOVAL OF SPACE DEBRIS

1. If the Institution determines that a space debris object constitutes a hazard, it shall immediately:
   a. inform the space debris object's owner and/or operator and the State of the space object registry; and
   b. proceed to consult the space debris object's owner and /or operator and its State of Registry and other States likely to be affected by the space debris object, regarding measures to be taken in relation to the space debris.
2. The owner and/or operator shall remove a space debris object determined to constitute a hazard, using the space object's own maneuver capabilities.
3. In a case where the space object is not capable of being maneuvered, State parties have to mandatorily require the owner and/or operator of the space object to contract an external removal arrangement with an institution that possesses such competence.
4. The owner and/or operator may contract with any space removal service provider, to remove the space debris determined to constitute a hazard on behalf of the owner and/or operator. Before such removal commences, the institution shall be informed and it may lay down conditions for such removal only to the extent necessary to ensure that the removal proceeds in a manner that is

consistent with considerations of safety and protection of functional space objects in the area and the Earth's orbital environment.

5. When the removal referred to in paragraphs 2 and 4 has commenced, the Institution may observe the removal to ensure that the removal proceeds effectively in a manner that is consistent with considerations of safety and protection of functional space objects in the area and the Earth's orbital environment.
6. The Institution shall:
   a. set a reasonable deadline within which the owner and/or operator must remove the space debris object taking into account the nature of the hazard determined in accordance with Article 7 (Determination of Hazard);
   b. inform the registered owner and/or operator and its State of Registry in writing of the deadline it has set and specify that, if the owner and/or operator does not remove the space debris object within that deadline, the Institution will proceed to contract a space removal service provider, at the expense of the owner's and/or operator's insurance; and
   c. inform the owner and/or operator and its State of Registry in writing that the institution intends to intervene immediately in circumstances where the hazard becomes particularly severe.
7. If the owner and/or operator does not remove the space debris object within the deadline set in accordance with paragraph 6(a), the Institution shall open a competition among qualified space removal service providers in order to select the appropriate one for the removal of that particular space debris object.
8. In circumstances where immediate action is required and the Institution has informed the State of Registry and the owner and/or operator accordingly, but there is no possibility of expeditious removal arranged by the space debris owner and/or operator, the Institution may immediately contract a space removal service provider with the most practical and expeditious means available, consistent with considerations of safety and protection of functional space objects and the Earth's orbital environment.
9. State parties shall take appropriate measures under their national law to ensure that the registered owners and/or operators comply with paragraphs 2 and 3.
10. State parties give their consent to the Institution to act under paragraphs 4–8, where required.

ARTICLE 11: LIABILITY OF THE OWNER AND/OR OPERATOR

1. Subject to Article 12 (Compulsory Removal Insurance or other Financial Security), the registered owner and/or operator shall be liable for the costs of locating and removing the hazardous space debris under Articles 8 (Locating Hazardous Space Debris Objects), 9 (Promulgation of Information about Hazardous Space Debris) and 10 (Measures to Facilitate the Removal of Space Debris), unless the registered owner and/or operator proves that:
   a. an operational space object of one State party was transformed into space debris as a direct or indirect result of an anti-satellite attack or an anti-satellite test by another state;
   b. the debris was wholly caused by an act or omission done with intent to cause damage by a third party; or
   c. the debris was wholly caused by the negligence or other wrongful act of any other State, international organization or private company [7];

2. Nothing in this Convention shall affect the application of the Convention on International Liability for Damage Caused by Space Objects, of 1972.
3. No claim for the costs referred to in paragraph 1 may be made against the registered owner and/or operator otherwise than in accordance with the provisions of this Convention.
4. Nothing in this article shall prejudice any right of recourse against third parties.

ARTICLE 12: COMPULSORY REMOVAL INSURANCE OR OTHER FINANCIAL SECURITY

1. The registered owner and/or operator of a space object shall be required by its State of Registry to maintain removal insurance or other financial security, such as a guarantee of a bank or similar institution, to cover liability under this convention, for at least a minimum amount as stipulated in Annex 3.
2. A certificate attesting that removal insurance or other financial security is in force in accordance with the provisions of this Convention shall be issued to each space object by the appropriate authority of the State of the space object's registry.
3. In absence of a national institution to issue the certificate in a State party, it may be issued or certified by the appropriate authority of any other State party. This compulsory insurance certificate shall be in the form of the model set out in Annex 4 to this Convention, and shall contain the following particulars:
   a. name of the space object and the State of Registry;
   b. size and shape of the space object;
   c. name and principal place of business of the registered owner and/or operator;
   d. COSPAR international identification number;
   e. type and duration of security;
   f. name and principal place of business of insurer or other person giving security and, where appropriate, place of business where the insurance or security is established;
   g. period of validity of the certificate, which shall not be longer than the period of validity of the insurance or other security;
   h. insured amount.
4. State parties shall mandate owners and/or operators under their registry to keep a permanent valid insurance or financial security certificate. In case of change of ownership of the space object, all parties involved shall make arrangements to keep a valid insurance or other financial security and the corresponding certificate in order to assure insurance coverage continuity.
5. States of Registry shall keep a certificate register. State parties shall mandate each registered owner and/or operator, or other interested party to deposit a copy of the insurance certificate of the space object in the institution, at least 6 months prior to the space object's launching. The institution will keep a certificate register. State parties and the Institution shall maintain certificate records in an electronic format, accessible to all State parties, attesting the existence of the certificates.
6. Nothing in this Convention shall be construed as preventing a State party and the Institution from relying on information obtained from other States or the Institution or other international

organizations relating to the financial standing of providers of insurance or financial security for the purposes of this Convention. In such cases, the State party relying on such information is not relieved of its responsibility as a State issuing the certificate required by paragraph 2.

7. Certificates issued and certified under the authority of a State party shall be accepted by other State parties for the purposes of this Convention and shall be regarded by other State parties as having the same force as certificates issued or certified by them, even if issued or certified in respect of a space object not registered in a State party. A State party and the Institution may at any time request consultation with the issuing or certifying State should it believe that the insurer or guarantor named in the certificate is not financially capable of meeting the obligations imposed by this Convention.
8. Any claim for removal costs arising under this Convention may be brought directly against the insurer or other person providing financial security for the registered owner's and/or operator's liability. In such a case the defendant may invoke the defence that the space debris was caused by the willful misconduct of the registered owner and/or operator [8].
9. A State party shall not authorize any space object to operate under its registry, unless a certificate has been issued according to this article.
10. Subject to the provisions of this article, each State party shall ensure, under its national law, that insurance or other security to the extent required by paragraph 1 is in force in respect of any space object, wherever registered:
    a. launched from its territory or any facility registered in such State party outside its territory;
    b. controlled landing in its territory.

FINAL CLAUSES [9]

Signatures

Annex 1: Definitions.
Annex 2: Technical issues.
Annex 3: Insurance Issues.
Annex 4: Model of certificate of insurance or other financial security in respect of liability for the removal of hazardous space debris.

Notes

[1] Some definitions of the IADC Space Debris Mitigation Guidelines (UN document A/AC. 105/C.1/L.260) have been modified here to fit the requirements of the removal self-enforcing mechanisms. An annex may include more definitions (Annex 1, not included here).

[2] The IADC has proposed to declare as protected regions two sensitive areas where most present-day satellites are operating. These two areas are the low Earth orbit (LEO) and the geostationary orbit (GEO). International Academy of Astronautics (IAA), "Position Paper on Space Debris Mitigation" (2006), p. 9. In the future, new space traffic demands and technology development may introduce new protected regions. The characteristics of new regions may be established in Annex 2 to this Convention (not included here).

[3] These orbits shall avoid long-term interference with LEO and GEO regions. United Nations General Assembly Resolution 62/217 "International Cooperation in the Peaceful Uses of Outer Space", 21 December 2007 (para. 26), Official Records of the General Assembly, 62nd Session, Supplement No. 20 (A/A/62/20), 2007, paras 117 and 118 and annex (IADC Guidelines), Guideline 6. Location and number of disposal regions may be set according to the development of technology and best practices in Annex 2 (not included here). Instead of determining a fixed period for acceptable orbital life (e.g. 25 years), it should gradually be reduced according to the development of the technology and best practices and shall be updated in Annex 2 (not included here).

[4] The IADC definition is: "3.2.1. Spacecraft. A spacecraft is an orbiting object designed to perform a specific function or mission (e.g. communications, navigation or Earth observation). A spacecraft that can no longer fulfill its intended mission is considered non-functional. (Spacecraft in reserve or standby modes awaiting possible reactivation are considered functional.)". *Supra* note 1.

[5] The IADC definition is: "3.1. Space debris. Space debris refers to all man-made objects, including fragments and elements thereof, in Earth orbit or re-entering the atmosphere, that are non-functional." *Supra* note 1.

[6] For example, by contracting with surveillance systems operators.

[7] For example, fault of the hired space control service.

[8] For example, the owner failed to remove the space object while it had the capabilities to do so.

[9] Provisions on Time limits, Amendment provisions, Settlement of disputes, Signature and Ratification, Entry into force, Denunciation, Depositary and Languages are omitted in this Draft Proposal.

The Proposed Space Preservation Treaty

The States party to this Treaty

Recognizing the common interest of all humankind in the exploration and non-weapons use of outer space for peaceful purposes,

Reaffirming that outer space plays an ever-increasing role in the future development of humankind,

Emphasizing the rights to explore and use outer space freely for peaceful purposes,

Keeping outer space from turning into an arena for military weapons confrontation, to assure security for all in outer space and safe functioning of space technology,

Recognizing that prevention of the placement of weapons and of an arms race in outer space would avert a grave danger for international peace and security,

Desiring to keep outer space as an environment where no weapon of any kind is placed,

Recalling the obligations of all States to observe the provisions of the Charter of the United Nations regarding the use or threat of use of force in their international relations, including in their space activities,

Reaffirming the importance and urgency of preventing an arms race in outer space and of approving concrete proposals on confidence building that could prevent such an arms race, as set out in United Nations General Assembly Resolutions on the prevention of an arms race in outer space,

Reaffirming the will of all States that the exploration and use of outer space, including the Moon, the planets and other celestial bodies, shall be for weapons-free, peaceful purposes and shall be carried out for the benefit and in the interest of all countries, irrespective of their degree of economic or scientific development,

Affirming that it is the policy of States party to this Treaty to permanently ban all space-based weapons,

Have agreed as follows:

ARTICLE I: PERMANENT BAN ON ALL SPACE-BASED WEAPONS AND WAR IN SPACE

A. Summary of Treaty—The Space Preservation Treaty (the "Treaty") bans war in space, and bans any and all covert space-based weapons programs, including those funded under unacknowledged secret access space ("Black Budget") programs, established, funded or operated by any State party to this Treaty, any State, any organization, corporation, partnership, group, individual or syndicate, or by any other means. This Treaty establishes an independent, international ombudsman agency—the Outer Space Peacekeeping Agency—to monitor any and all possible war in space and/or space-based weapons activity on land, sea, in the atmosphere, and in outer space, and to enforce this permanent ban.

B. Space-based weapons:

1. Each State party to this Treaty shall implement a permanent ban on research, development, testing, manufacturing, production, and deployment of any and all space-based weapons.

Space Safety Regulations and Standards. DOI: 10.1016/B978-1-85617-752-8.10041-8

2. Such permanent ban shall extend to any and all research, development, testing, manufacturing, production, and deployment of any and all space-based weapons, whether performed directly by an agency of a State party to this Treaty; by any organization, corporation, partnership, group, individual, or syndicate acting under contract to a State party to this Treaty; by any organization, corporation, partnership, group, individual, or syndicate; or by any organization, corporation, partnership, group, or syndicate acting in any unacknowledged secret access programs of any kind; or by any other entity or means, whether covert or overt, classified or unclassified, on land, sea, in the atmosphere, or in outer space.

C. Anti-satellite and missile defense weapons—Each State party to this Treaty shall implement a ban on research, development, testing, manufacturing, production, and deployment of weapons to destroy or damage objects in space that are in orbit. This shall include a ban on research, development, testing, manufacturing, production, and deployment of space-based anti-satellite weapons, space-based anti-ballistic missile systems, space-based missile defense systems, or space-based other anti-satellite systems, including anti-satellite systems placed in orbit or installed on structures or bodies in outer space.

D. Dual-use technologies (space hazards)—Each State party to this Treaty shall implement a ban on research, development, testing, manufacturing, production, and deployment of space-based dual-use technologies that permit such technologies, intended for commercial, civil, or scientific purposes, to be used also for space-based weapons, space-based anti-satellite, or space-based missile defense purposes. The Outer Space Peacekeeping Agency shall establish licensing requirements for any research, development, testing, manufacturing, production, and deployment of space-based technologies for use against space hazards such as asteroids and near Earth objects.

E. Weaponization of space objects—Each State party to this Treaty shall implement a ban on research, development, testing, manufacturing, production, and deployment of space objects, such as global positioning systems, space platforms and satellites, and spacecraft that are specifically mandated to conduct or enhance war on ground, sea, air or in space.

F. General ban—State parties undertake not to place in orbit around the Earth any objects carrying any kind of weapons, not to install such weapons on celestial bodies, and not to station such weapons in outer space in any other manner; not to resort to the threat or use of force against outer space objects or beings in outer space; not to assist or encourage other States, groups of States, or international organizations to participate in activities prohibited by this Treaty.

G. War in space—State parties undertake not to plan, research, develop, test, implement, mandate, or engage in any act of aggressive war in space, and not to threaten or use force in space against any other State party, State, organization, group, or being in space.

H. Denial of access to space—State parties undertake to guarantee the common interest of all humankind in the exploration and non-weapons use of outer space for peaceful purposes, and not to engage in denial of access to space, or the act of attempting to deny the lawful use of outer space to any other State, organization, or being. State parties agree that any State, organization, or being that engages in denial of access to space shall be subject to the jurisdiction of the Outer Space Peacekeeping Agency, which shall enjoin such denial of access to space.

I. Decommissioning of space-based weapons—State parties undertake to immediately and verifiably decommission any and all space-based weapons that they have caused to be placed in space prior to the effective date of this Treaty. State parties agree that any State, organization, or individual that has placed a space-based weapon in space prior to the effective date of this Treaty shall be subject to the jurisdiction of the Outer Space Peacekeeping Agency, which shall verifiably decommission such space-based weapons.

J. Tactical and strategic nuclear weapons and weapons of mass destruction in space—Each State party to this Treaty undertakes not to:
   1. Base in space any object carrying nuclear weapons or any other kinds of tactical or strategic weapons, including weapons of mass destruction.
   2. Install such space-based objects or weapons on celestial bodies, or station such weapons in outer space in any other manner.

K. Moon, planets, and celestial bodies—The Moon, the planets, and other celestial bodies shall be used by all States party to this Treaty exclusively for non-weapons, peaceful purposes. The establishment of weapons-related military bases in space, installations and fortifications for the testing of any type of military space-based weapons and the conduct of military maneuvers on celestial bodies or with space-based objects that are to be used as space-based weapons shall be forbidden.

L. Peaceful, scientific exploration and habitation—The use of military personnel for scientific research or for any other non-space-based-weapons, peaceful purposes shall not be prohibited. The use of any equipment or facility necessary for peaceful exploration or habitation of the Moon, the planets or other celestial bodies, or on objects in space shall also not be prohibited.

ARTICLE II: DEFINITION OF TERMS IN THIS TREATY

For purposes of this Treaty:

a. The terms "space" and "outer space" mean all space extending upward from an altitude greater that 100 kilometers above Earth's sea level.

b. The term "outer space object" or "space-based object" means any device, designed for functioning in outer space, being launched into an orbit around any celestial body, or being in the orbit around any celestial body, or on any celestial body, or leaving the orbit around any celestial body towards this celestial body, or moving from any celestial body towards another celestial body, traveling or placed in outer space by any other means.

c. The "use of force" or "threat of force" means any hostile actions against outer space objects or beings in outer space including, *inter alia*, those aimed at their destruction, damage, temporarily or permanently injuring, normal functioning, deliberate alteration of the parameters of their orbit, or the threat of these actions.

d. The term "space-based weapon" means and includes, without limitation:
   i. any device placed in outer space, based on any physical principle, specially produced or converted to eliminate, damage or disrupt normal function of objects in outer space, on the Earth or in its air, as well as to eliminate population, components of biosphere critical to human existence or any form of life or beings, or to inflict damage on them. A weapon will be considered as "placed" in outer space if it orbits the Earth at least once, or follows a section of such an orbit before leaving this orbit, or is stationed on a permanent basis somewhere in outer space.

- **ii.** any device in space that is capable of any of the following:
  - **1.** damaging or destroying an object (whether in outer space, in the atmosphere, or on Earth) by
    - **A.** firing one or more projectiles to collide with that object;
    - **B.** detonating one or more explosive devices in close proximity of that object;
    - **C.** directing a source of energy against that object, including molecular or atomic energy, subatomic particle beams, electromagnetic radiation, directed energy, plasma, or extremely-low-frequency (ELF) or ultra-low-frequency (ULF) energy radiation or some other source;
    - **D.** employing magnetic energy in any way, including to propel projectiles; or
    - **E.** employing any other unacknowledged or as yet undeveloped means of damaging or destroying.
- **iii.** any device placed in space capable of, or intended for, inflicting death or injury on, or damaging or destroying, a being or a population, or the biological life, bodily health, mental health, or physical and economic well-being of a being or a population,
  - **1.** through the use of space-based weapons systems using radiation, electromagnetic, sonic, laser, directed energy or other energies directed at individual beings or targeted populations for the purpose of electronic harassment, mood management, or mind control of beings or populations;
  - **2.** by expelling chemical or biological agents in the vicinity of a person or being in space;
  - **3.** through the use of any of the means described elsewhere in this Article I(d).
- **iv.** any device placed in space and intended or mandated as a weapon, including but not limited to the following:
  - **1.** electronic, magnetic, electromagnetic, or information weapons;
  - **2.** high-altitude, ultra-low-frequency weapons systems;
  - **3.** plasma, electromagnetic, magnetic, sonic, or ultrasonic weapons;
  - **4.** laser and directed energy weapons systems;
  - **5.** strategic, theater, or tactical weapons or weapons;
  - **6.** chemical, biological, environmental, climate, or tectonic weapons; and
  - **7.** weapons originating from off-planet cultures.

e. Space-based environmental weapons—The term "space-based weapon" shall also include, and not be limited to:

- **i.** any weapons system which weaponizes a natural ecosystem in space;
- **ii.** any weapons system one of whose components is placed in space and one of whose components weaponizes a natural ecosystem such as the ionosphere, upper atmosphere, climate, weather or tectonic systems;
- **iii.** any weapons systems using "scalar wave interferometry", a technology where two or more longitudinal, ultra-low-frequency waves are "aimed" at an intersecting point, at which time they interact in a unique way, using this scalar energy to weaponize the Ionosphere (60–800 km in altitude), which is in space. Such space-based "scalar wave interferometry" weapons systems weaponize the ionosphere in order to carry out the following types of space-based weapons attacks:
  - **1.** SDI (Strategic Defense Initiative) space-based radiofrequency weapon;
  - **2.** Space-based environmental weapon—Weather, climate and earthquake (tectonic) warfare;

3. Space warfare weapons system;
4. Space-based missile defense system;
5. Space-based scalar energy warfare against land and population targets, including cities, industrial sites, buildings, populations, and beings;
6. Space-based extremely-low-frequency (ELF) weapon for electromagnetic harassment and mood manipulation of target populations and beings;
7. Space-based biological and binary weapons system against populations and beings;
8. Damage or destruction upon a space object, a target population, being, municipality or region on land, air, or in space.

f. The term "being" shall include and not be limited to any human being, animal, and any living being on Earth or in space.

ARTICLE III: PUBLIC SAFETY—VERIFICATION AND ENFORCEMENT OF PERMANENT BAN ON WAR IN SPACE AND SPACE-BASED WEAPONS

Outer Space Peacekeeping Agency

A. The State parties find that banning war in space and the weaponization of space is hereby considered an urgent matter of public safety of the human population on land, sea, in the atmosphere and in outer space.

B. After the signing and ratification of this Treaty by five States, the provisions of this Treaty shall apply to all States, regardless of whether such States are signatories to this Treaty.

C. Each State party to this Treaty agrees to the establishment, funding, equipping, and deployment of an independent international ombudsman agency, the Outer Space Peacekeeping Agency.

D. The Outer Space Peacekeeping Agency shall have the exclusive jurisdiction to monitor any location on land, sea, air, and space for activities related to research, development, testing, manufacturing, production, and deployment of any and all space-based weapons, and to verify and enforce the permanent bans of space-based weapons, space-based weapons system decommissions, and prohibitions of space-based weapons under this Treaty. The jurisdiction of the Outer Space Peacekeeping Agency shall extend to the territory of all States party to this Treaty, to the territory of all States, and to all land, sea, atmosphere, and space.

E. The Outer Space Peacekeeping Agency shall have the exclusive jurisdiction to monitor and verify any space-based weapons-related activities in violation of this Treaty undertaken by any agency of any State party to this Treaty, any State, any organization, corporation, partnership, group, individual or syndicate, regardless of the location of such activities, on land, sea, atmosphere, or space.

F. Outer Space Peacekeeping Agency Conference—6 months from the date of entry into force of this Treaty under Article VIII hereof, the then signatories of this Treaty shall convene an Outer Space Peacekeeping Agency Conference. All of the States signatory to this Treaty at the time of the opening of the Conference shall be entitled to participate in and approve the formation of the Outer Space Peacekeeping Agency, including its legal statute, jurisdiction, mission, functions, administration, funding, regulations, operation, staffing, and collaborative forms of work.

ARTICLE IV: DUTIES OF STATE PARTIES AND OF THE UN SECRETARY GENERAL

A. Each State party to this Treaty undertakes to contact and urge other non-signatory State parties to sign, ratify, and implement this Space Preservation Treaty.

B. The Secretary General of the United Nations shall submit to the General Assembly of the United Nations within 90 days of the date that five State parties have signed and ratified this Treaty, and every 90 days thereafter, a report on:

 i. The implementation of the bans and prohibitions of this Treaty; and

 ii. Progress toward negotiating, signing, ratifying, and implementing this Treaty.

C. Each State party to this Treaty undertakes to take any measures it considers necessary in accordance with its constitutional processes to prohibit and prevent any activity in violation of the provisions of this Treaty anywhere under its jurisdiction or control.

D. Each State party to this Treaty undertakes not to assist, encourage or induce any State, group of States, international organization, or other entity or program to engage in activities contrary to the provisions of this Treaty.

E. Nothing in this Treaty can be interpreted as impeding the rights of the State parties to collaborate, explore, and use outer space for peaceful purposes in accordance with international law, which includes but is not limited to the Charter of the United Nations and the Outer Space Treaty of 1967.

ARTICLE V: AMENDMENTS TO THIS TREATY

A. Any State party to this Treaty may propose amendments to this Treaty. The text of any proposed amendment shall be submitted to the Depositary who shall promptly circulate it to all State parties.

B. An amendment shall enter into force for all States party to this Treaty upon the deposit with the Depositary of instruments of acceptance by a majority of State parties. Thereafter it shall enter into force for any remaining State party on the date of deposit of its instrument of acceptance.

ARTICLE VI: DURATION OF THIS TREATY

This Treaty shall be of unlimited duration.

ARTICLE VII: REVIEW CONFERENCE

A. Two years after the entry into force of this Treaty, a conference of the States party to this Treaty shall be convened by the Depositary at Geneva, Switzerland or another designated location. The conference shall review the operation of this Treaty with a view to ensuring that its purposes and provisions are being realized, and shall in particular examine the effectiveness of the provisions of this Treaty in eliminating the dangers of an arms race in space, confrontations in space or from space, and of space-based weapons.

B. At intervals of not less than 2 years thereafter, a majority of the States party to this Treaty may obtain, by submitting a proposal to this effect to the Depositary, the convening of a conference with the same objectives.

C. If no conference has been convened pursuant to paragraph (B) of this article within 2 years following the conclusion of a previous conference, the Depositary shall solicit the views of all States party to this Treaty, concerning the convening of such a conference. If one-third or 10 of the State parties, whichever number is less, respond affirmatively, the Depositary shall take immediate steps to convene the conference.

ARTICLE VIII: ENTRY INTO FORCE OF TREATY

A. This Treaty shall be open to all States for signature. Any State that does not sign this Treaty before its entry into force in accordance with paragraph (C) of this article may accede to it at any time.

B. This Treaty shall be subject to ratification by signatory States. Instruments of ratification or accession shall be deposited with the Secretary General of the United Nations.

C. Entry into force—This Treaty shall enter into force upon the deposit of instruments of ratification by five governments in accordance with paragraph (B) of this article, and in accordance with Article 102 of the Charter of the United Nations.

D. For those States whose instruments of ratification or accession are deposited after the entry into force of this Treaty, it shall enter into force on the date of the deposit of their instruments of ratification or accession.

E. The Depositary shall promptly inform all signatory and acceding States of the date of each signature, the date of deposit of each instrument of ratification or accession and the date of the entry into force of this Treaty and of any amendments thereto, as well as of the receipt of other notices.

F. This Treaty shall be registered by the Depositary in accordance with Article 102 of the Charter of the United Nations.

G. The authorized representative of each Signatory State party to the Space Preservation Treaty shall sign and date the Space Preservation Treaty and deposit a copy of the Signed Treaty with the UN Secretary General as Treaty Depositary as soon as possible.

ARTICLE IX: OFFICIAL LANGUAGES

This Treaty, of which the English, Arabic, Hebrew, Japanese, Chinese, French, Italian, Russian, and Spanish texts are equally authentic, shall be deposited with the Secretary General of the United Nations, who shall send certified copies thereof to the governments of the signatory and acceding States.

IN WITNESS WHEREOF, the undersigned, being duly authorized thereto by their respective governments, have signed this Treaty, opened for signature at

__________ on the ____day of __________, 20_______.

DONE at _____________, ______________ on ____________, ____, in the year 20__.

European Parliament Resolution of 10 July 2008 on Space and Security

The European Parliament,

- having regard to the European Security Strategy entitled "A secure Europe in a better world", adopted by the European Council on 12 December 2003,
- having regard to the EU Strategy against proliferation of Weapons of Mass Destruction, likewise adopted by the European Council on 12 December 2003,
- having regard to Council resolution of 21 May 2007 on the European Space Policy [1],
- having regard to the Treaty on the Functioning of the European Union (TFEU) and the Treaty on European Union (TEU), as amended by the Treaty of Lisbon, and their relevant provisions on European space policy (Article 189 of the TFEU), permanent structured cooperation on security and defence matters (Articles 42(6) and 46 of the TEU and Protocol 10) and enhanced cooperation in the civilian area (Part Six, Title III of the TFEU), as well as the solidarity clause (Article 222 of the TFEU) and mutual assistance provisions in the event of armed aggression against a Member State or States (Article 42(7) of the TEU),
- having regard to its resolution of 29 January 2004 on the action plan for implementing the European space policy [2],
- having regard to its resolution of 14 April 2005 on the European Security Strategy [3],
- having regard to the 1967 Treaty on Principles Governing the Activities of States in the Exploration and Use of Outer Space, including the Moon and Other Celestial Bodies ("the Outer Space Treaty"),
- having regard to the EU—Russia cooperation on space policy, which in 2006 created the Tripartite Space Dialogue between the European Commission, the European Space Agency (ESA) and Roscosmos (the Russian Federal Space Agency),
- having regard to Rule 45 of its Rules of Procedure,
- having regard to the report of the Committee on Foreign Affairs and the opinion of the Committee on Industry, Research and Energy (A6-0250/2008),

A. whereas freedom from space-based threats and secure sustainable access to, and use of, space must be the guiding principles of the European Space Policy,

B. whereas the various political and security challenges which the European Union is increasingly facing make an autonomous European Space Policy a strategic necessity,

C. whereas the lack of a common approach to space policy between EU Member States results in overly costly programmes,

D. whereas the crisis management operations within the framework of the European Security and Defence Policy (ESDP) suffer from a lack of interoperability between space assets operated by EU Member States,

E. whereas the European Union is lacking a comprehensive European space-based architecture for security and defence purposes,

Space Safety Regulations and Standards. DOI: 10.1016/B978-1-85617-752-8.10042-X

F. whereas the development of a new generation of launchers takes approximately 15 years and the present generation of launchers will need replacing in the next 20 years,
G. whereas development of space assets by the USA, Russia, Japan and other emerging spacefaring states, most notably China, India, South Korea, Taiwan, Brazil, Israel, Iran, Malaysia, Pakistan, South Africa and Turkey, is rapidly advancing,
H. whereas the French Presidency of the European Union during the second semester of 2008 sets out an advancement of the European Space Policy as one of its priorities,
I. whereas one of the most cost-effective elements of a space architecture and of achieving a sustainable fleet of space assets is on-orbit servicing, using *in situ* means.

GENERAL CONSIDERATIONS

1. Notes the importance of the space dimension to the security of the European Union and the need for a common approach necessary for defending European interests in space.
2. Underlines the need for space assets in order that the political and diplomatic activities of the European Union may be based on independent, reliable and complete information in support of its policies for conflict prevention, crisis management operations and global security, especially the monitoring of proliferation of weapons of mass destruction and their means of transportation and verification of international treaties, the transnational smuggling of light weapons and small arms, the protection of critical infrastructure and of the European Union's borders, and civil protection in the event of natural and man-made disasters and crises.
3. Welcomes the endorsement of the European Space Policy by the "Space Council" as proposed by a joint communication presented by the Commission and the European Space Agency (COM (2007)0212), especially the chapter on security and defence, while regretting the absence of any reference to the threat of weaponization of space within the "key issues to be considered in the development of a strategy for international relations" (as mentioned in Annex 3 to the above-mentioned Council Resolution of 21 May 2007); recommends, therefore, that the revised European Security Strategy should take this policy appropriately into account, and is of the view that space matters should be reflected in the possible White Paper on Security and Defence Policy.
4. Notes the inclusion of a legal basis for the European Space Policy in the Treaty of Lisbon; welcomes the opportunity given to it and to the Council to lay down, under the ordinary legislative procedure, the measures needed to shape a European Space Programme; calls on the Commission to submit to it and to the Council an appropriate proposal for such measures, together with a Communication relating to the establishment of appropriate relations with the European Space Agency; also welcomes the possibilities of permanent structured cooperation in security and defence matters and enhanced cooperation in the civilian area.
5. Encourages the Member States of the European Union, the European Space Agency and the various stakeholders to make greater and better use of the existing national and multinational space systems and to foster their complementarity; notes in this respect that common capabilities are needed for ESDP in at least the following areas: telecommunications, information management, observation and navigation; recommends the sharing and exchange of these data in line with the EU concept for Network Centric Operations Architecture.

6. Applauds the efforts of the International Academy of Astronautics and the International Association for the Advancement of Space Safety to promote remediation, understanding and measures in respect of space debris.

AUTONOMOUS THREAT ASSESSMENT

7. Calls on the EU Member States to pool and exchange the geospatial intelligence necessary for autonomous EU threat assessment.

EARTH OBSERVATION AND RECONNAISSANCE

8. Urges that the European Union Satellite Centre (EUSC) be fully developed to make full use of its potential; moreover, recommends the urgent conclusion of agreements between the European Union Satellite Centre and the EU Member States to provide imagery available to ESDP operation and force commanders while ensuring complementarity with Global Monitoring for Environment and Security (GMES) observation capacities and derived security-related information; in this regard, welcomes the Tactical Imagery Exploitation Station project, run jointly by the European Defence Agency (EDA) and the European Union Satellite Centre.
9. Recommends that the EU develop a common concept for geospatial intelligence, creating conditions for involvement of the EUSC in the planning for each ESDP operation requiring space-based observation and space-based intelligence; recommends that the EUSC establish a secure communication link in support of ESDP operations not only with the Operations Headquarters (OHQ) based in the EU but also with the Force Headquarters (FHQ) in the deployment region; furthermore, suggests that the EU explore the possibility of a financial contribution to the EUSC from the EU budget in order to provide sufficient funds to meet the increasing needs of ESDP operations.
10. Urges the EU Member States having access to the various types of radar, optical and weather observation satellites and reconnaissance systems (Helios, SAR-Lupe, TerraSAR-X, Rapid Eye, Cosmo-Skymed, Pleiades) to make them compatible; welcomes the bilateral and multilateral agreements between the leading EU countries (e.g. SPOT, ORFEO, the Helios cooperative framework, the Schwerin agreement and the future MUSIS); recommends that the MUSIS system be brought within a European framework and financed from the EU budget.
11. Emphasizes the importance of GMES for foreign as well as security and defence policies of the European Union; urges the creation of an operational budget line to ensure the sustainability of GMES services in response to users' needs.

NAVIGATION—POSITIONING—TIMING

12. Underlines the necessity of Galileo for autonomous ESDP operations, for the Common Foreign and Security Policy, for Europe's own security and for the Union's strategic autonomy; notes that, in particular, its public-regulated service will be vital in the field of navigation, positioning and timing, not least in order to avoid unnecessary risks.

13. Notes the first-reading agreement between Parliament and the Council on the proposal for a regulation on the further implementation of the European satellite radionavigation programmes (EGNOS and Galileo), which establishes that the Community is the owner of the system and that its deployment phase is fully financed by the Community budget.
14. Draws attention to its position adopted on 23 April 2008 on the European satellite radionavigation programmes (EGNOS and Galileo) [4], in particular, to the fact that the EGNOS and Galileo programmes should be considered as one of the major pillars of the future European Space Programme, and to the governance of these programmes, together with the Galileo Interinstitutional Panel, which may serve as a model in the development of a European Space Policy.

TELECOMMUNICATIONS

15. Underlines the need for secure satellite-supported communication for ESDP operations (EU Military Staff, EU Headquarters, deployable headquarters) and EU Member States' deployments under UN, NATO and other similar organizations.
16. Requests that the current and future satellite telecommunication systems at the disposal of the EU Member States (e.g. Skynet, Syracuse, Sicral, SATCOM Bw, Spainsat) be interoperable in order to provide for cost reduction.
17. Supports the cooperative development of a Software-Defined Radio (SDR) by the Commission and the European Defence Agency; notes that SDR will contribute to better interoperability of the ground segment of telecommunications systems.
18. Recommends that savings be achieved by shared use of the ground infrastructure supporting different national telecommunications systems.
19. Supports the possibility of funding future European satellite telecommunications systems supporting ESDP operations from the EU budget.

SPACE SURVEILLANCE

20. Supports the creation of a European space surveillance system leading to space situational awareness (including, for example, GRAVES and TIRA) to monitor the space infrastructure, space debris and, possibly, other threats.
21. Supports the possibility of funding the future European space situational awareness system from the EU budget.

SATELLITE-BASED EARLY WARNING AGAINST BALLISTIC MISSILES

22. Deplores the fact that EU Member States do not have access to instant data on ballistic missile launches around the world; expresses support, therefore, for projects leading towards satellite-based early warning against ballistic missile launches (such as the French "Spirale"); furthermore, calls for information acquired through these future systems to be available to all EU Member States in order to protect their population and to support possible

countermeasures, as well as to serve in the verification of compliance with the Nuclear Non-Proliferation Treaty, and for the purposes of ESDP operations and safeguarding Europe's security interests.

SIGNAL INTELLIGENCE

23. Supports the exchange of signal intelligence (electronic intelligence such as the French "Essaim" and communications intelligence) at European level.

AUTONOMOUS ACCESS TO SPACE AND INTERNATIONAL ENVIRONMENT

24. Supports secure, independent and sustainable access to space for the European Union as one of the preconditions of its autonomous action.

25. Recommends that the European non-commercial satellites be carried into orbit by European launchers, preferably from the territory of the European Union, bearing in mind the aspects of security of supply and protection of the European Defence Technological and Industrial Base.

26. Points out that it is necessary to increase the development effort in order for an enhanced *Ariane 5* to be available before 2015.

27. Recommends that strategic long-term investment in new European launchers be initiated as soon as possible, in order to keep up with the rising global competition; demands a greater degree of discipline for this project, in budgetary and time-frame terms.

28. Recommends that on-orbit servicing be established as a means of support to enhance the endurance, persistence, availability and operational efficiency of operational space assets and, at the same time, to reduce asset deployment and maintenance costs.

GOVERNANCE

29. Encourages strong inter-pillar cooperation for space and security, involving all the relevant actors (i.e. the Commission, the Council, the European Defence Agency and the European Union Satellite Centre), in order to safeguard the security policy and data security linked with the ESDP.

30. Strongly recommends the promotion of equal access for all EU Member States to operational data gathered using space assets under a reinforced ESDP framework.

31. Recommends that administrative and financial capacities for the management of space-related activities be developed by the European Defence Agency.

FINANCING

32. Points out that the EU budget commits expenditure amounting to approximately EUR 5250 million in the years 2007–2013 on common European space activities, resulting in an average expenditure of EUR 750 million per year over that period.

33. Calls on the European Union to set up an operational budget for space assets that serve to support the ESDP and European security interests.
34. Is alarmed by the fact that the lack of coordination among Member States results in a scarcity of resources due to unnecessary duplication of activities; therefore supports the idea of the launching of joint programmes by the Member States, which will provide costs savings in the longer term.
35. Furthermore, notes that the cost of the absence of a common European approach to the procurement, maintenance and functioning of space assets is estimated to amount to hundreds of millions of euros.
36. Points out that, as experience has shown, large-scale common projects cannot be properly managed when 27 different national budget authorities applying the principle of "fair return" are involved; therefore strongly recommends that these projects and programmes be financed from the EU budget.
37. Notes that the estimates of available expertise suggest that the level of investment needed to address the European security and defence needs in terms of satellite telecommunications, and the appropriate expenditure of the European Union on Earth observation and intelligence gathering, including signal intelligence, should be substantially increased in order to provide for the needs and ambitions of a comprehensive space policy.
38. Takes the view that the European Union, the European Space Agency, the European Defence Agency and their Member States should provide for reliable and adequate funding for the space activities envisaged and the research connected therewith; attaches great importance to the financing from the budget of the EU, such as on the Galileo project.

PROTECTION OF SPACE INFRASTRUCTURE

39. Underscores the vulnerability of strategic space assets as well as the infrastructure allowing access to space, e.g. launchers and space ports; therefore stresses the need for them to be adequately protected by ground-based theatre missile defence, planes and space surveillance systems; furthermore supports the sharing of data with international partners in the event that satellites are rendered inoperable by enemy action.
40. Calls for the vulnerability of future European satellite systems to be reduced through anti-jamming, shielding, on-orbit servicing, high-orbit and multi-orbital constellation architectures.
41. Emphasizes that the protective measures must be fully compliant with international standards regarding peaceful uses of outer space and commonly agreed transparency and confidence-building measures (TCBMs); asks EU Member States to explore the possibility of developing legally or politically binding "rules of the road" for space operators, together with a space traffic management regime.
42. Stresses that, as a result of this vulnerability, advanced communication should never be made fully dependent on space-based technologies.

INTERNATIONAL LEGAL REGIME FOR USES OF SPACE

43. Reiterates the importance of the principle of the use of space for peaceful purposes expressed in the above-mentioned 1967 Outer Space Treaty; is therefore concerned by the possible future weaponization of space.

44. Urges that under no circumstances should European space policy contribute to the overall militarization and weaponization of space.
45. Calls for the international legal regime to be strengthened so as to regulate and protect non-aggressive space uses and for the strengthening of TCBMs, within the framework of the drafting by the UN Committee on the Peaceful Uses of Outer Space (COPUOS) of space debris mitigation guidelines consistent with those of the Inter-Agency Debris Coordination Committee as well as the development by the UN Conference on Disarmament of a multilateral agreement on the prevention of an Arms Race in Outer Space; furthermore, asks the EU Presidency to represent the EU proactively in COPUOS; calls on the EU institutions to promote a conference to review the Outer Space Treaty, with the aim of strengthening it and expanding its scope to prohibit all weapons in space.
46. Calls on all international actors to refrain from using offensive equipment in space; expresses its particular concern about the use of destructive force against satellites, such as the Chinese anti-satellite system tested in January 2007, and the consequences of the massive increase in debris for space security; recommends, therefore, the adoption of legally binding international instruments focusing on banning the use of weapons against space assets and the stationing of weapons in space.
47. Calls on all space users to register their satellites, including military satellites, by way of a space security confidence-building measure promoting transparency; supports the Council's pursuit of a comprehensive EU Code of Conduct on Space Objects; demands that this Code be transformed into a legally binding instrument.
48. Urges the United Nations and the European Union to engage in the active diminution of, and protection from, space debris harmful to satellites.

TRANSATLANTIC COOPERATION ON SPACE POLICY AND MISSILE DEFENCE

49. Urges the European Union and the North Atlantic Treaty Organization to launch a strategic dialogue on space policy and missile defence, while bearing in mind the legal imperative of avoiding any action that might be incompatible with the principle of the peaceful use of space, especially on the complementarity and interoperability of systems for satellite communications, space surveillance, and early warning of ballistic missiles, as well as protection of European forces by a theatre missile defence system.
50. Calls on the European Union and the United States of America to engage in a strategic dialogue on the use of space assets and to take the global lead within and outside the UN to make sure that outer space is preserved for peaceful policies only.

OTHER INTERNATIONAL COOPERATION

51. Welcomes the strengthened cooperation between the European Union and the Russian Federation within the framework of the above-mentioned Tripartite Space Dialogue set up in 2006 between the European Commission, the European Space Agency and Roscosmos, including space

applications (satellite navigation, Earth observation and satellite communications) as well as access to space (launchers and future space transportation systems).

52. Instructs its President to forward this resolution to the Council, the Commission, the European Space Agency, the parliaments of the Member States and the Secretaries General of the United Nations, the North Atlantic Treaty Organization and the Organization for Security and Cooperation in Europe.

Notes

[1] OJC 136, 20.6. 2007, p. 1.
[2] OJC 96 E, 21.4.2004, p. 136.
[3] OJC 33 E, 9.2.2006, p. 580.
[4] Texts adopted, P6_TA(2008)0167.

UN COPUOS Space Debris Mitigation Guidelines

INTRODUCTORY NOTE

Ever increasing space debris poses the single most serious risk to the safety and sustainability of space operations of all nations and to human life and property on the surface of the Earth.

As early as 1982, the issue of space debris surfaced at international level during the second UN Conference on Space (UNISPACE II). After a brief discussion, the Conference recommended that: studies be undertaken to seek viable methods to minimize the probability of collision between active space objects and space debris; some of these methods could include the removal of inactive satellites from their orbits; based on such studies, appropriate regulatory measures should be taken. Though the United Nations Committee on the Peaceful Uses of Outer Space (COPUOS) acknowledged the seriousness of the problem in 1983, it was only in 1987 that its Scientific and Technical Subcommittee recommended that COSPAR and IAF undertake a study on space debris and in 1988 the UN General Assembly recognized that space debris is a serious potential hazard. It took 5 years for the COPUOS to agree in 1993 to include an item dealing with space debris in its agenda and in 1994, for the first time the Scientific and Technical Subcommittee considered, on a priority basis, the problem of space debris—a major positive step in international policy and lawmaking on space debris. From then onward, the Subcommittee continued deliberating on the issue of space debris. In July 1999, the third UN Conference on Space (UNISPACE III), after consideration of the issue of space debris, decided to recommend that in future, an action should be taken by the UN, to "improve the protection of the near and outer space environments through further research in and implementation of mitigation measures for space debris" [1].

It is interesting to note that (a) the Legal Subcommittee of COPOUS has never been allowed to address any regulatory matters related to space debris and (b) the Scientific and Technical Subcommittee, after debating the issue since 1994, adopted space debris mitigation guidelines in 2007. These guidelines were approved by the COPUOS, at its 50th session in 2007, as Space Debris Mitigation Guidelines of the Committee on the Peaceful Uses of Outer Space. In terms of binding enforceability and international legal status, this statement was one of the lowest forms ever put forth by a UN body on a highly serious, space-related international concern. The COPUOS Guidelines were subsequently simply endorsed (not adopted) by the General Assembly in its resolution 62/217 of 21 December 2007 (referred to as UN COPUOS Guidelines, as reproduced below). They are voluntary and based on, and consistent with, the Inter-Agency Space Debris Coordination Committee (IADC) guidelines, and present general recommendations in the form of seven guidelines to be implemented through national legislation. COPUOS, in 2007, agreed that its approval of the voluntary guidelines for the mitigation of space debris would increase mutual understanding on acceptable activities in space and thus enhance stability in space-related matters and decrease the likelihood of friction and conflict.

Space Safety Regulations and Standards. DOI: 10.1016/B978-1-85617-752-8.10043-1

The international space community, and even the general public, is becoming aware of the danger of the space debris problem. A UN circular, distributed extensively around the world, recently highlighted the "Space Debris Problem as One of the Ten Stories the World should Hear More About" [2]. This growing awareness of the negative impact of space debris upon the safety and sustainability of space operations has encouraged some spacefaring nations to take steps to mitigate the production of new debris through the development of national space debris mitigation measures, though there are some differences between their respective debris mitigation efforts. Such initiatives are useful in the short term, but the effectiveness of national and even multilateral regulatory initiatives could be limited since a single major accident could create hazards for the space activities of all States. The international community, through the UN, took 25 years to agree and adopt some basic principles and goals (without any specific international implementation mechanism) to reduce the production of space debris; however, according to Dave Finkleman, "[w]e are discovering that developing space debris mitigation goals was easier than achieving them". Moreover, the Guidelines do not deal with the extensive amount of debris currently in orbit that will automatically continue multiplying due to the "domino effect", also known as "Kessler Syndrome". Therefore, it is imperative that the Guidelines, irrespective of their inherent weaknesses, are implemented effectively nationally, and the international community must continue seriously searching for the means (a) to coordinate national space debris mitigation efforts and (b) to safely remove the existing debris, especially from those regions of space that are highly used.

Ram S. Jakhu

SPACE DEBRIS MITIGATION GUIDELINES OF THE COMMITTEE ON THE PEACEFUL USES OF OUTER SPACE [3]

1. Background

Since the Committee on the Peaceful Uses of Outer Space published its Technical Report on Space Debris in 1999 [4], it has been a common understanding that the current space debris environment poses a risk to spacecraft in Earth orbit. For the purpose of this document, space debris is defined as all man-made objects, including fragments and elements thereof, in Earth orbit or re-entering the atmosphere, that are non-functional. As the population of debris continues to grow, the probability of collisions that could lead to potential damage will consequently increase. In addition, there is also the risk of damage on the ground, if debris survives Earth's atmospheric re-entry. The prompt implementation of appropriate debris mitigation measures is therefore considered a prudent and necessary step towards preserving the outer space environment for future generations.

Historically, the primary sources of space debris in Earth orbits have been (a) accidental and intentional break-ups that produce long-lived debris and (b) debris released intentionally during the operation of launch vehicle orbital stages and spacecraft. In the future, fragments generated by collisions are expected to be a significant source of space debris.

Space debris mitigation measures can be divided into two broad categories: those that curtail the generation of potentially harmful space debris in the near term and those that limit their generation over the longer term. The former involves the curtailment of the production of mission-related space debris and the avoidance of break-ups. The latter concerns end-of-life procedures that remove decommissioned spacecraft and launch vehicle orbital stages from regions populated by operational spacecraft.

2. Rationale

The implementation of space debris mitigation measures is recommended since some space debris has the potential to damage spacecraft, leading to loss of mission, or loss of life in the case of manned spacecraft. For manned flight orbits, space debris mitigation measures are highly relevant due to crew safety implications.

A set of mitigation guidelines has been developed by the Inter-Agency Space Debris Coordination Committee (IADC), reflecting the fundamental mitigation elements of a series of existing practices, standards, codes, and handbooks developed by a number of national and international organizations. The Committee on the Peaceful Uses of Outer Space acknowledges the benefit of a set of high-level qualitative guidelines, having wider acceptance among the global space community. The Working Group on Space Debris was therefore established (by the Scientific and Technical Subcommittee of the Committee) to develop a set of recommended guidelines based on the technical content and the basic definitions of the IADC space debris mitigation guidelines, and taking into consideration the United Nations treaties and principles on outer space.

3. Application

Member States and international organizations should voluntarily take measures, through national mechanisms or through their own applicable mechanisms, to ensure that these guidelines are implemented, to the greatest extent feasible, through space debris mitigation practices and procedures.

These guidelines are applicable to mission planning and the operation of newly designed spacecraft and orbital stages and, if possible, to existing ones. They are not legally binding under international law.

It is also recognized that exceptions to the implementation of individual guidelines or elements thereof may be justified, for example, by the provisions of the United Nations treaties and principles on outer space.

4. Space debris mitigation guidelines

The following guidelines should be considered for the mission planning, design, manufacture and operational (launch, mission, and disposal) phases of spacecraft and launch vehicle orbital stages:

Guideline 1: Limit debris released during normal operations

Space systems should be designed not to release debris during normal operations. If this is not feasible, the effect of any release of debris on the outer space environment should be minimized.

During the early decades of the space age, launch vehicle and spacecraft designers permitted the intentional release of numerous mission-related objects into Earth orbit, including, among other things, sensor covers, separation mechanisms and deployment articles. Dedicated design efforts, prompted by the recognition of the threat posed by such objects, have proved effective in reducing this source of space debris.

Guideline 2: Minimize the potential for break-ups during operational phases

Spacecraft and launch vehicle orbital stages should be designed to avoid failure modes that may lead to accidental break-ups. In cases where a condition leading to such a failure is detected, disposal and passivation measures should be planned and executed to avoid break-ups.

Historically, some break-ups have been caused by space system malfunctions, such as catastrophic failures of propulsion and power systems. By incorporating potential break-up scenarios in failure mode analysis, the probability of these catastrophic events can be reduced.

Guideline 3: Limit the probability of accidental collision in orbit

In developing the design and mission profile of spacecraft and launch vehicle stages, the probability of accidental collision with known objects during the system's launch phase and orbital lifetime should be estimated and limited. If available orbital data indicate a potential collision, adjustment of the launch time or an on-orbit avoidance maneuver should be considered.

Some accidental collisions have already been identified. Numerous studies indicate that, as the number and mass of space debris increase, the primary source of new space debris is likely to be from collisions. Collision avoidance procedures have already been adopted by some Member States and international organizations.

Guideline 4: Avoid intentional destruction and other harmful activities

Recognizing that an increased risk of collision could pose a threat to space operations, the intentional destruction of any on-orbit spacecraft and launch vehicle orbital stages or other harmful activities that generate long-lived debris should be avoided.

When intentional break-ups are necessary, they should be conducted at sufficiently low altitudes to limit the orbital lifetime of resulting fragments.

Guideline 5: Minimize potential for post-mission break-ups resulting from stored energy

In order to limit the risk to other spacecraft and launch vehicle orbital stages from accidental break-ups, all on-board sources of stored energy should be depleted or made safe when they are no longer required for mission operations or post-mission disposal.

By far the largest percentage of the cataloged space debris population originated from the fragmentation of spacecraft and launch vehicle orbital stages. The majority of those break-ups were unintentional, many arising from the abandonment of spacecraft and launch vehicle orbital stages with significant amounts of stored energy. The most effective mitigation measures have been the passivation of spacecraft and launch vehicle orbital stages at the end of their mission. Passivation requires the removal of all forms of stored energy, including residual propellants and compressed fluids, and the discharge of electrical storage devices.

Guideline 6: Limit the long-term presence of spacecraft and launch vehicle orbital stages in the low-Earth orbit (LEO) region after the end of their mission

Spacecraft and launch vehicle orbital stages that have terminated their operational phases in orbits that pass through the LEO region should be removed from orbit in a controlled fashion. If this is not possible, they should be disposed of in orbits that avoid their long-term presence in the LEO region.

When making determinations regarding potential solutions for removing objects from LEO, due consideration should be given to ensuring that debris that survives to reach the surface of the Earth does not pose an undue risk to people or property, including through environmental pollution caused by hazardous substances.

Guideline 7: Limit the long-term interference of spacecraft and launch vehicle orbital stages with the geosynchronous Earth orbit (GEO) region after the end of their mission

Spacecraft and launch vehicle orbital stages that have terminated their operational phases in orbits that pass through the GEO region should be left in orbits that avoid their long-term interference with the GEO region.

For space objects in or near the GEO region, the potential for future collisions can be reduced by leaving objects at the end of their mission in an orbit above the GEO region such that they will not interfere with, or return to, the GEO region.

5. Updates

Research by Member States and international organizations in the area of space debris should continue in a spirit of international cooperation to maximize the benefits of space debris mitigation initiatives. This document will be reviewed and may be revised, as warranted, in the light of new findings.

6. Reference

The reference version of the IADC space debris mitigation guidelines at the time of the publication of this document is contained in the annex to document A/AC.105/C.1/L.260.

For more in-depth descriptions and recommendations pertaining to space debris mitigation measures, Member States and international organizations may refer to the latest version of the IADC space debris mitigation guidelines and other supporting documents, which can be found on the IADC website (www.iadc-online.org).

Notes

[1] The Space Millennium: Vienna Declaration on Space and Human Development (1999). *Third United Nations Conference on the Exploration and Peaceful Uses of Outer Space*, 19–23 July, UN Doc. A/CONF.184/L.16/Add.2

[2] http://www.un.org/en/events/tenstories/08/spacedebris.shtml (accessed: 10 October 2009).

[3] *Official Records of the General Assembly, 62nd Session, Supplement No. 20* (A/62/20), paras 117 and 118 and Annex. The UN General Assembly in its Resolution endorsed the Space Debris Mitigation Guidelines of the Committee on the Peaceful Uses of Outer Space in 2007. See: United Nations General Assembly, 62nd session, Agenda item 31, Document A/RES/62/217 (10 January 2008), paragraph 26.

[4] United Nations publication, Sales No. E.99.I.17.

APPENDIX 8

List of abbreviations

ADC Interagency Space Debris Coordination Committee
AIAA American Institute of Aeronautics and Astronautics
ALIP Alternating linear induction pump
AMC Applicable means of compliance
AMTEC Alkali metal thermal and electric energy conversion
ANC Air Navigation Commission (of the ICAO)
AOC Air Operator Certificate
ASAP NASA Aerospace Safety Advisory Panel
ASAT Anti-satellite weapon
ASD Aero Space Defense Group
ASI Agenzia Speziale Italiana
AST Associate Administrator for Commercial Space Transportation (USA)
ATC Ait Traffic Control
ATM Air Traffic Management
ATS Ait Traffic Services
ATV All-terrain vehicle
BNSC British National Space Centre
BOL Beginning-of-life phase of satellite
BSI British Standards Institute
CAMO Continuous Airworthiness Management Organization Agreement
CATEX Categorical exclusion
CCSDS Consultative Committee for Space Data Systems
CD Geneva-based Conference for Disarmament
CDI US Center for Defense Information
CDR Critical Design Review
CEN Comité Européen de Normalisation
CENELEC Comité Européen de Normalisation Electrotechnique
CEOS Committee on Earth Observation Satellites
CFE Commercial and foreign entities
CFR US Code of Federal Regulations
CHSS Commercial Human Space-flight Services
CHSV Commercial Human Space-flight Vehicles
CIL Critical Items List
CNCA Certification and Accreditation Administration—China
CNES Centre National d'Etudes Spatiales
CoC Code of Conduct
CoFR Certification of Flight Readiness
COLA Collision avoidance
COMSTAC Commercial Space Transportation Advisory Committee
COSPAR COPUOS Committee for Space and Atmospheric Research
COSTIND Chinese Commission of Science, Technology and Industry for National Defense
COTS US Commercial Orbital Transportation Services
CRD Comment Response Document
CRI Certification Review Items
CS Certification specification
CSF Commercial Spaceflight Federation
CSG Centre Spatial Guyanais
CSLA US Commercial Space Launch Act of 1984
CSLAA US Commercial Space Launch Amendments Act of 2004 (also CLSAA-2004)
CSSI Center for Space Standards and Innovation
CST Commercial Space Transportation
DARPA US Defense Advanced Research Projects Agency

Space Safety Regulations and Standards. DOI: 10.1016/B978-1-85617-752-8.10045-5

DDESB	DoD Explosives Safety Board
delta-V	Satellite orbital velocity
DIP	Dynamic isotope power system
DLR	German Deutsches Zentrum für Luft und Raumfahrt
DMEA	Damage modes and effects analysis
DoD	US Department of Defense
DOT	US Department of Transportation
EA	Environmental assessment
EASA	European Aviation Safety Agency
EC	European Commission or Expected casualty
ECLSS	Environmental Control and Life Support Systems
ECSS	European Cooperation on Space Standardization
EELV	Evolved Expendable Launch Vehicle
EEZ	Exclusive Economic Zone
EIRP	Effective isotropically radiated power
EIS	Environmental Impact Statement
ELDO	European Launcher Development Organization
ELoS	Equivalent level of safety
EN	European standards
EOL	End of life
ESA	European Space Agency
ESF	Equivalent safety findings
ESPI	European Space Policy Institute
ESRO	European Space Research Organization
ETSI	European Telecommunications Standards Institute
ETR	Eastern Test Range
EW	East–west direction
EWR	Eastern and Western Range
EU	European Union
EVA	Extravehicular Activity
FAA	US Federal Aviation Administration
FAA/AST	US FAA Office of Commercial Space Transportation
FCC	US Federal Communications Commission
FD	Flight dynamics
FIR	Flight Information Region
FL	Flight level
FMEA	Failure Modes and Effects Analysis
FMEA/CIL	Reliability analyses: Failure Mode Effects/Critical Elements List
FPSE	Free-piston Stirling engine
FRD	Flight Requirements Document
FSOA	French Space Operations Act
FSTS	Future Space Transportation System
FTS	Flight Termination System
GEO	Geostationary orbit
GMES	Global Monitoring for Environment and Security
GOST	Russian state standards
GP	General perturbation
GPS	Global Positioning System
GSC	Guyana Space Center
GSRP	Ground Safety Review Panel
HBP	Human behavior and performance
HCOC	Hague Code of Conduct against Ballistic Missile Proliferation
IA	Independent assessment
IAA	International Academy of Astronautics
IAASS	International Association for the Advancement of Space Safety
ICAO	International Civil Aviation Organization
ICD	Interface Control Document
ICT	Information and communication technical systems
IDMRD	Intermediate and Depot Maintenance Requirements Document
IFFAS	International Financial Facility for Aviation Safety
IFR	Instrument Flight Rule
IGA	Intergovernmental Agreement
IOT	In-orbit test
ISON	Russian International Scientific Observation Network
ISS	International Space Station
ISSB	Independent Space Safety Board
ISSF	International Space Safety Foundation
ITU	International Telecommunications Union

IV&V	Independent Verification and Validation
JAA	Joint Aviation Authorities in Europe
JARSWG	Joint American–Russian Safety Working Group
JspOC	Joint Special Operation Center
KE-ASAT	Kinetic-energy ASAT
LAE	Liquid apogee engine
LEO	Low Earth orbit
LEOP	Launch and early orbit phases
LOC	Loss of cooling
LSO	Launch Safety Officer
MA-OD	Mechanically alloyed oxide dispersed steels
MEM	Meteoroid Engineering Model
MEO	Middle Earth orbit
MIT	Massachusetts Institute of Technology
MOU	Memorandum of Understanding
MRTFB	Major Range and Test Facility Base
NAS	National Airspace System
NASA	US National Aeronautics and Space Administration
NATO	North Atlantic Treaty Organization
NextGen	Next Generation Air Transportation System
NIVR	Nederlands Instituut voor Vliegtuigontwikkeling en Ruimtevaart
NPA	Notice of Proposed Amendment
NPR	NASA Procedural Requirements
NPRM	Notice of Proposed Rule Making
NS	North–south direction
NSTS	National Source Tracking System
ODS	Oxide dispersed steels
OEM	Orbital ephemeris message
OMRSD	Operations and Maintenance Requirements and Specifications Document
OPM	Orbital Parameter Message
ORR	Operational Readiness Review
OSHA	Occupational Safety and Health Agency of the United States
OSI	Open Systems Interconnection
OST	Russian space industry standards
OSTC	Belgian Office for Science, Technical and Cultural Affairs
PAROS	Prevention of an Arms Race in Outer Space
PASO	ESA Products Assurance and Safety Offices
PCA	Power conversion assembly
PDR	Preliminary Design Review
PHTS	Primary heat transfer system
PLA	Chinese People's Liberation Army
PMA	Permanent magnet alternator
PPE	Personal protection equipment
PPWT	Treaty of the Prevention of the Placement of Weapons in Outer Space
PRA	Probabilistic Risk Assessment
PRACA	Problem Reporting and Corrective Action
PSF	Personal Spaceflight Federation
PSRP	Payload Safety Review Panel
PtF	Permit to Fly
RALCT	Risk and Lethality Commonality Team
RCC	Range Commander's Council
RCofA	Restricted Certificate of Airworthiness
RDT	Raketno-Kosmicheskaya Tekhnika
RF	Radio frequency
RFI	Radio frequency interference
RHU	Radioisotope heater units
RLV	Reusable launch vehicle
RMB	Renminbi: Chinese currency whose prinicipal unit is the yuan
RPS	Radioisotope power system
RTC	Restricted type certificate
RTG	Radioisotope thermoelectric generator
S^4	Submersion Subcritical Safe Space
SAIRS	Scalable AMTEC Integrated Reactor Space
SARPs	International Standards and Recommended Practices
SAS	Special Airworthiness Specifications
SATMS	Space and Air Traffic Management System
SCs	Special conditions

SCORE	Sectored compact fission reactor
SCS	Spacecraft survivability
SDR	System Definition Review
SEC	Space Enterprise Council
SES	Single European Sky
SESAR	Single European Sky ATM Research
SFP	Space-flight participant
SHEL	Model Licensing for Aviation Safety Software and Hardware and Environment and Liveware
SHO	Sufficiently high orbit
SIR	System Integration Review
SLASO	Australian Space Licensing and Safety Office
SLE	Space link extension
SMA	Safety and Mission Assurance
SMART	Safety and Mission Assurance Review Team
SOE	State-owned enterprise
SOPSO	Space Operation Support Office
SOT	US Secretary of Transportation
SP	Special perturbation
SRB	Solid Rocket Booster
SRP	Safety Review Panel
SRR	System Requirements Review
SSA	Space Situational Awareness
SSC	Satellite control center
SSE	Systems Safety Engineering
SSN	US Space Surveillance Network
SSP	Space Shuttle Program
SSTS	Single stage to space
STA	Space Transportation Association
STC	Space Transition Corridor
STE	Segmented thermoelectric
STP	Russian company standards
SUA	Special Use Airspace
SUIRG	Satellite Users Interference Reduction Group
SWF	Secure World Foundation
TA	Technical Authority
TBCM	Transparency and confidence-building measures
TC	Type certificate
TCA	Thermoelectric conversion assembly
TE	Thermoelectric
TEM	Thermoelectric magnetic
TFE	Thermionic fuel elements
TLE	Two-line element
TPV	Thermal photovoltaic
TR	Technical regulations
TRL	Technology readiness level
TS-D	Russian Technical Specifications for Description
TS-EQ	Russian Technical Specifications for Equipment
TS-OM	Russian Technical Specifications for Operations and Maintenance Manual
TTS	Thrust Termination System
UAS	Unmanned aerial system
UATCC	Upper Air Traffic Control Centers
UN COPUOS	United Nations Committee for the Peaceful Uses of Outer Space
UNEP	United Nations Environmental Program
USC	United States Code
USOAP	Universal Safety Oversight Audit Programme
	USSPACECOM
US Space Command	
	USSTRATCOM
US Strategic Command	
UV	Ultraviolet
VLJ	Very light jet
VR	Vulnerability reduction
WFF	Wallops Flight Facility
WMO	World Meteorological Organization
WSO	World Space Organization
WTR	Western Test Range

Index